PROBLEMS IN MIDDLE AND HIGH SCHOOL MATHEMATICS

PROBLEMS IN MIDDLE AND HIGH SCHOOL MATHEMATICS

Techniques of Solving

Mihai Rosu   **Alexandru Rosu**

Library of Congress Control Number:		2014921854
ISBN:	Hardcover	978-1-5035-2493-4
	Softcover	978-1-5035-2495-8
	eBook	978-1-5035-2496-5

Print information available on the last page.

Corrections by
Mihai Rosu, professor
Emanoil Theodorescu, PhD., Mathematics, University of Kansas, USA
Alexandru Rosu, Student, Univ. of Waterloo, Ontario, Canada
Vlad Rosu, Student, St. Theresa of Lisieux Catholic High School, Richmond
Hill, Ontario, Canada

Rev. date: 05/23/2016

To order additional copies of this book, contact:
Xlibris
1-888-795-4274
www.Xlibris.com
Orders@Xlibris.com
701494

Preface

We have developed this book - *Problems in Middle and High School Mathematics: Solving Techniques* - to help middle and high school students acquire a deeper understanding of mathematics, and prepare themselves for Olympiads and contests in this field. The wish is to instill in our students the art of reasoning and the power of evaluating and solving mathematics problems at a competitive level.

However, this work can be counted on with assisting many more students (even those apparently less inclined towards mathematics) with understanding the proper treatment of mathematical problems. We believe this work to be filling a gap in the area of high school mathematics in North-America, in terms of the material it covers, as well as the depth and rigor of treatment.

While common elements with other works of this type do exist, we have tried to enrich its content through the addition of special features, important problems, and diversified solutions. Intending to often present to students multiple approaches in solving a problem, each chapter presents necessary core knowledge, problems grouped on thematic and solving ideas (result of a lengthy pedagogical career), with examples solved in great detail (sometimes with multiple solutions), followed by similar problems with answers at the back.

Each chapter contains exercises that increase in difficulty, from the simplest ones to problems given in major mathematics competitions. At the end of each chapter, we have included problems I have published in the "*Mathematical Olympiads Correspondence Program*" (*http://www.math.ca/Competitions/MOCP/*) in Ontario, Canada.

Occasionally, some problems at the end of each chapter may appear very hard to solve. However, my thirty-year experience in teaching mathematics and my interactions with students enabled me to demonstrate the solutions in a clear, easy to understand manner.

We are confident that any students who pick up this book and solve one set of exercises can expand their knowledge, fuel their interest, and excel in their mathematics courses. We also hope this work will become instrumental to mathematics teachers working both in the private and public education systems.

Mihai Rosu

rosumihai@yahoo.ca

Alexandru Rosu

alexrosu97@gmail.com

===

===

Foreword

High school mathematics professor Mihai Rosu's volume "Methods of Solving Problems in Middle and High School Mathematics: Techniques of Solving" is a result of over three decades of professional activity in teaching high school mathematics. A mathematics professor formed – as a mathematics major - in the Romanian education system, at standards set earlier in the 20th century by a number of eminent Romanian mathematicians with doctorates from French and German universities, Mihai Rosu and his son Alexandru offer to the interested public – parents and students – a solid piece of work, in the strong tradition of the Romanian school of mathematics, well aligned to the mathematics curricula in North America.

The present volume is addressed to high school students (and middle school students, to a certain extent), with an offer of over 3,500 problems and exercises, most with full solutions, all with final answers. The chosen material varies in difficulty level from "introductory", all the way to "very advanced", facilitating a gradual evolution of the student from novice to really skillful. The rich bibliography – Romanian mathematical publications for high school students, and occasionally even university professor authors, North-American and Russian (Soviet) sources - indicates the broad reach of this work.

This volume is, at the same time, a concrete response to the call of many Canadians (university professors at "WISE Math", petitioners at "change. org" from multiple Canadian provinces (Alberta, Manitoba, British Columbia, Ontario) and many, many others, for raising the standards of Canadian mathematical (though not only that) primary and secondary education, both of students and of their teachers.

Those students who will be putting in the effort of carefully going through the material (including carefully reading the given solutions at the back of the book) will benefit enormously, towards gaining of a good, clear image of what modern high school mathematics might be. Most importantly, this work (together with others of the same type and similar caliber, covering other branches of high school mathematics) will prepare students for the significantly harder time of university mathematics, on their path to becoming highly qualified experts in their fields of interest.

Em. Theodorescu, Ph.D.

CONTENTS

Chapter I

NATURAL AND INTEGER NUMBERS

$N = \{0,1,2,...\}$ Natural number, $N^* = \{1,2,3,...\}$

$Z = \{-2,-1,0,1,2,...\}$ Integer numbers,

Properties of powers $a^n = \underbrace{a \cdot a \cdot ... \cdot a}_{n\ times}$, $n \in N^*$.

a) $a^0 = 1$;   b) $(ab)^n = a^n b^n$;

c) $\left(\dfrac{a}{b}\right)^n = \dfrac{a^n}{b^n}$;   d) $a^n a^m = a^{n+m}$;

e) $\dfrac{a^n}{a^m} = a^{n-m}$;   f) $\left(a^m\right)^n = a^{mn}$, $a,b \in Z$, $b \neq 0$, $m,n \in N$.

Base ten numeration system $\overline{abcd} = a \cdot 10^3 + b \cdot 10^2 + c \cdot 10 + d$.

Integer numbers. Order of operations

I. 1. Replace the □ with "=", "<", ">" such that the following relations are true:
a) $4 + 3 \square 2 + 3$;   b) $7 - 3 \square 2 + 3$;
c) $9 - 3 \square 3 + 3$;   d) $9 - 4 \square 5 + 2$;
e) $8 - 1 \square 5 + 2$;   f) $10 - 3 \square 6 + 3$;
g) $8 + 1 \square 8 + 1$;   h) $7 + 2 \square 6 + 3$;
i) $9 - 1 \square 5 + 3$;   j) $7 + 3 \square 6 + 4$.

I. 2. Calculate:
a) $1 + 3$;   b) $22 + 30$;
c) $17 - 2$;   d) $50 - 22$;
e) $65 + 47$;   f) $57 - 15$;
g) $55 + 34$;   h) $88 - 55$.

i) From the product of the numbers 27 and 22 subtract their difference;

j) Add 20 to one number, subtract 43 from the result so that 10 is the result. What is the number?

I. 3. Complete the following tables:

x	18	59	76
y	22	12	15
$x+y$			

a	b	c	$a+b$	$a+c$	$a+b+c$
32	14	19			
4	15	13			

I. 4. Find the unknown terms in the operations:

a) $x + 3 < 5$;

b) $4 + x = 9$;

c) $7 - x < 4$;

d) $3 + 5 = x + 3$;

e) $5 + x > 8 - 2$.

f) $45 - 12 + b = 56 + 17$;

g) $50 + 0 = x + 49$;

h) $65 - 22 + b = 58 + 37$;

i) $45 + 32 - a = 6 + 31$;

j) $a + 85 - 12 = 86 + 27$.

I. 5. a) Calculate:

$43 + 18 =$	$55 - 25 - 8 =$
$35 + 29 =$	$84 - 56 - 8 =$
$83 - 27 =$	$38 + 29 + 17 =$
$84 - 36 =$	$46 + 9 - 16 =$

b) Calculate by grouping the factors: $2 + 13 + 7 + 18 =$
$3 + 24 + 6 + 17 =$.

c) Find the unknown numbers:

$n - 26 = 25$	$35 + a = 71$
$75 + d = 79$	$35 - a = 711$
$12 + 13 - x = 14$	$78 - (x + 21) = 46$

d) From the sum of the numbers 9, 18, 27 and 36, subtract the difference between 36 and 17.

e) In a garden there are 48 roses, a second garden has 24 more roses than the first one. How many roses are in both gardens?

f) Vlad has 48 stamps and bought another 56 stamps. He gave to Alexandru 29 stamps. How many stamps does Vlad have left?

g) Fill in: 0, 5, 10, ___, ___, ___, ___, ___, ___, ___, 50;
0, 3, 6, ___, ___, ___, ___, ___, ___, ___,30;
0, 10, 20, ___, ___, ___, ___, ___, ___, ___, 100.

h) i) Multiply 767 by 18.
ii) Find the number 6 times larger than 97.
iii) Add 25 to the triple of the number 182.
iv) Substract 100 from the triple of the number 99.

i) Vlad is 7 years old, his mother is 6 times older than him and his father is 5 years older than his mother. How many times is Vlad's age less than his father's age?

j) Find „a" from the equalities: $a \div 6 = 7$; $9 \cdot a = 54$; $27 \div a = 3$.

I. 6. Fill out the table

a	b	c	d	c-b	$3a$-b	$3c$-d+1	a+$3b$-c	$3b$+$2c$-d+2
7		63	189	42				
10		16	48	0				
7		15	30	0				
5	15		135	30				
4		36	144	17				

I. 7. Calculate the expressions below when
$a = 3$, $b = 9$, $c = 27$, $d = 81$:
a) $d - 3c + b - 3a$;   **b)** $ad - 3ac + ab - 3a^2$;
c) $ad - c + 1$;   **d)** $a^2 b - ac + a$;
e) $bc - 3d + 1$;   **f)** $a^3 b - 3ac + 27 - a^3$;
g) $b^2 - d$;   **h)** $b^3 - bd$;
i) $a^2 b - d$;   **j)** $a^3 b - ad$.

I. 8. Calculate:
a) $-8 + 5 - 2$; **b)** $13 + (-8) + (-7) - (-3)$; **c)** $(-8) + (+6) - (-3)$;
d) $-5 - (-4) + 3 - 10$; **e)** $8 - (7-9) - 2$; **f)** $2 - 3 - 5 + 6 - 7 + 8 - 9$;
g) $(-10+8) + (4+6) - (-5-3)$.

I. 9. Calculate:
a) $(-4) + (-25) + 10 + (-25) + (+7) + 10 + 8 + (-30)$;
b) $(-27) + 32 + (-70) + (+40) + (-40)$;
c) $(-35) + 86 + 104 + 18 + (-20) + (+51) + (-15)$;
d) $45 + 31 + (-16) + (-31) + 22$;
e) $11 + 2 + 3 + (-4) + (-5) + (-6)$;
f) $1 + (-2) + 3 + (-4) + 5 + (-6)$;
g) $-8 - 7 - 6 - 5 - 4 - 3 - 2 - 1 + 0 - 1 + 2 - 3 + 4 - 5 + 6 - 7 + 8$.

I. 10. Calculate:

a) $(2 + 3 - 5 + 7 - 5) + (3 - 5 + 8 - 12) - (-3 + 5 - 7)$;

b) $(4 + 5 - 7) + (-4 - 9 + 13) - (-7 + 12 - 75)$;

c) $(42 + 23) - (23 + 42) + (-41 + 40 - 0)$;

d) $2 - [3 - (5 + 6 + 9 - 20 + 1)]$;

e) $(3 - 4 + 7) - [(5 - 4 + 9) - (-3 + 9) - 2]$;

f) $2 - \{3 - [+9 - (4 - 5)] + 1\}$;

g) $(-5) - [(-34) + (+16) - (-19) + (-1)] - (+11)$;

h) $(-11 + 13) - \{(-15) + (+7) - [-(-23) - (+29) \text{-}14 - (+13)] - (-32 + 45)\}$.

I. 11. Calculate:

a) $-39 - [-17 - (-15 + 39 - 26) - 20] - (-14 + 22)$;

b) $-(-52) - \{-24 + 31 - [(-37) + (+26) - (+15) + (-51)] - (-31)\}$;

c) $-32 + \{-33 - [-(-41) + (-19) + (-6)] - (-15 - 8)\}$;

d) $6 - \{-27 - [-4 - (-18 + 37) - (43 - 50)]\}$.

I. 12. Fill out the table

a	b	c	a-b	b-c	$(a$-$b)$+$2c$	$(2a$-$2b)$+$(b$-$c)$
26			11	1		
	25		1	1		
125			25	0		
45			5	10		
60		6	30	10		
	7		1	6		

I. 13. Calculate:

a) $13 - \{(-32) + [(-45) + (+43) + (-24)] - (+71)\} - 141$;

b) $44 - \{-43 - [-17 + (-59) + (+38)] + (+77)\} - 18$;

c) $-26 - \{-49 + [-37 + (+55) + (-32)] + (-33)\} - 30$;

d) $-35 + \{-22 - [-39 + (-52) + (+65)] + (-69)\} + 61$;

e) $-21 - \{-15 + [-30 + 3 + (-50)] + 16\} + 10$.

I. 14. Calculate:

a) $-99 - \{-93 - [-21 + 65 + (-50)]\} + 18$;

b) $24 - \{(-58) + 67 - [(-19) + 43] + 54\} + 3$;

c) $35 + \{-44 - [-24 + (-35 + 50) + 16] + (-40)\} + 62$;

d) $-32 + \{-10 - [71 - (-71 + 80) + (-40)] + (-50)\} + 83$;

e) $69 - \{-43 - [-35 + (-39) + 52] + 88\} + 33$;

f) $-5 - (-2) - \{(-5) - [(-8) + (-3) - (-7)] + 5\}$.

16

I. 15. Calculate:

a) $(-9) \times (+2) \times (-2) \times (-4)$; **b)** $(+4) \times (+1) \times (+3) \times (-2) \times (-2)$;

c) $-(-2) \times (+3) \times (-1) \times (-8) \times (-4)$; **d)** $(108) \div (-3)$; **e)** $(-35) \div (-5)$;

f) $45 \div (-9)$; **g)** $(-75) \div (+25)$; **h)** $(-92) \div (-23)$.

I. 16. Calculate:

a) $(-2) \times [27 + 108 \div (-9)]$; **b)** $24 + (-81) \div (+3) - (-2) \times (+6)$;

c) $21 + (-8) \times (+4) \div (-16)$; **d)** $-55 \div (-11) \times (-7) \div 35$;

e) $-4 \times \{(18 - 10) \div [6 - (-2)]\}$; **f)** $34 \div (-17) + (-24) \div (-8)$;

g) $64 \div 32 \div (-2)$; **h)** $54 \div [27 \div (-3)]$;

i) $(74 - 102 \div 3) \times 2 - 42 \div 6$; **j)** $(945 \div 9) \div (-5) - (-32) \div (+8)$.

I. 17. Calculate:

a) $-2 \times [(-3) \times (+8) + (-3) \times (+2)]$; **b)** $-2 \times [(-6 \times 8) + (-6 \times 2)]$;

c) $(-13) + (-6) \div 2 - \{(-5) \times (-4) - [2 - (3 - 5)]\}$;

d) $(-3) \times [(-15) + 4 \div (5 - 3)]$; **e)** $(-5) \times [17 + (-3) \times (-5) - 6]$;

f) $(-2) \times [(-40) \times (-1) + (-28) \div (+7)$;

g) $-[(-5) - (-48) \div (-16) + (-2) \times (+3)]$.

I. 18. Calculate:

a) $\{80 + [10 - 5 + (7 - 3)] - 9\} \div (-40)$;

b) $50 \times \{59 - [7 \times 8 + (13 - 2 \times 5)]\}$;

c) $22 \div \{17 - [15 - 3(7 - 4)]\}$;

d) $3 \times \{18 + [3 + 4 \times (4 - 2)]\}$;

e) $(15 - 8) \times \{8 + [3 + (7 - 1)]\}$;

f) $10 + [1 + (15 - 2) \times 4]$;

g) $30 - \{121 \div (-11) \times (-11) - 4 \times (-4) - 3 \times [3 - (9 - 2)]\}$;

h) $(-1) \times [(-7) + (-23)] \times (-7) - (8 - 10) \times (-4)$;

i) $5 - (-3)[4 - (7 - 3)] \div 3 \times [5 + (-3) \times (-6)]$;

j) $(-3)(-4) + [(-5) - (-6)] \times (-7)$;

k) $(-3) + (-4)[(-5) - (-6)] - (-7)$;

l) $[(-1) + (-6) + (-5)] \div (-6) - (+7)$.

I. 19. Calculate:

a) $(-5) \times \{(-14) \times (-3) - (-25) \times (-2) - [(-4) \times (+3) - (-5) \times (+1)]\}$;

b) $(-3) \times \{(-24) \times (-3) - (-5) \times (-13) - [-(-34) - 3 \times (-21 + 36)]\}$;

c) $(-4) \times [(-32) \div (-8) - (-48) \div (-4) + (-25 + 10)]$;

d) $(-9) \times \{-5 - [(-24) \div (-2) - (-54) \div (-6)] \times (-4)\}$.

I. 20. Calculate:

a) $[(-2) \times (+27) - (-9) \times (-4) - (-7) \times (+16)] \div (-10 + 8)$;

b) $[(-2) \times (-32) + (+3) \times (-24) - (-5)(-9)] \div (-65 + 12)$;

c) $\{-36 - [(-8) \times (+7) - (-23) \div (-1) + (-36) \div (-3) - (-38 + 19)]\} \div (-4)$;

d) $[(-6) \times (-5) - (-24) \div (-3) - (+27) \div (-3)] \times [(-2) \times (-7) - (-2) \times (-9)]$;

e) $[(+18) \div (-3) - (-45) \div (-3) - (-32) \div (+2)] \times [-24 + (-3) \times (-6)]$;

f) $[(-36) \div (-12) \times (+3) - (-18) \div (+2) \times (-2)] \times [-52 + (-25) \times (-2)]$.

I. 21. Fill out the table

a	b	c	d	a²c-3a²b+a+b	a²+3ab²-abc
6	18	54	162		
3	9	27	81		
11	33	99	297		
9	27	81	243		
17	51	153	459		

I. 22. Calculate:

a) $(-2)^3$; **b)** $(-5)^1$; **c)** $(-15)^2$; **d)** $(-1)^{203}$; **e)** $(-1)^{600}$; **f)** 17^1; **g)** 1^{500};

h) 451^0; **i)** 0^{801}; **j)** $(-2)^3 \cdot (-3)^3 \cdot (-1) + 2^3 \cdot 3^2$;

k) $(-1)^3 \cdot (-5)^2 + 2^3 \cdot (-3)^2 \cdot 3 - 5^0$; **l)** $2^{23} \cdot (-5)^{25} + 2^{25} \cdot (-5)^{23}$; **m)** 2^{3^4};

n) $(-3)^{2^3}$; **o)** $\left(2^1 \cdot 2^3 \cdot 2^5\right)^9 + [(-2)^1 \cdot (-2)^3 \cdot (-2)^5]^9$;

p) $(-2) \cdot (2^2)^3 \cdot (-2^4)^4 \div [2 \cdot (-2)^6 \cdot 2^{16}]$.

I. 23. Calculate:

a) $\left[(-2)^3 \cdot 2^3 \cdot 2 \cdot (-2)^2\right]^3 \div 8^5$; **b)** $\left[(-3)^4 \cdot 3^2 \cdot 3^3 \cdot (-3)^5\right]^2 \div 9^8$;

c) $2^3 \cdot \left[(-8)^1 \cdot (-4)^2 \cdot (-2)^3\right]^2 \div \left[(-2)^2\right]^9$;

d) $\left[(-9)^2 \cdot 27^4 \cdot (-81)^2\right] \div (-81^2)^3$.

I. 24. Calculate:

a) $\left\{\left[(-2)^4 \cdot 2^3 \cdot (-2)^3\right] \div \left[2 \cdot (-2)^2 \cdot (-2)^3\right]\right\} \div 4^2;$

b) $\left\{\left[3^3 \cdot (-3)^4 \cdot (-3)^5\right] \div \left[(-3)^4 \cdot (-3)^2 \cdot (-3)^3\right]\right\} \div (-3)^2;$

c) $\left\{\left[(-2)^6 \cdot 4^2 \cdot (-8)^5\right] \div \left[(-4)^3 \cdot (-64)^2\right]\right\} \div (-8);$

d) $\left\{\left[(-3)^4 \cdot 3^2 \cdot (-27)^2\right] \div \left[(-9)^3 \cdot (+81)\right]\right\} \div (-3.)$

I. 25. Calculate:

a) $-3^9 \div (-3)^9 + (-3)^5 \div (-3^5) + (-5 \cdot 4)^7 \div \left(2^{11} \cdot 5^3\right);$

b) $(-2)^{45} \div (2^8)^5 \cdot \left[(-3)^{21} + 5^{21} \div (-5)^{10} - (-8)^2 \cdot (-2)^9\right] \div \left(-3^5 + 5^{11} + 2^{15}\right);$

c) $\left[3 + (-3)^2 \cdot (-3^4)^6 + 3^3 \cdot 81^3 + 3 \cdot (2^4)^{11}\right] \div (1 + 3^{25} + 3^{14} + 2^{44}) - 2.$

I. 26. Calculate:

a) $(-3^3)^3 \cdot (-3^3)^2 \div (-3^{3^2}) + 2^{3^4} \div (-2)^{9^2} + (-2^4)^{2^2} \div 2^{2^{2^2}};$

b) $\left[(-8)^4 \cdot 32^2 \cdot 2^2 \div (4^7)\right]^3 \div 8^{10} - \left[8^5 \cdot (-4)^2 \div (-8)^5\right]^3 \div (-4)^5;$

c) $\left[(-8^2)^2 \div (4^{10} \div 32^2)\right]^5 \div (2^3)^3 - \left[(-81^2)^2 \div (9^3 \cdot 3^4)\right]^3 \div \left[81^2 \div (-3^3)^2\right]^9;$

d) $\left\{\left[(-25)^3 \cdot 5^7 \div (5^3)^4\right]^5 \div (-25)^2\right\}^4 \div \left[-625^2 \div 25^3\right]^2;$

e) $(-27)^6 \div \left[(-3)^5 \cdot (-3)^3\right]^2 + (-3)^2 \cdot (-3)^8 \div (-3^3)^3 + 3^3 \cdot 3 \div 3^4;$

f) $\left[(-2) \cdot (-2)^2 \cdot (-2)^3 \cdot \ldots \cdot (-2)^{111}\right] \div \left[(-2) \cdot (-2)^2 \cdot (-2)^3 \cdot \ldots \cdot (-2)^{110}\right].$

I. 27. Calculate:

a) $2^3 \cdot \left\{10 + 1^{20} \cdot \left[\left(2^2\right)^2 \cdot 2^{2^2} \div 2^4 - 5 \cdot \left(3^3 - 5^2\right)\right] \div 3 + 2\right\};$

b) $\left[5^{62} \div \left(5^2\right)^{30} + \left(3^2\right)^3 \div 3^{2^2}\right] \div 2;$ c) $\left(2^{n+1} \cdot 3^{n+2}\right) \div \left(2^n \cdot 3^n\right) + 1^{n+1}, n \in N;$

d) $\left(3^2 + 2 \cdot 2^3 + 5^9\right) \div \left[9 + 2^{12} \div 2^8 + \left(5^3\right)^3\right].$

I. 28. Calculate:

a) $(-1)^7 + (-1)^{86} - (-1)^{101} + (-1)^{37};$ b) $(-1)^n + 2(-1)^{n+1} - 3(-1)^{n+2};$

c) $(-1)^n + 3 \cdot (-1)^{2n+1} - 2(-1)^{2n};$

d) $1 \cdot (-1)^{2n} + 2(-1)^{2n+1} + 3(-1)^{2n+2} + 4(-1)^{2n+3} + \ldots + 100 \cdot (-1)^{2n+99}.$

19

I. 29. Calculate:

a) i) $1 - 2^3 + 3^2$, ii) $\left(5^2 + 5 + 1\right) \div \left(3^3 + 2^2\right)$;

b) $\left[10^3 \div 5^3 + \left(2^2\right)^3 \div 2^{2^2} - 3^3\right] \cdot 2^{1^2} + 1^{2010} + 2010^0$;

c) $968 - \left\{14 \cdot 13 - 5 \cdot \left[2^3 \cdot 5 \div \left(3^2 \cdot 5 \div 15 + 5\right) + 2 \cdot \left(13^4 \div 13^3 - 12\right)\right]\right\} \cdot 5$;

d) i) $4^2 \cdot 15 - 4^2 \cdot 14$, ii) $11^2 \cdot 20 - 11^2 \cdot 18 - 11^2$; iii) $30 \cdot 11^2 - 11^2 \cdot 28$;

   iv) $\left(2^{2007} + 2^{2007}\right) \div 2^{2006}$; v) $\left(3^{2007} + 3^{2005}\right) \div 3^{2005}$;

   vi) $2^7 - 2^6 - 2^5 - 2^4 + 2^3$;

e) $\left\{\left[\left(7^2 - 14\right) + 2^5 \div 4\right] \div \left(3^3 + 2^4\right) + 3^2\right\} \cdot 11 - \left[\left(4 + 3^2\right) \cdot 5 - 55\right]$;

f) $\left\{\left[\left(5^4 - 5^2\right) \div 5^2 + \left(2^5 + 4^3\right) \cdot 2^2\right] - 3^4\right\} - 2010^0$.

I. 30. Calculate:

a) $3x^3 \cdot (-3x^2)$; b) $-4m^4 \cdot (-3m^2)$; c) $2x^4 \cdot (-4x^3)$; d) $5a^2b \cdot (8a^2b^3)$;

e) $2x^3y \cdot (7x^3y^2)$; f) $3x \cdot 4xy \cdot (2 \cdot 3x^2)$; g) $2ab^2c \cdot (-6a^2b^2)$;

h) $5m^2n \cdot (2m^2n^5)$; i) $5a^2b \cdot (-a^2b^3) \cdot (4a^4b^3)$.

I. 31. The decreasing order of the numbers
aaa, bbb, aba, and *bab* when *a < b* is:
a) *bbb, aba, bab, aaa*   b) *aba, aaa, bbb, bab*
c) *bbb, bab, aba, aaa*   d) *bab, aaa, bbb, aba*
e) *aaa, bbb, aba, bab.*

I. 32. Collect similar terms in the expressions below:

a) $5a + 6a - 3a$; b) $4x^2y - 3x^2y + 2x^2y$; c) $4mn - 3mn - 6mn$;

d) $m^2 - 4 + 3m^2 + 1$; e) $2x^3 + 6x^3y - 4x^3 - 2x^3y$;

f) $2rs^2 + rs^2 + rs + 5rs - 4rs^2$; g) $-4a^2b + 2a^2b + 2ab - 5a^2b$;

h) $6x^2yz - 3x^2yz - 2x^2yz$; i) $2x^2y - 4x^2y + 5xy^2$.

I. 33. Fill out the table

a	b	c	d	6a³c÷a–18a³b÷a+b	3a³c²÷b–a³cd÷b+b÷a
5	15	45	135		
3	9	27	81		
1	3	9	27		
9	27	81	243		
17	51	153	459		

Equations and inequalities in the set Z (integer numbers)

I. 34. Prove that for any $k \in \mathbf{Z}$,

a) $3 \mid 3k$; **b)** $7 \mid 14k$; **c)** $5 \mid 5x + 5y$ for any $x, y \in \mathbf{Z}$;
d) $3 \mid a$ and $5 \mid a$, then $15 \mid a$, $a \in \mathbf{Z}$; **e)** $3 \mid (12a + 21b)$, for any $a, b \in \mathbf{Z}$;
f) $31 \mid 3^{n+3} + 4 \cdot 3^n$, for any $n \in \mathbf{N}$;

g) $17 \mid 2^{2n+3} \cdot 3^{n+1} + 3^{n+2} \cdot 4^{n+1} + 2^{n+3} \cdot 6^n$, for any $n \in \mathbf{N}$.

I. 35. Prove that if $7 \mid (5x + y)$, then $7 \mid (2x - y)$, where $x, y \in \mathbf{Z}$.

I. 36. Solve for a in the equations:

a) $5 + a = 8$; **b)** $55 + a = 59$; **c)** $5 - a = 2$; **d)** $78 - a = 71$; **e)** $a - 3 = 6$;
f) $38 - a = 24$; **g)** $a - 54 = 23$; **h)** $31 + a = 69$; **i)** $120 - 2a = 40$;
j) $279 - 3x = 222$.

I. 37. Solve in $\mathbf{Z}$ the equations:

a) $x + 4 = 11$; **b)** $y - 8 = 1$; **c)** $z - 2 = 3$; **d)** $x + 6 = 2$; **e)** $b + 4 = 1$; **f)** $-1 = x + 9$; **g)** $-20 + x = 19$; **h)** $(x + 6) - 3 = -2$; **i)** $(y - 5) + 19 = 14$; **j)** $(z - 4) + 6 = 21$;
k) $(b + 1) + 9 = 10$; **l)** $x - 8 = |2 - 9|$; **m)** $x - 7 = |4 - 11|$;

n) $x + |5 - 12| = -4$.

I. 38. Solve in $\mathbf{Z}$ the equations:

a) $12x = 144$; **b)** $8y = -32$; **c)** $7a = -28$; **d)** $-4x = 64$; **e)** $-7y = -35$;
f) $-35z = -105$; **g)** $-5x = 40$; **h)** $-2t = 22$; **i)** $3x = -3$.

I. 39. Solve in **Z** the equations:
a) $2x + 5 = 11$;   b) $11y - 5 = 17$;   c) $-13 + 4z = -37$;   d) $12 + 5t = -78$;
e) $19 = 5 - 2x$;   f) $17 = -3 - 5y$;   g) $-17 = -5 + 3x$;   h) $2x + 82 = 0$;
i) $4y + 180 = 0$.

I. 40. Solve in **Z** the equations:
a) $5x - x + 2 = -30$;   b) $5x - 4x + 2 = 5$;   c) $y - 7 - 4y = -4$;   d) $7x - 5x + 10 = 0$;
e) $4t - 3t - 2 = 0$;   f) $x - 14 - 3x = 0$;   g) $0 = a + 25 + 4a$;   h) $5 \times (x - 2) = -25$;
i) $6 \times (y + 9) = -36$;   j) $-3 \times (x - 3) = 12$;   k) $-8 \times (a + 5) = 56$;
l) $3 \times (t + 1) + 2 = -7$;   m) $2 \times (y - 3) - 7 = 5$;   n) $4 \times (x + 8) - 13 = -1$;
o) $5 \times (z - 2) + 30 = 20$;   p) $8 + 3 \times (x + 3) = -4$;   q) $13 + 4 \times (x - 4) = -3$;
r) $-4 - 3 \times (2x + 1) = -13$;   s) $-5 + 4 \times (3x - 6) = 7$;   t) $(2x - 10) + (x - 2) + x = 0$;
u) $(3 - x) + (4 - x) + (5 - x) = 0$;   v) $3(x + 4) - 2x = -5$;
w) $-(1 - 3y) + 2(y - 3) - y = -4 + 3y + 1$;   x) $-x - 2[6 - (1 - 2x)] = 0$;
y) $-2 \times [z + 3 \times (5 - z)] - 5 \times (z - 7) = 0$;
z) $5 \times a + 2 \times [3 \times (1 - a) - 2 \times (1 + a)] = 22$.

I. 41. Solve in **Z** the equations:
a) $4x = x + 21$;   b) $8x = 6x + 24$;   c) $12y = 80 + 8y$;   d) $a = 51 - 2a$;
f) $11x - 91 = 4x$;   g) $99x + 54 = 93x$;   h) $41y - 28 = 34y$;   i) $49 - 5x = 2x$;
j) $2x + 6 = 1 + x$;   k) $3y = 9 + 2y$;   l) $13a = 55 + 2a$;   m) $7x + 5 = 5x - 7$;
n) $4y - 3 = y + 6$;   o) $6t - 2 = t + 13$;   p) $3(x + 4) = 2x$;   q) $2(x - 5) = x - 1$;
r) $5 \times (1 + x) = 2 \times (7 + x)$;   s) $5x + 4 \times (1 - x) = -x$;   t) $2 \times (2 + x) - 3x = x + 10$;
u) $7x - 2 = 3x + 2$;   v) $4x - 2 \times (1 - x) = 4$;
w) $2 \times [3 \times (2 + x) - (-x)] = 5 \times (x + 3) - 2(1 - x)$;
x) $x - 1 - [2 \times (3 - 2x) - 3 \times (3 - x)] = x + 1$.

I. 42. Solve in **Z** the inequalities:
a) $x + 3 > 6$;   b) $x - 3 \leq 4$;   c) $-2 > x - 6$;   d) $x + 8 \leq 1$;   e) $3x < 12$;   f) $2x \leq -16$;
g) $-4x < 20$;   h) $-2x \geq -10$;   i) $-x + 4 \geq 1$;   j) $x - 5 < 13$;   k) $x + 3 \leq 12$.

I. 43. Solve in **Z** the inequalities:
a) $5x > 4x + 5$;   b) $7x - 3 > 6x + 6$;   c) $2x - 1 < 4x + 3$;   d) $3 + 3x > 7 - x$;
e) $9 - 2x \geq 6 - x$;   f) $3(x - 4) \leq 6$; g) $8 < 4 \cdot (3 - x)$;   h) $2 \cdot (x - 6) < 2 \cdot (x - 2)$;
i) $6 \cdot (1 - y) \geq 3 \cdot (3 - y)$.

I. 44. Solve in **Z** the inequalities:
a) $3 \cdot (3 - x) - 5 \cdot (5 + x) > 0$;   b) $2 \cdot (x - 3) > 3 \cdot (x + 2) - 3$;
c) $2 \cdot (x - 2) - 4 \leq 3(x + 1) + 3$;   d) $2 \cdot (y - 1) - 3(y + 2) \leq 3 - 2y$;
e) $3 - a + 2 \cdot (a - 3) \geq 3 \cdot (a + 1)$;
f) $3 \cdot (2y - 1) - 2 \cdot (y + 1) > 3y + 8$.

I. 45. Solve in **Z** the inequalities:

a) $-3 < x \le 4$; b) $4 \le 2x \pounds 8$; c) $-2 \le x + 2 < 3$; d) $-3 < x - 4 \le 2$;

e) $-3 < -2 + x \le 2$; f) $-4 < -x \le 2$; g) $-1 \le -x + 1 < 3$; h) $-9 \le -3x \le 3$;

i) $-9 \le -2x + 5 \le 3$; j) $-1 < -3x + 2 \le 5$.

TEST I. 1

1) The first three terms of two patterns are respectively 1,4,7 and 3,7,11 find the first six terms in each pattern.

2) Write each number in standard form.
   a) 800,000+60,000+5,000+30.
   b) seven hundred five thousand eight hundred six.

3) Calculate:
a) 50×40; b) $5 \times 7,000$; c) 37×78; d) $513 \div 9$; e) $632 \div 8$; f) $581 \div 7$.

4) Alex has 736 marbles to share among 8 people. How many marbles will each person get?

5) Vlad played video games for 329 minutes. His brother played 7 times less than Vlad. How long did his brother play?

6) Alex, Matthew, and Vlad worked together to cut their   neighbors' grass. At the end of the day, they jointly had earned exactly 9 of each of the 6 kinds of coins.
   a) How much money did they earn?

   b) The children shared the money equally. How much money did each child get?

7) Diana had $10.47. She spent $4.69. How much money does she have left?

8) At school, Alex bought chocolate ice cream for $1.19, a hot dog for $1.39, and a can of juice for $0.50. On the way home, he found a quarter on a bench. At the end of the day he had $2.45 in his wallet. How much money did Alex start with?

9) Matthew walks 80 m every minute. Alex walks 85 m every minute. How far will each person walk in 5 minutes? 20 minutes? 1 hour?

10) Vlad received $10 from his father. On the first day he bought a can of juice

for $0.69 and a cup of vanilla ice cream for $0.99. On the second day his mother gave him $2 and he bought a ball for $2.99. How much money does he have left?

TEST I. 2

Find out the value of *a* such that:
1) $18 + a = 49$; 2) $a - 31 = 49$; 3) $7 \cdot 4 + a = 5 \cdot 8$;
4) $a - 3 \cdot 7 = 5 \cdot 7$; 5) $a + 3 \cdot 7 = 5 \cdot 8$; 6) $4 \cdot 5 - a = 3 \cdot 6$;
7) $3 \cdot 7 + 2 \cdot 5 + a = 5 \cdot 9$; 8) $a - 3 \cdot 7 = 3 \cdot 9 + 4 \cdot 7 - 3 \cdot 4$;
9) $5 \cdot 9 + 4 \cdot 8 - a = 1 \cdot 9 + 9 \cdot 3$;
10) $4 \cdot 7 + 5 \cdot 5 + a = 5 \cdot 8 + 7 \cdot 7 + 8 \cdot 0$.

TEST I. 3

1) Find *a* from this equation
$(100 - 1) + (101 - 2) + (102 - 3) + (103 - 4) + (104 - 5) = 500 - a$.

2) Multiply a number by 4, add 24, divide by 2, and then subtract twice the original number. What is the result?

3) Calculate $2^7 - 2^6 - 2^5 - 2^4$.

4) The average of six numbers is 12. Subtract 6 from one of these numbers. What would the average be after you subtract 6 from one of these numbers?

5) The length of each side of a hexagon is a whole number. What number(s) cannot be the perimeter of this hexagon 5, 36, 77, or 78?

6) The sum of the squares of the first 20 positive integers is 2870. What is the sum of the squares of the first 19 positive integers?

7) Calculate $(2 + 4 + 6 + 8 + 10) \div (10 + 8 + 6 + 4 + 2)$.

8) How many edges does a cube have?

9) Alex has $2.50. He bought 13 gummy bears at 7¢ each. How many gumballs at 14¢ each can he buy with the money he has left?

10) Diana has just begun to read a 270-page book. If she reads 30 pages every day, how many days will it take her to read the whole book?

TEST I. 4

1) Calculate: $455 + \{455 + [455 + (455 - 455 \div 5)]\} - 7 \cdot 13 \cdot 16$.

2) Solve the equations:

i) $x \div 16 = 9$ and remainder 9; ii) $320 \div x = 80$;

iii) $5x + 8 = 53$; iv) $3 \cdot (x + 6) = 138$;

v) $(5x + 40) \div 5 = 17$.

3) If $ab - ac = 138$ and $a = 23$, then calculate $b - c$.

4) If $ab + ac = 432$ and $b + c = 12$, then calculate a.

5) If $a + b + c + d = 30$ and $a + d = 10$, then calculate $7a + 9b + 9c + 7d$.

6) If $x = 10$ and $a + b = 8$, then calculate $3x + 5a + 5b$.

7) Knowing that the difference of two numbers is 152, and one of them is 5 times bigger than the other, find the numbers.

8) The sum of two numbers is 265, and one of them is 4 times bigger than the other. Find the numbers.

9) The sum of two numbers is 513 . Divide one by the other one to get the quotient of 50 and the reminder of 3. Find the numbers.

10) Vlad buys 5 kg of grapes for $2 per kg and 20 kg of tomatoes for $3 per kg. How much money does he get back if he gives the vendor a 100 dollar bill?

TEST I. 5

1) Find a:

   a) $a \times 0.45 = 2.70$ b) $a \times 68 = 612$

   c) $2.3 \times a = 1.84$ d) $a \times 96 = 864$.

2) Find the missing number a in $(1.5 + 1 + a + 2.5) \div 2 = 9$.

3) Find the x in the following equation:

$5 \times [200.31 + (x + 15.45)] = 2005$

4) Calculate $255 \div (24.5 - 9.5) - (120 \div 6 - 159 \times 0) \div 2$

5) The sum of three numbers is 140. One of the numbers is 45 and another one is 70. Find out what the third number is.

6) In a farm there are pigs and hens. All together they have 40 heads and 100 legs. How many pigs and hens are in the farm?

7) The sum of two numbers a and b is 345, and a is 15 less than b.

Find the numbers.

8) Calculate $44.5 + 5 \times 32 \div 8 + 5 \times [40 + 8 \times (200 \div 5 - 72 \div 2)]$

9) The sum of three numbers is 120. One of them is 30, and the second one is 5 times more than the third number. Find the numbers.

10) Calculate $3 \times 1000 - [(5 \times 10 \times 2 - 8 \times 10)] \times 25 \div 10$.

TEST I. 6

1) Find a:

    a) $a \times 45 = 270$,   **b)** $a \times 63 = 567$,

    c) $a \times 33 = 264$,   **d)** $a \times 86 = 774$.

2) Find the missing number a in $(2 + 1 + a + 3) \div 2 = 9$.

3) Find x in the following equation $2 \times [500 + (x + 500)] = 2006$.

4) Compute $255 \div (25 - 10) - (240 \div 12 - 144 \times 0) \div 2$.

5) The sum of three numbers is 161. One of the numbers is 35 and the second one is 71. Find out what the third number is.

6) The last digit of a six-digit number is 6. If the digit 6 is exchanged with the digit to the left of it, then the new number is four times greater than the initial number. Find the number.

7) Compute $5 \times 1100 - \left[(25 \times 10 \times 2 - 9 \times 10)\right] \times 15 \div 10.$

8) The sum of two numbers a and b is 224, and a is 46 bigger than b. Find the numbers.

9) Compute $33 + 5 \times 64 \div 16 + 5 \times \left[44 + 8 \times (220 \div 5 - 93 \div 3)\right].$

10) The sum of three numbers is 212. One of them is 30, and the second one is 6 times more than the third one. Find the numbers.

TEST I. 7

1) Calculate: **a)** $2^3 \cdot 2^5 \cdot 2$;   **b)** $3^4 \cdot 3 \cdot 3^2 \cdot 3^3$;
c) $a^2 \cdot a^3 \cdot a^4 \cdot a^6$;   **d)** $7 \cdot 7 \cdot 7^5 \cdot 7^7$;   **e)** $\left(5^{10} \div 5^3\right) \div 5^2$;
f) $2^{10} \div 2^8$;   **g)** $\left(3^2\right)^4$;   **h)** $\left[\left(5^{12} \div 5^1 \cdot 5\right)^2\right]^3$.

2) a) $2^3 \cdot 2^5$; **b)** $3^4 \div 3^2$; **c)** $\left(2^3\right)^4$; **d)** $\left(5^2 \cdot 7^2\right)^4$.

3) Compare the numbers:
  a) 2^{10} & 8^3 ; **b)** 32^{20} & 16^{25}; **c)** 2^{15} & 3^{30} ;

  d) 3^{100} & 2^{150} **e)** 5^{30} & 3^{45} .

4) a) $\left(2^5 \cdot 3^7 \cdot 5^6\right)^2 \div \left(2^{10} \cdot 9^7 \cdot 5^1\right)$; **b)** $\left(2^{3^4} \div 2^{4^3}\right) \div \left(2^3\right)^4$;
  c) $32^{10} \div 4^{25}$;   **d)** $\left[\left(5 - 2^2\right)^{10} + \left(17 - 2^4\right)^{13}\right]^5$;

  e) $\left(25^2 \div 5^4\right)^{2010}$; **f)** $\left\{\left[\left(3^7 + 3^8 + 3^9 + 3^{10}\right) \div 3^7 - 1\right] \div 3 - 1\right\} \div 3$;

  g) $\left[2^3 \cdot 2^8 + 5^{70} \div 5^{10} - \left(3^2\right)^{25}\right] \div \left[4^2 \cdot 2^6 \cdot 2^1 + \left(5^5\right)^{2^2} - \left(3^2\right)^{5^2}\right]$.

5) Solve the equations:

  a) $2^x \cdot 2^3 = 2^7$; **b)** $x \cdot 2^{16} = 2^{19}$; **c)** $3^9 \div x = 3^7$; **d)** $\left(2^3\right)^x = 2^{18}$.

6) a) $\left(5^3 - 10^2\right)^2 = 5^x$; **b)** $\left(2^{10}\right)^4 \cdot 2^{5^2} \cdot 2^{12^0} = 2^x$; **c)** $\left[\left(5^{2^3} \cdot 5^{3^2}\right) \div 25^2\right]^5 = 5^x$.

7) Compare the numbers $a = 2^5 \cdot 3^7 \cdot 5^{26}$ and $b = 2^{30} \cdot 5^5 \cdot 7^7$.

8) Compare the numbers

$a = 4^{n+2} \cdot 9^{n+1} - 4^{n+1} \cdot 3^{2n+3}$ and $b = 3^{2n+3} \cdot 4^{2n+3} - 2^{2n+1} \cdot 6^{2n+3}$.

9) a) Find the natural number n if

$8^{(n-4)(n+4)} - 8 = 7 \cdot \left(8 + 8^2 + 8^3 + ...8^{2008}\right)$;

b) Prove that $25^{n+1} \cdot 3^{n+1} - 3^{n+2} \cdot 5^{2n+1} + 5^n \cdot 15^n$ is divisible by 31 for any $n \in N^*$.

10) If $a = 2^{100} + (2^{50})^2 + (2^{25})^4 + (2^5)^{20}$ and

$b = 2^{99} + (2^{33})^3 + (2^{11})^9 + (2^3)^{33} + 2^{99} \cdot x$, find x such that $a = b$.

TEST I. 8

1) Calculate: **a)** $1 - 2^3 + 3^2$; **b)** $\left(5^2 + 5 + 1\right) \div \left(3^3 + 2^2\right)$.

2) Calculate $\left[10^3 \div 5^3 + \left(2^2\right)^3 \div 2^{2^2} - 3^3\right] \cdot 2^{1^2} + 1^{2010} + 2010^0$.

3) Calculate

$22 \cdot 44 - \left\{14 \cdot 13 - 5 \cdot \left[2^3 \cdot 5 \div \left(3^2 \cdot 5 \div 15 + 5\right) + 2 \cdot \left(13^4 \div 13^3 - 12\right)\right]\right\} \cdot 5$.

4) Calculate, taking to common factor: **a)** $4^2 \cdot 15 - 4^2 \cdot 14$,

b) $11^2 \cdot 20 - 11^2 \cdot 18 - 11^2$; **c)** $30 \cdot 11^2 - 11^2 \cdot 28$; **d)** $\left(2^{2007} + 2^{2007}\right) \div 2^{2006}$;

e) $\left(3^{2007} + 3^{2005}\right) \div 3^{2005}$; **f)** $2^7 - 2^6 - 2^5 - 2^4 + 2^3$.

5) Calculate

$\left\{\left[\left(7^2 - 14\right) + 2^5 \div 4\right] \div \left(3^3 + 2^4\right) + 3^2\right\} \cdot 1 - \left[\left(4 + 3^2\right) \cdot 5 - 55\right]$.

6) Calculate $\left\{\left[\left(5^4 - 5^2\right) \div 5^2 + \left(2^5 + 4^3\right) \cdot 2^2\right] - 3^4\right\} - 2010^0$.

7) Find x from the equality

$$15 \cdot \left(3 \cdot x - 7 \cdot 2 + 3^2\right) + \left(2 \cdot x + 5^2 \cdot 2 - 2^2 \cdot 3^2\right) \cdot 5 = 5 \cdot \left(2 \cdot 4\right)^2 \div 2$$

8) Find x: $\left\{\left[(x - 4) \cdot 5 + 4^5 \div 2^{3^2} - 3^{10} \div (27)^3\right] \cdot 5 - 4^5 \div \left(2^3\right)^3\right\} \div 3 = 1.$

9) Calculate:

a) $1400 + 350 \div 5 + 15 \div 3 \cdot \left[265 - (50 \div 25 + 2) \cdot 65\right] \cdot 150;$

b) $7^{100} \div \left[7^{40} \cdot 7^{58} - \left(7^{10} \cdot 7^{15}\right)^5 \div 7^{27} + \left(8^{27} \div 8^{26} - 1^{120}\right)^{97} \cdot 7\right];$

c) Find the number k such that:

$64 + 2 \cdot 64 + 3 \cdot 64 + \ldots + 49 \cdot 64 = k^2.$

10) For any n a natural number we consider the numbers a, b and c:
$$a = \left[(3+4)^2 - \left(3^2 + 4^2\right)\right] \div \left(4^{n+1} \div 2^{2n}\right) - 2^2 \cdot 5 \div \left(3^2 - 2^2\right) - 1;$$

$$b = \left(25^n \div 5^{2n} + 1\right)^n; \quad c = \left[\left(2^n + 2^n\right) \div 2\right]^2.$$

Compare a, b, c. and find the number n such that $2ab = c$.

TEST I. 9

1) Show that the sum of three consecutive natural numbers is divisible by three.

2) Calculate:

a) $[(2 + 4 + 6 + \ldots + 100) + (1 + 3 + 5 + \ldots + 99)] \div 505;$

b) $2 + 4 + 6 + \ldots + 100 - (1 + 3 + 5 + \ldots + 99).$

3) Consider the numbers:

$a = 10 \cdot \{5 + 10 \cdot [362 - 10 \cdot (2^3 \cdot 3 + 24 \div 2^2)]\} + 2^2;$

$b = 2^3 + 2 \cdot [(3 \cdot 5^2 - 2^2 \cdot 18 \div 3^2 + 3 \cdot 11) \div 10^2].$ Prove that $a = 5^4 \cdot b + 4.$

4) Find $x \in N$ from the equality $(x - 2)^{2x-4} \cdot 2^3 = 8^{17}.$

5) Consider $A = 2^{2n+1} \cdot 3^{n+5}$, $B = 16^n \cdot 9^{n+4}$; $a = 4^n$, and $b = 3^{n+4}$, $n \cdot N^*$.
   a) Write A in terms of a and b.

   b) Show that B $= (a \cdot b)^2$.

6) Find the numeric value of each number.
   $x = 3^{1999} - (4 \cdot 9^{999} - 729^{333})$, $y = (8^{651} - 1024^{195}) \div (2^3 - 1)$, y^x, x^y, and show
   that the number $3^5 - 3^3$ is a perfect cube.

7) If $\overline{ac} \cdot b = \overline{bc} \cdot a + 1$, find the number $\overline{abc}$.

8) Compute the value of the expression
   $E = 1 \cdot (a - b) + 2 \cdot (b - c) + 3 \cdot (c - 4 \div 3)$ for $a + b + c = 4$.

9) Find the digit a such that the number $\left(\overline{a2} \cdot \overline{a0} + 1\right) \div \overline{aa}$ is a natural number.

10) Show that the number $a = 75 \cdot 5^{2n-1} - 8^n \div 2^{3n}$, for $n \cdot N$ is not
divisible by 5.

TEST I. 10

1) Calculate $804 - 4\{502 - [58 + 4(31 \cdot 9 - 234 \div 3)] \div 2\}$.

2) We add 46 to a number and the result is multiplied by 9 and we obtain 477.
What is the number?

3) If $a = 8 \cdot 6 - 5 \cdot 4 - 3 \cdot 2$;
       $b = 8 \cdot 7 + 6 \cdot 7 + 4 \cdot 3 + 2 \cdot 1$;
       $c = 3 \cdot 4 + 5 \cdot 6 + 6 \cdot 8 + 8 \cdot 10 + 11 \cdot 10$, then find $c \cdot (b - a)$.

4) Place the parentheses in the correct spot $7 \cdot 8 + 4 \cdot 11 - 50 = 14$.

Compute:

5) $9 \cdot (77 - 70) - [3 + (46 - 39) \cdot 3]$;

6) $102 - 10 - [3 \cdot (29 - 20) - (99 - 96) \cdot 8]$;

7) $8 \cdot (14 - 6) - [(16 - 9) + 14] + 2$;

8) $9 \cdot (20 - 11) - [2 + 6(17 - 8)]$;

9) $12.5 + 6 \times 32 \div 8 + 5 \times [40 + 8 \times (600 \div 15 - 144 \div 4)]$;

10) $6 + 5 \times 64 \div 16 + 4 \times [40 + 8 \times (200 \div 5 - 33)]$.

TEST I. 11

1) Compute:

a) $-2 \cdot \{13 - 2 \cdot [14 - 5(-13 + 20 - 6)]\}$;

b) $|2 \cdot (-5) - (-4)| - 6 + 1|-(-5)^2|$;

c) $\{[(-12) \div 4 - (8 - 9 + 3 - 5) - (-2)^3 \div 4] \cdot 3 - 2 \cdot 10\} \div (3 - 17)$;

d) $-[3 \cdot (-2)^3 + 56 \div (-2)^2 - (-2^5) \div 2^4]$;

e) $3.75 \div 0.2\overline{7} \cdot 0.\overline{1} + 0.5$; f) $[0.\overline{7} + 7.\overline{7}] \div (0.7 + 0.7)$;

g) $5.6 \cdot (4.5 \div 0.9 + 1.25) \div 1.4 - (2.127 + 5.873) \div 4$;

h) $0.1 \cdot \{2.5 - [0.4 \cdot (28.2 \cdot 1.5 - 28.1) - 0.08] \div 2.8\}$.

2) Solve the equations and inequations:

a) $x + 25.17 = 32.15$; b) $37.5 - x = 17.03$; c) $3.1 \cdot x = 1.24$;

d) $x - 1.2 < 3$; e) $[7.23 - (0.5x - 2.7)] \cdot 1.5 = 2.25$; f) $0.4x + 1.4 \le 9.6$;

g) $(2.4x + 5) \cdot 1.9 = 27.74$.

3) Show that the following numbers are perfect squares:

a) 100; b) 9^7; c) $2 \cdot (1 + 2 + 3 + ... + 1000) - 1000$; d) $(2^7 \div 2^3) \cdot 2^4$.

4) Consider the numbers $a = c^2 + cd$ and $b = c^3 - 5cd$, where $c = -4$ and $d = 3$.

a) Find a and b;

b) Solve the equation $a(x + c) + b(x + d) = x(a + b + c + d)$;

c) Calculate $S = c + 2c + 3c + ... + 100c$.

5) Prove that the following numbers are perfect squares:

a) 121; b) 4^5; c) $2 \cdot (1+2+3+...+2010) - 2010$; d) $\left(2^{2^5} \div 2^{5^2}\right) \cdot 2^{3^2}$.

6) Consider the number $\overline{ab}$, with $a \neq 0$ and $b \neq 0$.
 a) Prove that $\overline{ab} + \overline{ba} = 11 \cdot (a + b)$;

 b) Find all the numbers $\overline{ab}$ if $\overline{ab} + \overline{ba} = 99$.

7) If x and y are digits such that $\overline{23xy} + \overline{xy23} + \overline{xyxy} = 11413$, then find $x + y$.

8) a) Show that the number $10^{40} + 455$ is divisible by 5;

 b) Find the sum of the digits of the number $10^{40} + 455$;

 c) Show that $15^{15} - 5^5$ is divisible by 10;

 d) Prove that $7^n - 3$ is divisible by 2, $n \in \mathbb{N}^*$.

9) Let $N = 3^{n+2} \cdot 2^{n+3} + 3^{n+1} \cdot 2^{n+4}$; $n \in \mathbb{N}^*$ be a number. Show that this number is divisible by 720.

10) If a and b are digits, $a \neq 0$. Show that the number

$$\left(\overline{a1b} + \overline{a2b} + ... + \overline{a9b}\right) \div \left(\overline{a5b}\right)$$ is a perfect square.

TEST I. 12

1) Prove the equality $\overline{xx2} + \overline{x4} = \overline{xx1} + \overline{x2} + 3$, where x is a digit between zero and nine.

2) Determine the non-zero digit number a from the equality,
 $a = 0.\overline{1a} + 0.\overline{2a} + 0.\overline{3a} + ... + 0.\overline{9a}$.

3) Determine the non-zero digit a such that $2.\overline{1a} + 2.\overline{a1} = \overline{a.12} + 1.\overline{a2}$.

4) Show that the number $N = \underbrace{111...1}_{n\ times}\underbrace{222...2}_{n\ times}$ is a product of two consecutive numbers.

5) If a, b, c, d are non-zero digits such that $\overline{ab} + \overline{bc} + \overline{ca} = \overline{dd}$, show that $\overline{abc} + \overline{bca} + \overline{cab} = \overline{ddd}$.

6) Show that $a = \left(\overline{x1x} + \overline{x2x} + ... + \overline{x9x}\right) \div \overline{x5x}$ is a perfect square.

7) Prove that the number $a = 10^3 \cdot \left(\overline{x.x} + \overline{x.0x} + \overline{x.00x}\right)$ is divisible by 61 for any non-zero digit x a digit number.

8) Find the number $\overline{abc}$ if $\overline{abc0} + \overline{abc} = 2057$.

9) Find the digits a, b, c such that $\overline{abc2} + 3 \cdot \overline{abc} = \overline{4abc} + 1234$.

10) Find the non-zero digits a and b, such that $\overline{abb}^{\overline{ab}} = \overline{ab}^{20 \cdot a + 2 \cdot b}$.

TEST I. 13

1) Prove that the number $A = 2 + 4 + 6 + ... + 4002 + 2002$ is a perfect square.

2) Prove that the number $A = 3^{2n+3} \times 4^{2n+3} - 2^{2n+1} \times 6^{2n+3}$ is a perfect square, $n \in N$.

3) Prove that the number $n = 2^{2013} - 2^{2012} - 2^{2011} - 2^{2010}$ is a perfect square.

4) Prove that the number $A = 2^{\overline{n0}} + 3^{\overline{n1}} + 5^{\overline{n2}} + 6^{\overline{n3}}$ is not a perfect square, where n is a number verifying the equality $2^{6n+10} + 4^{3n+5} + 8^{2n+3} = 5 \cdot 2^{3(n+3)+15}$.

5) Prove that the number $n = 2^0 + 2^1 + 2^2 + 2^3 + ... + 2^{98}$ is divisible by 7.

6) Show that the number $n = 3^1 + 3^2 + ... + 3^{2010}$ is divisible by 363.

7) Prove that the number $N = 2013^{2013} + 2014^{2014} + 2015^{2015} + 2016^{2016} + 2017^{2017}$ is not a perfect square.

8) Prove that the number $a = 11^{2012} + 22^{2012} + 33^{2012} + ... + 99^{2012}$ is not a perfect square.

9) Prove that the number $a = 4^{n+1} \cdot 9^{n+2} + 6^{2n+1} \cdot 9 - 2^{2n} \cdot 9^{n+1}$ is divisible by 369, for any n a natural number.

10) Compare the numbers x, y, and z, where n is a natural number.
$x = \{1^2 + 2^3 \cdot [3^4 - 3 \cdot (2^5 - 5)]\} \cdot 2^n$;
$y = [(1 + 2 + 3 + \dots + 10) \div 11 - 2]^{n-1}$;
$z = [(1 + 2^2 \div 4) - (3^2 - 2^3)]^3 \cdot (1002 \div 3 - 331)$;

Find the last digit of the number $b = x \cdot y \cdot z$.

TEST I. 14

1) Calculate:

a) $1234 \cdot 56 + 45 \cdot 1234 - 1234$;

b) $c^a + c^b + c^c$ where

$$a = \left\{[10 \cdot 1000 - (300 - 100) \cdot 50]^{2008} + 1\right\}^{2009},$$
$$b = \left(2^2 \cdot 2^3 \cdot 2^4\right)^3 - 8 \cdot \left(4^3 \cdot 4^4 \cdot 4^5\right),$$
$$c = \left\{10^3 + 10^2 \cdot \left[10 + \left(5^2 - 5^3\right) \div 5\right]\right\} \div 20^2 + 1.$$

2) Compare the numbers:

a) 9^{60} and 27^{40}; b) $7 \cdot 7^2 \cdot 7^3 \cdot 7^4 \cdot \dots \cdot 7^{50}$ and 5^{1275}.

3) If $a = 2^{100} + (2^{50})^2 + (2^{25})^4 + (2^5)^{20}$ and
$b = 2^{99} + (2^{33})^3 + (2^{11})^9 + (2^3)^{33} + 2^{99} \cdot x$ find x such that $a = b$.

4) Consider the numbers

$$x = \left[2^{30^2} \cdot \left(2^6\right)^{100} \cdot 2 + \left(32^8\right)^{50} \div 2^{500}\right]^2 + 2^{3004} \text{ and } y = 5 \cdot \left(9^{1001} - 9^{1000} - 3^{2001}\right).$$

a) Show that x and y are perfect squares;

b) Show that x is divisible by 10;

c) Compare the numbers x and y.

5) Prove that the number $x = \overline{ab0} + \overline{ba0} + \overline{ab} + \overline{ba}$ is divisible by 11 for any digits a and b.

6) Prove that the number $x = \overline{ab0} + \overline{ba0}$ is divisible by 110 for any digits a and b.

7) Prove that the number $x = \overline{ab0} + \overline{ba0} + \overline{a0b} + \overline{b0a}$ is divisible by 211 for any digits a and b.

8) Find the number $\overline{xyz}$ knowing that $\overline{xyz0} + \overline{xyz} = 2002$.

9) Prove that the number $\dfrac{\overline{abcabcabc...abc}}{2010 \ digets}$, is divisible by 7 for any digits a and b.

10) Prove that the number $y = \overline{abc} + \overline{acb} + \overline{bac} + \overline{bca} + \overline{cab} + \overline{cba}$ is divisible by 37 for any digits a and b.

TEST I. 15

1) What is sum of the digits of the number
$x = 10^{2006} + 9 \cdot 10^{2002} - 2$?

2) Prove that the number $2\underbrace{99...99}_{n-1 \ times}82\underbrace{00...00}_{n-2 \ times}29$ can be written as a sum of three squares of three consecutive numbers.

3) Show that the number $\overline{ab0ab0}$ is divisible by 1001.

4) Find the digit x satisfying the equality $\overline{2x31} + \overline{24x2} + \overline{5x3} = \overline{5x16}$.

5) Find the number $\overline{xyz}$ such that $\overline{xyz0} + \overline{xyz} = 2002$.

6) Find x from the equalities:

a) $15 \cdot \left(3 \cdot x - 7 \cdot 2 + 3^2\right) + \left(2 \cdot x + 5^2 \cdot 2 - 2^2 \cdot 3^2\right) \cdot 5 = 5 \cdot (2 \cdot 4)^2 \div 2;$

b) $\left\{\left[(x-4) \cdot 5 + 4^5 \div 2^{3^2} - 3^{10} \div (27)^3\right] \cdot 5 - 4^5 \div \left(2^3\right)^3\right\} \div 3 = 1.$

7) Find the number $\overline{abc}$ such that $\overline{abc} + \overline{bc} = 196$.

8) Prove that the number $N = \underbrace{111...11}_{n\ times}\underbrace{222...22}_{n\ times}$ can be written as a product of two consecutive numbers.

9) Find the natural number n such that $2011^{n} - n^{2011} = 2010$.

10) Show that the number $444^{2010} + 9$ is divisible by 5.

Chapter II

RATIONAL NUMBERS AND FRACTIONS

The set of *rational numbers* $\mathbf{Q} = \left\{ \dfrac{a}{b} \mid a \in \mathbf{Z}, b \in \mathbf{Z}, b \neq 0 \right\}$.

Finite decimal numbers $a.b = \dfrac{ab}{10}$, $a.bc = \dfrac{abc}{100}$ (b and c digits).

Ratio and proportions $\dfrac{a}{b}$, $b \neq 0$ is a *ratio*. $\dfrac{a}{b} = \dfrac{c}{d}$ is a *proportion*.

The fundamental property of a proportion is $ab = cd$.

Denominator Addition/Subtraction property $\dfrac{a+b}{b} = \dfrac{c+d}{d}$ or $\dfrac{a-b}{b} = \dfrac{c-d}{d}$,

similarly for nominator *Addition/Subtraction property*

$\dfrac{a}{b \pm a} = \dfrac{c}{d \pm c}$. Also $\dfrac{a}{b} = \dfrac{c-a}{d-b}$ or $\dfrac{a^2}{b^2} = \dfrac{c^2}{d^2}$.

II. 1. Write the numbers as decimals or periodic numbers:

a) $\dfrac{3}{4}$; b) $\dfrac{3}{5}$; c) $\dfrac{3}{100}$; d) $\dfrac{5}{6}$; e) $\dfrac{2}{3}$; f) $\dfrac{7}{20}$; g) $\dfrac{34}{35}$; h) $\dfrac{13}{15}$; i) $\dfrac{3}{25}$.

II. 2. Write each number as a fraction:

a) 0.7; b) 2.31; c) 0.01; d) 0.013; e) $0.\overline{34}$; f) $0.\overline{7}$; g) $2.\overline{3}$; h) $1.\overline{43}$;

i) $0.9\overline{3}$; j) $0.5\overline{6}$; k) $2.1\overline{3}$; l) $1.41\overline{6}$.

II. 3. Simplify the fractions:

a) $\dfrac{35}{105}$; b) $-\dfrac{54}{88}$; c) $\dfrac{99}{8181}$; d) $-\dfrac{35}{49}$; e) $\dfrac{2010}{5 \cdot 12 \cdot 67}$; f) $\dfrac{3^3 \cdot 7^3}{3^3 \cdot 7^4}$; g) $\dfrac{256}{2^7 \cdot 5}$;

h) $\dfrac{5^2 - 4^2}{2^2 \cdot 3^2}$; i) $\dfrac{3^2 \cdot 2}{105 - 69}$; j) $\dfrac{972}{963}$.

II. 4. Compare the numbers:

a) $-\dfrac{3}{4}$ and -0.75; b) $-\dfrac{3}{4}$ and $-\dfrac{2}{4}$; c) $-\dfrac{2}{5}$ and $\dfrac{1}{7}$; d) 0.47 and $\dfrac{47}{100}$;

e) $1.\overline{4}$ and $1\dfrac{4}{9}$; f) -0.35 and 0.34; g) $-\dfrac{3}{8}$ and $-\dfrac{1}{3}$;

h) $-\dfrac{2}{3}$ and $-\dfrac{7}{8}$; i) $\dfrac{2}{3}$ and $\dfrac{3}{4}$.

II. 5. Write the numbers in an increasing order:

a) $-2,\ -\dfrac{4}{3},\ -\dfrac{1}{3},\ 0,\ \dfrac{2}{3},\ \dfrac{4}{3},\ 2$; b) $-1\dfrac{1}{5},\ -\dfrac{4}{5},\ -\dfrac{2}{5},\ 0,\ \dfrac{2}{5},\ 1\dfrac{1}{5},\ 1\dfrac{3}{5}$;

c) $-2\dfrac{1}{4},\ -\dfrac{1}{4},\ \dfrac{1}{2},\ \dfrac{3}{4},\ \dfrac{1}{4}$; d) $\dfrac{1}{4},\ \dfrac{5}{7},\ \dfrac{3}{3},\ \dfrac{8}{2},\ \dfrac{9}{5},\ \dfrac{2}{9}$;

e) $\dfrac{3}{5},\ 5\dfrac{3}{2},\ \dfrac{4}{1},\ 2\dfrac{5}{7},\ \dfrac{4}{7},\ 1\dfrac{9}{3}$.

II. 6. Calculate:

a) $\dfrac{3}{8}-\dfrac{5}{8}$; b) $-\dfrac{4}{9}+\dfrac{7}{9}$; c) $-\dfrac{4}{7}+\left(-\dfrac{1}{7}\right)$; d) $-\dfrac{4}{7}+\dfrac{5}{7}$; e) $\dfrac{2}{3}+\left(-\dfrac{2}{3}\right)$;

f) $0.9+0.7$; g) $0.46+0.75$; h) $0.36+0.9$; i) $-0.6+0.2$;

j) $0.3+(-0.14)$; k) $-0.4+(-0.3)$; l) $2.6+(-5.82)$;

m) $-0.9+6.15$; n) $-0.29+(-0.9)$; o) $6.5+(-0.019)$.

II. 7. Calculate:

a) $-\dfrac{4}{7}+\dfrac{1}{2}$; b) $\dfrac{4}{7}+\left(-\dfrac{1}{2}\right)$; c) $-\dfrac{5}{7}+\dfrac{4}{8}$; d) $-\dfrac{3}{5}+\left(-\dfrac{1}{2}\right)$; e) $\dfrac{2}{5}+\left(-\dfrac{2}{3}\right)$;

f) $-\dfrac{3}{5}-\dfrac{1}{3}$; g) $-\dfrac{2}{5}+\dfrac{1}{2}$; h) $-\dfrac{1}{2}+\dfrac{1}{4}$; i) $\dfrac{1}{2}+\left(-\dfrac{3}{6}\right)+\left(-\dfrac{1}{4}\right)+\left(-\dfrac{2}{3}\right)$;

j) $\dfrac{1}{2}-\dfrac{3}{4}+\dfrac{5}{6}-1\dfrac{3}{6}$; k) $1\dfrac{3}{4}+\dfrac{4}{3}-1\dfrac{1}{2}-0.75$; l) $0.45+2.3-4.43-10$;

m) $2.31-0.35$; n) $0.42-(-0.5)$; o) $-1.5-(-1.4)$; p) $0.25-(+0.26)$.

II. 8. Calculate:

a) $\dfrac{5}{7}-\left(+\dfrac{2}{7}\right)$; b) $\dfrac{3}{8}-\left(-\dfrac{5}{8}\right)$; c) $-\dfrac{2}{7}+\left(-\dfrac{1}{2}\right)$; d) $\dfrac{1}{5}-\left(-\dfrac{1}{9}\right)$; e) $-\dfrac{1}{8}-\left(-\dfrac{1}{3}\right)$;

f) $-\dfrac{1}{6}-\left(-1\dfrac{1}{2}\right)$; g) $-\dfrac{3}{4}-\left(+\dfrac{2}{5}\right)$; h) $-1\dfrac{1}{3}+\left(-1\dfrac{1}{4}\right)-\left(-1\dfrac{5}{6}\right)$;

i) $\left(-1\dfrac{1}{20}\right)+1\dfrac{2}{15}-\left(-\dfrac{5}{6}\right)$; j) $-\left[-\dfrac{1}{16}+\left(-\dfrac{1}{32}\right)\right]$.

38

II. 9. Calculate:

a) $\dfrac{2}{5} - \left(\dfrac{2}{6} - \dfrac{1}{5}\right)$; b) $-\dfrac{1}{7} - \left(\dfrac{2}{5} - \dfrac{4}{7}\right)$; c) $-2\dfrac{1}{2} - \left(-\dfrac{3}{5} + 1\dfrac{1}{4}\right)$;

d) $-\dfrac{5}{12} - \left(-\dfrac{7}{18} - \dfrac{13}{24}\right)$; e) $1\dfrac{1}{3} - \left(-\dfrac{5}{6} + 1\dfrac{1}{4}\right)$; f) $-\dfrac{1}{6} - \left(-\dfrac{5}{12} - \dfrac{7}{18}\right)$;

g) $-\dfrac{5}{9} - \left[-\dfrac{11}{24} + \left(\dfrac{5}{36} - \dfrac{7}{48}\right)\right]$; h) $1\dfrac{5}{16} - \left[\left(\dfrac{11}{36} - \dfrac{7}{48}\right) - \left(-\dfrac{5}{18} + \dfrac{11}{24}\right)\right]$;

i) $-2\dfrac{5}{4} - \left\{\left(-\dfrac{3}{4} + 5.6\right) - \left[-\dfrac{1}{2} + \left(-\dfrac{1}{6}\right)\right]\right\}$;

j) $\left[\dfrac{1}{3} - \left(-\dfrac{1}{6}\right)\right] - \{(-1.2) - [0.4 - (-0.9)]\}$;

k) $2\dfrac{3}{4} + \left\{-1\dfrac{2}{3} + \left[0.8 + (-7) + \left(+1\dfrac{1}{2}\right)\right] - 3\dfrac{1}{5}\right\}$;

l) $-3 + \left(-\dfrac{3}{4}\right) + \left[\left(-1\dfrac{5}{6}\right) + 2.5 + 2\right] - \left[3\dfrac{1}{2} + (-3)\right]$;

m) $-2\dfrac{1}{4} - \left\{\dfrac{3}{8} - \left[1\dfrac{1}{6} - \left(2\dfrac{1}{2} - 1\dfrac{2}{3}\right)\right] - 1\dfrac{3}{4}\right\}$.

Multiplication, Division, Order of Operations.

II. 10. Calculate:

a) $3 \cdot \dfrac{2}{3}$; b) $-\dfrac{4}{9} \cdot 9$; c) $-5 \cdot \left(-\dfrac{1}{5}\right)$; d) $-0.2 \cdot \dfrac{3}{2}$; e) $-6 \cdot (-2.5) \cdot \left(+\dfrac{1}{4}\right)$;

f) $10 \cdot (-0.25) \cdot \left(-\dfrac{3}{5}\right)$; g) $\dfrac{10}{3} \cdot \left(-\dfrac{7}{52}\right) \cdot \dfrac{26}{35}$; h) $0.12 \cdot (-0.7) \cdot (-2.5)$;

i) $-3 \cdot \dfrac{4}{3} \cdot \left(-\dfrac{7}{44}\right) \cdot \dfrac{11}{35} \cdot \left(-\dfrac{25}{75}\right)$.

II. 11. Calculate:

a) $\dfrac{2}{9} \div \dfrac{1}{3}$; b) $\dfrac{3}{25} \div 1\dfrac{1}{4}$; c) $-\dfrac{8}{7} \div \left(-3\dfrac{1}{2}\right)$; d) $-\dfrac{24}{35} \div \left(-1\dfrac{5}{7}\right)$;

e) $-\dfrac{33}{4} \div \left(-2\dfrac{1}{6}\right)$; f) $-3\dfrac{3}{7} \div \dfrac{48}{14}$; g) $14.4 \div 1.2$; h) $-1.19 \div 0.17$;

i) $17.5 \div (-0.25)$; j) $4.59 \div \left(-\dfrac{9}{10}\right)$; k) $6.5 \div \left(-\dfrac{5}{2}\right)$; l) $-9.6 \div \dfrac{3}{10}$;

m) $\dfrac{8}{3}\cdot\left(-\dfrac{5}{6}\right)\div\left(-\dfrac{15}{18}\right)$; n) $-\dfrac{5}{6}\div\dfrac{2}{6}\cdot(-0.2)$.

II. **12.** Calculate:

a) $\dfrac{5}{7}+\dfrac{2}{7}\cdot\dfrac{1}{3}$; b) $\dfrac{3}{5}-\dfrac{3}{5}\cdot\dfrac{1}{2}$; c) $\dfrac{5}{3}\cdot\dfrac{3}{8}-\dfrac{3}{4}\cdot2$; d) $\dfrac{4}{5}+\dfrac{1}{5}\cdot\left(1-\dfrac{3}{2}\right)$;

e) $\left(\dfrac{1}{5}+\dfrac{1}{6}\cdot\dfrac{1}{5}\right)\cdot6$; f) $\left(-1\dfrac{1}{4}+2\dfrac{1}{2}\right)\cdot(-2)+1$; g) $\dfrac{12}{20}+\dfrac{6}{5}\div\dfrac{6}{10}+\dfrac{1}{2}$;

h) $\dfrac{6}{4}-\dfrac{1}{3}\div\left(\dfrac{1}{5}+\dfrac{1}{15}\right)$; i) $0.5\div\dfrac{3}{5}+\dfrac{-6}{4}\cdot\dfrac{-1}{-2}$; j) $\dfrac{7}{12}\cdot\dfrac{3}{14}-\dfrac{7}{12}\cdot\dfrac{3}{14}$;

k) $-\dfrac{4}{10}\cdot\dfrac{4}{9}-\dfrac{2}{5}\cdot\dfrac{14}{9}$; l) $-\dfrac{5}{12}\cdot\dfrac{5}{27}-\dfrac{7}{12}\cdot\left(-\dfrac{7}{27}\right)$; m) $-\dfrac{3}{7}\cdot\dfrac{4}{5}-\dfrac{3}{7}\cdot\left(-\dfrac{5}{12}\right)$;

n) $\dfrac{11}{12}+\dfrac{2}{3}\div\left(-\dfrac{36}{27}\right)$; o) $\left(-\dfrac{8}{10}+\dfrac{5}{6}\right)\div\dfrac{4}{10}$; p) $\left(-\dfrac{44}{3}\right)\cdot\left(\dfrac{5}{21}-\dfrac{3}{14}\right)$;

q) $\left(\dfrac{4}{18}-\dfrac{7}{15}\right)\div\left(-\dfrac{2}{75}\right)$; r) $\dfrac{39}{21}\cdot\left(-\dfrac{7}{15}\right)-\dfrac{3}{7}\div\left(-\dfrac{9}{35}\right)$;

s) $\left(-3\dfrac{2}{10}\right)\div\left(-1\dfrac{2}{3}+0.4\right)\cdot\left(-1\dfrac{3}{16}\right)$; t) $10\cdot\left(2\dfrac{1}{7}-1\dfrac{1}{2}\right)\cdot[-0.1\overline{5}]$.

II. **13.** Calculate:

a) $-\left(2+\dfrac{7}{9}\right)\cdot\left[-\dfrac{1}{3}+\dfrac{1}{7}\cdot\left(\dfrac{2}{5}-\dfrac{1}{6}\right)\right]$; b) $\dfrac{1}{4}\div\left[-\dfrac{1}{3}-\dfrac{1}{2}\cdot\left(\dfrac{3}{4}-\dfrac{1}{2}\right)\right]$;

c) $\dfrac{3}{2}\div\left[\dfrac{1}{2}+\left(\dfrac{5}{12}-\dfrac{2}{3}\right)\div\left(-\dfrac{1}{4}\right)\right]$; d) $2\dfrac{13}{15}\div\left[-1\dfrac{1}{6}+1.16\div\left(3\dfrac{4}{15}-1\dfrac{1}{3}\right)\right]$;

e) $\left(\dfrac{1}{4}\cdot\dfrac{6}{2}+\dfrac{3}{4}\right)\div\left[\dfrac{1}{2}+\dfrac{3}{5}\cdot\left(-\dfrac{5}{4}\right)\right]$; f) $8\cdot\left[\dfrac{1}{4}+\left(\dfrac{7}{36}-\dfrac{1}{24}\right)\div\dfrac{1}{18}\right]$;

g) $2\dfrac{3}{4}-0.75\cdot\dfrac{11}{5}\div\left(2+0.5\div2\dfrac{1}{2}\right)$; h) $\left(1.91+0.09+\dfrac{3}{4}\right)\div\left(1\dfrac{2}{5}+2.84+0.16\right)$.

II. **14.** Calculate:

a) $\left(\dfrac{8}{5}\cdot\dfrac{5}{8}+1-\dfrac{6}{10}\cdot\dfrac{5}{6}\right)\div\left(\dfrac{4}{5}+\dfrac{18}{50}\cdot\dfrac{15}{12}-\dfrac{7}{10}\right)$;

b) $\left[-3\dfrac{1}{4}-\dfrac{3}{4}\cdot\left(-\dfrac{20}{9}\right)\right]\div\left(-\dfrac{38}{2}\right)+\left(\dfrac{5}{12}-\dfrac{7}{30}+\dfrac{2}{5}\right)\div\left(-\dfrac{1}{4}-\dfrac{4}{5}\right)$;

c) $\dfrac{4}{6}-1\dfrac{2}{3}\cdot\left(2\dfrac{9}{20}\cdot2\dfrac{13}{21}-7\dfrac{7}{15}\right)\div\left(1\dfrac{1}{35}\div1\dfrac{13}{14}-1\dfrac{7}{10}\right);$

d) $\left[-\dfrac{9}{14}\cdot2\dfrac{1}{3}+\left(\dfrac{4}{10}-\dfrac{3}{4}\cdot\dfrac{8}{9}\right)\div1\dfrac{3}{5}\right]\div\left(-2\dfrac{1}{12}\right);$

e) $\left(\dfrac{10}{5}\div\dfrac{3}{2}-\dfrac{3}{2}\div\dfrac{6}{2}\right)\cdot\left(\dfrac{1}{2}+2\right)-\dfrac{5}{4}\cdot\left(3\dfrac{3}{4}-2\dfrac{3}{4}\right);$

f) $\left[\left(3\dfrac{1}{9}\div3\dfrac{4}{9}+\dfrac{6}{62}\right)\div\dfrac{1}{4}-3\cdot\dfrac{2}{7}\right]\div\left[\left(6\dfrac{3}{5}\div1\dfrac{3}{8}-3\dfrac{4}{5}\right)\div\dfrac{1}{2}\right];$

g) $\dfrac{1}{4}+\dfrac{1}{4}\div\left[\dfrac{1}{4}\cdot\left(\dfrac{1}{4}+\dfrac{1}{4}\right)\right];$ h) $2\dfrac{1}{3}+\left[\left(1\dfrac{1}{2}+\dfrac{4}{9}\cdot2\dfrac{1}{4}\right)\div\dfrac{1}{9}+2\dfrac{3}{4}\right]\div\dfrac{1}{4};$

i) $\left[\left(1-1\dfrac{4}{5}\right)\div\left(2\dfrac{1}{2}-3\dfrac{1}{3}\cdot\dfrac{3}{5}+\dfrac{3}{4}\right)+2\div\left(1\dfrac{3}{4}-\dfrac{1}{4}\right)\right]\div\dfrac{2}{9}.$

II. 15. Simplify the expressions:

a) $34-8\cdot\left\{\dfrac{\left(15-9\dfrac{1}{3}\right)\div5\dfrac{2}{3}}{\left(19\dfrac{2}{3}-11\dfrac{7}{9}\right)\cdot\dfrac{9}{71}}\cdot\left[\left(\dfrac{79}{100}\div\dfrac{1}{4}\right)\div0.79+\dfrac{1}{4}\right]\right\};$

b) $1+4\cdot\left[\dfrac{(1-0.1)\cdot1\dfrac{2}{3}}{0.625+\dfrac{3}{8}+0.0625\div\dfrac{1}{16}}+2.04\div\dfrac{3}{25}\right]\div\dfrac{71}{4};$

c) $10-4\cdot\left[\dfrac{0.0625+\dfrac{2}{5}+2.04-\dfrac{1}{25}}{\left(2^3\cdot3^2\cdot5^4\right)\div\left(5^3\cdot3^2\cdot2^2\right)\cdot\dfrac{0.1}{3}+\dfrac{33.\overline{3}}{5^2\cdot2^2}}\right];$

d) $\dfrac{\left[\left(3\dfrac{2}{5}+5\dfrac{4}{15}\right)\cdot0.25\cdot\dfrac{7}{13}-\left(\dfrac{15}{2}-0.5\right)\div7\right]\cdot6}{\left(5.01-4\dfrac{7}{50}\right)\div0.03}+\dfrac{28}{29};$

e) $\dfrac{2\dfrac{3}{20}-1\dfrac{1}{2}\left(4\dfrac{1}{6}+\dfrac{0.003}{0.25}\div\dfrac{0.004}{0.2}-\dfrac{1}{2}\cdot9\dfrac{1}{3}\right)+\left(\dfrac{2}{5}\right)^2\cdot1.25}{\left(2.625+\dfrac{3}{8}+0.0625+\dfrac{15}{16}\right)-\left(0.03125+\dfrac{31}{32}+1.25+\dfrac{3}{4}\right)}+\dfrac{4}{5};$

f)
$$\dfrac{\left(5\dfrac{3}{5}-5\dfrac{4}{15}\right)\cdot 0.25\div\dfrac{1}{19}+\left(\dfrac{15}{2}-0.5\right)\div 7}{\left(\sqrt{26.4196}-4\dfrac{7}{50}\right)\div 0.125-6\dfrac{2}{3}}\ ;$$

g)
$$\dfrac{\left(\dfrac{7}{15}+\dfrac{14}{45}+\dfrac{2}{9}\right)\cdot 10\dfrac{1}{3}-1\dfrac{1}{1}\cdot\left(2\dfrac{2}{3}-1\dfrac{3}{4}\right)}{\left(\dfrac{3}{7}-\dfrac{1}{4}\right)\div\dfrac{3}{28}-1}\cdot 2+2;$$

h)
$$\dfrac{\left(6\dfrac{5}{9}-3\dfrac{1}{4}\right)\cdot 2\dfrac{2}{17}+40.5\cdot\dfrac{2}{9}\div 9}{4\div 6.25-1\div 5+\dfrac{1}{7}\cdot 1.96+0.28}\div 24;$$

i)
$$\dfrac{9-4.7\div\left(5-0.8\div 2\dfrac{4}{6}\right)}{\left(5\dfrac{3}{9}-3\dfrac{3}{4}\right)\div 1\dfrac{7}{12}+2}\cdot\dfrac{3}{8};$$

j) $50.05-\left(1.5+3\dfrac{2}{3}\right)\cdot 7\dfrac{1}{2}\div 3.15\cdot\dfrac{63}{31}-\left(1-2\dfrac{1}{5}\div 7\right)\cdot 1\dfrac{11}{24}-\dfrac{1}{20}.$

II. 16. Simplify the following expressions:

a)
$$\dfrac{\left(4\dfrac{3}{10}\div 3\dfrac{7}{12}-0.2\right)\cdot 5}{2\dfrac{7}{12}-0.25\cdot 9\dfrac{1}{3}\cdot\dfrac{5}{7}}\cdot 0.1+\dfrac{3.04}{2}+14\dfrac{12}{25};$$

b)
$$\dfrac{\left(8\dfrac{4}{45}-7\dfrac{1}{15}\right)\cdot 30}{1\dfrac{1}{3}}+\dfrac{4\dfrac{1}{4}\div 0.85+1\div 0.5}{(5.56-4.06)\div 3}\ ;$$

c)
$$\dfrac{\left(1\dfrac{16}{75}+2.46\right)\div(5\,.1\div 5)}{1\dfrac{2}{3}\div 1\dfrac{8}{9}\cdot\left(\dfrac{2}{15}+0.15\right)}+\dfrac{9.72-6\dfrac{13}{25}}{40.5\cdot\dfrac{2}{9}\div 9};$$

d)
$$\dfrac{\left(15-9\dfrac{1}{3}\right)\div 2\dfrac{5}{9}}{\left(20\dfrac{2}{3}-12\dfrac{7}{9}\right)\cdot\dfrac{9}{71}}\cdot\left(0.71-\dfrac{1}{4}\right)\div\left(0.71+\dfrac{1}{4}\right);$$

e) $2+0.125\cdot\dfrac{5\dfrac{8}{3}\cdot\left(4.2-3\dfrac{7}{11}+\dfrac{9}{5}\right)\cdot\dfrac{3}{23}}{4.8\div 5\dfrac{7}{10}\cdot 1\dfrac{3}{16}-\left(3\dfrac{1}{7}-2.8\right)\cdot 1\dfrac{1}{6}}-\dfrac{20}{11};$

f) $\dfrac{\left(5.225 - \dfrac{5}{9} - 3\dfrac{5}{6}\right) \cdot \dfrac{36}{43} + 1.3}{\left(2\dfrac{23}{50} + 1\dfrac{16}{75}\right) \div (5 \ .1 \div 5) - 0.09 + \dfrac{227}{300}};$

g) $\dfrac{15}{48} \div 0.125 + 1.456 \div \dfrac{91}{250} + 4\dfrac{1}{2} \cdot \dfrac{1}{3} + \dfrac{2.652 \div 1.3 - 1\dfrac{17}{30} + \dfrac{3}{50}}{\left(1.34 + 1\dfrac{1}{10} \div \dfrac{11}{15}\right) \div 5\dfrac{13}{40}};$

h) $\left(\dfrac{2\dfrac{1}{7}}{2\dfrac{1}{7} - 1\dfrac{2}{7}} + \dfrac{1}{2}\right) \div \dfrac{\left(1.\overline{7} - \dfrac{1}{9}\right) \cdot \left(2.\overline{3} - \dfrac{1}{2}\right)}{2.\overline{3} - \dfrac{1}{3}} \cdot \dfrac{\sqrt{\dfrac{1.\overline{7} - 1}{1.\overline{7} + 1} + \dfrac{18}{25}}}{3 - 7 \cdot 0.(142857)};$

i) Find A, B, C, and x from this equation: $\dfrac{B}{C} = \dfrac{A}{x}$

$A = \dfrac{1 + 1\dfrac{1}{2} \cdot \left(0.5 + 2\dfrac{1}{3}\right) + 0.6 \div \dfrac{1}{5}}{2 \cdot \sqrt{0.0121}}$; $B = \sqrt{\dfrac{0.01 + 0.001 + 0.0001 + 1 \div 1000}{10^{-4} \cdot \left(1 - \dfrac{1}{2}\right) \cdot \left(1 - \dfrac{1}{3}\right) \div \left(1 - \dfrac{1}{4}\right)}}$;

$C = \dfrac{1^2 + 2^2 + 3^2 + 4^2}{4^2 - 3^2 + 2^2 - 1^2}.$

Integer exponents. Properties of powers. The order of operations and use of brackets.

II. 17. Calculate:

a) $\left(+\dfrac{3}{5}\right)^3$; b) $\left(-\dfrac{2}{3}\right)^3$; c) $\left(-\dfrac{3}{5}\right)^2$; d) $\left(-2\dfrac{1}{3}\right)^4$; e) $\left(-\dfrac{11}{13}\right)^0$; f) $\left(-\dfrac{3}{22}\right)^1$;

g) $\left(+\dfrac{137}{219}\right)^0$; h) $\left(\dfrac{23}{11}\right)^1$; i) $(-1.1)^3$; j) $[-0.\overline{4}]^2$; k) $-(1.3)^2$;

l) $-(-0.5 + 0.2)^3$; m) $(-1 - 0.6)^3$; n) $\left(\dfrac{1}{6} - \dfrac{5}{4}\right)^3$.

II. 18. Calculate:

a) 3^{-2}; b) 5^{-3}; c) 4^{-2}; d) $(-6)^{-3}$; e) $(-5)^{-1}$; f) $\left(-\dfrac{1}{4}\right)^{-1}$; g) $\left(-\dfrac{2}{5}\right)^{-1}$;

h) $\left(-\dfrac{3}{4}\right)^{-2}$; i) $\left(-2\dfrac{1}{3}\right)^{-3}$; j) $\left(-0.\overline{6}\right)^{-1}$; k) $(-0.2)^{-3}$; l) $(-1.5)^{-2}$; m) $[-0.\overline{5}]^{-3}$.

II. 19. Use positive powers:

a) $32,\ 16,\ 8,\ 4,\ 2,\ 1,\ \dfrac{1}{2},\ \dfrac{1}{4},\ \dfrac{1}{8},\ \dfrac{1}{16},\ \dfrac{1}{32}$;

b) $\dfrac{1}{625},\ \dfrac{1}{125},\ \dfrac{1}{25},\ \dfrac{1}{5},\ 1,\ 5,\ 25,\ 125,\ 625$;

c) $\dfrac{1}{10000},\ \dfrac{1}{1000},\ \dfrac{1}{100},\ \dfrac{1}{10},\ 1,\ 10,\ 100,\ 1000,\ 10000$.

II. 20. Use negative powers for the numbers:

a) $\dfrac{1}{10^2}$; b) $\dfrac{1}{7^3}$; c) $\dfrac{1}{10000}$; d) $\dfrac{1}{27}$; e) $\dfrac{1}{16}$; f) $\dfrac{1}{121}$; g) $\dfrac{1}{729}$;

h) $\dfrac{1}{216}$; i) $\dfrac{1}{3125}$.

II. 21. Using properties of powers, calculate:

a) $2^4 \cdot 2^5$; b) $7^4 \cdot 7^2$; c) $(-8)^5 \cdot 8^3$; d) $(-3)^3 \cdot (-3)^2$; e) $4^5 \cdot 4^{-2}$;

f) $7^4 \cdot 7^{-5}$; g) $(-3)^{-2} \cdot (-3)^{-3}$; h) $\left(-\dfrac{2}{5}\right)^5 \cdot \left(-\dfrac{2}{5}\right)^3$; i) $\left(-\dfrac{2}{5}\right)^2 \cdot \left(-\dfrac{2}{5}\right)^3$;

j) $\left(-\dfrac{3}{7}\right)^3 \cdot \left(-\dfrac{3}{7}\right)$; k) $\left(-\dfrac{1}{2}\right)^{-2} \cdot \left(+\dfrac{1}{2}\right)^{-1}$; l) $(-3)^3 \cdot \left(-\dfrac{1}{3}\right)^{-2}$;

m) $\left(-\dfrac{2}{3}\right)^2 \cdot \left(-\dfrac{3}{2}\right)^{-3}$; n) $(-3)^{-4} \cdot (-3)^3 \cdot (-3)^2$;

o) $\left(-\dfrac{3}{7}\right)^4 \cdot \left(+\dfrac{3}{7}\right)^3 \cdot \left(+\dfrac{3}{7}\right)^{-4} \cdot \left(-\dfrac{3}{7}\right)^0$; p) $\left(-\dfrac{5}{11}\right)^{-3} \cdot \left(-\dfrac{5}{11}\right)^4 \cdot \left(-\dfrac{5}{11}\right)^5$;

q) $\left(-\dfrac{2}{3}\right)^7 \cdot \left(-\dfrac{3}{2}\right)^{-3} \cdot \left(-\dfrac{2}{3}\right)^{-6}$; r) $(-2.3)^3 \cdot (-2.3)^2 \cdot (+2.3)^{-4}$;

s) $(-2.7)^5 \cdot (-2.7)^{-7} \cdot (-2.7)^2$.

II. 22. Calculate:

a) $5^6 \div 5^2$; b) $2^4 \div 2^{-2}$; c) $(-2)^{-5} \div (-2)^{-3}$; d) $4^{-4} \div 4^{-1}$; e) $\left(\dfrac{2}{5}\right)^6 \div \left(\dfrac{2}{5}\right)^2$;

f) $\left(-\dfrac{5}{6}\right)^9 \div \left(-\dfrac{5}{6}\right)^8$; g) $\left[-\left(\dfrac{3}{4}\right)^{12}\right] \div \left(-\dfrac{3}{4}\right)^9$; h) $\left(-\dfrac{2}{3}\right)^6 \div \left(-\dfrac{3}{2}\right)^{-5}$;

i) $\left(+\dfrac{3}{5}\right)^{-5} \div \left(-\dfrac{5}{3}\right)^4$; j) $\left[\left(-\dfrac{4}{5}\right)^{13} \div \left(-\dfrac{4}{5}\right)^8\right] \div \left(+\dfrac{5}{4}\right)^{-4}$; k) $(-1.9)^8 \div (+1.9)^5$;

l) $(-0.\overline{4})^{21} \div (-0.\overline{4})^{19}$; m) $(-3.8)^{-5} \div (-3.8)^3$; n) $(-1.3)^2 \div (1.3)^{-5}$;

o) $\left[(-0.8)^3 \div (-0.8)^{-7}\right] \div (-0.8)^6$.

II. 23. Calculate:

a) $-(6^2)^8$; b) $\left[(-4)^2\right]^4$; c) $(-3^3)^4$; d) $(-2^4)^3$; e) $\left[\left(-\dfrac{2}{5}\right)^3\right]^2$; f) $\left[-\left(\dfrac{3}{4}\right)^2\right]^3$;

g) $\left[-\left(\dfrac{-3}{4}\right)^3\right]^2$; h) $\left(-\dfrac{2}{3}\right)^{2^3}$; i) $\left\{\left[\left(-\dfrac{3}{4}\right)^3\right]^2\right\}^5$; j) $\left[(-0.5)^2\right]^2$; k) $\left[(-1.8)^3\right]^2$;

l) $\left[-(0.\overline{3})^2\right]^4$; m) $(-0.5)^{3^4}$; n) $\left\{\left[(-1.5)^2\right]^3\right\}^3$; o) $\left[(-2)^{-3}\right]^{-2}$; p) $\left[(-4)^{-3}\right]^{-1}$;

q) $\left[\left(-\dfrac{2}{5}\right)^{-3}\right]^4$; r) $\left[\left(-\dfrac{3}{4}\right)^{-1}\right]^{-4}$; s) $\left[\left(-\dfrac{3}{5}\right)^2\right]^{-3}$; t) $\left\{\left[\left(-\dfrac{2}{3}\right)^{-4}\right]^{-3}\right\}^2$;

u) $\left\{\left[\left(-\dfrac{1}{2}\right)^0\right]^3\right\}^{-4}$; v) $\left\{\left[(-3.28)^{-2}\right]^3\right\}^0$.

II. 24. Using the properties of powers, calculate:

a) $\left(-\dfrac{2}{3}\right)^3 \cdot \left(-\dfrac{3}{2}\right)^2 \cdot \left(-\dfrac{3}{2}\right)^2$; b) $\left(\dfrac{1}{0.2}\right)^{-3} \cdot (-0.2)^3 \div \left(-\dfrac{1}{10}\right)^3$;

c) $(0.3)^{25} \cdot (0.3)^{12} \div \left\{\left[(0.3)^3\right]^9 \div \left[\left(\dfrac{3}{10}\right)^7\right]^2\right\}$;

d) $\left[\left(-\dfrac{1}{2}\right)^4 \cdot \left(-\dfrac{1}{2}\right)^3 \div \left(-\dfrac{1}{2}\right)^6\right]^2 \div \left(-\dfrac{1}{2}\right)^3$.

II. 25. Calculate:

a) $(-3)^5 \cdot (-27)^3 \cdot (-3)^{-10}$;   b) $2^{-3} \cdot (-2^7) \cdot 4^3$;   c) $10^{-6} \cdot 5^5 \cdot (-10)^3$;

d) $\left[(-3)^2\right]^3 \cdot \dfrac{1}{27}$;   e) $4^{-3} \div (-2)^7 \cdot 2^2$;   f) $\left[3^{12} \div (-9^2)\right] \div 0.\overline{3}$;

g) $-3^{11} \cdot (-2)^{-7} \div 6^{-7}$;   h) $\left[5^{-3} \div (-5)^{-6}\right] \div 25^2$;   i) $2^{-4} \cdot \dfrac{1}{4^{-20}} \cdot 16^{-4} \div \dfrac{1}{2^{12}}$;

j) $\left(\dfrac{1}{27}\right)^3 \cdot \left[(-3)^2\right]^3 \cdot 3^3 \div \left(\dfrac{1}{3^{-1}}\right)^6$;   k) $36^5 \cdot \left(-\dfrac{1}{6}\right)^6 \div \left[(-6)^{-1}\right]^2 \cdot \dfrac{1}{216}$;

l) $\left(\dfrac{1}{3}\right)^{-3} \cdot \dfrac{1}{3^8} \cdot 9^5 \cdot 9^{-6}$;   m) $\dfrac{2^3 \cdot 4^{-2} \cdot (-2)^4 \cdot 8^{-3}}{2^{-2} \cdot (-2)^3}$;

n) $(-3)^{-2} \cdot \left(-\dfrac{1}{3}\right)^{-3} \cdot (-3)^4 \div \left(\dfrac{1}{3}\right)^{-5}$;   o) $\left(\dfrac{1}{2} - \dfrac{2}{3}\right)^{-5} \cdot \left(\dfrac{2}{3} - \dfrac{1}{2}\right)^4 \div \dfrac{1}{27^{-1}}$;

p) $\left[\left(1\dfrac{1}{2}\right)^{-4} \div \left(0.25 + 1\dfrac{1}{4}\right)\right] \div \left(\dfrac{4}{9}\right)^{-2}$;   q) $\left[\left(-\dfrac{1}{2}\right)^{-3} \cdot \left(2 - \dfrac{5}{2}\right)^{-2}\right]^2 \div \left(4\dfrac{1}{2} - 3\right)^{-5}$.

II. 26. Calculate:

a) $\left(\dfrac{2}{3}\right)^2 \cdot \left(\dfrac{5}{7}\right)^2 \cdot \left(\dfrac{3}{2}\right)^2$;   b) $\left(\dfrac{1}{0.2}\right)^3 \cdot \left(\dfrac{1}{5}\right)^3 \cdot (0.1)^3$;   c) $\left(\dfrac{1}{2}\right)^4 \cdot (1)^4 \cdot \left(\dfrac{1}{0.5}\right)^4$;

d) $\left[\left(\dfrac{1}{2}\right)^5 \cdot \left(\dfrac{1}{2}\right)^3 \div \left(\dfrac{1}{2}\right)^6\right]^2 \div \left(\dfrac{1}{2}\right)^3$;   e) $\left(\dfrac{1}{27}\right)^4 \cdot \left(3^2\right)^3 \cdot 9 \div \left(\dfrac{1}{3}\right)^6$;

f) $36^3 \cdot \left(\dfrac{1}{6}\right)^6 \div \dfrac{1}{216} \cdot \left(6^1\right)^2$;   g) $\left(\dfrac{1}{3}\right)^3 \cdot \dfrac{1}{3^8} \cdot 9^5 \cdot 9^6$;   h) $\dfrac{4^3 \cdot 2^2 \cdot 2^4 \cdot 8^3}{\left(2^{-1}\right)^2 \cdot 2^3}$;

i) $5^2 \cdot \left(\dfrac{1}{5}\right)^3 \cdot 5^4 \div \left(\dfrac{1}{5}\right)^5$;   j) $\left[\left(\dfrac{1}{2}\right)^3 \cdot \left(2 - \dfrac{3}{2}\right)^2\right]^2 \div \left(6\dfrac{1}{2} - 6\right)^7$;

k) $\left(\dfrac{2}{3} - \dfrac{1}{2}\right)^6 \cdot \left(\dfrac{2}{3} - \dfrac{1}{2}\right)^4 \div \dfrac{1}{36}$.

II. 27. Calculate:

a) $\dfrac{5}{6^3} \cdot 6^4 - 1 \div 5^{-2}$;   b) $20^{-2} \cdot (-10)^2 - 5^2 \cdot (-4)^{-2}$;   c) $3 \cdot (-3)^{-3} + 3^{-2} - 9^{-2}$;

d) $(-2)^{-2} \cdot 3^1 + (-5)^2 \cdot 2^{-2} + (-4)^{-1} \cdot 3^2$;

e) $(-2.12)^0 \cdot (0.5)^{-2} - 10^2 \cdot (0.3)^2 + 3 \cdot (0.5)^{-2}$;

f) $\left[\left(-\dfrac{3}{7}\right)^6 \div \left(\dfrac{4}{7}-1\right)^4 - \dfrac{1}{7}\right]\cdot(-7)^2$; g) $\dfrac{15}{(-2)^3}+\left[(-1.25)^2-\left(-1\dfrac{3}{6}\right)^2\right]^2 \div \left(-2\dfrac{2}{3}\right)$;

h) $\left(5-\dfrac{3}{8}\right)^2 \cdot \left[(-2.5)^2-\left(-2\dfrac{1}{2}\right)^2\right]^{10}$; i) $-\left(\dfrac{1}{2}\right)^{-3}\cdot 0.25+\left(\dfrac{1}{3}\right)^{-2}\cdot 3^{-1}+0.3^{-1}\cdot\dfrac{1}{5}$;

j) $\left[\left(-2\dfrac{3}{4}\right)^3 \div (-1-1.75)^3\right]^2 \div \left(2+\dfrac{3}{4}\right)^2$;

k) $\left[(19-3^3)\div(-2)^2+45-(-8)^2\right]^0\cdot(-1.5)^3$;

l) $3^{-1}\div\left[2^{-1}\div(-2)^2\right]+(-0.1)^3\cdot(0.1)^{-2}$;

m) $0^5\cdot\left(-\dfrac{1}{5}\right)^{15}+\left\{\left[\left(-\dfrac{2}{3}\right)^3\right]^{-2}\right\}^0+\left[\left(-\dfrac{1}{2}\right)\cdot 4\right]^{-2}$;

n) $\left(-\dfrac{3}{4}\right)^5 \div \left(-\dfrac{3}{4}\right)^3-\left(-2+\dfrac{1}{2}\right)^6 \div \left(2-\dfrac{1}{2}\right)^4-\left(-\dfrac{3}{4}\right)^2$;

o) $\left[0.5^2\cdot\left(1+\dfrac{1}{2}\right)-\left(\dfrac{1}{3}\right)^2\right]\div\dfrac{19}{8}-\left(1.0\overline{5}-1\dfrac{3}{4}\right)\div\left(-1\dfrac{1}{5}\right)^{-2}$.

II. 28. Calculate:

a) $\left(\dfrac{7}{3}\right)^{-3}\cdot\left[\left(\dfrac{3}{7}\right)^{-2}\right]^{-3}\div\left[\left(\dfrac{3}{7}\right)^{-4}\right]^{-2}$; b) $\left[\left(-\dfrac{2}{3}\right)^6\cdot\left(-\dfrac{2}{3}\right)^{-3}\cdot\left(-\dfrac{2}{3}\right)^2\right]^{-2}$;

c) $\left(-\dfrac{2}{3}\right)^6\cdot\left(-\dfrac{2}{3}\right)^{-4}+\left(2-\dfrac{2}{3}\right)^2\div\left(2-\dfrac{2}{3}\right)^2-\left(-\dfrac{3}{2}\right)^{-2}$;

d) $\left(-\dfrac{3}{2}\right)^{-6}\div\left(-\dfrac{2}{3}\right)^4+\left(1-\dfrac{1}{3}\right)^2\div\left(2-\dfrac{4}{3}\right)^2+\left(-\dfrac{2}{3}\right)^3\cdot\left(1-\dfrac{1}{3}\right)$;

e) $\dfrac{1}{5^2}\cdot\left(-\dfrac{5}{3}\right)^3+(-3)^5\cdot(-3)^{-6}+\left(-\dfrac{2}{3}\right)^2$;

f) $(-0.4)^{-2}\div\left[(-2)^5\cdot(-2)^{-6}\right]^3-(-7)^3\cdot7^{-6}\div7^{-5}$;

g) $(-0.2)^{-3}\cdot\left[5^9\cdot(-5)^{-6}\right]^{-1}-\left[-0.\overline{3}+(-3)^{-2}\right]\div7^0$;

h) $\left\{\left[\left(\dfrac{6}{5}\right)^2\right]^{-3}\div\left[\left(\dfrac{6}{5}\right)^2\div\dfrac{125^{-2}}{216^{-3}}\right]+\left(\dfrac{125}{48}\right)^2\div\left(\dfrac{32}{15}\right)^{-3}\dfrac{2^{-7}\cdot(-5)^{-2}}{(-3)^{-5}\cdot5^{-1}}\right\}\cdot\left(\dfrac{5}{2}\right)^{-2}$;

i) $\left\{\left[\left(-\dfrac{9}{8}\right)^{-5} \div \left(\dfrac{27}{16}\right)^{-2} \cdot \dfrac{2^{-7}}{(-9)^{-3}}\right] \cdot \dfrac{48^{-1}}{4^{-2}} - \left[\left(\dfrac{9}{16}\right)^{-12} \cdot \left(\dfrac{9}{8}\right)^{16} \div \dfrac{4^{-2}}{9^{-1}}\right] \cdot \dfrac{4^{-2}}{9^{2}}\right\} \div 2^{-2}.$

Linear equations in the set of Q

II. 29. Solve the equations:

a) $6x - 2 = 7$; b) $6x - 7 = 1$; c) $5x + 2 = 0$; d) $3x - 3 = 8$; e) $2x - 1 = 8$;

f) $x + \dfrac{1}{4} = \dfrac{1}{3}$; g) $3x + 0.3 = \dfrac{1}{2}$; h) $-x + \dfrac{3}{4} = 0.5$; i) $-2x - \dfrac{2}{4} = \dfrac{1}{3}$;

j) $\dfrac{1}{5}x = -3$; k) $0.7 \cdot x = 7$; l) $-0.7x = -\dfrac{4}{5}$; m) $\dfrac{x}{2} = -3$; n) $\dfrac{a}{5} = -4$;

o) $\dfrac{y}{2} = -3.5$; p) $\dfrac{z}{0.6} = 5$; q) $-\dfrac{t}{6} = 0.5$; r) $\dfrac{x}{3} = 0.\overline{6}$; s) $\dfrac{5x}{9} = 0.\overline{3}$.

II. 30. Solve the equations:

a) $\dfrac{14}{3}x - 0.\overline{3} = 0$; b) $\dfrac{2}{5}x + \dfrac{1}{3} = 0$; c) $0.\overline{7}z - \dfrac{6}{7} = 5$; d) $0.7x - 0.2 = 7$;

e) $1.\overline{5}y - \dfrac{15}{8} = 0$; f) $0.4a + 0.4 = 0$; g) $\dfrac{4}{5}x + \dfrac{2}{3} = 0$; h) $2.4u - \dfrac{2}{7} = 1$;

i) $-1.9x - \dfrac{1}{7} = 8$; j) $-\dfrac{3}{5}y - \dfrac{3}{15} = 1$; k) $21 - \dfrac{19}{3}x = 2$; l) $2 - \dfrac{x}{3} = 5$;

m) $7 + \dfrac{x}{3} = -6$; n) $3 = \dfrac{3}{2}x + 6$; o) $21\dfrac{1}{2}b - \dfrac{2}{5} = -9$.

II. 31. Solve the equations:

a) $2x - 19x = 17$; b) $-5x + (-2+x) = 30$; c) $2x + 4 = x + 26$; d) $3x - 7 = 2x + 23$;

e) $-2 = -10x + 16x$; f) $\dfrac{3}{4} = -2x + 8x$; g) $\dfrac{1}{4}x + \dfrac{2}{3}x = 11$; h) $\dfrac{4}{5}x + \dfrac{1}{4}x = 42$;

i) $0.4x + 0.9x = 6.5$; j) $\dfrac{4}{9}y - \dfrac{5}{9} = -\dfrac{7}{27}$; k) $\dfrac{4}{3}x + \dfrac{1}{5}x = 2.3$; l) $4 = 4.3x + \dfrac{1}{11}$;

m) $-\dfrac{2}{5}y + \dfrac{1}{5}y - 2 = 8$; n) $7x - 8 = \dfrac{1}{3}x - 0.5$; o) $2(x + 7) - 4(x - 4) = 7$;

p) $4(x + 9) - 5(x + 12) = 8$; q) $4 - 3x - 3(8 - x) = 2(3 - x)$;

r) $3x - (1 - 2x) = 5x - 3$; s) $\dfrac{x}{4} + \dfrac{x}{3} = 7$.

II. 32. Solve the equations:

a) $-\dfrac{8}{3}x+\dfrac{1}{6}x=\dfrac{15}{2}$; b) $10x+\dfrac{5}{4}-3x=\dfrac{3}{8}$; c) $\dfrac{3}{4}x-\dfrac{2}{5}-\dfrac{5}{8}x=-\dfrac{1}{10}$;

d) $\dfrac{3}{5}x+\dfrac{3}{10}-\dfrac{1}{2}x=-\dfrac{1}{5}$; e) $\dfrac{3}{8}x-\dfrac{1}{2}-\dfrac{1}{4}x=\dfrac{3}{16}$; f) $\dfrac{3}{4}x-\dfrac{3}{5}-\dfrac{1}{8}x=-\dfrac{1}{10}$;

g) $-\dfrac{2}{3}+\dfrac{3}{5}x-\dfrac{1}{2}x=-\dfrac{3}{5}$; h) $-\dfrac{1}{9}-\dfrac{2}{3}x+\dfrac{1}{18}+\dfrac{1}{2}x=\dfrac{5}{6}$.

II. 33. Find x in the equations:

a) $3x-\dfrac{1}{4}=x+\dfrac{7}{4}$; b) $x+\dfrac{2}{5}=-x-\dfrac{8}{5}$; c) $4x-\dfrac{5}{3}=2x-\dfrac{2}{3}$;

d) $2\left(x-\dfrac{1}{2}\right)=3\left(x-\dfrac{1}{3}\right)-2$; e) $4\left(x-\dfrac{1}{4}\right)=5\left(x-\dfrac{1}{5}\right)-2$;

f) $4\left(2x-\dfrac{1}{4}\right)=8\left(4x-\dfrac{1}{8}\right)-12$; g) $\dfrac{2y}{5}+\dfrac{6y}{5}+\dfrac{12y}{5}+4y=40$;

h) $\dfrac{2+a}{5}+\dfrac{a}{3}-\dfrac{a}{5}+4a=\dfrac{2}{5}$; i) $2z+3\left(z-\dfrac{1}{6}\right)=4\left(z-\dfrac{1}{8}\right)-2$.

II. 34. Solve the equations:

a) $\dfrac{x}{2}-\dfrac{x-3}{3}=\dfrac{x}{4}+\dfrac{1}{6}$; b) $\dfrac{x}{3}-\dfrac{4-x}{2}=8\dfrac{5}{6}-\dfrac{x}{4}$;

c) $\dfrac{-2x+9}{3}-\dfrac{3x-3}{4}=\dfrac{x}{2}-x-1$; d) $\dfrac{-x-7}{2}-\dfrac{3x-5}{4}=\dfrac{x-2}{3}-2x+1$;

e) $\dfrac{2x+1}{2}-\dfrac{4x+1}{3}=\dfrac{2x+9}{4}-\dfrac{25}{6}$; f) $\dfrac{1}{5}\left\{\dfrac{1}{5}\left[\dfrac{1}{5}\left(\dfrac{1}{5}x-5\right)-5\right]-5\right\}-5=0$;

g) $2x+0.\overline{6}(x-3)=4-0.\overline{3}(5x-12)$; h) $0.8(x-0.\overline{3})+1.1\overline{6}(7x-5)=4.\overline{8}$;

i) $0.\overline{3}\cdot\left\{0.\overline{3}\cdot\left[0.\overline{3}\cdot\left(0.\overline{3}\cdot x+0.\overline{3}\right)\right]\right\}-0.\overline{3}=0$;

j) $\left[(x\div5-42)\cdot2-30\cdot14\right]\div6+29=79$.

II. 35. Solve în **Q** the equations:

a) $\dfrac{x-2}{3}+\dfrac{x-3}{4}=\dfrac{x-4}{6}+1$; b) $\dfrac{3x-1}{4}-\dfrac{x+1}{3}=x-1$;

c) $\dfrac{3x+1}{5}-\dfrac{4x-1}{2}=\dfrac{x+5}{10}-\dfrac{1}{2}$; d) $\dfrac{2x+3}{5}+\dfrac{5x-4}{3}-2=\dfrac{x}{15}$;

e) $\dfrac{a+1}{3}+\dfrac{a+3}{9}=0.\overline{3}$; f) $\dfrac{3+b}{7}+\dfrac{3b-1}{3}=b+\dfrac{1}{3}$;

49

g) $\dfrac{2(3x+5)}{15} - \dfrac{2(4x-1)}{3} = -\dfrac{2(3x+1)}{5} - \dfrac{2(7x-1)}{3}$;

h) $\dfrac{3x+1}{3} + \dfrac{3}{4} = \dfrac{x}{6} - \dfrac{3-x}{12}$; i) $\dfrac{3(x-1)}{2} - \dfrac{2(3-x)}{3} - \dfrac{1}{6} = \dfrac{x+1}{4}$;

j) $\dfrac{3(x-1)}{4} - \dfrac{5-x}{2} - \dfrac{3-2x}{6} = 1 - \dfrac{x}{12}$; k) $\dfrac{x}{3} - \dfrac{2-x}{5} = \dfrac{4(1-x)}{15}$;

l) $\dfrac{11x}{3} - \dfrac{3x+4}{5} = 3x - 2$; m) $\dfrac{4(2x-3)}{5} = 4 - \dfrac{3(x-3)}{4} - x$;

n) $\dfrac{x}{2} - \dfrac{x}{12} + 3 = \dfrac{2x}{9} - \dfrac{3-x}{6}$; o) $\dfrac{3x}{2} - \dfrac{5}{4} = 2x - \dfrac{2-2x}{3} - \dfrac{11}{6}$;

p) $\dfrac{5x+7}{2} - \dfrac{4x+7}{3} = 3x - \dfrac{1}{6}$; q) $\dfrac{5x+1}{3} - \dfrac{4-x}{2} = \dfrac{5x-8}{2} - 1$;

r) $\dfrac{5-x}{4} - \dfrac{1}{8} + 5x = 4 - \dfrac{3x}{2} - \dfrac{5}{4}$; s) $\dfrac{21x}{2} - \dfrac{5-x}{2} + \dfrac{3x}{8} = \dfrac{5x}{8} - \dfrac{3x}{2}$;

t) $\dfrac{t+5}{3} - \dfrac{t-9}{5} = 3 + \dfrac{3t-1}{15}$; u) $\dfrac{4a-1}{9} - \dfrac{5a-1}{6} - 3a = \dfrac{1}{3}$;

v) $\dfrac{3(x+8)}{4} - \dfrac{2(x-3)}{3} = \dfrac{3(x+5)}{3} - \dfrac{x}{4}$; w) $\dfrac{x-7}{4} + \dfrac{1}{3} - x + \dfrac{2x-1}{3} = \dfrac{x}{6}$;

x) $\dfrac{4(1-x)}{7} - \dfrac{2x-3}{3} - \dfrac{4x}{21} = 2(x-4)$; y) $\dfrac{x+3}{3} - \dfrac{3(x-2)}{4} = 1 + \dfrac{x-12}{2}$;

z) $\dfrac{2(5x-7)}{5} - \dfrac{3x-4}{3} = \dfrac{x+2}{10} - 5x$.

II. 36. Solve the equations:

a) $3(a-2) - \dfrac{1}{2} = 5\left(\dfrac{a}{2} - 1\right) + 2a$; b) $\dfrac{7}{3}(x-5) = -2\left(\dfrac{5}{3} - \dfrac{7}{3}x\right)$;

c) $5(x-4) = 6\left(\dfrac{x}{3} - 1\right) + 1$; d) $4(2.5x-1) + \dfrac{1}{3} = -2(3.5 - 3x) + \dfrac{1}{2}$;

e) $-0.\overline{5}(3x-2) - 0.5(4-2x) = 4(1-0.25x)$;

f) $0.4(3x-2) + 1 + 0.3 \cdot (5x-10) = 0$;

g) $0.5 \cdot (2x+1) - 0.4 \cdot (4x-1) = 2.3(x-4)$;

h) $0.5b + 12 = 0.\overline{3}(b+12)$; i) $4 \cdot \left(4x - 0.2\overline{3}\right) - 3\left(4x - 0.2\overline{3}\right) = 7x - 7$;

j) $\left(0.5\overline{3} - 5x\right) - 5\left(3 - 0.1\overline{3}x\right) = 11$.

II. 37. Solve the equations:

a) $a + 80 = \dfrac{9}{2}(10a + 8)$; b) $x + \dfrac{9x}{50} + 24\left(\dfrac{x+3}{100}\right) = 4.2$;

c) $0.80 = \dfrac{10 + a}{30 + a}$; d) $\dfrac{x-3}{2} - \dfrac{2(1-3x)}{0.2} + 1 = \dfrac{3(1-0.5x)}{2} - 7$;

e) $\dfrac{1}{2}\left\{\dfrac{1}{2}\left[\dfrac{1}{2}\left(\dfrac{1}{2}x - 1\right) - 1\right] - 1\right\} - 1 = 0$;

f) $x + \dfrac{1}{2}\left[\dfrac{2(x+1)}{3} + \dfrac{4(x+2)}{5}\right] = \dfrac{x-1}{2} + \dfrac{2}{3}\left[\dfrac{3(x-1)}{4} - \dfrac{6(x-1)}{5}\right]$;

g) $\dfrac{1}{2}\left\{\dfrac{1}{2}\left[\dfrac{1}{2}\left(\dfrac{1}{2}x - 1\dfrac{1}{2}\right) - 1\dfrac{1}{2}\right] - 1\dfrac{1}{2}\right\} - 1\dfrac{1}{2} = 0$;

h) $\dfrac{1}{2}\left\{\dfrac{1}{2}\left[\dfrac{1}{2}\left(\dfrac{1}{2}x - 3\dfrac{1}{2}\right) - 3\dfrac{1}{2}\right] - 3\dfrac{1}{2}\right\} - 3\dfrac{1}{2} = 0$;

i) $\dfrac{1}{2}\left\{\dfrac{1}{2}\left[\dfrac{1}{2}\left(\dfrac{1}{2}x + 2\right) + 2\right] + 2\right\} + 2 = 0$; j) $[(2x - 4.7)\cdot 3 - 1.9 + 9] \div 5 - 3.9 = 0.1$.

II. 38. Solve the equations:

a) $9(5x - 3) - 7(3x + 5) = 5(3x + 1) - 4(2x + 8) - 18$;

b) $(x + 5)^2 + (x - 6)(x + 6) = 2(x - 3)^2 - 7$;

c) $(5x + 7)(5x - 7) - (5x - 1)^2 = -70$;

d) $3(4x - 11) + 5(7 - 2x) = 16 - 3(3x - 10)$;

e) $(3x + 8)(x - 1) - 3(x + 1)^2 = -13$.

II. 39. Solve the equations:

a) $(3x + 1)(-4x + 3) = 0$; b) $\left(\dfrac{4}{7}x + 3\right)\left(\dfrac{1}{3}x + \dfrac{1}{2}\right) = 0$;

c) $\left(-\dfrac{5}{8}x - \dfrac{3}{7}\right)(0.\overline{3}x - 0.\overline{4}) = 0$; d) $\left(\dfrac{1}{13}x - \dfrac{7}{6}\right)\cdot\left(-\dfrac{8}{9}x + \dfrac{4}{7}\right) = 0$;

e) $\left(-\dfrac{1}{7}x + \dfrac{7}{8}\right)\cdot\left(-\dfrac{5}{9}x - \dfrac{4}{5}\right) = 0$.

II. 40. Solve the equations:

a) $\dfrac{2}{x} + \dfrac{3}{4} = \dfrac{5}{x} + \dfrac{1}{2} + 1$, $x \in \mathbf{Q} \setminus \{0\}$; b) $\dfrac{3x + 3}{2x + 5} = \dfrac{3x + 1}{2x + 1}$, $x \in \mathbf{Q} - \left\{-\dfrac{5}{2}; -\dfrac{1}{2}\right\}$;

c) $\dfrac{4}{x + 2} = \dfrac{5}{3 - x}$, $x \in \mathbf{Q} - \{-2; 3\}$; d) $\dfrac{2}{x + 2} + \dfrac{3}{x + 3} = \dfrac{5}{x}$, $x \in \mathbf{Q} - \{-3; -2; 0\}$;

e) $\dfrac{3}{x-2}+\dfrac{1}{x-3}=\dfrac{4}{x-1}$, $x \in \mathbf{Q}-\{1;2;3\}$; f) $\dfrac{5x+2}{3x-1}=\dfrac{4}{7}$, $x \in \mathbf{Q}-\left\{\dfrac{1}{3}\right\}$;

g) $\dfrac{5-3x}{2-3x}=5$, $x \in \mathbf{Q}-\left\{\dfrac{2}{3}\right\}$; h) $\dfrac{3}{4}-\dfrac{7x+2}{3x}=\dfrac{1}{6x}$, $x \in \mathbf{Q}\setminus\{0\}$;

i) $\dfrac{2x+3}{2x-3}=\dfrac{x+5}{x-4}$, $x \in \mathbf{Q}-\left\{\dfrac{3}{2};\ 4\right\}$; j) $\dfrac{2}{5x}+\dfrac{5}{4x}=\dfrac{3}{8}$, $x \in \mathbf{Q}\setminus\{0\}$;

k) $\dfrac{2x+1}{3x+1}=\dfrac{2x-1}{3x-1}$, $x \in \mathbf{Q}-\left\{-\dfrac{1}{3};\dfrac{1}{3}\right\}$; l) $\dfrac{3x+2}{3x-2}=\dfrac{x-1}{x+4}$, $x \in \mathbf{Q}-\left\{-4;\dfrac{2}{3}\right\}$.

II. 41.

a) If $a=\dfrac{2}{5}$ and $b+c=-\dfrac{25}{2}$ what is the value of $ab+ac$?

b) If $ab+ac=-4\dfrac{1}{6}$ and $b+c=\dfrac{36}{5}$ what is the value of a?

c) If $ab=\dfrac{2^2}{3^{-3}}$ and $ac=\dfrac{4}{3^{-4}}$ and $b+c=108$ what is the value of a?

d) If $ab=\dfrac{3^2}{4^{-3}}$ and $ac=\dfrac{12}{5^{-1}}$ what is the value of $a(b-c)$?

e) If $a=b+\dfrac{3^{-3}}{5^{-2}}$ and $c=-\dfrac{25}{27}$ what is the value of $ac-bc$?

f) If $ab=-\dfrac{4}{25}$ and $ac=\dfrac{16}{5}$ what is the value of $\dfrac{b}{c}$ and $\dfrac{c}{b}$?

II. 42.

a) Find the value of a such that the equation $ax+5=13+3x$ has the solution $x=5$.

b) Find the value of a such that the equation $ax-6=2x-5$ has the solution $x=2$.

c) Find the value of m such that the equation $2x+mx=m+3$ has the solution $x=3$.

d) Find the value of m such that the equation $3mx+1=x-2m$ has the solution $x=4$.

e) Find the value of m such that the equation $3m(x+3)+1-5m=-(2x-4)-12m$ has the solution $x=6$.

II. 43. Solve the equations, a and m are real parameters:

a) $2mx-1=x-2m$; b) $2x+a=ax+3$; c) $ax-5=4x+7$;

d) $ax+2x=9+3ax$; e) $ax+4=3a-3ax$;

f) $(2+x)(a+1)=2(a-2x)+3a$; g) $2ax+1+x=(a+4)x+2x$;

h) $2a(2 - x) = (1 - a)x + 3(1 - 3x)$;

i) $2(mx - 3m) + 1 + x = (m + 4)x - (2x - 1)$.

II. 44. Solve the equations:

a) $\dfrac{1}{x(x + 1)} + \dfrac{1}{(x + 1)(x + 2)} + \ldots + \dfrac{1}{(x + 2009)(x + 2010)} = \dfrac{2009}{2010}$

b) $\dfrac{1}{1 \cdot 2} + \dfrac{2}{1 \cdot 2 \cdot 3} + \dfrac{3}{1 \cdot 2 \cdot 3 \cdot 4} + \ldots + \dfrac{x}{1 \cdot 2 \cdot \ldots \cdot (x + 1)} = \dfrac{1 \cdot 2 \cdot 3 \cdot \ldots \cdot 2010 - 1}{1 \cdot 2 \cdot 3 \cdot \ldots \cdot 2010}$;

c) $\dfrac{2x}{x + 1} + \dfrac{2x + 1}{x + 2} + \dfrac{2x + 2}{x + 3} + \ldots + \dfrac{2x + 2013}{x + 2014} = 2014$;

d) $1 + \dfrac{x + 1}{2} + \dfrac{x + 2}{3} + \ldots + \dfrac{x + n - 1}{n} = n, \ n \in \mathbf{N}, n \geq 2$.

Linear equations, word problems

II. 45. A 12-foot board is cut into two pieces so that one piece is 4 feet longer than the other. How long is each piece?

II. 46. A 36-foot rope is cut into two pieces so that one piece is twice as long as the other. How long is each piece?

II. 47. A clothing store sells suits at \$140 and \$210. The store owners observes that they sold 40 suits at \$6,650. How many suits of each type did the owner sell?

II. 48. Find five consecutive positive integers if the product of the first three numbers is 216 less than the product of the last three numbers.

II. 49. Two gears have a total of 96 teeth and one gear has 18 less teeth than the other. How many teeth are on each gear?

II. 50. A number is 27.8 less than another number. If their sum is 66.7, find the numbers.

II. 51. A number is three times less than another number. If their sum is $\dfrac{34}{9}$, find the numbers.

II. 52. The same result is obtained when a number is multiplied by 5.8 and when it is lowered by 4. Find the number.

II. 53. The length of a rectangular field is 4 yards more than twice the width of the field. If the semiperimeter is 610 yards, what are the dimensions of the field?

II. 54. The side of a square is x cm and the dimensions of a rectangle are $(x + 3)$ cm and $(x + 5)$ cm. Find the sides of the square if the area of the rectangle is 83cm² larger than the area of the square.

II. 55. The length of a room is 2 feet less than twice its width. If the perimeter is 62 feet, what are the dimensions?

II. 56. A triangle ABC has $\angle B = 2x$ and $\angle C = 5x$. Find x such that the triangle ABC is isosceles.

II. 57. Adding the same number to the numerator and the denominator of the fraction $\dfrac{4}{7}$ we obtain $\dfrac{19}{31}$. Find the number.

II. 58. The perimeter of a rectangle is $4\dfrac{1}{2}$ cm and the length is 6.2 larger than the double of the width. Find the dimensions of the rectangle.

II. 59. One of the angles of a triangle is 72°, and another angle is three times larger than the third angle. Find the angles of the triangle.

II. 60. 3 kg of apples and 2 kg of oranges cost $5.05. If 1 kg of oranges costs $0.30 cents more than 1 kg of apples, find the prices of the apples and oranges.

II. 61. 10 math textbooks and 15 physics textbooks cost $2,000. The price of one math textbook and one physics texbook is $163. Find the price of each textbook.

II. 62. The sum of two numbers is 150. If the first number is divided by the second number, we obtain the quotient 4 and the remainder 15. Find the numbers.

II. 63. The difference between the length and width of a rectangle is 13 cm. Find the dimensions of the rectangle if the semiperimeter is 63cm.

Ratios, proportions, procents, equal proportions.

II. 64. Find the ratio of the length and width of a rectangle and the ratio of the width and length of the same rectangle knowing that:
a) $l = 17$cm, $w = 10$cm; b) $l = 7$dm, $w = 51$cm;
c) $l = 3$m, $w = 25$dm; d) $l = 4$m, $w = 360$cm.

II. 65. Find the ratio for:

a) the sum and difference of $\dfrac{2}{5}$ and 1.4;

b) the product and difference of 13 and 7.5.

II. 66. A circle has the circumference 16π cm and the radius 8 cm. Find the ratio of:
a) the circumference and the radius of the circle;
b) the area and the radius of the circle.

II. 67. The ratio of the lengths of the sides of two equilateral triangles is $\dfrac{3}{4}$.
Find the ratio of:
a) the perimeters;
b) the areas.

II. 68. Simplify the ratios:

a) $\dfrac{24}{36}$; b) $\dfrac{72}{42}$; c) $\dfrac{25}{45}$; d) $\dfrac{6a}{18a}$; e) $\dfrac{32b}{64b}$; f) $\dfrac{15a^3}{25a}$; g) $\dfrac{10\,\text{min}}{1\text{h}}$; h) $\dfrac{15\text{cm}}{4\text{dm}}$;

i) $\dfrac{15\text{m}}{4\text{hm}}$; j) $\dfrac{251}{4\text{dl}}$; k) $\dfrac{15\text{mm}}{4\text{cm}}$; l) $\dfrac{150\text{s}}{3\text{min}}$.

II. 69. Calculate the ratio:

a) $\dfrac{7a+3b}{5a-7b}$ if $\dfrac{a}{b}=\dfrac{4}{5}$; b) $\dfrac{2y-3x}{9x+5y}$ if $\dfrac{x}{y}=5\dfrac{1}{3}$; c) $\dfrac{3x+4y}{7x+31y}$ if $\dfrac{x}{y}=\dfrac{11}{23}$;

d) $\dfrac{m}{n}$ if $\dfrac{5m-2n}{3m-4n}=\dfrac{4}{7}$; e) $\dfrac{2x+7y}{3x-9y}$ if $\dfrac{3x}{5y}=\dfrac{4}{15}$; f) $\dfrac{a+3b}{3a+5b}$ if $\dfrac{a}{b}=\dfrac{1}{3}$;

g) $\dfrac{b}{a}$ if $\dfrac{2a+1}{b+2}=\dfrac{1}{2}$; h) $\dfrac{b}{a}$ if $\dfrac{2a-3b}{4a+5b}=\dfrac{5}{16}$;

i) $\dfrac{a}{x}$, $\dfrac{b}{x}$ and $\dfrac{c}{x}$ if $\dfrac{x}{a}+\dfrac{x}{b}+\dfrac{x}{c}=3$ and $\dfrac{b}{a}=\dfrac{0.\overline{2}}{0.7}$; $\dfrac{c}{b}=\dfrac{0.\overline{63}}{0.\overline{36}}$;

55

j) $\dfrac{x+3y}{2x-5y}$ if $(4x+3y)(x+y)=14xy$;

k) a, b, c, d, and e if $\dfrac{a}{7}=\dfrac{b}{2}=\dfrac{c}{6}=\dfrac{d}{4}=\dfrac{e}{5}$ and $7a+2b+6c+4d+5e=260$.

II. 70. Find the unknown term from the following proportions:

a) $\dfrac{x}{5}=\dfrac{12}{3}$; b) $\dfrac{x}{3.25}=\dfrac{0.54}{0.6}$; c) $\dfrac{14}{x}=\dfrac{35}{22}$; d) $\dfrac{4}{a}=\dfrac{19}{38}$; e) $\dfrac{38}{z}=\dfrac{72}{18}$;

f) $\dfrac{\frac{5}{35}}{z}=\dfrac{\frac{45}{20}}{2.5}$; g) $\dfrac{a}{18}=\dfrac{5}{36}$; h) $\dfrac{n}{85}=\dfrac{0.5}{17}$; i) $\dfrac{21}{x}=\dfrac{3.9}{1.3}$; j) $\dfrac{x}{4}=\dfrac{5}{2.8}$; k) $3=\dfrac{6}{x}$;

l) $\dfrac{x}{2^5 \div 2^3}=\dfrac{13^2}{3^0+3^1+3^2}$; m) $\dfrac{0.32}{3.2}=\dfrac{5.5}{2x}$; n) $\dfrac{x}{0.2\overline{3}}=\dfrac{0.1\overline{4}}{0.\overline{4}}$.

II. 71. Find x and y from the proportions:

a) $\dfrac{x}{y}=\dfrac{3}{7}$ and $x+y=50$; b) $\dfrac{x}{y}=\dfrac{13}{7}$ and $x-y=7$;

c) $\dfrac{x}{y}=\dfrac{2}{9}$ and $x-y=10$; d) $\dfrac{x}{y}=\dfrac{11}{27}$ and $x-2y=41$;

e) $\dfrac{3x}{2y}=\dfrac{5}{9}$ and $3x-4y=41$; f) $\dfrac{2x}{y}=\dfrac{3}{17}$ and $x+3y=5$.

II. 72. Find the variables from the proportions:

a) $\dfrac{28}{a+2}=7$; b) $\dfrac{2x-4}{7}=8$; c) $4=\dfrac{c-2}{5}$; d) $\dfrac{b+6}{2}=\dfrac{3b}{5}$; e) $\dfrac{6z}{5}=7z-4$;

f) $\dfrac{t+3}{6}=\dfrac{t}{10}$; g) $\dfrac{5x}{3}=\dfrac{2x-1}{4}$; h) $\dfrac{5x+1}{18}=\dfrac{6x-1}{21}$; i) $\dfrac{2x-1}{14}=\dfrac{3x+2}{28}$;

j) $\dfrac{2x-5}{2}=\dfrac{3x-7}{3}$; k) $\dfrac{4x+1}{8}=\dfrac{3x+3}{7}$; l) $\dfrac{6k+112}{24}=\dfrac{4k+8}{12}$;

m) $\dfrac{3x-4}{5}=\dfrac{10-3x}{20}$.

II. 73. Find the value of the ratio $\dfrac{x}{y}$:

a) $\dfrac{2x+3y}{5y}=\dfrac{3}{4}$; b) $\dfrac{5x-2y}{4y}=\dfrac{4}{3}$; c) $\dfrac{5x-y}{6}=\dfrac{4x-y}{4}$; d) $\dfrac{2x-y}{-5y}=\dfrac{3}{7}$;

e) $\dfrac{7x-2y}{2y}=\dfrac{4}{5}$; f) $\dfrac{3x-2y}{0.6}=\dfrac{5x-y}{0.4}$; g) $\dfrac{3x+2y}{5y-7x}=\dfrac{3}{8}$; h) $\dfrac{3y-2x}{4y-3x}=\dfrac{5}{7}$;

i) $\dfrac{2y}{6x+5y}=0.\overline{2}$.

II. 74. Write each percentage as a fraction:

a) 75%; **b)** 25%; **c)** 12%; **d)** 0.4%; **e)** 50%; **f)** 0.7%; **g)** 2.3%; **h)** 1.6%.

II. 75. Write each percentage as a decimal number:

a) 5%; **b)** 3%; **c)** 34%; **d)** 17%; **e)** 0.33%; **f)** 4.15%; **g)** 0.25%; **h)** 1.45%.

II. 76. Find the number which is:

a) 25% from 400; **b)** 8% from 75; **c)** $5\frac{1}{2}$% from 300; **d)** $\frac{1}{2}$% from 108;

e) 2.5% from 90; **f)** 120% from 65; **g)** 100% from 99; **h)** 200% from 20.3;

i) $4\frac{1}{3}$% from 250.

II. 77. Find the percentage:

a) 15 out of 90; **b)** 24 out of 30; **c)** 300 out of 400; **d)** 16 out of 25;

e) 12 out of 25; **f)** 24 out of 74; **g)** 450 out of 9; **h)** 180 out of 36;

i) 12 out of 15; **j)** 36 out of 54; **k)** 160 out of 250;

II. 78. Find the number x knowing that:

a) 15 is 20% of x; **b)** 16 is 60% of x; **c)** 4.5 is 20% of x; **d)** 16 is 15% of x;

e) 65% of x is 130; **f)** 3.5% of x is 21; **g)** 14% of x is 70; **h)** 300% of x is 190;

i) 250% of x is 180.

II. 79. Find the numbers x, y, z if $\dfrac{x}{3} = \dfrac{y}{4} = \dfrac{z}{5}$ and $x + z = 48$.

II. 80. Find the numbers x, y, z, t if $\dfrac{x}{2} = \dfrac{y}{3} = \dfrac{z}{5} = \dfrac{t}{7}$ and

$8x + 5y - 3z - 2t = 4$

II. 81. Find the numbers x, y, z if $\dfrac{x}{3} = \dfrac{y}{2} = \dfrac{z}{4}$ and $x \cdot y \cdot z = 192$.

II. 82. Let $\dfrac{x}{2} = \dfrac{y}{3}$ and $\dfrac{y}{4} = \dfrac{z}{5}, x, y \in Q_+^*$ be two proportions. Find the

numbers x, y, z if $x^2 + y^2 + z^2 = 433$.

II. 83. Find the numbers x, y, z if:

a) $\dfrac{x}{2} = \dfrac{y}{3} = \dfrac{4}{z}$ and $y = \dfrac{x+z}{2}$;

b) $\dfrac{x}{x+1} = \dfrac{y}{y+2} = \dfrac{z}{z+3}$ and $\dfrac{1}{x} + \dfrac{2}{y} + \dfrac{3}{z} = 54$.

II. 84. Find the numbers x, y, z if $\dfrac{x}{5} = \dfrac{y}{3}$ and $\dfrac{y}{2} = \dfrac{z}{4}$ and $x + 2y + 3z = 116$.

II. 85. Find the numbers x, y, z if $\dfrac{5}{x} = \dfrac{y}{3} = \dfrac{z}{4}$.

II. 86. Find the numbers x, y, z if $x + y + z = 4$, $\dfrac{x}{2} = \dfrac{y}{3}$ and $\dfrac{y}{6} = \dfrac{z}{8}$.

II. 87. Find the numbers a, b, c if $\dfrac{a}{3} = \dfrac{b}{4} = \dfrac{c}{6}$ and $a \cdot b \cdot c = 576$, $a, b, c \in \mathbf{N}$.

II. 88. Find the numbers $a, b, c, d \in \mathbf{Z}$ if $\dfrac{a}{b} = \dfrac{2}{3}$, $\dfrac{b}{c} = \dfrac{3}{4}$, $\dfrac{c}{d} = \dfrac{4}{5}$, and $abcd = 1920$.

II. 89. If $\dfrac{x}{a} = \dfrac{y}{b} = \dfrac{z}{c} = \dfrac{2}{3}$ calculate:

a) $\dfrac{x+y+z}{a+b+c}$; b) $\dfrac{x+3y+z}{a+3b+c}$; c) $\dfrac{x^2 + y^2 + z^2}{a^2 + b^2 + c^2}$.

II. 90. If $\dfrac{a}{3} = \dfrac{b}{4} = \dfrac{c}{5}$ calculate:

a) $\dfrac{2a-3b}{5c+a}$; b) $\dfrac{a+2b+3c}{3a+2b+c}$; c) $\dfrac{2a-3b+4c}{4a-3b+2c}$.

II. 91. If a, b, c are the sides of a triangle, show that the triangle is equilateral in both cases:

a) $\dfrac{(m+1)a+b+c}{a} = \dfrac{a+(m+1)b+c}{b} = \dfrac{a+b+(m+1)c}{c}$, $m \in \mathbf{R}$

b) $\dfrac{(m+1)a+b+c}{a+(m+1)b+c} = \dfrac{a+(m+1)b+c}{a+b+(m+1)c} = \dfrac{a+b+(m+1)c}{(m+1)a+b+c}$, $m \in \mathbf{R}$.

II. 92. Consider the numbers: $a = 2^{n+1} \cdot 5^n + 1$, $b = 2^n \cdot 5^{n+1} + 1$,

$c = 2^{n+3} \cdot 5^n + 7$, and $d = 2^{n+1} \cdot 5^{n+3} - 1$, where n is a natural number.

a) Show that a, b, c, d are not prime numbers.

b) Prove that $\dfrac{7d + c}{b - a}$ is a natural number.

c) Can c be a perfect square?

II. 93. Calculate the sums:

a) $S = 13 + 12 \cdot 13 + 12 \cdot 13^2 + 12 \cdot 13^3 + ... + 12 \cdot 13^{100}$;

b) $S = 3^{100} - 2 \cdot 3^9 - 2 \cdot 3^{98} - ... - 2 \cdot 3^2 - 2 \cdot 3 - 3$.

II. 94. Show that the number $\dfrac{\overline{abc} + \overline{bcd} + \overline{cda} + \overline{dab}}{a + b + c + d}$ is a natural number.

II. 95. Prove that the number $n = a(a+1)(a+2)(a+3) + 1$ is a perfect square.

II. 96. Show that the number $\underbrace{\overline{100...0200...01}}_{2n+3 \ times}$ is a perfect square.

II. 97. Show that the difference $\overline{aaa} - \overline{bbb}$ is divisible by 3.

II. 98. Find the five-digit number $abcd$ such that

$5 + 10 + 15 + ... + \overline{abcd} = \overline{abcd000}$

II. 99. Prove that the number $3^{3n+1} + 2 \cdot 3^{n+1}$ can be written as a sum of three cubes.

II. 100. Find the square of the number $A = \underbrace{33...34}_{n-1 \ times}$.

TEST II. 1

1) Alex has one-twentieth of a dollar in his pocket. What coins might he have?
2) Use base ten blocks to represent each fraction. Then write each fraction as a decimal. **a)** $\dfrac{3}{5}$; **b)** $\dfrac{3}{2}$; **c)** $2\dfrac{1}{5}$; **d)** $5\dfrac{4}{5}$.

3) Represent each fraction on a hundredths grid. Then write each number as a decimal. **a)** $\dfrac{1}{4}$; **b)** $\dfrac{3}{10}$; **c)** $\dfrac{2}{50}$; **d)** $\dfrac{2}{25}$.

4) Write each improper fraction as a decimal number.

    a) $7\dfrac{1}{2}$; **b)** $6\dfrac{2}{5}$; **c)** $3\dfrac{11}{25}$; **d)** $1\dfrac{1}{100}$.

5) Copy and complete. Replace each $\square$ with $<, >$ or $=$ to make the statement true.

    a) $\dfrac{75}{100}\ \square\ \dfrac{3}{4}$; **b)** $0.08\ \square\ \dfrac{8}{10}$; **c)** $0.6\ \square\ \dfrac{60}{100}$; **d)** $\dfrac{8}{5}\ \square\ 1\dfrac{3}{10}$.

6) Write two equivalent fractions for each decimal.
    a) 0.9; **b)** 0.40; **c)** 0.75, **d)** 0.5.

7) Write each division statement as an improper fraction and as a mixed number.
    a) $77 \div 8$; **b)** $84 \div 9$; **c)** $45 \div 7$; **d)** $79 \div 6$.

8) Vlad made 7 pizzas for his party. He invited 5 friends. How much pizza did Vlad think each person would eat?

9) Matthew's yard has a rectangular shape. It is 28.54 m wide and 31 m long. What is the area of the yard?

10) A snail traveled 0.94 m per hour. How far would the snail travel in 6 hours?

    TEST II. 2

Calculate:

a) $\dfrac{2}{5}\div\dfrac{1}{3}$; **b)** $\dfrac{6}{25}\div 1\dfrac{1}{4}$; **c)** $\dfrac{7}{34}\div 2\dfrac{5}{8}$; **d)** $3\dfrac{3}{5}\div\dfrac{3}{10}$; **e)** $14.4\div 1.2$;

f) $1.19\div 0.17$; **g)** $2.99\div\dfrac{23}{10}$; **h)** $4.5\div\left(-\dfrac{5}{2}\right)$; **i)** $-3.6\div\dfrac{3}{10}$;

j) $\dfrac{2}{3}\cdot\left(\dfrac{5}{6}\right)\div\left(\dfrac{15}{18}\right)$.

TEST II. 3

Find the solution for each of the following equations:
1) $2x - 3 = 5$; 2) $2x - 3 = 8 + 3x$; 3) $3x + 4 = 10$; 4) $3x + 7 = 8 - 2x$;
5) $2 - x = 6 + x$; 6) $4 - 2x = x + 7$; 7) $2(x-1) = 3x + 4$; 8) $3 - 4x = 5(3x+2)$;
9) $4 \cdot (x-2) = 6$; 10) $3\,(2-x) = 2x + 1$.

TEST II. 4

1) Calculate the arithmetic average of the numbers x and y, if:

$$x = [(-7+9)\cdot(-3)+(-2)^3 \div (-4)]^2 \div 8 \quad \text{and} \quad y = \left[0.\overline{3} - 1\frac{1}{2}\right]\cdot\left(-\frac{1}{4}\right)^{-1}\cdot 3\,.$$

2) Solve the equation $\dfrac{5}{6}(x+3) + 2\dfrac{3}{4} = \dfrac{3x+2}{3} - \dfrac{3-x}{2}$.

3) Find the number x, y, z if they are in proportion with the numbers 9, 10, and 12 and their sum is 217.

4) Simplify the expressions:

  a) $\left(2^3 \cdot 6 - 2^2 \cdot 5\right) \div 7 - \left(1002 \div 3 - 331\right)$;

  b) $\left[\left(7^2 - 3^2\right) \div 5 - 2^3\right] \cdot 2007 + 16 \div \left(7 \cdot 3^2 - 5\ \right).$

5) Calculate $a = \left[\left(0,\overline{3}+\dfrac{1}{4}\right)\cdot\left(1+\dfrac{5}{7}\right)\right] \div \left[\dfrac{2}{3} \div \left(\dfrac{1}{2}+\dfrac{1}{2}\cdot 0,\overline{3}\right)\right].$

6) Find the positive integer solution of the following equations:
  a) $(x+1)(x+2) = 72$; b) $x + 2 = 5(x-2)$;
  c) $\left(2^2 + 2^4 + 2^6\right)\cdot x = 2^3 + 2^5 + 2^7\,.$

7) Calculate:

  a) $\left[2.4^2 - 0.76 + 0.2 \cdot \left(5.4 + 0.144 \div 1.2^2\right)\right]$;

  b) $10 \cdot \left\{21^2 \div 441 + 2 \cdot \left[\left(2^{70} \cdot 3^{30}\right) \div \left(2^{69} \cdot 3^{30}\right) - 15^0\right]\right\}$;

  c) $\left\{20.5 \div \left[25 \cdot \left(1 - \dfrac{1}{5}\right) + \left(2\dfrac{1}{3} - 0.75\right)\cdot\dfrac{6}{19}\right]\right\}^2\,.$

8) Solve the equation $\left[\left(1 + 1.\overline{6}\right) - \dfrac{5}{3} + 1\dfrac{1}{3}\right] \div 1.\overline{16} \cdot \left(\dfrac{1}{3}\right)^2 - x = \dfrac{2}{3}\,.$

9) Calculate the area of a rectangle with the semiperimeter of 80 cm and the length $\dfrac{3}{10}$ of the perimeter.

10) Calculate $a = \dfrac{1.\overline{1}}{1} + \dfrac{2.\overline{2}}{2} + \dfrac{3.\overline{3}}{3} + \ldots + \dfrac{9.\overline{9}}{9}$;

$b = \dfrac{1}{1.\overline{1}} + \dfrac{2}{2.\overline{2}} + \dfrac{3}{3.\overline{3}} + \ldots + \dfrac{9}{9.\overline{9}}$ and $c = \dfrac{1.\overline{1}}{1.\overline{1} \cdot 2.\overline{2}} + \dfrac{1.\overline{1}}{2.\overline{2} \cdot 3.\overline{3}} + \ldots + \dfrac{1.\overline{1}}{8.\overline{8} \cdot 9.\overline{9}}$.

TEST II. 5

1) Find x in each of the following proportion $\dfrac{x}{1.5} = \dfrac{0.5}{0.2}$.

2) Find x in each of the following proportion:

a) $\dfrac{3}{9} = \dfrac{2}{x}$; b) $\dfrac{4}{9} = \dfrac{x}{5}$; c) $\dfrac{6}{15} = \dfrac{0.3}{x}$; d) $\dfrac{5.4 - 1.8}{x} = \dfrac{7.5 - 4.5}{2.4 - 1.4}$.

3) Find the rational numbers x, y such as $x + y = 10$ and $\dfrac{x}{0.2} = \dfrac{y}{0.05}$.

4) Find the rational numbers x, y, and z such as $x + y + z = 24$ and

$\dfrac{x}{3} = \dfrac{y}{4} = \dfrac{z}{5}$.

5) Find the rational numbers x, y, z such as $z - y = 4$ and $\dfrac{x}{5} = \dfrac{y}{2} = \dfrac{z}{6}$.

6) Find the rational numbers x, y, z such that $2x + 3y - 5z = -20$

and $\dfrac{x}{5} = \dfrac{y}{2} = \dfrac{z}{4}$.

7) Find the sides of a triangle if the perimeter is 24 and the sides are proportional with the numbers 3, 4, 5.

8) Find the rational numbers x, y, z such as $\dfrac{x}{12} = \dfrac{y}{15} = \dfrac{z}{30}$ and $xyz = 1600$.

9) Consider the proportions $\dfrac{x}{22} = \dfrac{y}{28} = \dfrac{3z}{40}$. Find x, y, and z if $\dfrac{x}{11} + \dfrac{y}{6} + \dfrac{z}{17} = 13$.

TEST II. 6

1) Calculate:

a) $[(9-4.5)\div 0.03]\div[(4.5-3.65)\cdot 4-0.4]$;

b) $[(4.3-4.15)\div 1.5]\div[(1.88+2.12)\cdot 0.0125]$;

2) Calculate:

a) $\{[2+(3+3\cdot 5)\div 2]+[1-(7-3\cdot 2)\div 2]\cdot 10\}\div 5$;

b) $\{[3+(1+1\cdot 5)\div 2]+[1-(4-0\cdot 2)\div 2]\cdot 10\}\div 5$.

3) If $a=\left(-\dfrac{2}{3}\right)^{-3}\cdot\left(+\dfrac{2}{3}\right)^{4}+\left(\dfrac{7}{6}\right)^{2}\cdot\left(-\dfrac{18}{98}\right)\div\dfrac{3}{4}$ and

$b=0.2\overline{3}\div 0.21\cdot 0.1^{2}-0.3\overline{1}$ then find x from the equation

$ax+b=1.\overline{3}$.

4) Calculate: $\left(1-\dfrac{1}{2}\right)^{2}\left(1-\dfrac{1}{3}\right)^{2}\left(1-\dfrac{1}{4}\right)^{2}........\left(1-\dfrac{1}{16}\right)^{2}$.

5) Find x such that $\dfrac{1}{9}\cdot\left\{\dfrac{1}{7}\cdot\left[\dfrac{1}{5}\cdot\left(\dfrac{1}{3}\cdot(x-2)-4\right)+6\right]+8\right\}=\dfrac{1}{3}$.

6) Calculate $(1-7^{-1})\cdot(1-8^{-1})\cdot(1-9^{-1})\cdot\cdots\cdot(1-2007^{-1})$.

7) Calculate: a) $\dfrac{1-\sqrt{2}}{\sqrt{1\cdot 2}}+\dfrac{\sqrt{2}-\sqrt{3}}{\sqrt{2\cdot 3}}+...+\dfrac{\sqrt{2010}-\sqrt{2009}}{\sqrt{2010\cdot 2009}}$;

b) $\dfrac{1}{2012}\left(2013-\dfrac{1007}{\sqrt{1+3+5+...+2013}}\right)^{2013}$.

8) Solve the equation $\dfrac{x}{2}+\dfrac{x}{3}+...+\dfrac{x}{n}=\dfrac{1}{2}+\dfrac{1}{3}+...+\dfrac{1}{n}$, $n\in\mathbb{I}$.

9) Let A, B, C, D be collinear points and $AD=12$cm. $AB=BD$ and $CD=$ 4cm. Find the length of the segments AB, BC, BD, and AM, where M is the midpoint of BC.

10) In the diagram we have:

A   B   C   D   F   G

$AG=57$cm, $AB=CD=DF$, $BC=FG=12$cm. Find the length of the segments AB, BF, CG, and the distance from A to the midpoint of CG.

63

TEST II. 7

1) Calculate the arithmetic average of the numbers x and y, where:

$$x = [(-7+9)\cdot(-3)+(-2)^3 \div(-4)]^2 \div 8 \text{ and } y = \left(0.\overline{3}-1\frac{1}{2}\right)\cdot\left(-\frac{1}{4}\right)^{-1}\cdot 3.$$

2) Solve the equation $\dfrac{5}{6}(x+3)+2\dfrac{3}{4} = \dfrac{3x+2}{3}-\dfrac{3-x}{2}$.

3) Find the numbers x, y, z if they are in proportion with the numbers 9, 10, and 12 and their sum is 217.

4) Calculate:

a) $\left(2^3 \cdot 6 - 2^2 \cdot 5\right) \div 7 - (1002 \div 3 - 331)$;

b) $\left[\left(7^2 - 3^2\right) \div 5 - 2^3\right] \cdot 2007 + 16 \div \left(7 \cdot 3^2 - 5\right)$.

5) Calculate $a = \left[\left(0.\overline{3}+\dfrac{1}{4}\right)\cdot\left(1+\dfrac{5}{7}\right)\right] \div \left[\dfrac{2}{3} \div \left(\dfrac{1}{2}+\dfrac{1}{2}\cdot 0.\overline{3}\right)\right]$.

6) Solve the equations:

a) $(x+1)(x+2) = 72$;

b) $3x+2 = 5(x-2)$;

c) $\left(2^2 + 2^4 + 2^6\right)\cdot x = 2^3 + 2^5 + 2^7$.

7) Calculate:

a) $\left[2.4^2 - 0.76 + 0.2 \cdot \left(5.4 + 0.144 \div 1.2^2\right)\right]$;

b) $10 \cdot \left\{16^2 \div 256 + 2\left[\left(2^{60} \cdot 3^{20}\right) \div \left(2^{59} \cdot 3^{20}\right)\cdot 2010^0\right]\right\}$;

c) $61.5 \div \left[25 \cdot \left(1-\dfrac{1}{5}\right)+\left(2\dfrac{1}{3}-0.75\right)\cdot\dfrac{6}{19}\right]$.

8) Solve the equation $\left(1+1.\overline{6}-\dfrac{2}{9}+5\dfrac{1}{3}\right) \div 1.\overline{16} \cdot \left(\dfrac{1}{2}\right)^2 - x = \dfrac{2}{3}$.

9) Calculate the area of a rectangle with the perimeter 80 cm and the length $\dfrac{3}{5}$ from the semiperimeter.

10) Calculate $a = \dfrac{1.\overline{1}}{1}+\dfrac{2.\overline{2}}{2}+\dfrac{3.\overline{3}}{3}+...+\dfrac{9.\overline{9}}{9}$,

$b = \dfrac{1}{1.\overline{1}}+\dfrac{2}{2.\overline{2}}+\dfrac{3}{3.\overline{3}}+...+\dfrac{9}{9.\overline{9}}$, and

$$c = \frac{1.\overline{1}}{1.\overline{1} \cdot 2.\overline{2}} + \frac{1.\overline{1}}{2.\overline{2} \cdot 3.\overline{3}} + \dots + \frac{1.\overline{1}}{8.\overline{8} \cdot 9.\overline{9}}$$

TEST II. 8

1) Calculate:

a) $\left[1.4^2 - 0.06 + 4.37 \cdot \left(3.4 + 0.144 \div 1.2^2\right)\right]^0$;

b) $10 \cdot \left\{21^2 \div 441 + 2 \cdot \left[\left(2^{101} \cdot 3^{207}\right) \div \left(2^{100} \cdot 3^{207}\right) \cdot 168^0\right]\right\}$;

c) $\left\{20.5 \div \left[25 \cdot \left(2 - 1\frac{1}{5}\right) + \left(3\frac{1}{3} - 1.75\right) \cdot \frac{6}{19}\right]\right\}^{-5}$.

2) Find x from the proportion $\dfrac{3^{21} + 3^2 + 3^{23}}{39} = \dfrac{3^{20}}{x}$.

3) Calculate

$$\left(\frac{2^2 \cdot 2^5 \cdot 2^1 - 2^3 \cdot 2^4 \cdot 2^5}{2^5 \cdot 2^3 \cdot 2^2 - 2^4 \cdot 2^5 \cdot 2^5} \div \frac{-3^3 \cdot 3^4 \cdot 3^5 - 3^7 \cdot 3^5 \cdot 3^3}{-3^3 \cdot 3^2 \cdot 3^4 - 3^3 \cdot 3^7 \cdot 3^2}\right) \div \frac{1}{36}.$$

4) Solve the equation $\dfrac{5}{6}(-x + 4) + 2\dfrac{3}{4} = \dfrac{3x - 2}{3} - \dfrac{3 - x}{2} - \dfrac{41}{12}$.

5) Solve the equation $\left(1.\overline{6} - \dfrac{2}{9} + 6\dfrac{1}{3}\right) \div \left(1.1\overline{6} \cdot 5^{-1}\right) - \dfrac{x}{3} = \dfrac{2}{3}$.

6) Calculate $\left(\dfrac{1}{0.\overline{2}} + \dfrac{1}{0.2} + \dfrac{1}{0.36\overline{3}}\right) \div \left(2\dfrac{3}{4} + 9\dfrac{1}{2}\right)$.

7) Calculate:

a) $\left[\left(3^6 \cdot 3^2 \cdot 3^4\right)^{10} \div 3^{118} + 1\right]^3 \div 1000$;

b) $[4^{17} \div 2^{18} \cdot (-8)^{12}]^2 \div (-8)^{20} \cdot 2^{-42}$.

8) Calculate $a = \left[\left(0.\overline{3} + \dfrac{2}{3}\right) \cdot \left(1 + \dfrac{2}{7}\right)\right] \div \left[\dfrac{2}{7} \div \left(\dfrac{5}{6} + \dfrac{1}{2} \cdot 0.\overline{3}\right)\right] \cdot \dfrac{2}{9}$.

9) a) Calculate $11 \cdot [-1 + 343 \div (441 \cdot 21 \div 27)]$;

   b) Solve the equation $5x - 4 = 2x + 8$;

   c) Find positive integers x such that $2x - 1 < 7$.

10) Solve the equation $\dfrac{2x-9}{3}-\dfrac{x-3}{4}=\dfrac{x-1}{2}-x+\dfrac{1}{12}$.

TEST II. 9

1) The sum of Mary's and her mother's age is 40. Her mother is three times older than Mary.

    a) How old is Mary?

    b) When does the mother's age become twice the age of Mary's ?

2) Calculate:

    a) $(2+4+6+...+84)\cdot\left(\dfrac{2}{7\cdot9}+\dfrac{2}{9\cdot11}+\dfrac{2}{11\cdot13}+...+\dfrac{2}{19\cdot21}\right)$;

    b) $a=\dfrac{4}{1}+\dfrac{7}{2}+\dfrac{10}{3}+\dfrac{13}{4}+...+\dfrac{301}{100}-\left(1+\dfrac{1}{2}+\dfrac{1}{3}+\dfrac{1}{4}+...+\dfrac{1}{100}\right)$.

3) a) Prove that $\dfrac{1}{n(n+1)}=\dfrac{1}{n}-\dfrac{1}{n+1}$, for any natural number;

    b) Calculate the sum $S=\dfrac{1}{2}+\dfrac{1}{6}+\dfrac{1}{12}+\dfrac{1}{20}+...+\dfrac{1}{600}$;

    c) Show that $S=\dfrac{1}{1^2}+\dfrac{1}{2^2}+\dfrac{1}{3^2}+...+\dfrac{1}{20^2}<\dfrac{39}{20}$;

    d) Find the natural number n such that

$$1+\dfrac{1}{1+2}+\dfrac{1}{1+2+3}+...+\dfrac{1}{1+2+...+n}=\dfrac{200}{101}.$$

4) a) Calculate $\dfrac{a}{b}$ if $a=\left(1+1.\overline{6}-\dfrac{2}{9}+5\dfrac{1}{3}\right)\div1.1\overline{6}$ and

$$b=\left(1-\dfrac{1}{2}\right)\cdot\left(1-\dfrac{1}{3}\right)\cdot\left(1-\dfrac{1}{4}\right)\cdot...\cdot\left(1-\dfrac{1}{20}\right).$$

    b) Show that the natural number

$n=7+7^2+7^3+...+7^{2009}+7^{2010}$ is divisible by 19.

5) Show that the number $N=10\cdot2^{6n+2}+3\cdot2^{6n+3}$, $n\in\mathbb{N}$ is both a perfect square and a perfect cube.

6) If $\dfrac{a}{4}=\dfrac{b}{6}=\dfrac{c}{8}$ prove that $a+c=2b$, and if $\dfrac{1}{a}+\dfrac{1}{b}+\dfrac{1}{c}=\dfrac{13}{144}$ find a, b, and c.

7) If $\dfrac{a}{b} = \dfrac{3}{2}$ and $\dfrac{c}{d} = \dfrac{4}{5}$ find $\dfrac{3ac - bd}{5ac - bd}$.

8) Find a, b, c such that a and b are in proportion with 5 and 8, b and c are in inverse proportions with 3 and 2 prove that $3a + 2b - c = 57$.

9) Calculate: a) $100 + 250 \div 5 + 15 \div 3 \cdot \left[265 - \left(50 \div 25 + 2\right) \cdot 65\right] \cdot 50$

b) $6^{100} \div \left[6^{40} \cdot 6^{58} - \left(6^{10} \cdot 6^{15}\right)^5 \div 6^{27} + \left(7^{57} \div 7^{56} - 1^{100}\right)^{97} \cdot 6 \right]$

c) Find k such that $64 + 2 \cdot 64 + 3 \cdot 64 + \ldots + 49 \cdot 64 = k^2$.

10) Calculate $a^2 + 3ab - 3ac + d^2$ knowing that $a = 3$, $b - c = 5$, and

$d = \left[2^{18} \cdot \left(2^3 \cdot 3\right)^{202} \right] \div \left(8 \cdot 9 \cdot 2^{309} \cdot 3^{99}\right)^2 + 3 \cdot (2 + 4 + 6 + \ldots + 2010)^0$

TEST II. 10

1) Find the integer x from the equalities:

a) $0.\overline{1x} + 0.\overline{2x} + 0.\overline{3x} + \ldots + 0.\overline{9x} = \dfrac{26x}{11}$;

b) $0.\overline{1x} + 0.\overline{2x} + 0.\overline{3x} + \ldots + 0.\overline{9x} = \dfrac{50 + x}{11}$.

2) Show that the number $a = 2^{2n+2} \cdot 3^{2n+2} + 4^{n+3} \cdot 9^{n+1} - 36^{n+1}$ is divisible by 1296, where n is a natural non-zero number.

3) Calculate the sums:

$S_1 = 2^n \cdot 3^{n+1} \cdot 5 + 2^{n+1} \cdot 3^n \cdot 7 + 2^n \cdot 3^n, n \in N$;

$S_2 = 3^{n+1} \cdot 5^{n+1} + 3^{n+2} \cdot 5^n + 15^n, \ n \in N$;

$S_3 = 3^n \cdot 7^{n+2} \cdot 11^{n+1} + 21^{n+1} \cdot 11^{n+2} - 40 \cdot 3^{n+3} \cdot 7^{\ n}, \ n \in N$;

$S_4 = 2 \cdot 4^n \cdot 2^n \cdot 3^{4n} + 8^n \cdot 16 \cdot 81^n \cdot 9 + 2^{3n+2} \cdot 27^n \cdot 3^n \cdot 243$.

4) Prove that the fraction $\dfrac{9^n + 21 \cdot 3^n + 68}{2 \cdot 3^n + 8}$ is a natural number for any n a natural number.

5) Prove that $A = \dfrac{36^n \cdot 5^{n+1} \cdot 7 - 2^{2n+1} \cdot 3^{2n+1} \cdot 5^n}{6^n \cdot 5^{n-1} \cdot 19 + 2^{n+1} \cdot 15^n}$ is a natural number.

6) Simplify the fractions:

a) $\dfrac{2 \cdot 7^{x+2} - 3 \cdot 7^{x+1} + 3 \cdot 7^x}{2^{x+4} \cdot 3^{x+1} - 2^{x+3} \cdot 3^x}$; b) $\dfrac{11^{x+2} - 11^{x+1} + 10 \cdot 11^x}{13^{x+1} - 13^x}$.

7) Consider the number $N = \overline{abc} + \overline{bca} + \overline{cab}$ where $a \ne b \ne c$.
   a) Find N, if a, b, c are consecutive digits such that $a + b + c = 6$;

   b) Show that $\dfrac{\overline{abc} + \overline{bca} + \overline{cab}}{185}$ is divisible by 37.

8) If the middle digit of a 3-digit number is removed, the result is a 2-digit number that is 6 times less than the initial number. Find the initial number.

9) Find the natural number $\overline{abc}$ if $\overline{a.b} + \overline{b.c} + \overline{c.a} = 3.\overline{3}$.

10) Find a and b, $a < b$ such that the number $A = \sqrt{0.\overline{ab2} + 0.\overline{b2a} + 0.\overline{2ab}}$ is an integer number.

11) Find the non-zero digits a, b, c, d such that $a + b + c = 2d$, $a < c < d < b < 9$, and $\overline{a.bcd} + \overline{b.cda} + \overline{c.dab} + \overline{d.abc} = 10$.

68

Chapter III

ALGEBRAIC EXPRESSIONS. COORDINATE GEOMETRY. SLOPES. LINEAR FUNCTIONS

$a^2 - b^2 = (a - b)(a + b);$

$(a + b)^2 = a^2 + 2ab + b^2;$

$(a - b)^2 = a^2 - 2ab + b^2;$

$a^3 - b^3 = (a - b)(a^2 + ab^2 + b^2);$

$a^3 + b^3 = (a + b)(a^2 - ab^2 + b^2);$

$(a + b)^3 = a^3 + 3a^2b + 3ab^2 + b^3;$

$(a - b)^3 = a^3 - 3a^2b + 3ab^2 - b^3;$

$(a + b + c)^2 = a^2 + b^2 + c^2 + 2ab + 2b + 2ca, \ a, b, c \in \mathbf{R}.$

III. 1. Collect like terms:

a) $\dfrac{1}{2}a^2b + 3b^2x - 4a^2b - \dfrac{5}{2}b^2x$; b) $2m^2n - 5an + 3m^2n - 7an$;

c) $0.6ax - 0.04ax - 0.24ax - 0.14ax$; d) $\dfrac{1}{2}x^2 - \dfrac{1}{2}x + x^2 + \dfrac{3}{4}x + 3x - 4x^2$;

e) $\dfrac{1}{0.5}bxy - \dfrac{0.4}{4}x^2y - 0.5x^2y - \dfrac{5}{2}bxy$; f) $\dfrac{1}{0.1}x + 0.2x - \dfrac{0.5}{2}x - 0.15x - 0.2x$;

g) $0.3a - \dfrac{0.5}{2}a + \dfrac{3}{0.2}b - \dfrac{3}{4}a + b$; h) $\dfrac{2}{3}ax^3 + \dfrac{1}{2}ax^3 - \dfrac{2}{3}ax^3 - \dfrac{3}{2}ax^3$;

i) $\dfrac{2}{5}by - \dfrac{3}{2}by - 3by + 12by$; j) $3a^2b - \dfrac{2}{3}a^2b + 3\dfrac{1}{2}ab^2 - \dfrac{1}{6}a^2b$;

k) $\left(3x^2 - \dfrac{2}{3}y^2z - xy\right) - \left(2x^2 - xy + \dfrac{1}{6}y^2z\right)$.

III. 2. Collect like terms:

a) $0.2 \ ab^2 + 0.28ab^2 - \dfrac{2}{5}ab^2 + \dfrac{1}{2}ab^2$; b) $3(m - n)^2 + 2(m - n)^2 - (m - n)^2$;

c) $2(a + b)^3 + 5(a + b)^3 - 3(a + b)^3$; d) $-2x^2 - (-5x^2 + 2) - 8x^2 + 10 + 4x^2$;

e) $\left(3x^3 - 4x^2 + x + 1\right) - \left(4x^3 - x^2 - x - 2\right) - \left(4x^3 - 2x^2 + 2x + 8\right)$;

f) $6a + \{4b + [6c - 2a - (2a - c)]\} - [8a - (7b + c)]$;

g) $\left(5x^2 - 3x + 4\right) - \left(4x^2 - 3x + 1\right) - \left(2x^2 + 4\right)$;

h) $(2a - b) - (b + c - 2d) + (b + 2c - d) + (2b - a)$;

i) $\left(\dfrac{4}{7}ab + \dfrac{1}{5}bc - \dfrac{2}{3}ac\right) - \left(-\dfrac{4}{7}ab + \dfrac{3}{10}bc - \dfrac{1}{5}ac\right)$;

j) $\left(\dfrac{1}{3}a^2x + \dfrac{1}{4}ax^2 + \dfrac{2}{3}x^2\right) - \left(-\dfrac{5}{6}a^2x + \dfrac{1}{3}ax^2 - \dfrac{3}{8}x^2\right)$.

III. 3. Calculate and collect like terms:

a) $\left(3a^3\right) \cdot \left(2a^3\right) + \left(3a^4\right) \cdot a^2$; b) $\left(4x^3\right) \cdot \left(3x^7\right) - \left(2x^6\right) \cdot \left(-x^4\right)$;

c) $\left(3x^3\right) \cdot \left(2x^2\right) \cdot (5x) - (3x) \cdot \left(2x^5\right) - \left(3x^2\right) \cdot \left(6x^4\right)$;

d) $\left(4x^2\right) \cdot (3x) \cdot \left(\dfrac{1}{4}x^6\right) - \left(\dfrac{2}{3}x\right) \cdot \left(-3x^6\right) \cdot \left(-2x^2\right)$;

e) $\left(10\,a^2b\right) \cdot \left(\dfrac{3}{5}a^4b^2\right) - \left(5ab^2\right) \cdot \left(2a^5b\right) + 2ab \cdot \left(-a^5b^2\right)$;

f) $\left(8a^2b^2\right) \cdot \left(0.5b^3\right) - (-3ab) \cdot \left(ab^4\right) + (-2b) \cdot \left(a^2b^4\right)$;

g) $\left(0.\overline{2}xy^2z^3\right) \cdot \left(9x^3yz\right) - (4xyz) \cdot \left(\dfrac{1}{2}x^3y^2z^3\right) + (12xyz) \cdot \left(\dfrac{1}{4}x^3y^2z^3\right)$;

h) $\left(\dfrac{2}{3}ab^2\right)^5 \div \left(\dfrac{2}{3}ab^2\right)^2 + \left(\dfrac{-2}{3}ab^2\right)^5 \div \left(-\dfrac{2}{3}ab^2\right)^2$.

III. 4. Calculate:

a) $(x+1)^2$; b) $(x-1)^2$; c) $(x-2)^2$; d) $(x+3)^2$; e) $(2x+1)^2$; f) $(3x-1)^2$;

g) $(2x-2)^2$; h) $(4x-2)^2$; i) $(5x+3)^2$; j) $(4x-3)^2$; k) $(-3x+1)^2$;

l) $(-x-3)^2$; m) $(6x-1)^2$; n) $(0.5x+4)^2$; o) $(-a^2-1)^2$; p) $(5u-2)^2$;

q) $(a+7)^2$; r) $(2b-3)^2$; s) $(-1-3x)^2$; t) $(2x^2-x)^2$; u) $(x+5x^2)^2$;

v) $(a^2-2x)^2$; x) $(t^2-5y^3)^2$.

III. 5. Calculate:

a) $\left(\dfrac{x}{2} - \dfrac{y}{3}\right)^2$; b) $\left(x - \dfrac{1}{3}\right)^2$; c) $\left(-\dfrac{b}{4} + \dfrac{4}{3}\right)^2$; d) $\left(\dfrac{3a}{2} - \dfrac{2b}{5}\right)^2$; e) $\left(2r - \dfrac{4}{3}\right)^2$;

f) $\left(\dfrac{2a}{5}+\dfrac{5}{4}\right)^2$; g) $\left(-\dfrac{3a}{8}-\dfrac{2}{3}\right)^2$; h) $\left(\dfrac{a^2}{6}+\dfrac{2a}{3}\right)^2$; i) $\left(\dfrac{3}{2}x^4y+\dfrac{2}{3}x^3y^2\right)^2$;

j) $\left(\dfrac{2}{7}a^5b+0.3a^3b^2\right)^2$; k) $\left(3x^2y-2y\right)^2$; l) $(2y-0.5x^3y^2)^2$;

m) $(2x^2y^2-0.2x^3y)^2$; n) $(\sqrt{2}+a)^2$; o) $(\sqrt{3}a+\sqrt{2}b)^2$; p) $(x-\sqrt{3})^2$;

q) $\left(\sqrt{2}x-\dfrac{1}{\sqrt{2}}y\right)^2$; r) $\left(\dfrac{2}{\sqrt{3}}x+\sqrt{3}\right)^2$; s) $\left(\dfrac{3}{\sqrt{2}}x+\sqrt{2}\right)^2$; t) $(2y-\sqrt{3})^2$.

III. 6. Calculate:

a) $(x-4)(x+4)$; b) $(2x+3)(2x-3)$; c) $(3x-1)(3x+1)$; d) $(3x-5)(3x+5)$;

e) $(x-0.4)(x+0.4)$; f) $(3-2x)(3+2x)$; g) $(0.2x-1)(0.2x+1)$;

h) $(3x-4)(3x+4)$; i) $\left(\dfrac{x}{3}+\dfrac{1}{2}\right)\left(\dfrac{x}{3}-\dfrac{1}{2}\right)$; j) $\left(\dfrac{1}{2}a+\dfrac{2}{3}b\right)\left(\dfrac{1}{2}a-\dfrac{2}{3}b\right)$;

k) $\left(3x^2+\dfrac{1}{3}\right)\left(3x^2-\dfrac{1}{3}\right)$; l) $\left(\dfrac{y}{3}-\dfrac{3}{2}\right)\left(\dfrac{y}{3}+\dfrac{3}{2}\right)$; m) $\left(\dfrac{a}{3}-3a^2\right)\left(\dfrac{a}{3}+3a^2\right)$;

n) $\left(\dfrac{5}{3}x+\dfrac{2}{5}y\right)\left(\dfrac{5}{3}x-\dfrac{2}{5}y\right)$; o) $\left(\dfrac{4}{3}x+\dfrac{3}{2}y\right)\left(\dfrac{4}{3}x-\dfrac{3}{2}y\right)$;

p) $(\sqrt{3}-x)(\sqrt{3}+x)$; q) $(\sqrt{5}x-1)(\sqrt{5}x+1)$.

III. 7. Calculate:

a) $x^2-(2x+1)^2$; b) $3(x+3)-2(x-3)^2$; c) $8-(2x+5)^2-2x(x-3)$;

d) $2(x^2+2)^2-3(x^2-2)^2$; e) $(x^2+2x)^2-(x^2-2x)^2$;

f) $3(x-3)(x+3)+(x+9)^2-4x^2$; g) $(x-3)(x+3)-(x-4)^2$;

h) $5x-2(x+3)^2-3(2x-5)(2x+5)+14x^2$;

i) $3x^2+7(2x+3)^2+4(-2x+5)(2x+5)-15x^2$.

III.8. Calculate:

a) $(x+y+z)^2$; b) $(x+y+2)^2$; c) $(x+y+0.1)^2$; d) $(a^2+a+1)^2$;

e) $(x^2+y^2+z^2)^2$; f) $(\sqrt{2}+\sqrt{3}+1)^2$; g) $(a+b-1)^2$;

h) $(x+2y-1)^2$; i) $(x-2y-2z)^2$; j) $(\sqrt{2}-2x+1)^2$.

III. 9. Factor:

a) $x(a+b) - y(a+b)$; b) $a(x-y) - b(x-y)$; c) $m(x+y) - n(x+y)$;

d) $x(a-b) + 2(a-b)$; e) $(xy+yz) + (xa+za)$; f) $(ab-ac) + (bd-cd)$;

g) $mx + xy - ny - mn$; h) $4x - 6 - 10ax + 15a$; i) $2ax^2 + bx^2 - 2ay^2 - by^2$;

j) $ay + a\sqrt{3} - y\sqrt{2} - \sqrt{6}$; k) $a^2x + bx - a^2y - by$; l) $x^3 - x^2 + x - 1$;

m) $x^3 + 3x^2 + 3x + 1$; n) $x^3 + x^2 - 2x - 2$;

o) $(4x-3)(3x-5) + (4x-3)(-2x+1) - 3 + 4x$;

p) $(5x-1)(-5x+2) - (5x+6)(2-5x)$;

q) $(x+2)(x+2\sqrt{3}) - (x+2\sqrt{3})$; r) $(1-2x)(x+\sqrt{3}) - (x-3)(x+\sqrt{3})$.

III. 10. Factor:

a) $2m^3 - 6m^5 + 6m^2n + 2m^2$; b) $a^2(a+3x) + x^2(a+3x)$;

c) $mn(m^2 + n^2) - 2n^2(m^2 + n^2)$; d) $a(2-x^2) + 3b(x^2 - 2) - 2 + x^2$;

f) $(2a-3b)(p+3m) + (3a-4b)(p+3m)$; g) $x^3 + x^2z + 3xz^2 + 3z^3$;

h) $x^3 - x^2z + 2xz^2 + 2z^3$; i) $(m^2 + 4m)^2 - 9$; j) $(2a-3b)^2 - 4b^2$;

k) $9(m-4p)^2 - 64m^2$; l) $100m^2 - 9(3m-2p)^2$; m) $(n+3q)^2 - 4(q-n)^2$;

n) $4p^{12} - 20p^6z^4 + 25z^8$; o) $-30ab^2 + 25b^4 + 9a^2$; p) $-6a^2b^4 + 1 + 9a^4b^8$;

q) $3x + 18x^3y^3 + 27x^5y^6$; r) $x^4 - 9x^2 - x^2 + 9$; s) $9a^2 - a^4 - 9b^2 + a^2b^2$;

t) $36x - 4xy^2 + 9 - y^2$; u) $4a - 16ac^2 - a^2 + 4a^2c^2 - 4 + 16c^2$;

v) $36a^{n+2} + 16a^{n-2}b^2 + 48a^nb$; x) $x^2 + 2xy + y^2 - z^2$;

y) $9 - y^2 - 6yz - 25z^2$; z) $3x^5 - 192x - 4x^4 + 256$.

III. 11. Simplify the fractions. Assume that no denominator is equal to zero.

a) $\dfrac{x^3}{y^3} \cdot \dfrac{y^2}{x^2}$; b) $\dfrac{-5x^3}{6y^2}\left(-\dfrac{18}{x}\right)$; c) $\left(\dfrac{-8x}{7y}\right)\left(\dfrac{-7}{x}\right)$; d) $-2x^2 \cdot \dfrac{5}{x^5} \cdot \dfrac{x^2}{4}$;

e) $\dfrac{2x^2}{6y}\left(-\dfrac{9xy}{y^2}\right)\left(\dfrac{-2}{6x^2}\right)$; f) $\dfrac{mx^2 + my}{x^2 - 3x} \cdot \dfrac{x-3}{x^2 + y}$; g) $\dfrac{ax^2}{ny^3} \div \dfrac{ax^2}{ny^3}$;

h) $\dfrac{x+y^2}{x-y^2} \div \dfrac{ax+ay^2}{bx-by^2}$; i) $\left(\dfrac{b}{2} - \dfrac{5b}{4} - \dfrac{b}{6}\right) \div \dfrac{b}{12}$; j) $\left(\dfrac{3}{a} + \dfrac{2}{a} - \dfrac{1}{a}\right) \div \left(\dfrac{5}{a} - \dfrac{1}{a} - \dfrac{4}{3a}\right)$.

III. 12. Perform the operations. Assume that no denominator is equal to zero.

a) $\dfrac{x}{3}+\dfrac{x}{3}-\dfrac{2x}{3}+\dfrac{5x}{3}$; b) $\dfrac{x}{y}+\dfrac{3x}{y}+\dfrac{4y-3x}{y}$; c) $\dfrac{x}{2}+\dfrac{3x-1}{3}+\dfrac{2x+5}{4}$;

d) $x+\dfrac{3x}{5}+\dfrac{1-2x}{2}$; e) $\dfrac{x}{x+2y+4z}+\dfrac{2y}{x+2y+4z}+\dfrac{4z}{x+2y+4z}$;

f) $x-\dfrac{x}{5}+\dfrac{2-x}{15}+\dfrac{1-x}{3}$; g) $2x-\dfrac{2-3x}{3}+\dfrac{1-2x}{2}$; h) $3x+\dfrac{2x-1}{2}-\dfrac{1-x}{4}$;

i) $\dfrac{1}{x}+\dfrac{1}{2x}-\dfrac{1}{3x}+\dfrac{1}{4x}$; j) $1-\dfrac{2}{7x}-\dfrac{5}{14x^2}$; k) $\dfrac{1}{2(x+1)}-\dfrac{1}{4(x+1)}-\dfrac{1}{6(x+1)}$;

l) $\dfrac{3}{5x-10}+\dfrac{1}{3x-6}-\dfrac{1}{x-2}$.

III. 13. Simplify the fractions. Assume that no denominator is equal to zero.

a) $\dfrac{xy+3y}{x^2+3x}$; b) $\dfrac{5x-20}{x^2-16}$; c) $\dfrac{7x^2+21xy}{9x^3-81xy^2}$; d) $\dfrac{40x^2-40}{5x+5}$;

e) $\dfrac{(2x+3)(x-2)-9(2x+3)}{(x-1)(4x^2-9)}$; f) $\dfrac{x^6-y^6}{x^9-x^6y^3}$; g) $\dfrac{x^2-8x+7}{x^2-49}$; h) $\dfrac{x^3-8y^3}{x^2-4y^2}$;

i) $\dfrac{x^2-4y^2-2x-4y}{4y^2-x^2-2y-x}$; j) $\dfrac{14x^3+14a^3}{7x+7a}$.

III. 14. Simplify the fractions. Assume that no denominator is equal to zero.

a) $\dfrac{72x^2y^3z}{288x^3y^2z^2}$; b) $\dfrac{5a^3y^3}{5a^2y^5}$; c) $\dfrac{24ax^2}{18abx^4}$; d) $\dfrac{(x-4)(x-2)}{(x-2)(x+3)}$; e) $\dfrac{3x^2-3x}{6x-6}$;

f) $\dfrac{x^2-4}{(x+2)^2}$; g) $\dfrac{x^2+x}{2x+2}$; h) $\dfrac{9x^2-1}{3x+1}$; i) $\dfrac{5x+1}{25x^2-1}$; j) $\dfrac{x^2-6x+9}{x^2-9}$; k) $\dfrac{a^2-25}{a+5}$;

l) $\dfrac{36x^2-25}{36x^2+60x+25}$; m) $\dfrac{4x^2+32x+64}{2x^2-32}$; n) $\dfrac{2x^3-5x^2-12x}{4x^3-20x^2+16x}$;

o) $\dfrac{3x^3-4x^2-4x}{4x^3-6x^2-4x}$; p) $\dfrac{a^3-a^2+2a-2}{a^3-a^2-2a+2}$; q) $\dfrac{x^2+3x-4}{x^2-5x+4}$;

r) $\dfrac{2x^4+x^3-6x^2}{2x^4-7x^3+6x^2}$; s) $\dfrac{(x^2-5x+2)(x^2-5x)-3}{(x^2-5x-2)(x^2-5x)+1}$;

t) $\dfrac{(x^2-4x+4)(x^2-4x)-5}{(x^2-4x+2)(x^2-4x)-15}$; u) $\dfrac{(2x^2-3x-4)(2x^2-3x)+4}{(2x^2-3x)(2x^2-3x-3)+2}$.

73

Algebraic rational expressions

III. 15. Calculate the expressions and simplify the fractions. Assume that no denominator is equal to zero.

a) $\dfrac{2x}{x^2+3x}+\dfrac{6}{x^2+3x}$; b) $\dfrac{2x+3}{x^2-4}+\dfrac{x+3}{x^2-4}$; c) $\dfrac{x^2(x-2)}{x^3+8}+\dfrac{4x}{x^3+8}$;

d) $\dfrac{3x-2y}{x-3y}-\dfrac{2x-y}{x-3y}-\dfrac{2y}{x-3y}$; e) $\dfrac{1}{x^2-1}+\dfrac{1}{x^2-1}\cdot\dfrac{1-x}{x+1}$;

f) $\dfrac{3a}{6a^2x^2}-\dfrac{x}{6a^3x}+\dfrac{a}{3a^3x^2}$; g) $\dfrac{3}{x^2-1}-\dfrac{3}{x^2+1}$; h) $\dfrac{3x+5}{x^2-1}+\dfrac{4}{x-1}+\dfrac{1}{x+1}$;

i) $\dfrac{x-1}{x^2-3x}+\dfrac{x+2}{x^2-3x}+\dfrac{1}{x^2-3x}$; j) $\dfrac{2x+17}{(x-2)(x+5)}+\dfrac{1}{(x-2)(x+5)}$;

k) $\dfrac{x-1}{x+1}+\dfrac{x+1}{x-1}-\dfrac{4}{x^2-1}$; l) $\dfrac{x^2}{x^2-3x}+\dfrac{9-6x}{x^2-3x}$; m) $\dfrac{2}{x-1}-\dfrac{x+5}{x^2-1}+\dfrac{5}{x+1}$;

n) $\dfrac{3}{x^2+2}+\dfrac{1}{x+1}-\dfrac{5-2x^2}{\left(1-x^2\right)\left(x^2+2\right)}$.

III. 16. Simplify the expressions. Assume that no denominator is equal to zero.

a) $\dfrac{x+1}{x-1}-\dfrac{x+3}{x-2}$; b) $3x+\dfrac{2x-3x^4}{x^3-1}$; c) $\dfrac{3(x-y)}{x(x+y)}-\dfrac{2(x-2y)}{x(x+y)}$;

d) $\dfrac{x-a+2b}{6(a-b)}-\dfrac{2x+4a-b}{2(a-b)}+\dfrac{2x+a+b}{3(a-b)}+\dfrac{2x+10a-2b}{12(a-b)}$;

e) $\dfrac{\left(2x^3-4x^2\right)\div 2x+x(3-x)-2}{x-2}-\dfrac{2-x}{2}$;

f) $\dfrac{6x^2+4xy}{2x+v}+\dfrac{2\left(x^2-2xy-y^2\right)}{2x+v}+2y$;

g) $\dfrac{x^2+xy}{x}+\dfrac{\left(2x^4y-6x^3y\right)\div 2x^2y}{x^4-3x^3}-\dfrac{2xy-x^2}{2x}$;

h) $\dfrac{x-y+1}{x-y}+\dfrac{x+y-2}{4(x-y)}-\dfrac{x+y+7}{3(x-y)}+\dfrac{x+y+22}{12(x-y)}$;

i) $\dfrac{1-2x}{x^2-3x}+\dfrac{2+x}{x^2+3x}+\dfrac{x+7}{x^2-9}$;

j) $\dfrac{-1-2x}{(x+1)(x-2)}+\dfrac{3+x}{(x+1)(x+4)}+\dfrac{x+8}{x^2+2x-8}$.

III. 17. Find the domain and simplify the following rational expressions:

a) $\left[\left(\dfrac{x}{x+1}-\dfrac{x^2}{x^2+2x+1}\right)\div\left(\dfrac{x}{x^2-1}-\dfrac{1}{x+1}\right)\right]\dfrac{x+1}{x}$;

b) $\left[\left(\dfrac{x}{x+2}+\dfrac{1}{x^2-4}\right)\left(\dfrac{x+1}{x-1}+\dfrac{2x+5}{1-x^2}\right)\right]\div\left(\dfrac{1}{x+1}-\dfrac{1}{2x}\right)$;

c) $\dfrac{x-1}{4}\cdot\left[\dfrac{x+3}{x+1}-\dfrac{x+1}{x+4}\left(\dfrac{2x+3}{x-1}-1\right)\right]$;

d) $\left(\dfrac{x^2+8}{x^3-8}+\dfrac{x}{x^2+2x+4}-\dfrac{1}{x-2}\right)\left(\dfrac{x^2}{x^2-4}-\dfrac{2}{2-x}\right)$;

e) $\left\{\left[\left(\dfrac{x}{x+1}-\dfrac{1}{1-x}-\dfrac{2x}{x^2-1}\right)\left(\dfrac{x-1}{x+1}\right)-\dfrac{2x-2}{x+1}+1\right]-1\right\}\div\dfrac{x+3}{x+1}$;

f) $\left(\dfrac{5}{2x+3}+\dfrac{2}{3-2x}+\dfrac{2x+9}{4x^2-9}\right)\dfrac{(2x+3)^2}{4}-\left(\dfrac{y}{x}-2+\dfrac{x}{y}\right)\div\left(\dfrac{y}{x^2+xy}-\dfrac{2}{x+y}+\dfrac{x}{y^2+xy}\right)$;

g) $\left[\left(\dfrac{2x^3+4x^2}{8x^2}-\dfrac{x^2+4}{4x}\right)\div\dfrac{x-2}{2x}+1\right]^2+\left(\dfrac{2x-1}{4}-\dfrac{x+1}{3}+\dfrac{x^2+8}{12}\right)\div\dfrac{(x+1)^2}{12}+x$

h) $E(x)=\left(\dfrac{2x+1}{x^2+3x}+\dfrac{2-x}{x^2-3x}-\dfrac{x-7}{x^2-9}\right)\div\dfrac{x+3}{x^3+2x^2-3x}$, also solve the

equation $E(x)=x-1$;

i) $\left[\dfrac{y-1}{(y+1)^2-y}-\dfrac{1-3y+y^2}{y^3-1}-\dfrac{1}{y-1}\right]\div\dfrac{y^2+1}{1-y}$;

j) $\left[\left(\dfrac{x^2-x}{x^2+1}+\dfrac{2x^2}{x^3-x^2+x-1}\right)\div\dfrac{x^2}{x^2-1}\right]\div\dfrac{2x^2+5x+3}{2x^2+3x}$;

k) $\dfrac{x+1}{x+2}+\left(\dfrac{x^2+7x+10}{x^2+x-6}-\dfrac{x^2+3x-10}{x^2+5x+6}\right)\div\dfrac{8x^2+40x}{x^2+x-6}.$

III. 18. Simplify the expressions. Assume that no denominator is equal to zero.

a) $\left[\dfrac{(x+y)^2 + 2y^2}{x^3 - y^3} - \dfrac{1}{x-y} + \dfrac{x+y}{x^2 + xy + y^2} \right] \div \left(\dfrac{1}{y} - \dfrac{1}{x} \right);$

b) $\left\{ \dfrac{x^2 - y^2}{(x+y)^2} + \left(\dfrac{x^2 - y^2}{(x+y)^2} \right)^3 + 2\left[\dfrac{x^2 - y^2}{(x+y)^2} \right]^2 \right\} \div \dfrac{x^2 - y^2}{(x+y)^4};$

c) $\dfrac{x^2 - 1}{xy} \div \left[\left(\dfrac{x^2 - xy}{x^2 y + y^3} - \dfrac{2x^2}{y^3 - xy^2 + x^2 y - x^3} \right) \cdot \left(1 - \dfrac{y-1}{x} - \dfrac{y}{x^2} \right) \right];$

d) $\left\{ \dfrac{1}{m^2 n} + \left[\dfrac{1}{n^4} + \dfrac{1}{m^2} + \dfrac{2}{m-n^2} \left(\dfrac{1}{m} - \dfrac{1}{n^2} \right) \right] \div \dfrac{(m-n^2)^2}{-mn^2} \right\} \left(\dfrac{m}{m+n} + \dfrac{n}{m-n} + \dfrac{2mn}{m^2 - n^2} \right);$

e) $\left(\dfrac{x}{y} + \dfrac{y}{x} + 2 \right) \left(\dfrac{x+y}{2x} - \dfrac{y}{x+y} \right) \div \left[\left(x + 2y + \dfrac{y^2}{x} \right) \left(\dfrac{x}{x+y} + \dfrac{y}{x-y} \right) \right];$

f) $\left(\dfrac{1}{2x-y} + \dfrac{3y}{y^2 - 4x^2} - \dfrac{2}{2x+y} \right) \div \left(\dfrac{4x^2 + y^2}{4x^2 - y^2} + 1 \right);$

g) $\left(\dfrac{1}{a - 2b} + \dfrac{6b}{4b^2 - a^2} - \dfrac{2}{a + 2b} \right) \div \left(\dfrac{a^2 + 4b^2}{a^2 - 4b^2} + 1 \right);$

h) $\left[\dfrac{x^2}{x^2 - y^2} - \dfrac{x^2 y}{x^2 + y^2} \left(\dfrac{x}{xy + y^2} + \dfrac{y}{x^2 + xy} \right) \right] \div \dfrac{x}{x-y};$

i) $\left(\dfrac{x^2 + y^2}{xy} - 2 \right) \div \left(\dfrac{2x^2 + 2xy}{x^2 + 2xy + y^2} - 1 \right) \cdot \left(\dfrac{1}{x-y} - \dfrac{1}{x+y} \right);$

j) $\left(\dfrac{a+b}{b} - \dfrac{2b}{b-a} \right) \cdot \dfrac{b-a}{a^2 + b^2} - \left(\dfrac{a^2 + 1}{2a - 1} - \dfrac{a}{2} \right) \div \dfrac{2+a}{1 - 2a}.$

III. 19. Consider the rational expression

$$E(x) = \dfrac{x^2 + 5x + 6}{x^2 + 4x + 4} - \dfrac{1}{x+2}.$$

a) Show that $E(x)$ is a positive integer for any $x \in \mathbf{R} \setminus \{-2\}$;

b) Find $x \in \mathbf{N}$ such that the number $\dfrac{2x+4}{x^2 - 4}$ is an integer number.

III. 20. Let $E(x) = \left(\dfrac{x+2}{x^2-x} + \dfrac{x-3}{x^2-1} - \dfrac{x-1}{x^2+x} \right) \div \dfrac{x^2+3x+2}{x^3-2x^2+x}$ be a

rational expression.

a) State the domain of $E(x)$;

b) Solve the equation $E(x) = \dfrac{3(x+1)}{x+2}$.

III. 21. Consider the algebraic expression

$$E(x) = \dfrac{9}{x-5} \cdot \left(\dfrac{x}{x+5} + \dfrac{5}{x-5} - \dfrac{10x}{x^2-25} \right).$$

a) Determine the domain of $E(x)$;

b) Find a simple form for $E(x)$;

c) Find the positive integer x such that $E(x)$ is an integer number.

III. 22. Consider the algebraic expression

$$E(x) = \left(\dfrac{x^2}{x+2} - \dfrac{x^3}{x^2+4x+4} \right) \div \left(\dfrac{x}{x+2} - \dfrac{x^2}{x^2-4} \right).$$

a) State the domain of $E(x)$;

b) Find a simple form for $E(x)$;

c) Find the positive integers x, such that $E(x)$ is an integer positive number.

d) Find the numeric value of x, such that $\dfrac{1}{x} \cdot E(x) = \dfrac{4}{x+2}$.

III. 23. Consider the rational expression

$$F(x) = \left(\dfrac{x^2+1}{x^2-2x+1} - \dfrac{x+1}{x-1} \right) \div \dfrac{-4}{1-x^2}.$$

a) Determine the domain of $F(x)$;

b) Find a simple form for $F(x)$;

c) Find the positive integer x such that $F(x)$ is an integer number.

III. 24. Consider the algebraic expression

$$E(x) = \left(\frac{1}{x+2} + \frac{2x}{2-x} + \frac{2x^2}{x^2-4} \right) \div \frac{3x+2}{3 \cdot (2-x)} \ .$$

a) Determine the domain of $E(x)$;

b) Show that $E(x) = \dfrac{3}{x+2}$;

c) Find the positive integer x such that $E(x)$ is an integer number.

III. 25. Consider $A = x^2 - x - 12$ and $B = x^2 - 12x + 36$.

a) Find $A - B$;

b) Find the numeric value of x such that $A = B$;

c) Prove that $\dfrac{A}{x-4} - \dfrac{B}{x-6}$ is constant for any $x \in \mathbf{R} \setminus \{4, 6\}$.

III. 26. Consider the rational expression

$$E(x) = \left(\frac{2}{x-5} + \frac{x}{x+5} \right) \div \frac{x^2 - 3x + 10}{x^2 - 6x + 5}.$$

a) Prove that $E(x) = \dfrac{x-1}{x+5}$, for any $x \in \mathbf{R} \setminus \{-5, 1, 5\}$;

b) Find the integer x such that $E(x)$ is an integer number.

III. 27. Consider the rational algebraic expression

$$E(x) = \left(\frac{x^4 + x^2}{x^6 + x^4 + x^2 + 1} + \frac{1}{x^4 + 1} \right) \div \left(\frac{1}{x^2 - 1} - \frac{2x^2}{x^6 - x^4 + x^2 - 1} \right) .$$

a) Determine the domain of $E(x)$;

b) Find a simple form for $E(x)$.

III. 28. Consider the rational algebraic expression

$$E(x) = \left(\frac{3x}{x-2} + \frac{4-6x}{x^2-4} - \frac{2x}{x+2} \right) \div \frac{x+2}{3}.$$ Find the domain D and the value of

$x \in D$ such that $E(x) \in \mathbf{Z}$.

III. 29. Consider the expressions $E_1(x) = \left(\dfrac{x}{x+1} - \dfrac{x}{x-1} \right) (x^2 - 1)$ and

$E_2(x) = \dfrac{x^3 - 2x^2 + x}{x^3 - x}$.

a) Determine the domain of $E_1(x)$ and $E_2(x)$;

78

b) Find a simple form for $E_1(x)$ and $E_2(x)$;

c) Prove that $\dfrac{E_1(x)}{x+1}+E_2(x)$ is an integer for any x from the domain.

III. 30. Consider the expression

$$E(x)=\left(\frac{1}{1-2x}-\frac{9x^2-18x^3}{1-4x+4x^2}\right)\div\left(1-\frac{5x^2}{1-4x^2}\right).$$

a) Determine the domain of $E(x)$;

b) Find the integer x such that $\dfrac{E(x)}{2x-1}$ is an integer number.

III. 31. Let $E(x)=\left(\dfrac{x}{x+3}-\dfrac{6x}{x^2-9}-\dfrac{3}{3-x}\right)\div\dfrac{x-3}{2x-1}$ be an algebraic

expression.

a) Determine the domain of $E(x)$;

b) Show that $E(x)=\dfrac{2x-1}{x+3}$;

c) Find the integer x such that $\dfrac{2E(x)}{2x-1}$ is an integer number.

III. 32. Solve the equations. Assume that no denominator is equal to zero.

a) $\dfrac{1}{x-3}=2$; **b)** $\dfrac{2x-1}{x}=\dfrac{4x}{2x-1}$; **c)** $3x+\dfrac{1}{x}=3x-\dfrac{1}{x}$; **d)** $\dfrac{2x-6}{x-3}=2$;

e) $\dfrac{3x-3}{x-1}=\dfrac{3x-4}{x-1}$; **f)** $\dfrac{2}{x}-\dfrac{2}{x+1}=0$; **g)** $x+1+\dfrac{x-1}{x-2}=3+\dfrac{x-1}{x-2}$;

h) $\dfrac{x}{x-3}+\dfrac{x-5}{x}=-\dfrac{2x}{3-x}$; **i)** $\dfrac{1}{x+1}-\dfrac{1}{2x+2}-\dfrac{2}{3x+3}-\dfrac{1}{6x-6}=0$;

j) $\dfrac{2}{1+x}-\dfrac{1}{x-3}=\dfrac{4}{(x+1)(3-x)}$; **k)** $\dfrac{x^2}{x-5}=\dfrac{25}{x-5}$; **l)** $\dfrac{a+b}{x}+\dfrac{a-b}{a}=1$;

m) $\dfrac{2x}{x-2}-\dfrac{1}{x+2}=\dfrac{2x^2}{x^2-4}$; **n)** $\dfrac{3(x+5)}{3x}=\dfrac{4x+1}{4x+2}$; **o)** $\dfrac{x^2-9}{x+5}=x+3$;

p) $\dfrac{3x-2}{x-5}=\dfrac{3x+1}{x+5}$; **q)** $\dfrac{x+\dfrac{1}{2}}{2}-\dfrac{2x-\dfrac{x+3}{4}}{5}=4$;

r) $\dfrac{2x+1}{2x-1} - \dfrac{3(2x-1)}{2x+1} - \dfrac{8x^2}{1-4x^2} = 0$; s) $\dfrac{a^2+x}{b^2-x} - \dfrac{a^2-x}{b^2+x} = \dfrac{4\left(abx + a^2 - b^2\right)}{b^4 - x^2}$,

$a \neq b$;

t) $\dfrac{x}{8} + \dfrac{1 - \dfrac{x+1}{2}}{2 - \dfrac{2}{2 - \dfrac{2}{2 + \dfrac{x}{2}}}} = -\dfrac{1}{4}$; u) $\dfrac{5+\sqrt{7}}{x} = \dfrac{9}{5-\sqrt{7}}$.

III. 33. Solve the equations.

a) $2x(3x-2) - 3\left[(2-x)(-2x+3) - \dfrac{x-3}{2}\right] = -4$;

b) $\dfrac{1}{(x+1)^2} + \dfrac{4}{x(x+1)^2} = \dfrac{5}{2x(x+1)}$; c) $\dfrac{2x+26}{5x^2-5} - \dfrac{12}{x^2-1} - \dfrac{2}{1-x} = 0$;

d) $\dfrac{10}{4-x} = \dfrac{4}{1-3x} - \dfrac{5}{x-4}$; e) $\dfrac{x+a}{a-x} + \dfrac{x-a}{a+x} = \dfrac{8a}{a^2-x^2}$;

f) $\dfrac{x}{a} + 1 \div \left(1 - \dfrac{a^2}{b^2}\right) = 1 \div \left(\dfrac{a}{b} - \dfrac{b}{a}\right)$; g) $\dfrac{x}{a}(3a+1) = \dfrac{3a}{1+a} + \dfrac{(2a+1)x}{a(a+1)^2} + \dfrac{a^2}{(a+1)^3}$;

h) $\dfrac{a^3+1}{a^3-1} = \dfrac{a(x-1) - a^2 + x}{a(x+1) + a^2 + x}$; i) $\dfrac{ax-x}{2a+2} + \dfrac{ax}{a^2-1} - \dfrac{a-x}{a-1} = \dfrac{x}{2} + \dfrac{a+x}{a+1}$;

j) $x \div \left[\dfrac{(a-1)^2}{4a} + 1\right] = \left[\dfrac{(a-1)^2 + a}{3}\right] \div \left[\dfrac{(a+1)^3}{3a} - a - 1\right]$;

k) $\dfrac{(x-1)^2 + 5(x-1) + 16}{x+3} - 5(x+2) = -4x - 7$;

l) $\dfrac{16a^2x^2 - 48ax + 36}{8ax - 12} - \dfrac{ax-5}{5} - \dfrac{9ax}{5} = 2 - x$;

m) $\dfrac{36-x^2}{x+6} - \dfrac{12x^2 + 36x + 27}{3(2x+3)} = \dfrac{2x-1}{5}$.

III. 34. Solve the equation $0.\overline{1x1} + 0.\overline{2x2} + 0.\overline{3x3} + \ldots + 0.\overline{9x9} = x$.

III. 35. Solve the equation $\dfrac{1}{3x} + \dfrac{1}{5y} = \dfrac{1}{6}$, where x and y are integer numbers.

III. 36. Solve the equations:

a) $-0.7x + 3.8 = x + 0.4$; b) $0.25x - 1.5 = 0$; c) $5x - 2\sqrt{3} = \sqrt{3}$;

d) $\sqrt{3}x - \sqrt{6} = 0$; e) $-\sqrt{3}x = \sqrt{3} - 3$; f) $2.\bar{1} - 2x = \dfrac{1}{4} + \dfrac{5}{12} + \dfrac{7}{18}$;

g) $x - \left|7 - \sqrt{5}\right| = \left|\sqrt{5} - 7\right|$; h) $3x - \sqrt{2} = 2x + \sqrt{18}$; i) $5\left(x + \sqrt{2}\right) + \dfrac{1}{2} = 0.25$;

j) $(x - 1)\sqrt{2} = \left(\sqrt{3} - 1\right) \cdot x + 1 - \sqrt{3}$;

k) $\left(\sqrt{5} - 2\right)\!\left(\sqrt{5} + 2\right) = \left(\dfrac{\sqrt{2}}{2} - x\right)\!\left(\sqrt{7} - \sqrt{5}\right)\!\left(\sqrt{7} + \sqrt{5}\right)$;

l) $\sqrt{2}x + 3 = 5\sqrt{2}(x - 1)$; m) $3(x - 1) = \dfrac{5x - 1}{2}$; n) $-\dfrac{3}{4}x + \dfrac{1 - \sqrt{3}}{3} = 0$;

o) $\dfrac{1 - 0.4x}{0.6} - \dfrac{1 - 3x}{2} = \dfrac{0.5x - 0.4}{0.3}$; p) $0.\bar{3}z - \sqrt{2} = 1.5z + 0.\bar{5}$;

q) $3\sqrt{2} + x = \sqrt{2}\left(x - \sqrt{2}\right) + 5$; r) $\dfrac{0.2(2 - 3x)}{0.01} - \dfrac{0.25 - 2x}{0.1} = \dfrac{0.6 - x}{0.02} + 7.5$;

s) $\dfrac{1}{2}\left[\dfrac{1}{3}\left(\dfrac{1}{4} \cdot x + 2\right) - 1\right] = 0$; t) $\dfrac{1}{9}\left\{\dfrac{1}{7}\left[\dfrac{1}{5}\left(\dfrac{1}{3} \cdot (x - 1) + 4\right) + 6\right] + 8\right\} = 1$;

u) $\dfrac{3x}{2} - \dfrac{x - 2}{3} - 1 = x - \dfrac{2 - x}{6}$;

v) Find $a \in \mathbf{R}$ such that $x = -2$ be a solution for the equations:

i) $ax + 5 = 6 - a$;

ii) $3x + a = 5$;

iii) $2a - 3x = 3(a + 1) - 1 - 4x$;

iv) $2x + 2a = 3a + x - 5$;

w) Find m such that in the equation $mx + 5 = 11$ have the solution
i) 1; ii) 2; iii) 3; iv) -1.

III. 37. Linear equations containing absolute value.

a) $\left|x + 2\right| = 6$; b) $\left|x - 3\right| = 5$; c) $\left|3x + 2\right| = 7$; d) $20 \div \left|x - 1\right| = 4$;

e) $2\left|x\right| + 1 = 7$; f) $\left|x\right| + \dfrac{1}{2} = \dfrac{3}{4} + \left|2x\right|$; g) $\left|2x + 1\right| - 6 = -3\dfrac{1}{3}$

h) $3\left(2\left|3x + 1\right| + 6\right) - 10 = 14$; i) $\left|x^2 - 9\right| + \left|2x - 6\right| = 0$; j) $\left|x\right| = m$;

k) $\left|x + 1\right| = a$; l) $\left|x\right| + \left|x + 1\right| = -1$; m) $\left|a\right| \cdot x = a$, $a \in \mathbf{R}$;

n) $|x| = -x$; o) $|2m + 1| \cdot x = 2m + 1$; p) $\sqrt{m^2} \cdot x = m(m - 1)$;

q) $\sqrt{m^2 + 4m + 4} \cdot x = m + 2$; r) $|m| \cdot x = m$; s) $|m| \cdot x + 2x = 2 + m$;

t) $|a| \cdot x + 2ax = 2$, $a \in \mathbf{R}$; u) $|x + |x + 1|| + |x - |x + 1|| = 2$;

v) $\sqrt{x^2 - 4x + 4} = 3x - 1$; w) $|2x - 1| = |2x + 3|$; x) $|x - 2| - 2|2x + 3| = 0$;

y) $||x| + |x + 2|| = x + 5$.

III. 38. Solve each equation and state the domain for their solutions.

a) $2x = 3 - mx$; b) $mx - m = 1 - x$; c) $mx - 4 = 3x$; d) $3(x + 1) = m + 2$;

e) $a^2 x = b(3a - bx)$; f) $m^2 x + mx + 3 = 2x + m$; g) $m^2 x + 3 - 3x = m + 2mx$;

h) $m^2 x - 1 = 3mx + 4x$; i) $mx - 4m = m^2 + 4x$; j) $m^2 x = 2m(1 - x)$;

k) $mx - 1 = x - m^2$; l) $2ax + 7 = x + 3a$; m) $mx + 3(x + m) = x + m - 1$;

n) $(m + 1)x + a = x$, $a, m \in \mathbf{R}$; o) $am - ax = mx - am$;

p) $3mx + 2a - 5 = 2(xm + a)$; q) $1 - (x - m) \cdot a = (a - x) \cdot m$;

r) $a \cdot (x - a) = m(x - m)$; s) $4\sqrt{2}x - \sqrt{2}m = 4mx$;

t) $\dfrac{m}{x - 3} = \dfrac{x + 2}{x^2 - 9}$, $m \in \mathbf{R} \setminus \{-3, 3\}$; u) $2a(x - 2a) = m(x - 3m)$.

III. 39. Compute, factor, and solve each equation.

a) $-2x = 4x^2$; b) $x^2 - 9 = 0$, $x \in \mathbf{N}$; c) $x^2 - 3 = 0$; d) $3x^2 + 5x = 0$;

e) $\dfrac{x^2 - 2x}{x} = 0$; f) $4x^2 - 25 = 0$; g) $(x + 1)^2 - 16 = 0$;

h) $4(x + 1)^2 - 9(x - 1)^2 = 0$; i) $(x - 1)(x - 2)(x + 3) = 0$;

j) $(x + 2)(x - 5) + (x + 2)(3x - 4) = 0$; k) $8x + 16 + x^2 = 0$; l) $3 - 2x - x^2 = 0$;

m) $8x^2 - 7x - 1 = 0$; n) $1 + 4x^2 = 5x$, $x \in \mathbf{N}$; o) $2x^2 - 3x - 9 = 0$;

p) $(x + 3)(5x + 9) - x^2 + 9 = 0$; q) $\dfrac{x^2 + 5x + 4}{x^2 + 3x + 2} = 1$;

r) $\dfrac{x^3 - 9x + 3x^2 - 27}{x^2 - 9} = x^2 + 3x$.

III. 40. Solve the inequations:

a) $-x + 2 \geq 0$; b) $-3x + 1 < 0$; c) $-7x - 1 \geq 0$; d) $-4x \geq -2x + 3$;

e) $-6x \geq -8x$; f) $4(x-1) - \dfrac{1-x}{2} \leq 3(x+2) - \dfrac{12-x}{5} - 2x$;

g) $2(2-3x) - \dfrac{1-x}{2} + \dfrac{12-x}{3} \geq 2(2x-1) - \dfrac{3-x}{6}$;

h) $(x-1)^2 - 5 > (x+4)^2$; i) $(x+1)(x+2) + 3(1-x) < (x-1)^2$;

j) $\dfrac{x-1}{3} - (1-2x) > \dfrac{1}{4}x - \dfrac{1-2x}{6}$; k) $\dfrac{3-x}{4} - \dfrac{3}{2}(x-1) > \dfrac{2}{3} - \dfrac{4x-3}{6}$;

l) $(x-1)(2x+3) \leq (2x-5)(x+4)$; m) $\dfrac{x-1}{4} - \dfrac{3-x}{2} \geq \dfrac{1}{3}x - \dfrac{5(x+3)}{6} + \dfrac{3}{4}$;

n) $\dfrac{2x+27}{35} < \dfrac{3x+5}{7} + 1 + \dfrac{1-3x}{5}$; o) $\dfrac{3-x}{5} - \dfrac{2-x}{2} \geq x + 2$;

p) $\dfrac{x+1}{21} - \dfrac{5-x}{3} - 8 \cdot \left(\dfrac{4-x}{7} - \dfrac{2-x}{3} \right) \geq 5x - \dfrac{3x - \dfrac{2}{3}}{7}$;

r) $(x+1)^3 - (x-2)^2 + x \geq (x+2)^2 + (x-1)^3 + 2(x\sqrt{2} - 1)(x\sqrt{2} + 1)$

III. 41. Solve the systems of equations.

a) $\begin{cases} 2x + 3y = -1 \\ 4x - 3y = 7 \end{cases}$; b) $\begin{cases} 4x + 3y = 5 \\ 2x - 5y = 9 \end{cases}$; c) $\begin{cases} 4x - 5y = 1 \\ 7x - 9y = 2 \end{cases}$;

d) $\begin{cases} 8a - 1\ b = 5 \\ 3a + 4b = 10 \end{cases}$; e) $\begin{cases} a - b = 2 \\ a + b = 4 \end{cases}$; f) $\begin{cases} 4a - 3x = 1 \\ 3a + 4x = 7 \end{cases}$;

g) $\begin{cases} 5m - 4n = -1 \\ 8m - 2n = -6 \end{cases}$; h) $\begin{cases} \dfrac{1}{5}x + \dfrac{1}{2}y = -2 \\ \dfrac{1}{4}x - \dfrac{1}{15}y = -\dfrac{71}{60} \end{cases}$; i) $\begin{cases} \dfrac{1}{7}x + \dfrac{2}{3}y = 5 \\ \dfrac{3}{7}x - \dfrac{1}{3}y = 1 \end{cases}$;

j) $\begin{cases} \dfrac{1}{3}a + \dfrac{2}{5}b = 6 \\ \dfrac{1}{6}a - \dfrac{3}{10}b = -2 \end{cases}$; k) $\begin{cases} \dfrac{1}{x} + \dfrac{2}{y} = 3 \\ \dfrac{3}{x} - \dfrac{1}{y} = 2 \end{cases}$.

III. 42. Solve the systems of linear equations:

a) $\begin{cases} x + y = -1 \\ 4x + 3y = -2 \end{cases}$;

b) $\begin{cases} \dfrac{x-1}{2} + \dfrac{y+1}{5} = \dfrac{1}{5} \\ \dfrac{x-4}{7} - \dfrac{1-y}{2} = -\dfrac{13}{14} \end{cases}$;

c) $\begin{cases} \dfrac{2y - x + 5}{2x + y - 5} = 4\dfrac{1}{2} \\ \dfrac{x-2}{3} - \dfrac{3-y}{2} + 2x - y = 1 \end{cases}$;

d) $\begin{cases} 1 - \dfrac{x + 2y}{7} = \dfrac{1}{3}[2 - 2(x-2) - y] \\ \dfrac{x-2}{4} - \dfrac{1+y}{3} + x + 5y = -3 \end{cases}$;

e) $\begin{cases} x\sqrt{2} + y\sqrt{3} = -1 \\ 5x\sqrt{2} - 2y\sqrt{3} = 16 \end{cases}$;

f) $\begin{cases} \left(\sqrt{5} + \sqrt{3}\right)x + \left(\sqrt{5} - \sqrt{3}\right)y = 4 \\ 2x - y = \sqrt{5} - 3\sqrt{3} \end{cases}$;

g) $\begin{cases} \dfrac{1}{x} - \dfrac{1}{y} = 6 \\ \dfrac{3}{x} + \dfrac{2}{y} = 3 \end{cases}$;

h) $\begin{cases} \dfrac{1}{x} + 10y = -3 \\ \dfrac{3}{x} + 4y = 4 \end{cases}$;

i) $\begin{cases} \dfrac{28}{x} = \dfrac{82}{y} \\ x\left(\dfrac{28}{x} - 2\right) = y\left(\dfrac{82}{y} - 2\right) + 54 \end{cases}$;

j) $\begin{cases} \dfrac{x-1}{5} + \dfrac{y+3}{3} = \dfrac{x}{4} - \dfrac{1}{4} \\ x + y = -2 \end{cases}$.

III. 43. Solve the systems:

a) $\begin{cases} \dfrac{2}{x} + \dfrac{3}{y} = -\dfrac{1}{6} \\ \dfrac{3}{x} - \dfrac{2}{y} = \dfrac{5}{6} \end{cases}$;

b) $\begin{cases} \dfrac{2x - 3y}{4} - \dfrac{2x - y}{3} = -\dfrac{3}{4} \\ (x-2)^2 - (3-y)^2 = (x-y)(x+y) - 7 \end{cases}$;

c) $\begin{cases} \dfrac{x + y - 9}{x - 2y - 3} = -2 \\ \dfrac{x+y}{4} - \dfrac{3x - 2y - 6}{6} = -\dfrac{7}{12} \end{cases}$;

d) $\begin{cases} x + 2 \cdot [y - 6 \cdot (x - 1)] = -1 \\ y - 3 \cdot [x - 6 \cdot (y + 1)] = -4 \end{cases}$;

e) $\begin{cases} \dfrac{x}{a} + \dfrac{y}{b} = 3 \\ \dfrac{x}{3a} + \dfrac{y}{4b} = \dfrac{5}{6} \end{cases}$, $a, b \in \mathbf{R} \setminus \{0\}$;

f) $\begin{cases} \dfrac{x}{a+b} + \dfrac{y}{a-b} = 2a \\ x - y = 4ab \end{cases}$;

g) $\begin{cases} \dfrac{2}{x + y - 1} + \dfrac{3}{x - y + 2} = 5 \\ \dfrac{1}{x - y + 2} - \dfrac{2}{x + y - 1} = -1 \end{cases}$;

h) $\begin{cases} \dfrac{3}{x+y} = \dfrac{4}{x+z} = \dfrac{5}{y+z} \\ (x+y)(x+z)(y+z) = 60 \end{cases}$;

i) $\dfrac{x + 2y - 7}{1} = \dfrac{2y - x + 15}{2} = \dfrac{2x + y + 19}{3}$.

III. 44. The sum of two by two sides of a triangle is 45 m, 52 m, and 48 m, respectively. Find the length of its sides.

III. 45. Two triangles are similar. The sides of the first one are equal to 7 cm, 10 cm, 11 cm respectively, and the perimeter of the second triangle is 70 cm. Find the sides of the second triangle.

III. 46. The sum of three natural numbers is 222. If we divide the first number by the second one, we will obtain 10, and if we divide the second number by the third one, we will also obtain 10. Find the numbers.

III. 47. Solve the systems of equations:

a) $\begin{cases} \dfrac{x-3}{y+1}=1 \\ \dfrac{x+1}{y-2}=3 \end{cases}$; b) $\begin{cases} \dfrac{a-1}{b-1}=\dfrac{1}{5} \\ \dfrac{a+14}{b+14}=\dfrac{1}{2} \end{cases}$; c) $\begin{cases} p+\dfrac{5q-1}{2}=14 \\ 5p+\dfrac{q-2}{2}=\dfrac{71}{2} \end{cases}$ d) $\begin{cases} \dfrac{5}{x-1}\div\dfrac{4}{y-1}=3\dfrac{3}{4} \\ \dfrac{6}{x+8}\div\dfrac{8}{y+12}=1\dfrac{1}{5} \end{cases}$;

e) $\begin{cases} \dfrac{3x+2y}{3}-x=6 \\ 4y-\dfrac{x-y}{3}=37 \end{cases}$; f) $\begin{cases} \dfrac{x+y}{2}-\dfrac{x-y}{2}=1 \\ \dfrac{x+y}{3}+\dfrac{x-y}{4}=\dfrac{2}{3} \end{cases}$; g) $\begin{cases} \dfrac{2x-y}{3}+\dfrac{2x+y}{4}=1 \\ \dfrac{6}{x+1}\div\dfrac{5}{y+2}=2.4 \end{cases}$;

h) $\begin{cases} 4x+7y+10z=21 \\ x+2y+2z=5 \\ 3x+5y-2z=6 \end{cases}$; i) $\begin{cases} 2x-3y-4z=-16 \\ 3x+y-z=2 \\ 5x-2y+3z=10 \end{cases}$;

j) $\begin{cases} 3x+y-3z=1 \\ x-3y+4z=2 \\ 2x+2y-z=3 \end{cases}$; k) $\begin{cases} 2x-y+5z=25 \\ 4x-3y+3z=17 \\ x-2y+z=8 \end{cases}$; l) $\begin{cases} 3x-y-5z=-6 \\ 2x+y-z=4 \\ 3x-5y+4z=4 \end{cases}$;

m) $\begin{cases} a-b+2c=-2 \\ 3a+2b-c=-3 \\ 2a+3b-5c=-3 \end{cases}$; n) $\begin{cases} x-4y+2z=-29 \\ 8x+2y-3z=2 \\ 2x+5y+z=40 \end{cases}$; o) $\begin{cases} -m-n+2p=0 \\ m+2n-4p=-1 \\ -m+3n-2p=0 \end{cases}$;

p) $\begin{cases} x^{-1}+y^{-1}=5 \\ y^{-1}+z^{-1}=7 \\ z^{-1}+x^{-1}=6 \end{cases}$.

III. 48. Find the sum of $\dfrac{7}{2^4-2}+\dfrac{13}{3^4-3}+\dfrac{21}{4^4-4}+...+\dfrac{n^2+n+1}{n^4-n}$.

85

Coordinate Geometry. Slopes. Linear Functions.

III. 49. Show that A, B, and C are the vertices of a right angled triangle and in each case state the right angle:
a) $A(2, 0)$, $B(2, 5)$, $C(-5, 0)$;
b) $A(-5, 3)$, $B(-6, -2)$, $C(-1, -3)$.

III. 50. Use the midpoint formula to find the coordinates of the midpoint of AB given:
a) $A(3, -1)$ and $B(1, 7)$; **b)** $A(4, 1)$ and $B(4, -3)$;
c) $A(5, 0)$ and $B(-3, 6)$; **d)** $A(4, 0)$ and $B(0, 3)$;
e) $A(0, -1)$ and $B(4, -5)$; **f)** $A(-2, 7)$ and $B(4, -3)$.

III. 51. Suppose M is the midpoint of PQ, find the coordinates of Q given:
a) $P(-2, 3)$ and $M(-4, 9)$; **b)** $M(-2, -1)$ and $P(4, 7)$;
c) $P(1, -2)$ and $M(1, 6)$; **d)** $M(1, -1)$ and $P(-1, 5)$.

III. 52. Suppose M is the midpoint of AB, find the coordinates of A given:
a) $M(4, -3)$ and $B(-2, 3)$; **b)** $B(0, 4)$ and $M(-3, -2)$;
c) $M(4, -4)$ and $B(-5, 3)$; **d)** $B(1, 4)$ and $M(-3, -4)$.

III. 53. The points A, B, C, and D are collinear, such that
$AB = BC = CD$. Find the coordinates of B and D, when:
a) $A(5, -3)$ and $C(-1, 3)$; **b)** $A(1, 4)$ and $C(-3, -2)$.

III. 54. Find the coordinates of B and D, when $AB = BC = CD$.

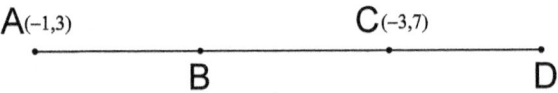

III. 55. $ABCD$ is a parallelogram.
Diagonals AC and BD bisect each
other at O.
Find the coordinates of O and C.

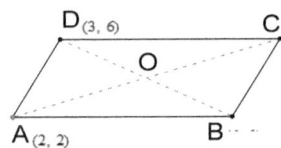

III. 56. Triangle ABC has vertices
$A(1, 2)$, $B(5, 0)$ and $C(3, -4)$. M is the midpoint of AB and N is the midpoint of BC.
a) Use the gradients to show that MN is parallel to AC;
b) Show that MN is half of AC.

III. 57. In a triangle ABC, M is the midpoint of AB and N is the midpoint of CM. Find the coordinates of N.

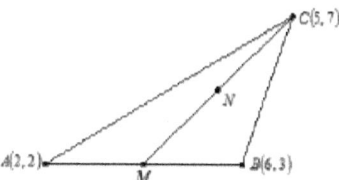

III. 58. Given the points $A(0, 4)$, $B(5, 4)$, $C(2, 0)$ and $D(-3, 0)$
a) Show that $ABCD$ is a rhombus;
b) Use the gradients to verify that its diagonals are perpendicular.

III. 59. Given the points $A(1, 4)$, $B(5, 6)$ and $C(3, 0)$:
a) Show that ABC is a right angle isosceles triangle;
b) Find the coordinate of M, the midpoint of BC;
c) Use the gradients to verify that AM and BC are perpendicular.

III. 60. Consider the diameter of the semicircle AB, the point $M(4, k)$ lies on the arc AB
a) Find k ;
b) Find the gradients of AM and BM.

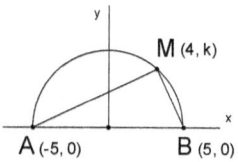

III. 61. Given the points $A(8, 7)$, $B(17, 7)$, $C(2, -2)$, and $D(-7, -2)$:
a) Use the gradients to show that $ABCD$ is a parallelogram;
b) Use the distance formula to check that $AB = DC$ and $BC = AD$;
c) Find the midpoints of AC and BD.

III. 62. Find the value of a given:
a) $A(5, 2)$ and $B(-2, a)$ are 7 units apart;
b) $C(1, -1)$, is 5 units from $D(-2, a)$;
c) $E(3, a)$ and $F(5, -2)$ are $\sqrt{13}$ units apart.

III. 63. Consider the triangle ABC with vertices $A(5, 4)$, $B(-1, 6)$, and $C(3, -4)$.
a) Calculate the length of the median from point A;
b) Find the coordinates of the points that divide the segment BC into eight equal parts.

III. 64. $A(4, 1)$, $B(3, -4)$ and $C(8, -5)$ are three points in the Cartesian plane.
a) Find the gradient of AB and BC;
b) What can be said about the length of segments AB and BC?

c) Classify the triangle ABC.

III. 65. Find the distance between the points $A(x, y)$, and $B\left(\dfrac{2(x+my)}{1+m^2}, \dfrac{2m(x+my)}{1+m^2}\right)$, where m is a real number.

III. 66. Consider the points $A(-1, 2)$, and $B(3, 4)$. Find the distance from the point O to the line AB.

III. 67. Find the distance from:

a) $A(0, 3)$ to the line $4x - 3y - 1 = 0$;

b) $A(-1, 3)$ to the line $x + y - 5 = 0$;

c) $O(0, 0)$ to the line $y = 2x - 10$;

d) $A(3, -1)$ to the line $y = 2x + 3$;

e) $A(0, -1)$ to the line $y = -\dfrac{5x}{2} - \dfrac{3}{2}$.

III. 68. Find the distance from the point:

a) $A(4, -2)$ to the line joining the points $B(1, 3)$ and $C(-4, -2)$;

b) $A(4, 1)$ to the line joining the points $B(1, -8)$ and $C(-4, -2)$.

III. 69. Find m, such that the distance from the point $A(m, m + 1)$ to the line $3x - 4y - 1 = 0$ is 1.

III. 70. Show that the triangle with vertices $A\left(1+\sqrt{3}, \dfrac{3}{2}\right)$, $B\left(\dfrac{1}{2}, \dfrac{1+\sqrt{3}}{2}\right)$, and $C\left(1, \dfrac{1}{2}\right)$ is right.

III. 71. Show that the triangle with vertices $A(1, 2)$, $B(3, -1)$, and $C\left(\dfrac{4+3\sqrt{3}}{2}, \dfrac{1+2\sqrt{3}}{2}\right)$ is equilateral.

III. 72. Consider the points $A(4, -1)$ and $B(2, 3)$. Find a point P on x-axis such that $PA = PB$.

III. 73. Consider the points $A(-3, 3)$ and $B(3, 7)$. Find a point P on the line $y = 2$ such that the angle APB be right.

III. 74. Consider the points $A\left(1, 1 + 2\sqrt{3}\right)$, $B(3, 5)$, and $C\left(6, 1 - \sqrt{7}\right)$. Find a point Q at the same distance to the points A, B, and C.

III. 75. Find the linear function $f : \mathbf{R} \to \mathbf{R}$ such that the graph contain the points:

a) $A(2, 0)$ and $B(-1, 3)$;   b) $C(-3, -9)$ and $D(5, 7)$;
c) $E(0, 5)$ and $F(2, 4)$;   d) $G(1, -6)$ and $H(-4, 14)$;
e) $I(-3, -10)$ and $J(4, -3)$;   f) $K(-1, -4)$ and $L(1, 6)$.

III. 76. Find the gradient (*slope*) and y-intercept of the lines with the equations:

a) $y = \dfrac{x - 2}{3}$;   b) $\dfrac{x}{3} + \dfrac{y}{5} = -7$;   c) $\dfrac{2x}{3} + \dfrac{y}{4} = -\dfrac{7}{6}$.

III. 77. Find the equation of the line:

a) having gradient $-\dfrac{1}{2}$ and cutting the y-axis at 4;

b) parallel to a line with slope 3, and passing through the point $(-1, -2)$;
c) cutting the x-axis at 4 and the y-axis at 3;
d) cutting the x-axis at -2 and passing through the point $(2, -5)$;

e) perpendicular to a line with the gradient $\dfrac{1}{2}$ and cutting the x-axis at -1;

f) perpendicular to a line with the gradient -3, and passing through the point $(4, -1)$;
g) with x and y-intercepts 2 and -5, respectively;
h) passing through the points $(-1, 3)$ and $(2, 1)$.

III. 78. Check if the following points are collinear:
a) $A(-2, 1)$, $B(1, 7)$, and $C(-1, 1)$;   b) $D(1, 6)$, $E(3, 8)$, and $F(-5, 0)$;
c) $G(2, -5)$, $H(-3, 10)$, and $I(4, 3)$;   d) $J(1, 0)$, $K(2, 7)$, and $L(3, 14)$.

III. 79. A line has slope $\dfrac{2m}{3}$. Find m given that the line is:

a) Parallel to a line with slope $\dfrac{1}{6}$;

b) Parallel to a line with slope $-\dfrac{2}{3}$;

c) Perpendicular to a line with slope 5;

d) Perpendicular to a line with slope $\dfrac{6}{5}$.

III. 80. A line passes through the points $A(2,4)$ and $B(5, a)$
a) Find the gradient of AB;

b) Find a if AB is parallel to a line of gradient of $\dfrac{1}{3}$;

c) Write the equation of the perpendicular bisector of the segment AB for $a = 5$.

III. 81. A line passes through the points $P(-1,6)$ and $Q(b, 4)$.
Find b if PQ is:

a) Parallel to a line with of gradient of $-\dfrac{2}{3}$;

b) Perpendicular to a line of gradient of $-\dfrac{4}{3}$.

III. 82. For $A(5,1)$, $B(1, -1)$, $C(a + 1, 2)$ and $D(0, 3)$, find a when:
a) AB is parallel to CD;
b) BC is perpendicular to AD.

III. 83. Consider the linear function $f : R \rightarrow R$,
$f(x) = 3x - 6a + 2$, where $a \in N$. Find a such that the point
$A(a, - a + 8)$ lie on the graph of f.

III. 84. The lines $3x - 4y - 1 = 0$ and $mx - 3y - 6 = 0$ are orthogonal to
one another. Find the value of m.

III. 85. Draw the graph of the function
$g : \{-1, 0, 1, 2\} \rightarrow R$, $g(x) = 2x + 2$.

III. 86. Consider the points $A(2, 3)$, $B(5, 5)$, $C(-2, 4)$ and $D(4, k)$. Find the
value of k if:
a) AB is parallel to CD;
b) AC is perpendicular to BD.

III. 87. Consider the function $f : R \rightarrow R$, $f(x) = ax + a + 2$, $a \in R$. Find
the value of the number a if the point $A(1, 4)$ lies on the graph of f.

III. 88. Find b if

a) $(2, b)$ lies on the line with the equation $3x - 2y = -4$;

b) $(-1, b)$ lies on the line with the equation $3x - 4y = 6$;

c) $(b, -4)$ lies on the line with the equation $5x + y = 1$;

d) $(b, -3)$ lies on the line with the equation $4x - y = 8$;

e) $(b, 3)$ lies on the line with the equation $3x + by = 12$.

III. 89. Consider the functions $f, g : \mathbf{R} \to \mathbf{R}$, $f(x) = 2x + 1$, $g(x) = -x + 2$. Find the coordinates of the point of intersection of their graphs. Are the points $A(0, 1)$, $B\left(\dfrac{1}{2}, 2\right)$, and $C(-1, -1)$ collinear?

III. 90. Find the function $f : \mathbf{R} \to \mathbf{R}$, $f(x) = ax + b$, where $a, b \in \mathbf{R}$, a, b < 0, if the area between the graph of f and Ox and Oy axes is 4 and the slope of f is –3.

III. 91. Consider the function $f : \mathbf{R} \to \mathbf{R}$, $f(x) = x - 3$.

a) Find $f(4)$;

b) Draw the graph of $f(x)$;

c) Find the points of intersection of f with the axes;

d) Which of the points $A(5, 2)$ and $B(-2, 5)$ lay on the graph?

e) Find the area of the triangle between the graph of f and the axes;

f) Find the distance from the origin to the graph of $f(x)$;

g) Find the measure of the angle between the graph of $f(x)$ and the x-axis ;

h) Solve the equation $f(x) + 2 = 1 - x$;

i) Solve the inequation $3 - 2f(x) \geq 3f(x) - 2$.

III. 92. Find the function $f : \mathbf{R} \to \mathbf{R}$, $f(x) = ax + b$ such that the graph pass through the points $A(1, 2)$ and $B(3, -2)$.

III. 93. Consider the functions $f, g : \mathbf{R} \to \mathbf{R}$, $f(x) = 2x - 8$ and $g(x) = x - 7$.

a) Find the coordinates of the points of intersection between G_f (*graph of the function* f) and the x-axis and the y-axis;

b) Draw the graph for the function f;

c) Find the point of intersection between the graphs of the functions f and g;

d) Find the perimeter and the area of the triangle formed by the graph of f and the x-axis and y-axis;

e) The distance from the origin to the graph of f;

f) The radius of the incircle and circumcircle of the triangle formed by the graph of f and the x-axis and y-axis;

g) Find the linear function $h : \mathbf{R} \to \mathbf{R}$, $h(x) = ax + b$ if
$h(x) + h(2 + 2x) = f(1 + x) + g(3 + x)$.

h) Solve the inequation $f(x + 2) + 2f(3 - 2x) \geq 6$;

i) Solve the system of equations $\begin{cases} f(x) + 3g(1 - 2y) = -40 \\ 3f(x + 3) - 2g(y - 1) = 0 \end{cases}$;

j) Prove that $\dfrac{f(a) + f(b)}{2} = f\left(\dfrac{a + b}{2}\right)$;

k) Which of the points $A(1, 3)$, $B(0, 4)$, $C(11, 14)$ lay on the graph of f?

III. 94. Find the point M which lies on the graph of the linear function:

a) $f : \mathbf{R} \to \mathbf{R}$, $f(x) = 8x - 7$, with equal coordinates;

b) $g : \mathbf{R} \to \mathbf{R}$, $g(x) = -3x + 6$, x intercept is 2;

c) $h : \mathbf{R} \to \mathbf{R}$, $h(x) = \dfrac{3}{7}x + 1$, y intercept is 1;

d) $i : \mathbf{R} \to \mathbf{R}$, $i(x) = -\dfrac{1}{3}x + 3$, y intercept is two thirds of x intercept.

III. 95. Find $m \in \mathbf{R}$ such that the point $A\left(\sqrt{2}, 6\right)$ lie on the graph of
$f : \mathbf{R} \to \mathbf{R}$, $f(x) = mx\sqrt{3} + 5\sqrt{3} - 3$.

III. 96. Find $m, n \in \mathbf{R}$ such that the points $A(2, -1)$ and $B(3, 1)$ lie on the
graph of the linear function $f : \mathbf{R} \to \mathbf{R}$, $f(x) = (m + 2n - 3)x - 3m - n$.

III. 97. Find $m, n \in \mathbf{R}$ such that the points $A(m, -10)$ and $B(0, -4)$ lie on the
graph of the linear function $f : \mathbf{R} \to \mathbf{R}$, $f(x) = (3n + 1)x - 2m + 2$.

III. 98. Find $a \in \mathbf{R}$ such that the points $A(1, -2)$, $B(2a, -2a)$,
and $C(3, 5)$ are collinear.

III. 99. Calculate the area and the perimeter for the triangle between the
graphs of the functions $f : \mathbf{R} \to \mathbf{R}$, $f(x) = 2x + 1$, $g : \mathbf{R} \to \mathbf{R}$, $g(x) = -x + 7$,
and $h : \mathbf{R} \to \mathbf{R}$, $h(x) = 3$.

III. 100. Write the equation of the line that has:

a) slope $-\dfrac{3}{4}$ and y-intercept 5;

b) slope $-\dfrac{3}{7}$ and passing through the point $A(1, -2)$;

c) y – intercept of 6 and passing through the point $B(1, -3)$;

d) same x – intercept as $x + y - 5 = 0$, parallel to equation $3x - y - 5 = 0$;

e) same y – intercept as $3x + y - 2 = 0$, perpendicular to the line of equation $x + 2y - 4 = 0$.

III. 101. Consider the function $f : \mathbf{R} \to \mathbf{R}$, $f(x) = 3x - 6$.
a) Draw the graph of f;

b) Find the area of the triangle formed by G_f, the x-axis and the y-axis;
c) Find $a \in \mathbf{R}$ knowing that the point $A(2a, a - 1)$ lies on the graph of f;
d) Calculate the sum $f(1) + f(2) + f(3) + \ldots + f(10)$.

III. 102. Consider the function $f : \mathbf{R} \to \mathbf{R}$, $f(x) = x - 3$.
a) Draw the graph of f;
b) Prove that the point $M(-2, -5)$ lie on the graph of f;
c) Find the area of the triangle formed by G_f and x-axis and y-axis;
d) Find the distance from the origin to the graph of f;
e) Calculate $E = f\left(\sqrt{3}\right) \cdot f\left(-\sqrt{3}\right)$.

III. 103. a) Find the linear function $f : \mathbf{R} \to \mathbf{R}$, $f(x) = mx + b$, such that:
a) $f(1) + f(2) + f(3) = 3m + b + 12$ and $f(2) = 7$;
b) $f(x - 3) = 2x + 1$.

III. 104. Consider the function
$f : \mathbf{R} \to \mathbf{R}$, $f(x) = (2a - 2)x + b - 3$.
a) Find a such that the graph of f be parallel to x-axis;
b) Calculate b such that the origin be on the graph of f;
c) Find $a \in \mathbf{N}$ such that the function be decreasing;

d) If $a = 2$ and $b = 1$, find $M \in G_f$ such that the coordinates are equal.

III. 105. Consider the points $A(a, 0)$ and $B(0, b)$. Find a point M lying on the line $y = 3x$, such that the angle AMB be a right angle.
If C is the fourth vertex of the rectangle $OACB$, show that $CM \perp OM$.

III. 106. Consider the function $f : \mathbf{R} \to \mathbf{R}$, $f(x) = 2mx + m + 3$.

a) Find $m \in \mathbf{R}$ such that the point $A(-1, 2)$ lie on the graph of f;
b) Draw the graph of f;
c) Find the area of the triangle formed by G_f and the x-axis and the y-axis.

III. 107. Consider the function $f : \mathbf{R} \to \mathbf{R}$, $f(x) = 2x - 4$.

a) Draw the graph of f;

b) Find x-axis and y-axis intercepts;

c) Find the area of the triangle AOB formed by G_f and the x-axis and the y-axis;

d) Prove that $Q(1, -2)$ is the circumcenter of $\triangle AOB$;

e) Find the length of the altitude of $\triangle AOB$.

III. 108. Consider the linear functions $f : \mathbf{R} \to \mathbf{R}$, $f(x) = x + 3$ and $g : \mathbf{R} \to \mathbf{R}$, $g(x) = 2x + 2$.

a) Draw the graphs of f and g;

b) Find the area of the triangle formed by G_f and G_g and the x-axis.

III. 109. Consider the function $g : \mathbf{R} \to \mathbf{R}$, $g(x) = 5 - 2x$. Find:

a) the value of a such that the point $R(2, a)$ lie on the graph of f;

b) the geometric mean of $g\left(\sqrt{6}\right)$ and $g\left(-\sqrt{6}\right)$;

c) x-axis and y-axis interception.

III. 110. Draw the graph of the function $f : \mathbf{R} \to \mathbf{R}$,

$$f(x) = \begin{cases} -x + 1 \ \text{ if } \ x < 2 \\ -2x + 3 \ \text{ if } \ x \geq 2 \end{cases}$$

III. 111. Draw the graph of the function $f : \mathbf{R} \to \mathbf{R}$,

$$f(x) = \begin{cases} -x + 1 \ \text{ if } \ x < 2 \\ 2x - 3 \ \text{ if } \ x \geq 2 \end{cases}$$ and solve the equation $f(x) = 4$.

III. 112. Draw the graph of the function $f : [-4, 1) \cup (3, \infty) \to \mathbf{R}$,

$$f(x) = \begin{cases} a - x \ \text{ if } \ -4 \leq x < -2 \\ 3 \ \text{ if } \ -2 \leq x \leq 1 \\ x + a \ \text{ if } \ x \geq 3 \end{cases}$$, where a is a solution of the equation

$$\frac{3a}{2} - \frac{a-2}{3} - 1 = a - \frac{2-a}{6}.$$

III. 113. Find the number $a \in \mathbf{R}$, if the graph of the function

$$f : \mathbf{R} \to \mathbf{R}, f(x) = \begin{cases} -2x + 1 \ \text{ if } \ x < -3 \\ ax - 3 \ \text{ if } \ x \geq -3 \end{cases}$$ passes through the point $A(-2, -5)$.

III. 114. Consider the function $f : \mathbf{R} \to \mathbf{R}$ such that

$f(x-1) + 3f(0) = 2x - 6$.

a) Show that $f(0) = -1$;

b) Show that $f(x) = 2x - 1$;

c) Calculate $\dfrac{1}{f(1)f(2)} + \dfrac{1}{f(2)f(3)} + ... + \dfrac{1}{f(2014)f(2015)}$.

III. 115. Consider the points $A(1,2)$, $B(5,4)$ and a mobile point L on x-axis. The lines LA, LB intersect the line $y=x$ at the points P, Q and the y-axis at the points M, N. Prove that:

a) The line NP passes through a fixed point situated on the line AB;

b) The line MQ is parallel to the line AB.

Chapter IV

POWERS AND RADICALS

$\mathbf{N} = \{0,1,2,...\}$ Natural numbers, $\mathbf{N}^* = \{1,2,3,...\}$

$\mathbf{Z} = \{-2,-1,0,1,2,...\}$ Integer numbers,

$\mathbf{Q} = \left\{ \dfrac{a}{b} \mid a \in \mathbf{Z}, b \in \mathbf{Z}, b \neq 0 \right\}$ Rational numbers,

$\mathbf{R} - \mathbf{Q}$ Irrational numbers, $\mathbf{R}$ The set of real numbers.

Properties of powers $\quad a^n = \underbrace{a \cdot a \cdot ... \cdot a}_{n \ times}, \quad n \in \mathbf{N}^*.$

a) $a^0 = 1$,

b) $(ab)^n = a^n b^n$

c) $\left(\dfrac{a}{b}\right)^n = \dfrac{a^n}{b^n}$

d) $a^n a^m = a^{n+m}$

e) $\dfrac{a^n}{a^m} = a^{n-m}$

f) $\left(a^m\right)^n = a^{mn}, \quad a,b \in \mathbf{R}, \ b \neq 0, \ m,n \in \mathbf{R}.$

Power function with integer
positive (natural) exponent

$f : \mathbf{R} \to [0, \infty), \ f(x) = x^2$

a similar behavior has the graph
of the function

$f : \mathbf{R} \to [0, \infty), \ f(x) = x^{2n}, n \in \mathbf{N}.$

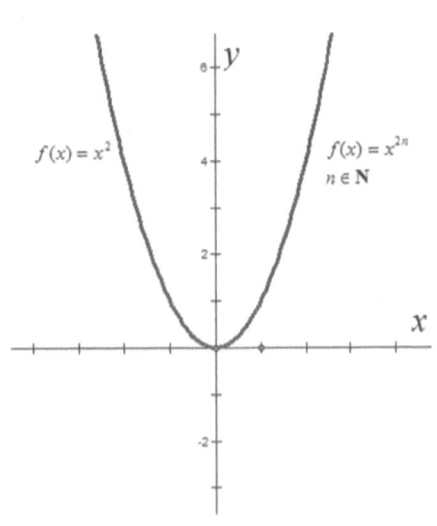

$f(x) = x^2$

$f(x) = x^{2n}$
$n \in \mathbf{N}$

Power function with integer positive
(natural) exponent

$$f : \mathbf{R} \to \mathbf{R}, \ f(x) = x^3$$

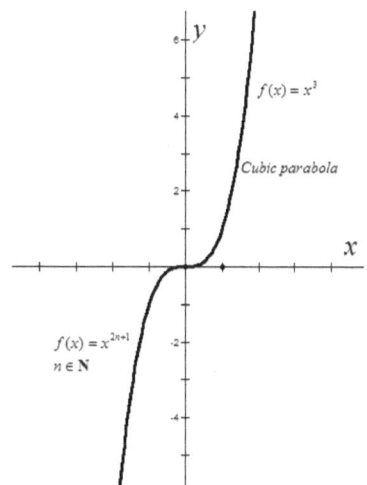

A similar behavior has the graph

$$f(x) = x^{2n+1}, \ n \in \mathbf{N}.$$

Power function with integer
negativ exponent

$$f : \mathbf{R} - \{0\} \to [0, \infty), \ f(x) = x^{-2}$$

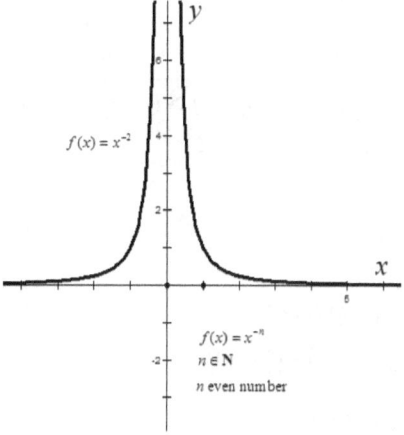

A similar behavior has
the graph of

$$f : \mathbf{R} - \{0\} \to [0, \infty), \ f(x) = x^{-n}, \ n \in \mathbf{N},$$
n even number.

$f : \mathbf{R} - \{0\} \rightarrow \mathbf{R} - \{0\},\ f(x) = x^{-3}$
and a similar behavior has
the graph of $f(x) = x^{-n}$, $n \in \mathbf{N}$,
n odd number.

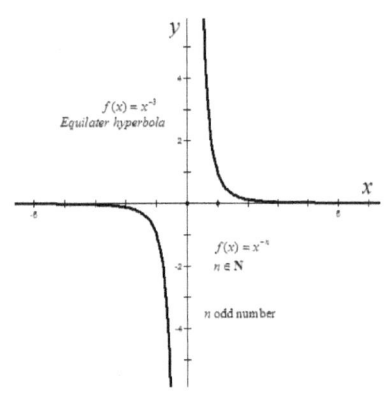

Properties of radicals $\sqrt[n]{(a)^n} = a$, $n \in \mathbf{N}^*$

a) $\sqrt[n]{ab} = \sqrt[n]{a}\sqrt[n]{b}$,

b) $\sqrt[n]{\dfrac{a}{b}} = \dfrac{\sqrt[n]{a}}{\sqrt[n]{b}}$,

c) $\sqrt[n]{a^{nm}} = a^m$,

d) $\left(\sqrt[n]{a}\right)^m = \sqrt[n]{a^m}$, $\sqrt[n]{a^m} = a^{\frac{m}{n}}$,

e) $\sqrt[n]{\sqrt[m]{a}} = \sqrt[m]{a}$, $a \geq 0$, $b > 0$, $m, n \in \mathbf{N}$.

Square root function
$f : [0, \infty) \rightarrow [0, \infty)$,

$f(x) = \sqrt{x}$

$f^{-1} : [0, \infty) \rightarrow [0, \infty)$

$f^{-1}(x) = x^2$

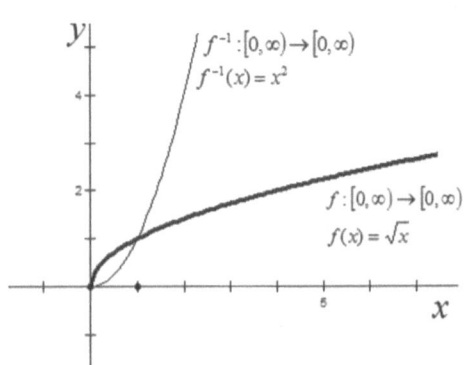

98

IV. 1. Write using powers:

a) $x \cdot x \cdot x \cdot x$; b) $y \cdot y \cdot y \cdot y$; c) $4 \cdot a \cdot a \cdot a$; d) $-3 \cdot b \cdot b \cdot b$; e) $-2 \cdot x \cdot x \cdot y$;
f) $3 \cdot a \cdot a \cdot b \cdot b \cdot b$; g) $(2a) \cdot (2a) \cdot (2a)$; h) $(a^2 b^3) \cdot (a^2 b) \cdot (a^2 b^2)$;
i) $a^4 \cdot a^5 \cdot a^6$; j) $(-3x^3 y^3)(-3x^2 y^3)(-3x^7 y^3)$; k) $a^{12} \cdot a^3 \cdot a^6$; l) $x^2 \cdot x^3 \cdot x^6$;
m) $x^n \cdot x$, $n \in \mathbf{N}^*$; n) $x^n \cdot x^n$, $n \in \mathbf{N}^*$; o) $a^m \cdot a^n \cdot a$, $m,n \in \mathbf{N}^*$.

IV. 2. Simplify the expressions bellow and eliminate any negative exponents.

a) $(3y^2)(2y^3)$; b) $\left(6x^2 y^4\right)\left(\frac{1}{2}x^5 y\right)$; c) $(2y)^3$; d) $\dfrac{x^9 (2x)^4}{x^4}$; e) $\dfrac{a^{-4} b^4}{a^{-5} b^5}$;

f) $b^{-4}\left(\frac{1}{3}b^7\right)(6b^{-8})$; g) $(rs)^3 (2s)^{-2}(2r)^4$; h) $(2u^2 t^4)^3 (3u^3 t)^{-2}$; i) $\dfrac{(2y^2)^4}{2y^5}$;

j) $\dfrac{(2x^4)^2 (3x^4)}{(x^3)^4}$; k) $\dfrac{(x^2 y^3)(xy^4)^{-3}}{x^2 y^3}$; l) $\left(\dfrac{c^4 d^3}{c^3 d^2}\right)\left(\dfrac{d^2}{c^3}\right)^3$; m) $\dfrac{(xy^2 z)^4}{(x^2 y^2 z)^3}$;

n) $\dfrac{(x^{-1})^3 (x^2)^3}{x^4 (x^{-2})^2}$; o) $\left(\dfrac{xy^{-2} z^{-3}}{x^2 y^2 z^4}\right)^{-3}$; p) $\left(\dfrac{q^{-1} rs^{-2}}{r^{-5} s^2 q^{-8}}\right)^{-1}$; q) $\left(3ab^2 c^2\right)\left(\dfrac{3a^2 b}{c^3}\right)^2$;

r) $\dfrac{x^2 \left(y^{-3}\right)^2}{\left(xy^{-1}\right)^2 \left(x^3\right)^2}$.

IV. 3. Simplify:

a) $2\sqrt{2} + 4\sqrt{2}$; b) $10\sqrt{3} - 8\sqrt{3}$; c) $6\sqrt{5} - 8\sqrt{5}$; d) $-3\sqrt{7} + 2\sqrt{7}$;
e) $3\sqrt{3} - 2\sqrt{3} - (1 - 5\sqrt{3})$; f) $2(\sqrt{5} + 1) + 4(1 - \sqrt{5})$;
g) $5(\sqrt{3} - \sqrt{2}) - 2(\sqrt{2} - \sqrt{3})$; h) $4(\sqrt{3} - 1) - 2(2 - 3\sqrt{3})$.

IV. 4. Simplify:

a) $(3\sqrt{2})^2$; b) $(2\sqrt{3})^2$; c) $(2\sqrt{6})^2$; d) $(-2\sqrt{5})^2$; e) $(2\sqrt{8})^2$; f) $(2\sqrt{7})^2$;
g) $(-4\sqrt{3})^2$; h) $(-3\sqrt{3})^2$; i) $\sqrt{2} \cdot (-5\sqrt{2})$; j) $(-2\sqrt{2})(-5\sqrt{3})$; k) $(-2\sqrt{6}) \cdot 3\sqrt{6}$;
l) $\sqrt{8} \cdot (-3\sqrt{8})$.

IV. 5. Simplify:

a) $\dfrac{\sqrt{75}}{\sqrt{5}}$; b) $\dfrac{\sqrt{3}}{\sqrt{48}}$; c) $\dfrac{\sqrt{27}}{\sqrt{3}}$; d) $\dfrac{\sqrt{3}}{\sqrt{60}}$; e) $\dfrac{3\sqrt{8}}{\sqrt{2}}$; f) $\dfrac{4\sqrt{15}}{\sqrt{5}}$; g) $\dfrac{4\sqrt{6}}{\sqrt{24}}$;

h) $\dfrac{3\sqrt{98}}{6\sqrt{7}}$; i) $\dfrac{2\sqrt{35}}{\sqrt{70}}$; j) $\dfrac{4\sqrt{34}}{\sqrt{68}}$.

IV. 6. Write in simplest radical form:

a) $\sqrt{250}$; b) $\sqrt{128}$; c) $\sqrt{40}$; d) $\sqrt{63}$; e) $\sqrt{72}$; f) $\sqrt{125}$;

g) $\sqrt{288}$; h) $\sqrt{175}$; i) $\sqrt{800}$; j) $\sqrt{450}$.

IV. 7. Write in simplest radical form $a + b\sqrt{n}$:

a) $\dfrac{3+\sqrt{8}}{2}$; b) $\dfrac{6-\sqrt{12}}{2}$; c) $\dfrac{3+\sqrt{18}}{3}$; d) $\dfrac{8-\sqrt{32}}{2}$; e) $\dfrac{6+\sqrt{72}}{3}$;

f) $\dfrac{9+\sqrt{27}}{6}$; g) $\dfrac{4-\sqrt{50}}{2}$; h) $\dfrac{5-\sqrt{200}}{5}$; i) $\dfrac{18-\sqrt{27}}{3}$.

IV. 8. Expand and simplify:

a) $\sqrt{3}\left(2+\sqrt{3}\right)$; b) $-\sqrt{3}\left(4-\sqrt{3}\right)$; c) $\sqrt{2}\left(2\sqrt{2}-\sqrt{5}\right)$; d) $-\sqrt{5}\left(2\sqrt{5}-\sqrt{3}\right)$;

e) $-2\sqrt{3}\left(1-2\sqrt{3}\right)$; f) $\left(\sqrt{5}+2\right)\left(\sqrt{5}-3\right)$; g) $\left(\sqrt{7}+\sqrt{2}\right)\left(\sqrt{7}-\sqrt{2}\right)$;

h) $\left(3\sqrt{3}+1\right)\left(3\sqrt{3}+3\right)$; i) $\left(6-2\sqrt{5}\right)\left(2-\sqrt{5}\right)$; j) $\left(2\sqrt{2}+1\right)\left(\sqrt{2}+3\right)$.

IV. 9. Expand and simplify:

a) $\left(2\sqrt{8}+1\right)\left(3\sqrt{8}+2\right)$; b) $\left(\sqrt{3}+\sqrt{6}\right)^2$; c) $\left(2\sqrt{2}+3\right)^2$; d) $\left(2\sqrt{1}+6\right)\left(2\sqrt{1}-6\right)$;

e) $\left(2\sqrt{2}-\sqrt{5}\right)\left(2\sqrt{2}+\sqrt{5}\right)$; f) $\left(2\sqrt{6}+3\right)\left(2\sqrt{6}-3\right)$;

g) $\left(\sqrt{3}-1\right)\left(1+\sqrt{3}-\sqrt{2}\right)\left(\sqrt{6}-\sqrt{2}\right)$.

IV. 10. Write each fraction with integer denominator:

a) $\dfrac{1}{1+\sqrt{2}}$; b) $\dfrac{2}{3-2\sqrt{2}}$; c) $\dfrac{1}{2+\sqrt{5}}$; d) $\dfrac{\sqrt{2}}{1-\sqrt{2}}$; e) $\dfrac{1+\sqrt{2}}{1-\sqrt{2}}$; f) $\dfrac{\sqrt{3}}{3-\sqrt{3}}$;

g) $\dfrac{-2\sqrt{3}}{1-\sqrt{3}}$; h) $\dfrac{1+\sqrt{3}}{2-\sqrt{3}}$; i) $\dfrac{4}{1+\sqrt{3}-\sqrt{2}}$; j) $\dfrac{3}{\sqrt[3]{2}-1}$; k) $\dfrac{1}{1+\sqrt[3]{2}+\sqrt[3]{4}}$;

l) $\dfrac{1}{2\sqrt[3]{4}+3\sqrt[3]{2}+1}$; m) $\dfrac{-23}{\sqrt[3]{2}-\sqrt{3}}$.

IV. 11. Write in a simple form the radicals:

a) $\sqrt{\left(3\sqrt{2}-5\right)^2}-\sqrt{\left(1-\sqrt{2}\right)^2}$; b) $\sqrt{\left(2\sqrt{3}-4\right)^2}-\sqrt{\left(1-\sqrt{3}\right)^2}$;

c) $4\sqrt{5}\cdot\left(\sqrt{5}-2\right)-20$; d) $4\sqrt{2}\cdot\left(\sqrt{2}-3\right)-8$; e) $2\sqrt{3}\cdot\left(\sqrt{3}-2\right)-6$;

f) $\dfrac{2-\sqrt{3}}{\sqrt{12}}-\dfrac{2-3\sqrt{3}}{\sqrt{27}}+\dfrac{3-\sqrt{3}}{\sqrt{3}}$; g) $\dfrac{1-\sqrt{2}}{\sqrt{2}}-\dfrac{3+2\sqrt{2}}{\sqrt{8}}-\dfrac{1-3\sqrt{2}}{\sqrt{18}}$;

h) $\dfrac{1-\sqrt{2}}{\sqrt{18}}+\dfrac{1+2\sqrt{2}}{\sqrt{8}}-\dfrac{1+\sqrt{2}}{\sqrt{2}}$; i) $\dfrac{2-\sqrt{3}}{\sqrt{12}}-\dfrac{1-3\sqrt{3}}{\sqrt{27}}+\dfrac{2-2\sqrt{3}}{2\sqrt{3}}$.

IV. 12. Calculate:

a) $-\sqrt{2}\cdot\left(-\sqrt{7}\right)$; b) $\sqrt{8}\div\sqrt{2}$; c) $\left(\sqrt{5}\right)^2$; d) $\sqrt{13}\cdot\left(\sqrt{3}-\sqrt{2}\right)$; e) $-2\sqrt{3}+4\sqrt{3}$;

f) $5\sqrt{2}+8-3\sqrt{2}-10$; g) $4\sqrt{2}-\sqrt{18}$; h) $3\sqrt{45}+4\sqrt{20}$; i) $3\sqrt{75}-5\sqrt{27}$;

j) $\left(2\sqrt{18}-3\sqrt{8}\right)+\left(3\sqrt{32}-\sqrt{50}\right)$; k) $\left(2\sqrt{20}-\sqrt{45}+2\sqrt{18}\right)-\left(2\sqrt{20}+3\sqrt{72}\right)$;

l) $\left(\sqrt{6}-3\sqrt{3}+5\sqrt{2}-\sqrt{8}\right)\cdot2\sqrt{6}$; m) $\left(5\sqrt{48}-6\sqrt{27}+4\sqrt{12}\right)\div\sqrt{3}$;

n) $\left(3\sqrt{50}+5\sqrt{200}-3\sqrt{450}\right)\div\sqrt{10}$; o) $\left(2\sqrt{8}+3\sqrt{5}-7\sqrt{2}\right)\left(\sqrt{72}-5\sqrt{20}-2\sqrt{2}\right)$.

IV. 13. Calculate:

a) $2\cdot\sqrt{2\sqrt{2\sqrt{2\sqrt{2\sqrt{2\sqrt{2\sqrt{4}}}}}}}$; b) $\sqrt{5\sqrt{5\sqrt{5\sqrt{5\sqrt{5\sqrt{5\sqrt{25}}}}}}}$;

c) $\sqrt{17+\sqrt{71-\sqrt{43+\sqrt{31+\sqrt{23+\sqrt{4}}}}}}$;

d) $\sqrt{33}\cdot\sqrt{6+\sqrt{3}}\cdot\sqrt{3+\sqrt{3+\sqrt{3}}}\cdot\sqrt{3-\sqrt{3+\sqrt{3}}}$;

e) $\sqrt{2+\sqrt{3}}\cdot\sqrt{2+\sqrt{2+\sqrt{3}}}\cdot\sqrt{2+\sqrt{2+\sqrt{2+\sqrt{3}}}}\cdot\sqrt{2-\sqrt{2+\sqrt{2+\sqrt{3}}}}$.

IV. 14. Calculate:

a) $\sqrt{3}+2\sqrt{3}-4\sqrt{3}$; b) $3\sqrt{5}-6\sqrt{5}+5\sqrt{5}$; c) $3\sqrt{6}+2\sqrt{6}-5\sqrt{6}+3\sqrt{6}$;

d) $3\sqrt{6}\cdot2\sqrt{3}\cdot(-2\sqrt{2})$; e) $2\sqrt{75}+\sqrt{50}+2\sqrt{32}-3\sqrt{12}$;

f) $\left(2\sqrt{24}-\sqrt{54}+3\sqrt{96}-\sqrt{150}\right)\cdot\sqrt{6}$; g) $-3\sqrt{25}\div\sqrt{5}$;

h) $\sqrt{2+\sqrt{4}} \div (-2)$; i) $\left(5\sqrt{32}+3\sqrt{8}-4\sqrt{18}\right) \div \sqrt{2}$;

j) $12+2\sqrt{5}\cdot\left[2\sqrt{500}+2(\sqrt{45}-\sqrt{125})\right]$;

k) $\left(1-\sqrt{2}\right)\left(1+\sqrt{2}\right)+\left(\sqrt{2}-\sqrt{3}\right)\left(\sqrt{2}+\sqrt{3}\right)$.

IV. 15. Calculate:

a) $\dfrac{2}{\sqrt{2}}+\dfrac{1}{\sqrt{3}}$; b) $\dfrac{3}{\sqrt{5}-\sqrt{2}}\cdot\dfrac{1}{\sqrt{2}-1}$; c) $\sqrt{2}+\left(\dfrac{1}{\sqrt{2}}\right)^{-2}$; d) $\dfrac{12}{\sqrt{13}-1}-\sqrt{13}$;

e) $\left(\dfrac{1}{2\sqrt{3}}-\dfrac{3}{\sqrt{75}}+\dfrac{7}{\sqrt{12}}\right)\cdot\sqrt{27}$; f) $\dfrac{2}{\sqrt{3}+\sqrt{5}}+\dfrac{2}{\sqrt{5}+\sqrt{7}}+\dfrac{2}{\sqrt{7}+\sqrt{9}}$;

g) $\left(\dfrac{\sqrt{3}-\sqrt{2}}{\sqrt{6}}+\dfrac{\sqrt{4}-\sqrt{3}}{\sqrt{12}}+\dfrac{\sqrt{5}-\sqrt{4}}{\sqrt{20}}\right)\div\dfrac{1}{\sqrt{10}}$;

h) $\left(\dfrac{\sqrt{4}-\sqrt{3}}{\sqrt{12}}+\dfrac{\sqrt{5}-\sqrt{4}}{\sqrt{20}}+\dfrac{\sqrt{6}-\sqrt{5}}{\sqrt{30}}\right)\div\dfrac{1}{\sqrt{6}}$;

i) $\dfrac{4-\sqrt{6}}{\sqrt{2}}+\dfrac{\sqrt{6}-1}{\sqrt{3}}-\dfrac{2-\sqrt{6}}{\sqrt{2}}-\dfrac{\sqrt{6}+5}{\sqrt{3}}$;

j) $\left(\dfrac{\sqrt{2}-\sqrt{1}}{\sqrt{2}}+\dfrac{\sqrt{3}-\sqrt{2}}{\sqrt{6}}+\dfrac{\sqrt{4}-\sqrt{3}}{\sqrt{12}}+\dfrac{\sqrt{5}-\sqrt{4}}{\sqrt{20}}\right)\div\dfrac{5-\sqrt{5}}{5}$;

k) $\dfrac{1}{1+\sqrt{2}}+\dfrac{1}{\sqrt{2}+\sqrt{3}}+....+\dfrac{1}{\sqrt{9}+\sqrt{100}}$.

IV. 16. Rationalize the denominators:

a) $\dfrac{2}{\sqrt{3}}$; b) $\dfrac{6}{\sqrt{12}}$; c) $\dfrac{\sqrt{27}}{3\sqrt{2}}$; d) $\dfrac{1-\sqrt{2}}{1+\sqrt{2}}$; e) $\dfrac{6-\sqrt{6}}{\sqrt{6}-1}$; f) $\dfrac{\sqrt{3}}{2+\sqrt{3}}$;

g) $\dfrac{37}{7-2\sqrt{3}}$; h) $\dfrac{4}{\sqrt{7}+\sqrt{3}}$; i) $\dfrac{6}{3+\sqrt{2}-\sqrt{5}}$; j) $\dfrac{1+3\sqrt{2}-2\sqrt{3}}{\sqrt{2}+\sqrt{3}+\sqrt{6}}$.

IV. 17. Rationalize the denominators:

a) $\dfrac{10}{\sqrt[3]{3}+\sqrt[3]{7}}$; b) $\dfrac{4}{\sqrt{2}\left(3+\sqrt{5}\right)}$; c) $\dfrac{2}{\sqrt{3+\sqrt{5}-\sqrt{13+4\sqrt{3}}}}$;

d) $\dfrac{\sqrt{2}}{\sqrt{3+2\sqrt{2}}}+\dfrac{\sqrt{2}}{\sqrt{3-2\sqrt{2}}}$; e) $\dfrac{2}{\sqrt{13+30\sqrt{2+\sqrt{9+4\sqrt{2}}}}}$.

IV. 18. Give a simple form for the expressions:

a) $\dfrac{\sqrt{7}-\sqrt{5}}{\sqrt{7}+\sqrt{5}}$; b) $\dfrac{\sqrt{7}+\sqrt{3}}{\sqrt{7}-\sqrt{3}}$; c) $\dfrac{5}{\sqrt{7}+\sqrt{2}}+\dfrac{3}{\sqrt{5}-\sqrt{2}}$;

d) $\dfrac{4}{\sqrt{7}+\sqrt{3}}+\dfrac{3}{\sqrt{6}-\sqrt{3}}$; e) $\left(\dfrac{\sqrt{2}}{2}-\dfrac{1}{2\sqrt{2}}\right)^{2}\left(\dfrac{\sqrt{2}-1}{\sqrt{2}+1}-\dfrac{\sqrt{2}+1}{\sqrt{2}-1}\right)$;

f) $\left(\dfrac{1+\sqrt{3}}{2}-\dfrac{2}{1+\sqrt{3}}\right)^{2}-\left(\dfrac{1-\sqrt{3}}{2}-\dfrac{2}{1-\sqrt{3}}\right)^{2}$; g) $\left(\sqrt{11+4\sqrt{7}}-\sqrt{11-4\sqrt{7}}\right)^{2}$.

IV. 19. Give a simple form for the expressions:

a) $\sqrt{\dfrac{2+\sqrt{3}}{2-\sqrt{3}}}+\sqrt{\dfrac{2-\sqrt{3}}{2+\sqrt{3}}}$; b) $\sqrt{2^{4}}+\sqrt{\dfrac{5}{0.02}}+\sqrt{\dfrac{5}{0.002}}+\sqrt{\dfrac{555}{0.0002}}$;

c) $\dfrac{1+\sqrt{2}+\sqrt{3}+\sqrt{6}}{1+\sqrt{2}+\sqrt{5}+\sqrt{10}}$; d) $\dfrac{\sqrt{10}+\sqrt{14}+\sqrt{15}+\sqrt{21}}{\sqrt{10}-\sqrt{14}+\sqrt{15}-\sqrt{21}}$.

IV. 20. Give a simple form:

a) $\sqrt{54}$; b) $\sqrt[3]{-108}$; c) $\sqrt[4]{4\cdot\sqrt[3]{4}}$; d) $\dfrac{a\sqrt{b\sqrt{c}}}{c\sqrt{b\sqrt{a}}}$, $a,b,c>0$;

e) $2\cdot\sqrt[3]{432}+3\cdot\sqrt[3]{-54}-6\cdot\sqrt[3]{-128}+5\cdot\sqrt[3]{16}+7\cdot\sqrt[3]{-250}$;

f) $4\cdot\sqrt[3]{-3}-\sqrt[3]{\dfrac{8}{9}}+\sqrt[3]{\dfrac{3}{8}}-\sqrt[3]{7\dfrac{1}{9}}-\sqrt[3]{-0,375}+\sqrt[3]{46\dfrac{7}{8}}$;

g) $\left(\sqrt[4]{81}-\sqrt[4]{2}\right)\left(\sqrt[4]{162}-\sqrt[4]{32}+3\right)\left(9+\sqrt{2}\right)$;

h) $\left(\sqrt[8]{5}-\sqrt[8]{4}\right)\left(\sqrt[8]{5}+\sqrt[8]{4}\right)\left(\sqrt[4]{5}+\sqrt[4]{4}\right)\left(\sqrt{5}+\sqrt{4}\right)$.

IV. 21. Simplify the expressions bellow and eliminate any negative exponents. Assume that all letters denote positive numbers.

a) $x^{\frac{2}{3}}\cdot\sqrt[5]{x}$; b) $(-2a^{\frac{3}{4}})(5\sqrt{a^{3}})$; c) $(4b)^{\frac{1}{2}}(8b^{\frac{2}{5}})$; d) $\left(8x^{6}\right)^{-\frac{2}{3}}$; e) $\left(c^{2}\cdot\sqrt[3]{d^{9}}\right)^{-\frac{1}{3}}$;

f) $\left(4\sqrt{x^{12}}\,y^{8}\right)^{\frac{3}{2}}$; g) $\left(\sqrt[4]{y^{3}}\right)^{\frac{2}{3}}$; h) $\left(\sqrt[3]{a^{8}}\right)^{-\frac{3}{4}}$; i) $\left(2x^{4}\cdot\sqrt[3]{y^{-4}}\right)^{3}(8^{-1}x^{-9}y^{3})^{\frac{2}{3}}$;

j) $(x^{-5}y^{15}\sqrt{z^{20}})^{-\frac{3}{5}}$; k) $\left(a^{\frac{1}{3}}\cdot\sqrt[5]{a^{4}}\cdot a^{-\frac{1}{10}}\right)^{30}$.

IV. 22. Simplify the expressions and eliminate any negative exponents. Assume that all letters denote positive numbers.

a) $\left(\dfrac{x^6 y^2}{y^4}\right)^{\frac{5}{2}}$; b) $\left(\dfrac{-2\sqrt[4]{x^3}}{y^{\frac{1}{4}}\sqrt{z}}\right)^4$; c) $\sqrt[3]{\left(-27a^{\frac{3}{4}}\right)^2\left(9a^{\frac{3}{2}}\right)^{\frac{1}{2}}}$;

d) $\left(8x^{\frac{3}{4}}\right)^{\frac{2}{3}}\left(9\cdot\sqrt[3]{x^2}\right)^{\frac{1}{2}}$; e) $\left(3\cdot\sqrt[4]{b^2}\right)^{-\frac{2}{3}}\sqrt{\left(9b^6\right)^{-1}}\left(3b^6\right)^{\frac{2}{3}}$; f) $\sqrt{\dfrac{\sqrt[5]{x^3}\cdot\sqrt[4]{x^5}}{\sqrt[3]{x^2}}}$;

g) $\sqrt{(x^{-2}y^{-6})^{-1}}\cdot(x^2y^3)^{-\frac{1}{6}}$; h) $\sqrt{\left(\dfrac{\sqrt[3]{x^{-1}}\cdot x^{-2}}{x^4}\right)^{-3}}$; i) $\left(4x^2y^4\right)^{\frac{3}{2}}\sqrt{\left(2^2x^{-2}y^{-4}\right)^3}$;

j) $\sqrt[6]{\left(y^{\frac{3}{4}}\right)^2}\cdot(y^{\frac{6}{4}})^{\frac{2}{3}}$; k) $\sqrt{\dfrac{(x^a)^4}{x^{2a+b}}\dfrac{(x^b)^4}{x^{a+2b}}}$.

IV. 23. Calculate:

a) $\sqrt{\dfrac{1-x}{1+x}}$ for $x=\dfrac{\sqrt{3}}{2}$; b) $\dfrac{\sqrt{1+x}}{1+\sqrt{1+x}}-\dfrac{\sqrt{1-x}}{1-\sqrt{1-x}}$ for $x=\dfrac{1}{2}$;

c) $\dfrac{\sqrt{2-x}}{8}\left(\sqrt{\dfrac{2-x}{2+x}}+\sqrt{\dfrac{2+x}{2-x}}\right)$ for $x=\dfrac{1}{4}$;

d) $\dfrac{(x+4)(x+2)(x-1)(x-3)}{46}$ for $x=\dfrac{\sqrt{3}-1}{2}$;

e) $\dfrac{x+\sqrt{3}}{\sqrt{x}+\sqrt{x+\sqrt{3}}}+\dfrac{x-\sqrt{3}}{\sqrt{x}-\sqrt{x-\sqrt{3}}}$ for $x=2$;

f) $\dfrac{\sqrt{x-2\sqrt{2}}}{\sqrt{x^2-4x\sqrt{2}+8}}-\dfrac{\sqrt{x+2\sqrt{2}}}{\sqrt{x^2+4x\sqrt{2}+8}}$ for $x=3$;

g) $\dfrac{\sqrt{1+x}-\sqrt{1-x}}{\sqrt{1+x}+\sqrt{1-x}}$ for $x=\dfrac{2a}{a^2+1}$, $a\in[-1,1)$;

h) $\dfrac{(a+x)(x+b)+(a-x)(x-b)}{(a+x)(x+b)-(a-x)(x-b)}$ for $x=\sqrt{ab}$.

IV. 24. Calculate:

a) $\dfrac{\sqrt[3]{1-x^2}-\sqrt[3]{1+x^2}}{\sqrt[3]{1-x^2}+\sqrt[3]{1+x^2}}$ for $x=\sqrt{\dfrac{b(3a^2+b^2)}{a(a^2+3b^2)}}$;

b $\dfrac{\sqrt[4]{x+1}+\sqrt[4]{x-1}}{\sqrt[4]{x+1}-\sqrt[4]{x-1}}$ for $x=\dfrac{a^2+b^2}{4ab}+\dfrac{ab}{a^2+b^2}$;

c) $\left[\dfrac{\left(x^2+a^2\right)^{\frac{1}{2}}+\left(x^2-a^2\right)^{\frac{1}{2}}}{\left(x^2+a^2\right)^{\frac{1}{2}}-\left(x^2-a^2\right)^{\frac{1}{2}}}\right]^{-2}$ for $x=a\left(\dfrac{m^2+n^2}{2mn}\right)^{\frac{1}{2}}$, $a>0$, $n>m>0$;

d) $\dfrac{\left(\sqrt{x+1}+\sqrt{x-1}\right)^n-\left(\sqrt{x+1}+\sqrt{x-1}\right)^n}{\left(\sqrt{x+1}+\sqrt{x-1}\right)^n+\left(\sqrt{x+1}+\sqrt{x-1}\right)^n}$ for $x=\dfrac{a^2+b^2}{2ab}$, where $a,b>0$.

IV. 25. Simplify:

a) $\left(x^{\frac{1}{3}}-1\right)\left(x^{\frac{2}{3}}+x^{\frac{1}{3}}+1\right)\left(x^{\frac{1}{3}}+1\right)\left(x^{\frac{2}{3}}-x^{\frac{1}{3}}+1\right)$;

b) $\left(x^2-x^{\frac{3}{2}}y^{\frac{1}{2}}-2x^{\frac{1}{2}}y^{\frac{1}{4}}+2y^{\frac{3}{4}}\right)\div\left(x^{\frac{1}{2}}-y^{\frac{1}{2}}\right)$;

c) $\dfrac{x-y}{x^{\frac{1}{3}}-y^{\frac{1}{3}}}-\dfrac{x+y}{x^{\frac{1}{3}}+y^{\frac{1}{3}}}$; **d)** $\dfrac{\left(9^n-9^{n-1}\right)^{\frac{1}{2}}}{\left(27^{n-1}-19\cdot27^{n-2}\right)^{\frac{1}{3}}}$.

IV. 26. Consider $a,b,c,d\in\mathbf{R}$, prove that :

a) $(a+b)^5-a^5-b^5=5ab(a+b)\left(a^2+ab+b^2\right)$;

b) $a^3-b^3=\left(a\dfrac{a^3-2b^3}{a^3+b^3}\right)^3+\left(b\dfrac{2a^3-b^3}{a^3+b^3}\right)^3$, $a+b\neq0$;

c) $\left(a^3+b^3+c^3+d^3\right)^2=9(bc-ad)(ca-bd)(ab-cd)$,
where $a+b+c+d=0$.

IV. 27. Draw the graph of each function and its base function:

a) $f_1:\mathbf{R}\to[0,\infty)$, $f_1(x)=4x^2$;

b) $f_2:\mathbf{R}\to\mathbf{R}$, $f_2(x)=x^3-8$;

c) $f_3:\mathbf{R}\to[0,\infty)$, $f_3(x)=(x-2)^2$;

d) $f_4 : \mathbf{R} \to [0, \infty)$, $f_4(x) = (x+1)^4$;

e) $f_5 : \mathbf{R} \to [0, \infty)$, $f_5(x) = |x-1|^3$;

f) $f_6 : \mathbf{R} \to [0, \infty)$, $f_6(x) = |x+2|^3$;

g) $f_7 : \mathbf{R} \setminus \{0\} \to \mathbf{R} \setminus \{-1\}$, $f_7(x) = x^{-2} - 1$;

h) $f_8 : \mathbf{R} \setminus \{0\} \to \mathbf{R} \setminus \{-1\}$, $f_8(x) = x^{-3} - 1$;

i) $f_9 : \mathbf{R} \setminus \{1\} \to \mathbf{R} \setminus \{0\}$, $f_9(x) = \dfrac{1}{x+1}$;

j) $f_{10} : \mathbf{R} \setminus \{-1\} \to (0, \infty)$, $f_{10}(x) = \dfrac{1}{(x+1)^2}$.

IV. **28.** For each function find its inverse:

a) $f(x) = \sqrt{x+1}$;   b) $f(x) = \sqrt{2-x}$;

c) $f(x) = \begin{cases} \dfrac{x-6}{2} & \text{if } x < 12 \\ \sqrt{x-3} & \text{if } x \ge 12 \end{cases}$;   d) $f(x) = \begin{cases} -\sqrt{-x} & \text{if } x < 0 \\ \sqrt{x} & \text{if } x \ge 0 \end{cases}$;

e) $f(x) = \sqrt{x + 2\sqrt{x-1}} - \sqrt{x - 2\sqrt{x-1}}$, $x \ge 2$;

f) $f(x) = \sqrt{1 + 2x\sqrt{1-x^2}} + \sqrt{1 - 2x\sqrt{1-x^2}}$, $x \ge 0$.

IV. **29.** Find the range of the functions:

a) $f(x) = \sqrt{x^2 + x + 1} - \sqrt{x^2 - x + 1}$, $x \in \mathbf{R}$;

b) $f(x) = \sqrt[3]{x^3 - 3x^2 + 3x - 1} + \sqrt{x^2 - 2x + 1}$, $x \in \mathbf{R}$.

IV. **30.** Calculate:

a) $2x - 1 + \sqrt{(x-4)^2}$ b) $x - 3 + \sqrt{(x-3)^2}$;

c) $(x^2 - 4) \dfrac{\sqrt{x^2 - 4x + 4} + 4}{\sqrt{x^2 - 4x + 4}}$;   d) $\dfrac{\sqrt{x^4 - 2x^3 + x^2}}{x^2 - \sqrt{x^2 - x + 1}}$;

e) $\dfrac{\sqrt{4x^2 - 12x + 9} + 6}{2x - 3} \sqrt{4 - 12x^{-1} + 9x^{-2}}$;

f) $2x + \sqrt{x^2} + \sqrt{x^2 + 2x + 1} + 1$;   g) $\dfrac{(x+1)\sqrt{x^2 - 6x + 9}}{(x-1)^2 - 4\sqrt{x^2 - 4x + 4}}$;

h) $4+\left|\sqrt{x^2-4x+4}-2\right|+\sqrt{x^2-8x+16}$;

i) $\dfrac{\sqrt{x^2-6x+9}+3x-3}{2\sqrt{x^2-8x+16}-\sqrt{x^2-10x+25}}$.

IV. 31. Find the value of x such that the radicals are defined:

a) $\sqrt{2x-1}$; **b)** $\sqrt[3]{x^2-3x+1}$; **c)** $\sqrt{-2-6x}$; **d)** $\sqrt[3]{x-2}$; **e)** $\sqrt[6]{1-3x}$;

f) $\sqrt{\left(x^2-4x\right)^2}$; **g)** $\sqrt[4]{x^2-7x+12}$; **h)** $\sqrt[6]{\left(4-x^2\right)x^2}$;

i) $\sqrt{(x+1)(x-2)(x+4)}$; **j)** $\sqrt{3-x}-\sqrt{x-1}$; **k)** $\sqrt{x^2-1}+\sqrt[3]{x^2-2}$;

l) $\sqrt{2x-4}+3\cdot\sqrt{5-x}+\sqrt[3]{5-4x}$; **m)** $\sqrt[3]{1+2x}+\sqrt{6-3x}$; **n)** $\sqrt{\dfrac{2-|x-2|}{2+|x-2|}}$.

IV. 32. Without using a calculator compare the radicals:

a) $4\sqrt{2}$ and $\sqrt{27}$; **b)** $3\sqrt[4]{5}$ and $\sqrt[4]{216}$; **c)** $21\sqrt{5}$ and $12\sqrt{14}$;

d) $2\cdot\sqrt{\dfrac{6}{7}}$ and $\sqrt{\dfrac{47}{14}}$; **e)** $-2\cdot\sqrt{\dfrac{1}{75}}$ and $-\sqrt{\dfrac{24}{450}}$;

f) $-3\sqrt[3]{14}$ and $-2\sqrt[3]{18}$; **g)** $2\cdot\sqrt[3]{\dfrac{1}{44}}$ and $\sqrt[3]{\dfrac{7}{325}}$.

IV. 33. Write in an increasing order the numbers (without calculators):

a) $\sqrt[3]{5}$, $\sqrt[4]{6}$, $\sqrt[12]{628}$; **b)** $\sqrt{4}$, $\sqrt[4]{15}$, $\sqrt[3]{11}$; **c)** $\sqrt{5}$, $\sqrt[3]{6}$, $\sqrt[12]{777}$;

d) $2\cdot\sqrt[3]{3}$, $\sqrt[6]{16}$, $\sqrt[12]{250}$; **e)** $\sqrt[5]{50}$, $\sqrt[10]{20}$, $\sqrt[15]{448}$;

f) $\sqrt[3]{7}$, $\sqrt[4]{17}$, $\sqrt[12]{4000}$; **g)** $\sqrt{2}$, $\sqrt[3]{3}$, $\sqrt[5]{5}$.

IV. 34. Prove that:

a) $\left(4+\sqrt{15}\right)+\sqrt{4-\sqrt{15}}=\sqrt{11}$; **b)** $\left(4+\sqrt{15}\right)-\sqrt{4-\sqrt{15}}=\sqrt{6}$;

c) $\sqrt{5+\sqrt{24}}+\sqrt{5-\sqrt{24}}=\sqrt{12}$;

d) $\sqrt{3+\sqrt{5}}+\sqrt{3-\sqrt{5}}+\sqrt{6-\sqrt{35}}=\sqrt{6+\sqrt{35}}$;

e) $\left(4+\sqrt{15}\right)\left(\sqrt{10}-\sqrt{6}\right)\cdot\sqrt{4-\sqrt{15}}=2$;

f) $\dfrac{1-2^{-2}}{2^{\frac{1}{2}}-2^{-\frac{1}{2}}} - \dfrac{2}{2^{\frac{3}{2}}} + \dfrac{2^{-2}-2}{2^{\frac{1}{2}}-2^{-\frac{1}{2}}} = -\dfrac{3\sqrt{2}}{2}$; g) $\dfrac{\sqrt{7+4\sqrt{3}}\cdot\sqrt{19-8\sqrt{3}}}{4-\sqrt{3}} = 2+\sqrt{3}$;

h) $\sqrt{10+\sqrt{24}+\sqrt{40}+\sqrt{60}} = \sqrt{2}+\sqrt{3}+\sqrt{5}$.

IV. 35. Prove that:

i) $\dfrac{2+\sqrt{3}}{\sqrt{2}+\sqrt{2+\sqrt{3}}} + \dfrac{2-\sqrt{3}}{\sqrt{2}-\sqrt{2-\sqrt{3}}} = \sqrt{2}$;

j) $\dfrac{3+\sqrt{3}+\sqrt{5}+\sqrt{15}}{3+\sqrt{2}+\sqrt{3}+\sqrt{6}} = 3-\sqrt{6}-\sqrt{10}+\sqrt{15}$;

k) $\left(\dfrac{\sqrt{3}+1}{1+\sqrt{3}+\sqrt{5}} + \dfrac{\sqrt{3}-1}{1-\sqrt{3}+\sqrt{5}}\right)\cdot\left(\sqrt{5}-\dfrac{2}{\sqrt{5}}+2\right) = 2\sqrt{3}$;

l) $\dfrac{\sqrt{5-2\sqrt{6}}\left(5+2\sqrt{6}\right)\left(49-20\sqrt{6}\right)}{3\sqrt{3}-3\sqrt{18}+3\sqrt{12}-\sqrt{8}} = 1$; m) $\dfrac{\sqrt{21+8\sqrt{5}}}{4+\sqrt{5}}\cdot\sqrt{9-4\sqrt{5}}+2 = \sqrt{5}$;

n) $\sqrt{8+2\sqrt{10+2\sqrt{5}}} - \sqrt{8-2\sqrt{10+2\sqrt{5}}} = \sqrt{20-4\sqrt{5}}$.

IV. 36. Show that:

a) $\sqrt[3]{7+5\sqrt{2}} - \sqrt[3]{7-5\sqrt{2}} = 2$;   b) $\sqrt[3]{45+29\sqrt{2}} - \sqrt[3]{45-29\sqrt{2}} = 2\sqrt{2}$;

c) $\sqrt[3]{20+14\sqrt{2}} + \sqrt[3]{20-14\sqrt{2}} = 4$;   d) $\sqrt[3]{44+18\sqrt{6}} + \sqrt[3]{44-18\sqrt{6}} = 4$;

e) $\sqrt[3]{5+2\sqrt{13}} + \sqrt[3]{5-2\sqrt{13}} = 1$;   f) $\sqrt[3]{77+20\sqrt{13}} + \sqrt[3]{7-20\sqrt{13}} = 7$;

g) $\sqrt[3]{9+\sqrt{80}} + \sqrt[3]{9-\sqrt{80}} = 3$;   h) $\sqrt[3]{6+\sqrt{\dfrac{847}{27}}} + \sqrt[3]{6-\sqrt{\dfrac{847}{27}}} = 3$.

IV. 37. Show that:

a) $\dfrac{7-4\sqrt{3}}{2-\sqrt{3}} = \sqrt[3]{26-15\sqrt{3}}$; b) $\dfrac{11-6\sqrt{2}}{3-\sqrt{2}} = \sqrt[3]{45-29\sqrt{2}}$;

c) $\dfrac{2\cdot\sqrt[3]{2}}{1+\sqrt{3}} = \dfrac{\sqrt[3]{20+12\sqrt{3}}}{2+\sqrt{3}}$; d) $\dfrac{\sqrt[3]{7+2\sqrt{6}}}{\left(\sqrt{6}+1\right)^{-1/3}} + \dfrac{\sqrt[3]{7-2\sqrt{6}}}{\left(\sqrt{6}-1\right)^{-1/3}} = 2\sqrt{6}$;

e) $5\cdot\sqrt[3]{3\sqrt{128}} - 3\cdot\sqrt[3]{9\sqrt{162}} - 1\cdot\sqrt[6]{18} + 2\cdot\sqrt[3]{75\sqrt{50}} = 0$;

f) $\sqrt{\dfrac{\sqrt{2}}{3} + \dfrac{3}{\sqrt{2}} + 2} - \dfrac{1}{3}\cdot\dfrac{9\cdot\sqrt[4]{2}-2\sqrt{3}}{\sqrt{6}-\sqrt[4]{8}} = -1$;

108

g) $\left(\dfrac{\sqrt[4]{8}-1}{\sqrt[4]{2}-1}+\sqrt[4]{2}\right)^{\frac{1}{2}}\left(\dfrac{\sqrt[4]{8}+1}{\sqrt[4]{2}+1}-\sqrt[4]{4}\right)\left(1-\sqrt{2}\right)^{-1}=1.$

IV. 38. Show that:

h) $\dfrac{\sqrt{2}\cdot\sqrt{\sqrt[4]{8}-\sqrt{2}+1}}{\sqrt{\sqrt[4]{8}+\sqrt{\sqrt{2}-1}}-\sqrt{\sqrt[4]{8}-\sqrt{\sqrt{2}-1}}}=1;$

i) $\dfrac{\sqrt[3]{\sqrt{3}+\sqrt{6}}\cdot\sqrt[6]{9-6\sqrt{2}}-\sqrt[6]{18}}{\sqrt[6]{2}-1}=-\sqrt[3]{3};$

j) $\dfrac{25\cdot\sqrt[4]{2}+2\sqrt{5}}{\sqrt{250}+5\cdot\sqrt[4]{8}}-\sqrt{\dfrac{\sqrt{2}}{5}+\dfrac{5}{\sqrt{2}}+2}=-1;$

k) $\sqrt[3]{\sqrt[5]{\dfrac{32}{5}}-\sqrt[5]{\dfrac{27}{5}}}=\sqrt[5]{\dfrac{1}{25}}+\sqrt[5]{\dfrac{3}{25}}-\sqrt[5]{\dfrac{9}{25}};$

n) $\left[\dfrac{\left(1+3^{-1/2}\right)^{1/6}}{\left(3^{1/2}+1\right)^{-1/3}}-\dfrac{\left(3^{1/2}-1\right)^{1/3}}{\left(1-3^{-1/2}\right)^{-1/6}}\right]^{-2}\cdot\dfrac{3^{5/6}\div2^{-1}}{\sqrt{3}+\sqrt{2}}=3.$

IV. 39. Calculate the sums

$S_1=\dfrac{1}{1+\sqrt{2}}+\dfrac{1}{\sqrt{2}+\sqrt{3}}+\dfrac{1}{\sqrt{3}+\sqrt{4}}+...+\dfrac{1}{\sqrt{n-1}+\sqrt{n}}\,;$

$S_2=\dfrac{1}{2\sqrt{1}+1\sqrt{2}}+\dfrac{1}{3\sqrt{2}+2\sqrt{3}}+\dfrac{1}{4\sqrt{3}+3\sqrt{4}}+...+\dfrac{1}{(n+1)\sqrt{n}+n\sqrt{n+1}};$

$S_3=\dfrac{\sqrt{2}-1}{\sqrt{2}}+\dfrac{\sqrt{3}-\sqrt{2}}{\sqrt{6}}+\dfrac{\sqrt{4}-\sqrt{3}}{\sqrt{12}}+...+\dfrac{\sqrt{n+1}-\sqrt{n}}{\sqrt{(n+1)n}};$

$S_4=\sqrt{2-\sqrt{3}}+\sqrt{4-\sqrt{15}}+\sqrt{6-\sqrt{35}}+...+\sqrt{2n-\sqrt{4n^2-1}};$

$S_5=\dfrac{1}{\sqrt{1!}\left(2+\sqrt{2}\right)}+\dfrac{2}{\sqrt{2!}\left(3+\sqrt{3}\right)}+...+\dfrac{n}{\sqrt{n!}\left(n+1+\sqrt{n+1}\right)}.$

IV. 40. Simplify the irrational expressions:

a) $\sqrt{x\cdot\sqrt[3]{x\cdot\sqrt{x}}}\,$; b) $\sqrt{\dfrac{\sqrt[3]{a^3}}{a^2}}\cdot\sqrt[3]{\dfrac{a^2\cdot\sqrt{a}}{\sqrt{a^3}}}\cdot\sqrt[4]{\dfrac{\sqrt{a}}{a\cdot\sqrt[3]{a^2}}}\cdot\sqrt[8]{\dfrac{\sqrt{a^1}}{\sqrt{a}}}\,$;

c) $\sqrt[4]{\dfrac{x}{2}}\cdot\dfrac{\left(\sqrt[8]{x}-\sqrt[8]{2}\right)^2+\left(\sqrt[8]{x}+\sqrt[8]{2}\right)^2}{\sqrt{x}-\sqrt[4]{2x}}\div\dfrac{\left(\sqrt[4]{x}+\sqrt[4]{2}-\sqrt[8]{2x}\right)\left(\sqrt[4]{x}+\sqrt[4]{2}+\sqrt[8]{2x}\right)}{2-\sqrt[4]{2x^3}};$

109

d) $\dfrac{a(a-2)-b(b+2)+\sqrt{ab}(b-a+2)}{a+b-\sqrt{ab}} \div \left(1+2\dfrac{a^2+b^2+ab}{b^3-a^3}\right)$, $a\neq b$;

e) $\dfrac{\left(\sqrt{a}+\sqrt{ax}+x+x\sqrt{x}\right)^2\left(1-\sqrt{x}\right)^2}{\left(x+x^{-1}-2\right)a^{-\frac{1}{4}}} - \dfrac{\left(x\sqrt{a}\right)^{\frac{3}{2}}}{\left(ax^{-1}+4\sqrt{a}+4x\right)^{\frac{1}{2}}}$, $x\in\mathbf{R}_+$.

IV. 41. Simplify the irrational expressions:

a) $\dfrac{\sqrt{a^2-b+\sqrt{c}}\cdot\sqrt{a-\sqrt{b+\sqrt{c}}}\cdot\sqrt{a+\sqrt{b+\sqrt{c}}}}{\sqrt{\dfrac{a^3}{b}-2a+\dfrac{b}{a}-\dfrac{c}{ab}}}$;

b) $\dfrac{\left(\sqrt[4]{a}+\sqrt[4]{b}-\sqrt[8]{ab}\right)\left(\sqrt[4]{b}+\sqrt[4]{a}+\sqrt[8]{ab}\right)}{\sqrt[4]{a^3b}-b}\div\dfrac{\left(\sqrt[8]{a}+\sqrt[8]{b}\right)^2+\left(\sqrt[8]{a}-\sqrt[8]{b}\right)^2}{b^{-\frac{1}{4}}\left(\sqrt{a}-\sqrt{b}\right)}$;

c) $\left(\dfrac{2(a+1)+2\sqrt{a^2+2a}}{3a+1-2\sqrt{2a^2+a}}\right)^{\frac{1}{2}}-\left(\sqrt{2a+1}-\sqrt{a}\right)^{-1}\sqrt{a+2}$, $a\in\mathbf{R}_+$;

d) $\dfrac{\sqrt{\sqrt{3}+2}\cdot\sqrt[4]{7-4\sqrt{3}}+\sqrt[3]{\sqrt{x}(x+27)}-9x-27}{\sqrt{x}-2-\sqrt{2-\sqrt{3}}\cdot\sqrt[4]{7+4\sqrt{3}}}$;

e) $\dfrac{\left(\sqrt[8]{x}+\sqrt[8]{y}\right)^2+\left(\sqrt[8]{x}-\sqrt[8]{y}\right)^2}{x-\sqrt{xy}}\div\dfrac{\left(\sqrt[4]{x}+\sqrt[8]{xy}+\sqrt[4]{y}\right)\left(\sqrt[4]{x}-\sqrt[8]{xy}+\sqrt[4]{y}\right)}{\sqrt[4]{x^3y}-y}$.

IV. 42. Simplify the expressions:

a) $\dfrac{\left[a^{\frac{2}{3}}\cdot(2a)^{\frac{1}{3}}\cdot2^{\frac{2}{3}}\right]^{\frac{1}{2}}}{\left[a^{-\frac{3}{2}}\cdot(2a)^{-\frac{1}{2}}\cdot2^{-\frac{3}{2}}\right]^{-\frac{1}{4}}}$; **b)** $\dfrac{\left[\sqrt{x^{-1}}\cdot\sqrt{2^3}+x\right]^{\frac{1}{4}}\cdot\sqrt[8]{(x-2)^{-3}}}{\left(\dfrac{\sqrt{2}}{\sqrt{x}-\sqrt{2}}-\sqrt{\dfrac{2}{x}+1}\right)^{\frac{1}{4}}}$;

c) $\dfrac{\left[\dfrac{\left(1+x^{-\frac{1}{2}}\right)^{\frac{1}{6}}}{\left(x^{\frac{1}{2}}+1\right)^{\frac{1}{3}}}-\dfrac{\left(x^{\frac{1}{2}}-1\right)^{\frac{1}{3}}}{\left(1-x^{-\frac{1}{2}}\right)^{\frac{1}{6}}}\right]^{-2}\cdot x^{\frac{1}{12}}}{x^{\frac{1}{2}}+(x-1)^{\frac{1}{2}}}$, $x>1$;

110

d)
$$\dfrac{\dfrac{1}{\sqrt{x-1}}-\sqrt{x+1}}{\dfrac{1}{\sqrt{x+1}}-\dfrac{1}{\sqrt{x-1}}}\div\dfrac{\sqrt{x+1}\cdot\sqrt{x^2-1}}{(x-1)\sqrt{x+1}-(x+1)\sqrt{x-1}}\ ;$$

e)
$$\dfrac{\sqrt{\left(\dfrac{a^4+16}{a^4-4a^2}+\dfrac{8}{a^2-4}\right)\left(a^3-4a\right)-4\sqrt{a}}}{(a-2)\sqrt{\dfrac{a}{a-2}-\dfrac{2}{a+2}-\dfrac{4a}{a^2-4}}}.$$

IV. 43. Simplify the expressions:

a) $\dfrac{\sqrt{x^2-4xy+4y^2}}{\sqrt{x^2+4xy+4y^2}}-\dfrac{8xy}{x^2-4y^2}+\dfrac{2y}{x-2y},\ 0<x<2y;$

b) $\left(\dfrac{1+\sqrt{x}}{\sqrt{1+x}}-\dfrac{\sqrt{1+x}}{1+\sqrt{x}}\right)^2-\left(\dfrac{1-\sqrt{x}}{\sqrt{1+x}}-\dfrac{\sqrt{1+x}}{1-\sqrt{x}}\right)^2,\ x\in[0,\infty);$

c) $\left(\sqrt{\dfrac{1}{x^2}-1}+\dfrac{1}{x}\right)\left(\dfrac{1-x}{\sqrt{1-x^2}+x-1}+\dfrac{\sqrt{1+x}}{\sqrt{1+x}-\sqrt{1-x}}\right),$

$x\in(-1,0)$;

d) $\dfrac{\sqrt{1+\sqrt{1-x^2}}\left((1+x)\sqrt{1+x}-(1-x)\sqrt{1-x}\right)}{2+\sqrt{1-x^2}},\ x\in(-1,1)$;

e) $\dfrac{\left(\sqrt{\sqrt{\dfrac{x-1}{x+1}}+\sqrt{\dfrac{x+1}{x-1}}}-2\right)\left(2x+\sqrt{x^2-1}\right)}{(x+1)\sqrt{x+1}-(x-1)\sqrt{x-1}},\ x>1.$

IV. 44. Simplify the expressions:

a) $\left[\left(\sqrt{xy}-\dfrac{y^2\sqrt{x}}{\sqrt{xy^2}+\sqrt{y^3}}\right)\div\dfrac{\sqrt[4]{xy}-\sqrt{y}}{x-y}-x\sqrt{y}\right]^2\div\sqrt[3]{xy\sqrt{xy}}-\left(\dfrac{x}{\sqrt{x^4-1}}\right)^{-2};$

b) $\left(\sqrt[3]{x^2}+\dfrac{2\sqrt[3]{x^2}}{\sqrt[3]{x}-2}\right)\dfrac{\sqrt[3]{x^2}-4}{\sqrt[3]{x^3}+2\sqrt[3]{x^2}}-\left(2+\dfrac{\sqrt[3]{x^2}}{\sqrt[3]{x}+2}\right)^{-1}\dfrac{x-8}{\sqrt[3]{x}+2},\ x\in\mathbf{R}\setminus\left\{-\sqrt[3]{2},0\right\};$

c) $\left(\dfrac{x^2y^3}{(x+y)^{\frac{5}{2}}}-\dfrac{2x^2y^2}{(x+y)^{\frac{3}{2}}}+\dfrac{x^2y}{(x+y)^{\frac{1}{2}}}\right)\div\left(\dfrac{x^3}{(x+y)^{\frac{5}{2}}}-\dfrac{x^3y}{(x+y)^{\frac{7}{2}}}\right);$

d) $\left(\sqrt[3]{\dfrac{8x^3+24x^2+18x}{2x-3}}-\sqrt[3]{\dfrac{8x^3-24x^2+18x}{2x+3}}\right)-\left(\dfrac{1}{2}\sqrt[3]{\dfrac{2x}{1}-\dfrac{1}{6x}}\right)^{-1}$,

$x\in\left(\dfrac{3}{2},\infty\right)$;

e) $\dfrac{\left(\sqrt[8]{x}+\sqrt[8]{y}\right)^2+\left(\sqrt[8]{x}-\sqrt[8]{y}\right)^2}{x-\sqrt{xy}}\div\dfrac{\left(\sqrt[4]{x}+\sqrt[8]{xy}+\sqrt[4]{y}\right)\left(\sqrt[4]{x}-\sqrt[8]{xy}+\sqrt[4]{y}\right)}{\sqrt[4]{x^3y}-y}$.

IV. 45. Find the value of the expressions for indicated value:

a) $\left[\dfrac{(x+2y)\div 8y^3}{x^2+2xy+2y^2}-\dfrac{(x-2y)\div 8y^3}{x^2-2xy+2y^2}\right]+\left(\dfrac{y^{-2}}{4x^2-8y^2}-\dfrac{1}{4x^2y^2+8y^4}\right)$

$x=\sqrt[4]{6},\ y=\sqrt[8]{2}$;

b) $\dfrac{a^{-1}-b^{-1}}{a^{-3}+b^{-3}}\div\dfrac{a^2b^2}{(a+b)^2-3ab}\left(\dfrac{a^2-b^2}{ab}\right)^{-1}$, $a=1-\sqrt{2},\ b=1+\sqrt{2}$;

c) $\left(\dfrac{a}{b}+\dfrac{b}{a}+2\right)\left(\dfrac{a+b}{2a}-\dfrac{b}{a+b}\right)\div\left[\left(a+2b+\dfrac{b^2}{a}\right)\cdot\left(\dfrac{a}{a+b}+\dfrac{b}{a-b}\right)\right]$,

$a=0.75,\ b=\dfrac{4}{3}$;

d) $\sqrt{a^3-b^3+\sqrt{a}}\cdot\dfrac{\sqrt{a^{3/2}+\sqrt{b^3}+\sqrt{a}}\cdot\sqrt{a^{3/2}-\sqrt{b^3}+\sqrt{a}}}{\sqrt{\left(a^3+b^3\right)^2-a\left(4a^2b^3+1\right)}}$, $a=4,\ b=\sqrt[3]{62}$;

e) $\dfrac{\sqrt{2}(x-a)}{2x-a}-\left[\left(\dfrac{\sqrt{x^2}}{\sqrt{2x^2}+\sqrt{ax}}\right)^2+\left(\dfrac{\sqrt{2x}+\sqrt{a}}{2\sqrt{a}}\right)^{-1}\right]^{1/2}$, $a=0.32,\ x=0.08$;

f) $\dfrac{\sqrt{x-2\sqrt{x+3}}+4}{x^{1/2}-(x-3)^{1/2}-\sqrt{3x+x^2}+\sqrt{x^2-9}}-\dfrac{1}{\sqrt{x}+\sqrt{x-3}}$, $x=9$;

g) $\dfrac{\sqrt{x+4\sqrt{x-4}}+\sqrt{x-4\sqrt{x-4}}}{\sqrt{1-\dfrac{8}{x}+\dfrac{16}{x^2}}}$, $x=5$;

h) $\dfrac{\sqrt{3}x^{3/2}-5x^{1/3}+5x^{4/3}-\sqrt{3x}}{\sqrt{3x+10\sqrt{3}x^{5/6}+25x^{2/3}}\sqrt{1-2x^{-1}+x^{-2}}}$, $x=-1$.

IV. 46. Simplify the expressions:

a) $\dfrac{\left[(3x+2)^{-1/2}+(3x-2)^{-1/2}\right]^{-1}+\left[(3x+2)^{-1/2}-(3x-2)^{-1/2}\right]^{-1}}{\left[(3x+2)^{-1/2}+(3x-2)^{-1/2}\right]^{-1}-\left[(3x+2)^{-1/2}-(3x-2)^{-1/2}\right]^{-1}}$;

b) $\dfrac{\left(\sqrt{a}+\sqrt{ax}+x+x\sqrt{x}\right)^2\left(1-\sqrt{x}\right)^2}{\left(x+x^{-1}-2\right)a^{-1/4}}-\dfrac{\left(x\sqrt{a}\right)^{3/2}}{\left(ax^{-1}+4\sqrt{a}+4x\right)^{-1/2}}$;

c) $\left(\sqrt{\dfrac{(1-n)\cdot\sqrt[3]{1+n}}{n}}\cdot\sqrt[3]{\dfrac{3n^2}{4-8n+4n^2}}\right)^{-1}\div\sqrt[3]{\left(\dfrac{3n\sqrt{n}}{2\sqrt{1-n^2}}\right)^{-1}}$;

d) $\dfrac{2a\left(a+2b+\sqrt{a^2+4ab}\right)}{\left(a+\sqrt{a^2+4ab}\right)\left(a+4b+\sqrt{a^2+4ab}\right)}$.

IV. 47. Simplify the expressions:

a) $\dfrac{\left(xy^{-1}+1\right)^2}{xy^{-1}-x^{-1}y}\cdot\dfrac{x^3y^{-3}-1}{x^2y^{-2}+xy^{-1}+1}\div\dfrac{x^3y^{-3}+1}{xy^{-1}+x^{-1}y-1}$;

b) $\left(\dfrac{\sqrt[4]{x^3}-y}{\sqrt[4]{x}-\sqrt[3]{y}}-3\sqrt[12]{x^3y^4}\right)^{-1/2}\left(\dfrac{\sqrt[4]{x^3}+y}{\sqrt[4]{x}+\sqrt[3]{y}}-\sqrt[3]{y^2}\right)$;

c) $\left(4x-1\right)\left\{\dfrac{1}{8x}\left[\left(\sqrt{8x-1}+4x\right)^{-1}-\left(\sqrt{8x-1}-4x\right)^{-1}\right]\right\}^{1/2}$;

d) $\dfrac{\sqrt{4\left(x-\sqrt{y}\right)+yx^{-1}}\cdot\sqrt{9x^2+6\cdot\sqrt[3]{2yx^3}+\sqrt[3]{4y^2}}}{6x^2+2\cdot\sqrt[3]{2x^3y}-3\cdot\sqrt{yx^2}-\sqrt[6]{4y^5}}$;

e) $\dfrac{1}{\sqrt{x}+\sqrt{x-4}}-\dfrac{\sqrt{x-2\sqrt{x+4}}+5}{x^{1/2}-(x-4)^{1/2}-\sqrt{4x+x^2}+\sqrt{x^2-16}}$.

IV. 48. Write in a simple form the radicals:

a) $\sqrt{5+2\sqrt{6}}$; b) $\sqrt{6-\sqrt{20}}$; c) $\sqrt{10-2\sqrt{21}}$; d) $\sqrt{4\sqrt{2}+2\sqrt{6}}$;
e) $\sqrt{17+\sqrt{288}}$; f) $\sqrt{7-4\sqrt{3}}$; g) $\sqrt{28-5\sqrt{12}}$;
h) $\sqrt{17-4\sqrt{9+\sqrt{90}}}$; i) $\sqrt{2-2a\sqrt{2-a^2}}$, $1\le a\le\sqrt{2}$.

IV. 49. Solve the irrational equations:

a) $\sqrt{2x-2}=4$; b) $\sqrt{3-4x}=4$; c) $\sqrt{x\sqrt{x\sqrt{x\sqrt{x\sqrt{x\sqrt{x\sqrt{x^2}}}}}}}=5$;

d) $\sqrt{24-10x} = 3-4x$; e) $\sqrt{2x^2-7} = 3+x$; f) $\sqrt{4x+1} = 3-3x$;

g) $\sqrt[3]{x^2+2x}-2 = 0$; h) $x = 6\left(\sqrt{x-2}-1\right)$; i) $\sqrt[3]{2x-1}-\sqrt[6]{x+1} = 0$;

j) $\dfrac{4-x}{\sqrt{x^2-8x+32}} = \dfrac{3}{5}$; k) $\sqrt[3]{1+7x} = 1+x$;

l) $\sqrt{(3-a)^2}-\sqrt{4-a+2\sqrt{3-a}} = 3-2a$, $a \in \mathbf{R}$.

IV. 50. Solve the irrational equations:

a) $\sqrt{x-2}+\sqrt{2x-3} = 5$;   b) $\sqrt{2x-1}+\sqrt{2-x} = 2$;

c) $\sqrt{3x+1}+\sqrt{4-3x} = 3$;   d) $\sqrt{x-1}+\sqrt{6-x} = 3$;

e) $\sqrt{x+8}-2 = \sqrt{x}$;   f) $\sqrt{2x+1}+\sqrt{1-x} = 2$;

g) $\sqrt{x+7}+\sqrt{x+2} = \sqrt{3x+19}$; h) $\sqrt{5x}-\sqrt{2x-1} = \sqrt{3x+1}$;

i) $\sqrt{3x+7}+\sqrt{x-3} = 2\sqrt{x+1}$.

IV. 51. Solve the irrational equations:

a) $\sqrt{x^2-\sqrt{2x+1}} = 2-x$;

b) $\sqrt{\dfrac{x+4}{x-4}}+2\sqrt{\dfrac{x-4}{x+4}} = \dfrac{1}{3}$;   c) $\sqrt{\sqrt{x+16}-\sqrt{x}} = 2$;

d) $\sqrt{2x+3}+\sqrt{10x-12} = \sqrt{5x-1}+\sqrt{7x-8}$;

e) $\dfrac{\sqrt{x^2-16}}{\sqrt{x-3}}+\sqrt{x+3} = \dfrac{7}{\sqrt{x-3}}$.

IV. 52. Solve the equations:

a) $\sqrt{x+2}+\sqrt{x+7} = 5$;   b) $\sqrt{15-x}+\sqrt{3-x} = 6$;

c) $\sqrt{x+1}-\sqrt{9-x} = \sqrt{2x-12}$; d) $\sqrt{x+9}+\sqrt{x+1} = \sqrt{4x+16}$;

e) $\sqrt{3x+4}+\sqrt{x-4} = 2\sqrt{x}$.

IV. 53. Solve the equations:

a) $\sqrt{x+3}+\sqrt{x-1} = \sqrt{4x-2+2\sqrt{5}}$;

b) $\sqrt{x+4}+\sqrt{x-1} = \sqrt{4x-5+4\sqrt{6}}$;

c) $\sqrt{x-5}+\sqrt{x-12} = \sqrt{1+\sqrt{2}}$.

IV. 54. Solve the equations:

a) $\left(\sqrt{x+2}+\sqrt{x+1}\right)^3 + \left(\sqrt{x+2}+\sqrt{x+1}\right)^2 = 2$;

b) $\left(\sqrt{x}-5\right)\left(\sqrt{x}-4\right)\left(\sqrt{x}-3\right)\left(\sqrt{x}-2\right) = 24$;

c) $x^2 - 4x - 6 = \sqrt{2x^2 - 8x + 12}$.

IV. 55. Solve the equations:

a) $\sqrt{x-1+2\sqrt{x-2}} - \sqrt{x-1-2\sqrt{x-2}} = 1$;

b) $\sqrt{x+7+4\sqrt{x+3}} + \sqrt{x+7-4\sqrt{x+3}} = 6$;

c) $\sqrt{x+3+2\sqrt{x+2}} + \sqrt{x+3-2\sqrt{x+2}} = 2$;

d) $\sqrt{x+1-4\sqrt{x-3}} + \sqrt{x+1+4\sqrt{x-3}} = 4$;

e) $\sqrt{x+3-4\sqrt{x-1}} + \sqrt{x+8-6\sqrt{x-1}} = 1$;

f) $\sqrt{x+27-10\sqrt{x+2}} + \sqrt{x+1\ -6\sqrt{x+2}} = 4$;

g) $\sqrt{x+2\sqrt{x-1}} + \sqrt{x-2\sqrt{x-1}} = 2$;

h) $\sqrt{x-a+1-2\sqrt{x-a}} + \sqrt{x-a+4+4\sqrt{x-1}} = 3$, $x \geq a, a \in \mathbf{R}$.

IV. 56. Solve the equations:

a) $2 \cdot \sqrt{5 \cdot \sqrt[8]{x+1}+4} - \sqrt{2 \cdot \sqrt[8]{x+1}-1} = \sqrt{20 \cdot \sqrt[8]{x+1}+5}$;

b) $\sqrt{10x+2\sqrt{25x^2-1}} - \sqrt{10x-2\sqrt{25x^2-1}} = 2\sqrt{5x-1}$;

c) $\sqrt{6x+2\sqrt{9x^2-4}} - \sqrt{6x-2\sqrt{9x^2-4}} = 2\sqrt{3x-2}$.

IV. 57. Solve the equations:

a) $\dfrac{\sqrt{4+x}+\sqrt{4-x}}{\sqrt{4+x}-\sqrt{4-x}} = \dfrac{4}{x}$;

b) $\dfrac{x+\sqrt{2a-x}}{x-\sqrt{2a+x}} = \dfrac{\sqrt{a}+1}{\sqrt{a}-1}$, $a>0$;

c) $\dfrac{\sqrt{a+x}+\sqrt{a-x}}{\sqrt{a+x}-\sqrt{a-x}} = \dfrac{\sqrt{a}+1}{\sqrt{a}-1}$, $a>0$;

d) $\dfrac{a+x+\sqrt{2ax+x^2}}{a+x-\sqrt{2ax+x^2}} = \dfrac{\sqrt{a}+1}{\sqrt{a}-1}$, $a>1$;

e) $\dfrac{1}{x-5\sqrt{x}+6}+\dfrac{1}{x-3\sqrt{x}+2}+\dfrac{1}{x-\sqrt{x}}=\dfrac{1}{x-3\sqrt{x}}$,

$x \in \mathbf{R}\setminus\{0,1,4,9\}$.

IV. 58. Solve the equations:

a) $\sqrt[3]{3x-1}+\sqrt{3x+7}=6$;　　b) $\sqrt[3]{1+x}-\sqrt{2+x}=-1$;

c) $\sqrt[3]{8x+1}-\sqrt{3x-2}=1$;　　d) $\sqrt[3]{12+x^2}+\sqrt[3]{12-x^2}=2\sqrt[3]{3}$;

e) $\sqrt[3]{\sqrt{x}+3}+\sqrt[3]{13-\sqrt{x}}=4$;　f) $\sqrt[3]{9-\sqrt{x+2}}+\sqrt[3]{7+\sqrt{x+2}}=4$.

IV. 59. Solve the equations:

a) $\sqrt[3]{x+1}+\sqrt[3]{x+2}+\sqrt[3]{x+3}=0$;

b) $\sqrt[3]{x+1}+\sqrt[3]{x+2}=\sqrt[3]{2x+3}$;　c) $\sqrt[3]{x+2}+\sqrt[3]{x-14}=\sqrt[3]{x-6}$;

d) $2\sqrt[3]{x+4-4\sqrt{x}}-\sqrt[3]{x+4+4\sqrt{x}}=3\cdot\sqrt[3]{x-4}$.

IV. 60. Solve the equations:

a) $\sqrt[3]{(1-x)^2}+\sqrt[3]{(8+x)^2}=\sqrt[3]{(8+x)(1-x)}+3$;

b) $\sqrt[3]{(x-1)^2}+\sqrt[3]{(3-x)^2}=\sqrt[3]{(x-1)(3-x)}+1$.

IV. 61. Solve the equations:

a) $\sqrt{x-b^2}-\sqrt{x-a^2}=a-b$;

b) $\sqrt{4x^2-5x+5}-\sqrt{4x^2-5x+2}=1$;

c) $\sqrt[3]{x^2+2}+\sqrt[3]{4x^2+3x-2}=\sqrt[3]{3x^2+x+5}+\sqrt[3]{2x^2+2x-5}$;

d) $\sqrt[3]{x^2+x}+\sqrt[3]{4x^2+7x+3}=\sqrt[3]{x^2+3x+2}+\sqrt[3]{4x^2+5x+1}$.

IV. 62. Solve the equations:

a) $\sqrt[4]{28+5x}+\sqrt[4]{54-5x}=4$;　　b) $\sqrt[4]{x-7}-\sqrt[3]{x+4}=-1$;

c) $\sqrt[4]{x+8}+\sqrt[4]{x-8}=2$;　　d) $\sqrt[4]{x-1}+\sqrt[4]{2-x}=1$;

e) $\sqrt[4]{x-5}-\sqrt[4]{5x-9}=-2$.

IV. 63. Solve the equations:

a) $\sqrt{x\sqrt[5]{x}}-26\cdot\sqrt[5]{x\sqrt{x}}=27$; b) $\dfrac{\sqrt{x+4}+\sqrt{x-4}}{2}+6=x+\sqrt{x^2-16}$;

c) $\dfrac{4}{\sqrt[5]{x}+2}+\dfrac{\sqrt[5]{x}+3}{5}=2$;   d) $\dfrac{1}{\sqrt{x}+\sqrt[3]{x}}+\dfrac{1}{\sqrt{x}-\sqrt[3]{x}}=\sqrt{2}$;

e) $\dfrac{\sqrt[3]{x^4}-1}{\sqrt[3]{x^2}-1}-\dfrac{\sqrt[3]{x^2}-1}{\sqrt[3]{x}+1}=4$;   f) $\dfrac{x\cdot\sqrt[5]{x}-1}{\sqrt[5]{x^3}-1}-\dfrac{\sqrt[5]{x^3}-1}{\sqrt[5]{x}-1}=2$;

g) $\dfrac{\sqrt[7]{x-\sqrt{2}}}{2}-\dfrac{\sqrt[7]{x-\sqrt{2}}}{x^2}=\dfrac{x}{2}\cdot\sqrt[7]{\dfrac{x^3}{x+\sqrt{2}}}$;

h) $\dfrac{\sqrt{1+x}+\sqrt{1-x}}{\sqrt{1+x}-\sqrt{1-x}}=\dfrac{8}{\sqrt[3]{225x}}$;   i) $\dfrac{4}{\sqrt[5]{(3x-4)^2}}=\dfrac{\sqrt[5]{(3x-4)^2}}{4}$;

j) $42\cdot\sqrt[12]{x^{-7}}-\sqrt[4]{x^{-1}}\cdot\sqrt[3]{x^2}=5\cdot\sqrt[12]{x^1}$;

k) $5\cdot\sqrt[3]{x\cdot\sqrt[5]{x}}+3\cdot\sqrt[5]{x\cdot\sqrt[3]{x}}=8$.

IV. 64. Solve the equations:

a) $\sqrt[n]{(1+x)^2}+\sqrt[n]{(1-x)^2}=4\cdot\sqrt[n]{1-x^2}$, $n\in\mathbf{N}^*$;

b) $\sqrt{1+\sqrt{x}}-2\sqrt{1-\sqrt{x}}=\sqrt[4]{1-x}$;

c) $2\sqrt{2+\sqrt[3]{x}}-2\sqrt{2-\sqrt[3]{x}}=3\sqrt[4]{4-\sqrt[3]{x^2}}$;

d) $6\cdot\sqrt[3]{2x+1}-\sqrt[3]{3x-2}=5\cdot\sqrt[6]{(2x+1)(3x-2)}$;

e) $5\cdot\sqrt[n]{1+\sqrt[5]{x}}-2\cdot\sqrt[n]{1-\sqrt[5]{x}}=3\cdot\sqrt[2n]{1-\sqrt[5]{x^2}}$;

f) $\sqrt[7]{2x+1}+\sqrt[7]{x+2}=2\cdot\sqrt[14]{(2x+1)(x+2)}$;

g) $3\cdot\sqrt[3]{x-6}+2\cdot\sqrt[3]{x-5}=7\cdot\sqrt[6]{(x-6)(x-5)}$.

IV. 65. Solve the equations:

a) $\dfrac{\sqrt{x^2+6x}}{\sqrt{x+1}}+\sqrt{x+5}=\dfrac{4+\sqrt{21}}{\sqrt{x+1}}$;

b) $\sqrt{\dfrac{1}{x}+\sqrt{1-\dfrac{x^2}{9}}}-\sqrt{\dfrac{1}{x}-\sqrt{1-\dfrac{x^2}{9}}}=\sqrt{x}$;

c) $\dfrac{\sqrt[3]{(10-x)^2}+\sqrt[3]{(10-x)(x-1)}+\sqrt[3]{(x-1)^2}}{\sqrt{\sqrt[3]{10-x}+\sqrt[3]{x-1}}}=\dfrac{7}{\sqrt{3}}$;

d) $\dfrac{(32-x)\sqrt[3]{x+3}-(x+3)\sqrt[3]{32-x}}{\sqrt[3]{x+3}-\sqrt[3]{32-x}}=-30$;

e) $x\cdot\sqrt[3]{\dfrac{x}{2-x}}+(2-x)\cdot\sqrt[3]{\dfrac{2-x}{x}}=2$.

IV. 66. Solve the equations:

a) $\sqrt{6+3\sqrt{2+\sqrt{2+x}}} + \sqrt{2-\sqrt{2+\sqrt{2+x}}} = 2x$;

b) $\sqrt{1+\sqrt{x(2-x)}} + \sqrt{1-\sqrt{x(2-x)}} = \sqrt{2(2-x)}$.

IV. 67. Consider $m, n \in \mathbf{N}^{\bullet}$. Express m in terms of n from the equalities:

a) $\left(\sqrt{5}-2\right)^{n} = \sqrt{m+1} - \sqrt{m}$; b) $\left(\sqrt{17}-4\right)^{n} = \sqrt{m} - \sqrt{m-1}$;

c) $\left(5\sqrt{2}+7\right)^{n} = \sqrt{m+1} - \sqrt{m}$; d) $\left(\sqrt{2}-1\right)^{n} = \sqrt{m} - \sqrt{m-1}$;

e) $\left(\sqrt{m+1}+\sqrt{m}\right)^{2} = \sqrt{n+1} - \sqrt{n}$.

IV. 68. Solve the equations:

a) $\sqrt[2x+2]{x^{3}\cdot\sqrt[x]{x^{3}}} = \sqrt[2x^{2}]{x^{9}}$; b) $\sqrt[4x-3]{10-4\sqrt[x]{(-x)^{x}}} = \sqrt[5x]{2x}$;

c) $\sqrt[2x-1]{15-4\sqrt[x]{(2-2x)^{4x-6}}} = \sqrt[3x-4]{2x^{2}-2}$; d) $\sqrt[6-x]{3x-3} = \sqrt[2x-5]{2x^{2}-5}$;

e) $\sqrt[11-x]{x-3} + \sqrt[16x-8]{10x-4} = \sqrt[4]{x-1}^{\,x+6}$;

f) $\sqrt[8x-4]{x-1} + \sqrt[5x-7]{1-x} = \sqrt[12-2x]{x^{2}-2x}$;

g) $3 + \sqrt[x]{7x-3} = \sqrt{x^{2}-7x+14}\sqrt[14]{6x-5}$.

IV. 69. Solve the inequations:

a) $\sqrt{x-1} < \sqrt{3-x}$; b) $2\sqrt{x^{2}-3x-10} < 39-2x$;

c) $\sqrt{x+2} < \sqrt{x-2} + \sqrt{x-3}$; d) $\sqrt{x^{2}-3x+2} > 2-x$;

e) $\sqrt{x^{2}-4x} > x-3$; f) $\sqrt{2-\sqrt{x+2}} < \sqrt{3+x}$;

g) $\dfrac{\sqrt{x}}{3-x} < 1$; h) $\dfrac{2\sqrt{x+2}}{1-2\sqrt{2-x}} < 1$; i) $\sqrt{x^{2}-53x+196} < x-13$;

j) $\sqrt[3]{x^{3}-3x^{2}+5x-6} > x-2$; k) $\sqrt{x-2} - \sqrt[3]{x-3} > 1$;

l) $\dfrac{3x}{x-2} - \sqrt{\dfrac{3x}{x-2}} > \sqrt[4]{\dfrac{12x}{x-2}}$; m) $\dfrac{\sqrt{2-x}+4x-3}{x} > 2$.

IV. 70. Solve the systems of equations:

a) $\begin{cases} x\sqrt{y} + y\sqrt{x} = 30 \\ x\sqrt{x} + y\sqrt{y} = 35 \end{cases}$; b) $\begin{cases} \dfrac{1}{\sqrt{x}} + \dfrac{1}{\sqrt{y}} = \dfrac{4}{3} \\ xy = 9; \end{cases}$; c) $\begin{cases} x\sqrt{y} + y\sqrt{x} = 6 \\ x^{2}y + y^{2}x = 20 \end{cases}$;

d) $\begin{cases} \sqrt{3x-y+12} + \sqrt{3x+y+5} = 4 \\ \sqrt[4]{3x-y+12} + \sqrt[4]{3x+y+5} = 2 \end{cases}$; e) $\begin{cases} \sqrt{\dfrac{3x-2y}{2x}} + \sqrt{\dfrac{2x}{3x-2y}} = 2 \\ x^2 - 8y^2 = 18(1-y) \end{cases}$;

f) $\begin{cases} \sqrt[3]{x} + \sqrt[3]{y} = 5 \\ \sqrt[3]{x^2} - \sqrt[3]{xy} + \sqrt[3]{y^2} = 7 \end{cases}$; g) $\begin{cases} \sqrt[4]{x} + \sqrt[4]{y} = 3 \\ x + y = 17 \end{cases}$;

h) $\begin{cases} \sqrt{x+\sqrt{y}} + \sqrt{x-\sqrt{y}} = 2 \\ \sqrt{y+\sqrt{x}} + \sqrt{y-\sqrt{x}} = 1 \end{cases}$; i) $\begin{cases} \sqrt{x+y} + \sqrt[4]{x-y} = 8 \\ \sqrt[4]{x^3 + x^2 y - xy^2 - y^3} = 15 \end{cases}$.

IV. 71. Find the real numbers $x_1, x_2,...,x_n$ such that

a) $\sqrt{x_1 - 1} + \sqrt{2(x_2 - 2)} + ... + \sqrt{n(x_n - n)} = \dfrac{1}{2}(x_1 + x_2 + ... + x_n)$;

b) $\sqrt{x_1 - 1} + 2\sqrt{x_2 - 2^2} + ... + n\sqrt{x_n - n^2} = \dfrac{1}{2}(x_1 + x_2 + ... + x_n)$.

IV. 72. Calculate the sum $S = \left(1 - \dfrac{1}{4}\right) + \left(2 - \dfrac{1}{4}\right) + \left(3 - \dfrac{1}{4}\right) + ... + \left(n - \dfrac{1}{4}\right)$

, were n is a positive integer and solve the equation

$\sqrt{x_1 - 1} + \sqrt{x_2 - 2} + \sqrt{x_3 - 3} + ... + \sqrt{x_n - n} + \dfrac{n(2n+1)}{4} = x_1 + x_2 + ... + x_n$.

IV. 73. In a triangle with the sides a, b, c,

$$\dfrac{\sqrt{a}}{\sqrt{b} + \sqrt{c} - \sqrt{a}} = \dfrac{\sqrt{b}}{\sqrt{c} + \sqrt{a} - \sqrt{b}} = \dfrac{\sqrt{c}}{\sqrt{a} + \sqrt{b} - \sqrt{c}}.$$

Show that the triangle is equilateral.

IV. 74. If a, b, c are the sides of $\triangle ABC$ such that

$\sqrt{2(a+b)} = \sqrt{a+c} + \sqrt{a-c}$, then show that the triangle is right.

IV. 75. Prove that the numbers

$A = \dfrac{\sqrt{56 + \sqrt{56 + \sqrt{56 + ...}}}}{\sqrt{12 + \sqrt{12 + \sqrt{12 + ...}}}} - \sqrt{42 + \sqrt{42 + \sqrt{42 + ...}}}$ and

$B = \dfrac{\sqrt{30 + \sqrt{30 + \sqrt{30 + ...}}}}{\sqrt{6 + \sqrt{6 + \sqrt{6 + ...}}}} - \sqrt{72 + \sqrt{72 + \sqrt{72 + ...}}}$ are integers.

Chapter V

QUADRATIC FUNCTIONS
AND APPLICATIONS

General form: $f : \mathbf{R} \to \mathbf{R}$, $f(x) = ax^2 + bx + c$, $a, b, c \in \mathbf{R}$, $a \neq 0$.

Factored form: $f(x) = a(x - x_1)(x - x_2)$, where x_1 and x_2 are the roots of the equation $ax^2 + bx + c = 0$.

Vertex form (standard form): $f(x) = a(x - h)^2 + k$, $V(h, k)$ or

$$f(x) = a\left(x + \frac{b}{2a}\right)^2 - \frac{\Delta}{4a}, \quad \Delta = b^2 - 4ac,$$

$$V\left(-\frac{b}{2a}, -\frac{\Delta}{4a}\right) \quad \text{or} \quad V\left(\frac{x_1 + x_2}{2}, f\left(\frac{x_1 + x_2}{2}\right)\right).$$

The roots of the equation $ax^2 + bx + c = 0$, $a \neq 0$ are

$x_{1,2} = \dfrac{-b \pm \sqrt{b^2 - 4ac}}{2a}$. The nature of the roots is given by the value of Δ.

a) If $\Delta = b^2 - 4ac > 0$, there are two different real roots;

b) If $\Delta = b^2 - 4ac = 0$, there are two equal real roots;

c) If $\Delta = b^2 - 4ac < 0$, there are two conjugate complex roots.

Vieta's formulas (formulas between coefficients and roots) $x_1 + x_2 = -\dfrac{b}{a}$ and $x_1 x_2 = \dfrac{c}{a}$.

V. 1. Factor:

a) $x^2 - 49$; b) $25x^2 - 1$; c) $81x^2 - 4$; d) $(x + 8)^2 - (x + 3)^2$;

e) $49(x - 2)^2 - 81(x - 3)^2$; f) $x^2 + 7x + 12$; g) $x^2 - 9x - 36$;

h) $x^2 + 6x - 16$; i) $6x^2 - 11x + 4$; j) $7x^2 + 13x - 2$.

V. 2. For the following functions, find y-intercept:

a) $y = (x + 3)(x - 2)$;

b) $y = (x - 5)\left(x + \sqrt{3}\right)$;

c) $y = -(x + 4)(x - 1)$;

d) $y = -x(x - 4)$;

e) $y = (x-4)(x+2)$; f) $y = -(x-1)(x-3)$.

V. 3. For the following quadratics, find the x-intercept and the axis of symmetry:

a) $y = (x-1)(x+3)$; b) $y = -(x+2)(x-3)$; c) $y = (x-3)(x+7)$;

d) $y = -(x-2)(x-5)$; e) $y = (x+5)^2$; f) $y = x^2 - 4x - 7$;

g) $y = (ab)x^2 + (b^2 - ac)x - b$, $a,b,c \in \mathbf{R}$, $ab \neq 0$.

V. 4. Write the quadratic functions in the vertex form:

a) $f(x) = x^2 - 2x + 8$; b) $f(x) = -x^2 + 5x + 2$;

c) $f(x) = \frac{2}{3}x^2 - 4x + 1$; d) $f(x) = 0.5x^2 - 0.2x + 0.3$;

e) $f(x) = -3x^2 + x + 0.1$; f) $f(x) = 5x^2 - \frac{2}{3}x + 1$.

V. 5. Determine the axis of symmetry and the vertex of each parabola associated to the functions:

a) $f(x) = -2x^2 + 3x + 1$; b) $f(x) = 3x^2 - x + 1$;

c) $f(x) = -4x^2 + 3x - 2$; d) $f(x) = 5x^2 - 2x - 1$;

e) $f(x) = 0.4x^2 - 0.1x + 0.3$; f) $f(x) = \frac{1}{3}x^2 - \frac{1}{2}x + \frac{1}{4}$.

V. 6. Write the vertex form of the quadratic equations:

a) with vertex $(-2, 7)$ and y-intercept of 3 units;

b) that is shifted up 3 units, compressed vertically by a factor of $\frac{1}{2}$, flipped upside down and shifted right by 5 units;

c) that is exactly the same shape as the basic parabola, but has vertex $(-1, 4)$ and opens down.

V. 7. Consider the equation $y = x^2$. Write the equation of the transformed function which is:

a) compressed vertically by a factor of $\frac{1}{4}$, reflected in the x-axis, and translated 4 units to the right, and 3 units upward;

b) stretched vertically by a factor of 5, reflected in the y-axis, and translated 4 units to the left, and 3 units downward.

121

V. 8. Find the roots of the equation $ax^2 + bx + c = 0$, $a,b,c \in \mathbf{R}$, $a \neq 0$ in the following cases:

a) $a + b + c = 0$; b) $a - b + c = 0$; c) $4a + 2b + c = 0$.

V. 9. a) The point $A(1, k)$ lies on $y = -x^2 + 4x - 7$. Find k;

b) The point $B(-2, l)$ lies on $y = -3x^2 - 5x$. Find l;

c) The point $C(m, 2)$ lies on $y = x^2 + 4x - 3$. Find m;

d) The point $D(n, 13)$ lies on $y = 2x^2 + 3x + 8$. Find n.

V. 10. For each function calculate the maximum or minimum:

a) $f(x) = -3x^2 + 2x - 1$; b) $f(x) = 2x^2 + x - 3$;

c) $f(x) = -x^2 + 4x + 2$; d) $f(x) = -2x^2 + 4x + 3$;

e) $f(x) = 5x^2 - 2x + 3$; f) $f(x) = x^2 - 4$.

V. 11. Draw the graph of the functions:

a) $f(x) = x^2 - 3x + 2$; b) $f(x) = -2x^2 - 7x - 5$;

c) $f(x) = -x^2 + 10x - 25$; d) $f(x) = 2x^2 - 3x + 1$;

e) $f(x) = -x^2 + 16$; f) $f(x) = -x^2 + 5x$.

V. 12. The graph

of $y = |f(x)|$, where

$f(x) = ax^2 + bx + c$,

$a < 0$

is shown.

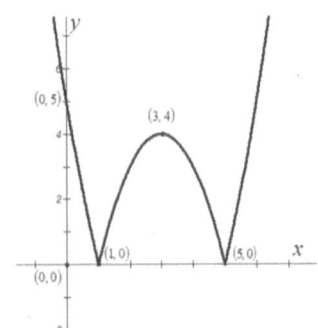

Find the equation of f.

V. 13. Let $a,b,c \in \mathbf{R}$ and $0 < a < b < c < 1$. Prove that the equation $ax^2 + 8ab(1 - c)x + b = 0$ does not have real roots.

V. 14. Find the real number a, if the length of the segment determined of the points $A(-1, 2)$, and $B(4 - a, 4 + a)$ is equal to 5.

V. 15. Find $m \in \mathbf{R}$ such that each linear function passes through the indicated point:

a) $f : \mathbf{R} \to \mathbf{R}$, $f(x) = (m + 2)x - 4m + 3$, $A(m, -m + 9)$;

b) $f : \mathbf{R} \to \mathbf{R}$, $f(x) = (m + 3)x - 2m^2 - 6$, $B(m + 2, 5m - 9)$.

V. 16. Given the sketch to the right, find the equation of the parabola, given the additional information:

a) the point $(3, -3)$ is on the graph;

b) the y-value of the vertex is -8;

c) the y-intercept is 36.

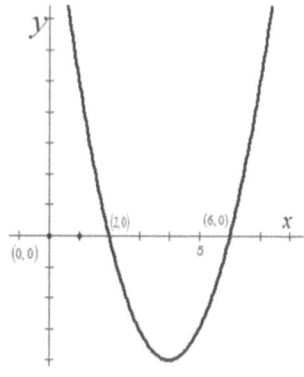

V. 17. Let

$f(x) = a(x - p)(x - q)$.

The graph is shown to the right.

Write down the value of p and q and find the coordinates of the vertex.

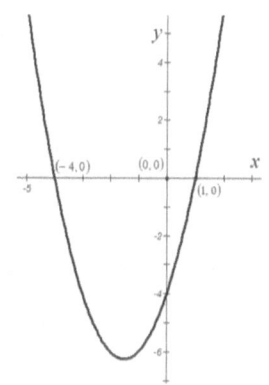

V. 18. Find the function $f : \mathbf{R} \to \mathbf{R}$, $f(x) = ax^2 + bx + c$, $a, b, c \in \mathbf{R}, a \neq 0$ such that

a) the graph crosses y-axis at $A(0, 1)$ and the vertex is $V(2, -3)$;

b) the vertex is $V\left(-\dfrac{5}{4}, -\dfrac{49}{8}\right)$ and passes through the point $A(0.5, 0)$.

V. 19. Consider the functions $f, g : \mathbf{R} \to \mathbf{R}$, $f(x) = x^2 - 2x + m$ and $g(x) = 2x^2 + 4x + 5$. Find $m \in \mathbf{R}$ such that f and g have the same range.

V. 20. Find the maximum for the algebraic expressions:

a) $E(x) = \dfrac{2x^2 - 8x + 17}{x^2 - 4x + 8}$; **b)** $F(x) = \dfrac{3x^2 - 6x + 5}{x^2 - 2x + 3}$.

V. 21. For the functions f, g: $\mathbf{R} \circledR \mathbf{R}$, calculate $f \circ g$ and $g \circ f$:

a) $f(x) = 2x^2$, $g(x) = 3x^2$; **b)** $f(x) = x^2 + 1$, $g(x) = 3x - 2$;

c) $f(x) = x^2 + 3x$, $g(x) = x^2 - 4x + 1$;

d) $f(x) = \begin{cases} 3x - 2 & \text{if } x \le -2 \\ -2x & \text{if } x > -2 \end{cases}$, $g(x) = 4x - 6$;

e) $f(x) = \begin{cases} x^2 + 1 & \text{if } x < 2 \\ 2x + 1 & \text{if } x \ge 2 \end{cases}$, $g(x) = \begin{cases} 3x - 2 & \text{if } x \le -1 \\ 2x^2 & \text{if } x > -1 \end{cases}$;

f) $f(x) = \begin{cases} x^2 + 3x & \text{if } x < -2 \\ -2x - 4 & \text{if } x \ge -2 \end{cases}$, $g(x) = \begin{cases} 3x - 2 & \text{if } x \le 1 \\ x^2 - 4x + 1 & \text{if } x > 1 \end{cases}$;

g) $f(x) = \begin{cases} 2x^2 & \text{if } x \in \left(0, \dfrac{1}{3}\right) \\ 5 & \text{if } x \in \left[\dfrac{1}{3}, 1\right) \end{cases}$, $g(x) = \begin{cases} 3x^2 & \text{if } x \in \left(0, \dfrac{1}{2}\right) \\ x & \text{if } x \in \left[\dfrac{1}{2}, 1\right) \end{cases}$;

h) $f(x) = x - 3 + |x - 3|$, $g(x) = 4 + ||x - 2| - 2| + |x - 4|$.

V. 22. Prove that the functions:

a) $f : (-\infty, 3] \to [-1, \infty)$, $f(x) = x^2 - 6x + 8$;

b) $f : [3, \infty) \to [-1, \infty)$, $f(x) = x^2 - 6x + 8$;

c) $f : \left(-\infty, \dfrac{5}{2}\right] \to \left(-\infty, \dfrac{9}{4}\right]$, $f(x) = -x^2 + 5x - 4$;

d) $f : \left[\dfrac{5}{2}, \infty\right) \to \left(-\infty, \dfrac{9}{4}\right]$, $f(x) = -x^2 + 5x - 4$ are bijections and find

their inverses. Draw the graph for the inverse and parent function.

Notice: The function $f:\mathbf{R} \to \mathbf{R}$, $f(x) = x^2 - 6x + 8$ is not an inversible function, it has an inverse, but the inverse is *not a function*.

V. 23. Find the inverses and draw the graphs of the bijective (inversible) functions:

a) $f:(-\infty, -1] \cup [0, 3) \to [0, \infty)$, $f(x) = x^2 - 2x + 1$;

b) $f:\mathbf{R} \to \mathbf{R}$, $f(x) = \begin{cases} x^2 - 4x & \text{if } x < 0 \\ -x & \text{if } x \geq 0 \end{cases}$;

c) $f:\mathbf{R} \to \mathbf{R}$, $f(x) = \begin{cases} 2x - 8 & \text{if } x < 4 \\ x^2 - 6x + 8 & \text{if } x \geq 4 \end{cases}$;

d) $f:\mathbf{R} \to \mathbf{R}$, $f(x) = \begin{cases} \dfrac{x+6}{3} & \text{if } x < 6 \\ x^2 - 6x + 4 & \text{if } x \geq 6 \end{cases}$;

e) $f:\mathbf{R} \to \mathbf{R}$, $f(x) = \begin{cases} -x^2 + 6x - 9 & \text{if } x < 2 \\ x^2 - 4x + 3 & \text{if } x \geq 2 \end{cases}$.

V. 24. The graph of the quadratic function $f(x) = -5(x - 1)(x + 7)$ intersects x-axis at the points A and B. Find the length of the segment AB.

V. 25. Determine the equation for each quadratic function.

a) x intercept is -3 and 1, containing the point $(4, -21)$;

b) x intercept is $-1 \pm 2\sqrt{2}$, containing the point $(1, -6)$.

V. 26. Find the point(s) of intersection between the parabola $y = x^2 - 6x + 8$ and the lines:

a) $y = 4x - 16$; b) $y = 4x - 17$; c) $y = 4x - 18$.

V. 27. Determine the value of m in $y = -x^2 - x - 3 + m$ that will result in the intersection of the line $y = 3x - 1$ with the quadratic at:
a) two points; b) one point; c) no point.

V. 28. Let $f:\mathbf{R} \to \mathbf{R}, f(x) = |ax^2 - 3x + 2c|$ a function. If $0 < a < 2$ and $c < 0$, determine a and c such that $f(0) = 4$ and $f(1) = 6$.

125

V. 29. Determine the quadratic function $f(x) = ax^2 + bx + c$

$a, b, c \in \mathbf{R}, a \neq 0$, such that the graph passes through the points:
a) $A(1, 0)$, $B(-1, 6)$, and $C(2, 3)$;
b) $A(-2, 12)$, $B(-1, 6)$ and crosses x-axis at the point $C(2, 0)$;
c) $A(1, 0)$, $B(-1, 10)$, and $C(2, -2)$;
d) $A(-2, 0)$, $B(1, 6)$, and the maximum is 6;
e) $V(1, 2)$ is the vertex and crosses y-axis at the point $C(0, -3)$;
f) $A(-1, 13)$, $B(2, 10)$, and the minimum is 9.

V. 30. If $f(x)$ is a quadratic function such that $f(-1) = 6$, $f(0) = 1$, and $f(1) = 0$, find $f(2)$.

V. 31. Consider the function $f : \mathbf{R} \to \mathbf{R}$,

$$f(x) = \begin{cases} \left(a^2 - 2\right)x + 4a & \text{if } x < 2 \\ x^2 - 2(b+1)x + 1 & \text{if } x \geq 2 \end{cases}.$$

Determine $a, b \in \mathbf{R}$ such that the graph of f passes through the points $(1, -6)$ and $(3, 4)$.

V. 32. Determine the quadratic function $f(x) = ax^2 + bx + c$ $a, b, c \in \mathbf{R}, a \neq 0$, such that the point $V(3, 5)$ is the vertex and it crosses the y-axis at the point $A(0, -4)$.

V. 33. Two parabolas have the vertices $V_1(2, a)$ and $V_2(2, b)$. If one of their common points of intersection is $A_1(7, 5)$, find the second common point of intersection.

V. 34. The parabola $f(x) = x^2$ meets the parabola

$f(x) = -x^2 + 6x - 4$ at the points P and Q. Find the equation of the line passing through P and Q.

V. 35. The quadratic equation $5x^2 + bx + c = 0$ has one of the roots 2 and the quadratic equation $5x^2 + cx + b = 0$ has one of the roots 3. Find the real numbers b and c.

V. 36. The inequality $4\left(x^2 + ax + 2\right) - 3\left(x^2 + 2x + b\right) > 0$ has the solution $x \in (-\infty, -2) \cup (4, \infty)$. Find $a, b \in \mathbf{R}$.

V. 37. Find $a,b \in \mathbf{R}$ such that the inequality

$(x-3)^2 + 2(x-1)^2 + ax + b \leq 0$ hold for any $x \in [-1,4]$.

V. 38. Let A, B be the points of intersection of the graph of the function

$f : \mathbf{R} \to \mathbf{R}$, $f(x) = -x^2 + (m+1)x - m$ with the x-axis and the point V the

vertex of the parabola. Find $m \in (1, \infty)$ such that the area of the triangle VAB be 1.

V. 39. Find the value of x such that the function

$f(x) = (x - a_1)^2 + (x - a_2)^2 + ... + (x - a_n)^2$ have the minimum value, where

$a_1, a_2, ..., a_n \in \mathbf{R}$.

V. 40. The quadratic function $f(x) = ax^2 + bx + c$, $a,b,c \in \mathbf{R}$, $a \neq 0$ has a maximum of five units and passes through the points $A\,(1, 1)$ and $B\,(-1, -11)$. Find a, b, c.

V. 41. Let x_1 and x_2 be the roots of the equation $x^2 - 2mx + m - 1 = 0$. Calculate in terms of the parameter $m \in \mathbf{R}$, the expressions:

a) $x_1 + x_2$; b) $x_1 x_2$; c) $x_1^2 + x_2^2$; d) $x_1^3 + x_2^3$; e) $x_1 - x_2$; f) $x_1^4 + x_2^4$;

g) $x_1^{-1} + x_2^{-1}$; h) $x_1^{-2} + x_2^{-2}$; i) $x_1^{-3} + x_2^{-3}$; j) $\sqrt{x_1} + \sqrt{x_2}$; k) $\sqrt[4]{x_1} + \sqrt[4]{x_2}$.

V. 42. Determine the parameter $m \in \mathbf{R}$ such that the roots of the equations fulfil the indicated condition.

a) $mx^2 + 2(m-1)x + 5 = 0$, $x_1 - x_2 = 2$;

b) $(m-1)x^2 + (3m-2)x + 3 = 0$, $x_1 = x_2$;

c) $(m+2)x^2 + (m-1)x + m = 0$, $x_1 = 3x_2$;

d) $3x^2 + (m-2)x + m + 6 = 0$, $x_1^2 + x_2^2 = \dfrac{8}{3}$;

e) $(m+4)x^2 - (m+6)x - m + 4 = 0$, $x_1 x_2 = x_1 + x_2$;

f) $mx^2 - (m-1)x - m + 1 = 0$, $x_1^3 + x_2^3 = 4$;

g) $(m+3)x^2 + (m-2)x + m + 1 = 0$, $x_1 + x_2 + 2x_1 x_2 = \dfrac{3}{2}$;

h) $x^2 - 2mx + m + 3 = 0$, $(x_1 - 1)^2 + (x_2 - 1)^2 = 6$;

i) $x^2 - 2(3m+1)x + 3m + 4 = 0$, $\sqrt{\dfrac{1+x_1}{1+x_2}} + \sqrt{\dfrac{1+x_2}{1+x_1}} = \dfrac{5}{2}$;

j) $x^2 + (1-2m)x + m^2 - 2m - 3 = 0$, $\dfrac{x_1}{x_2} + \dfrac{x_2}{x_1} \leq 2$.

127

V. 43. Let x_1 and x_2 be the roots of the equation $x^2 - 3x - 1 = 0$.

Denote $S_n = x_1^n + x_2^n$, for any $n \in \mathbf{N}^*$.

a) Solve the equation;

b) Calculate S_2, S_3;

c) Calculate the value of the expressions $E = \dfrac{x_1}{x_2 + 1} + \dfrac{x_2}{x_1 + 1}$ and

$$F = \frac{x_1^2 - 3x_1 + 1}{x_1^2 - 4x_1 + 2} + \frac{x_2^2 - 3x_2 + 1}{x_2^2 - 4x_2 + 2};$$

d) Find $m \in \mathbf{R}$ from the equation

$$\frac{x_1^2 - (m+1)x_1 + 2}{x_1^2 - (m+1)x_1 + 3 + m} + \frac{x_2^2 - (m+1)x_2 + 2}{x_2^2 - (m+1)x_2 + 3 + m} = -6;$$

e) Prove the identity $S_{n+2} = 3S_{n+1} + S_n, \forall n \in \mathbf{N}^*$.

V. 44. The equation $(1 - 2m)x^2 + 2x - 3 = 0$ has real roots. Find the largest numerical value that m may have.

V. 45. Find the quadratic equation with the roots:

a) $x_1 = 3$, $x_2 = -3$; b) $x_1 = -2$, $x_2 = 0$; c) $x_1 = -4$, $x_2 = 2$;

d) $x_1 = \dfrac{1 + \sqrt{3}}{2}$, $x_2 = \dfrac{1 - \sqrt{3}}{2}$; e) $x_1 = \dfrac{1 + i\sqrt{2}}{3}$, $x_2 = \dfrac{1 - i\sqrt{2}}{3}$.

V. 46. Consider the equation $x^2 + mx + 2m - 1 = 0$ with the roots x_1 and x_2. Find a quadratic equation in terms of y with the roots y_1 and y_2 such that:

a) $y_1 = \dfrac{1}{x_1}$, $y_2 = \dfrac{1}{x_2}$; b) $y_1 = 1 + \dfrac{1}{x_1}$, $y_2 = 1 + \dfrac{1}{x_2}$;

c) $y_1 = \dfrac{x_1 + x_2}{x_1}$, $y_2 = \dfrac{x_1 + x_2}{x_2}$; d) $y_1 = \dfrac{x_1}{x_2}$, $y_2 = \dfrac{x_2}{x_1}$;

e) $y_1 = x_1^{-2}$, $y_2 = x_2^{-2}$; f) $y_1 = x_1^{-3}$, $y_2 = x_2^{-3}$.

V. 47. Consider the equation $x^2 - ax + 3 = 0$, $a \in \mathbf{R}$ with the roots x_1 and x_2.

a) For $a = 2$ find the quadratic equation in terms of y with the roots

$$y_1 = \frac{1}{x_1 + 1} \quad \text{and} \quad y_2 = \frac{1}{x_2 + 1};$$

b) Find $a \in \mathbf{R}$ such that $x_1^2 + x_2^2 = a(a-2)$;

c) Find $a \in \mathbf{R}$ from the ecuation $\dfrac{x_1^2 - ax_1 + 2}{x_1^2 - ax_1 + 3 + a} + \dfrac{x_2^2 - ax_2 + 2}{x_2^2 - ax_2 + 3 + a} = -6$.

V. 48. Solve the following equations in the set of **R**.
Assume that no denominator is equal to zero.

a) $\dfrac{1}{x^2 - 6x + 1} + \dfrac{2}{x^2 - 6x + 2} + \dfrac{3}{x^2 - 6x + 3} + \dots + \dfrac{2013}{x^2 - 6x + 2013} = 2013$.

b) $x^2 + (x+2)^2 + \dots + (x + 2012)^2 = (x+1)^2 + (x+3)^2 + \dots + (x + 2013)^2$.

V. 49. Solve the equations:

a) $(x+1)(x+2) + 18 = (x+3)(x+4)$; b) $(x+1)^2 + (x+2)^2 = 2x^2 + 3x + 11$;

c) $\dfrac{(x+2)(x-1)}{2} - \dfrac{(x+1)(x+3)}{6} = \dfrac{(x+5)(x-1)}{3}$;

d) $\dfrac{(x-2)(2x-1)}{3} + \dfrac{(x+2)(x+1)}{4} = \dfrac{1}{12}\dfrac{x(x+3)}{12} - \dfrac{x-7}{3}$;

e) $\dfrac{2}{3} + \dfrac{(x+2)(x-5)}{6} + \dfrac{x^2 - 5x + 4}{2} = \dfrac{(x+3)(4x-1)}{6}$;

f) $\dfrac{(x+4)(x+1)}{6} + \dfrac{(1-2x)(3+x)}{4} = \dfrac{1 - 2x - 4x^2}{12}$;

g) $\dfrac{(x-1)^2 - 3(x-1) + 4}{x+1} - 5(x+3) = -4x - 14$;

h) $\dfrac{2x^2 + 2x + 1}{(x+1)(x+2)} + \dfrac{2x^2 + 2x + 3}{(x+1)(x+3)} = \dfrac{2x^2 + 2}{(x+2)(x+3)} + 2$.

V. 50. Solve the equations:

a) $\dfrac{a+1}{5a} - \dfrac{1}{a} = 1$; b) $\dfrac{1}{a^2} - \dfrac{4}{a} = \dfrac{3}{a^2}$; c) $\dfrac{4}{x} - \dfrac{1}{x^2} = \dfrac{1}{5x^2}$; d) $\dfrac{x-5}{x^2} + \dfrac{1}{x} = \dfrac{2}{x} - 5$;

e) $\dfrac{1}{x+5} - \dfrac{1}{x^2 + 5x} + \dfrac{2}{3} = \dfrac{4}{x^2 + 5x}$; f) $\dfrac{5}{p+6} - \dfrac{1}{p^2 + 6p} - \dfrac{2}{7} = \dfrac{2}{p^2 + 6p}$;

g) $\dfrac{1}{2v} = \dfrac{5v + 15}{v^2 - 6v} - \dfrac{v+6}{2v^2 - 12v}$; h) $\dfrac{5}{x+1} - \dfrac{8}{3} = \dfrac{6}{x^2 - 2x - 3} + \dfrac{2x - 5}{x - 3}$;

i) $\dfrac{n^2 + 7n + 6}{n^2} = \dfrac{1}{6} - \dfrac{1}{6n^2}$; j) $\dfrac{x^2 - 3x - 4}{x^3 - x^2} - \dfrac{1}{x^2} = \dfrac{x-2}{x^2}$;

k) $\dfrac{v-6}{2v^2 + 2v - 4} + \dfrac{v}{2v - 2} = \dfrac{1}{2}$; l) $\dfrac{x-3}{2x + 10} + 2x - 12 = \dfrac{x^2 + 3x - 18}{2x + 10}$.

129

V. 51. Solve the equations:

a) $\dfrac{1}{x-1}+\dfrac{1}{x-4}=\dfrac{5}{4}$;

b) $\dfrac{x+3}{x^2-4}-\dfrac{1}{x+2}=\dfrac{2x}{2x-x^2}$;

c) $\dfrac{1}{x^2+3}-\dfrac{1}{x^2+6}=\dfrac{1}{18}$;

d) $\dfrac{1}{x^3+3}-\dfrac{1}{x^3+4}=\dfrac{1}{20}$;

e) $\dfrac{x-2}{x-1}+\dfrac{x+2}{x+1}=-\dfrac{x+1}{x+2}-\dfrac{x-1}{x-2}$;

f) $3\left(x+\dfrac{1}{x^2}\right)-7\left(1+\dfrac{1}{x}\right)=0$;

g) $4\left(x^2+\dfrac{1}{x^2}\right)+8\left(x-\dfrac{1}{x}\right)=29$;

h) $\dfrac{x+b}{a+b}-\dfrac{a-b}{x-b}=\dfrac{x+c}{a+c}-\dfrac{a-c}{x-c}$;

i) $\dfrac{x+a}{x+b}+\dfrac{x+b}{x+a}=2.5$;

j) $\dfrac{x+6}{x+3}+\dfrac{x+10}{x+5}+\dfrac{x+14}{x+7}=6$;

k) $\dfrac{(x-k)^2+x(x-k)+x^2}{(x-k)^2-x(x-k)+x^2}=\dfrac{7}{3}$;

l) $\dfrac{x}{a-b}+\dfrac{2a-x}{a+b}-\dfrac{a-b}{x}=1$;

m) $\dfrac{1-a^2}{ax+1}+\dfrac{a+x}{a}=1$;

n) $\dfrac{ax^2}{x+1}=(a-1)^2$;

o) $\dfrac{3a}{x-3a}+\dfrac{2a}{x-2a}+\dfrac{6a}{x-a}=6$.

V. 52. Solve the equations in the set of **R**:

a) $\dfrac{x^2+1}{x+1}+\dfrac{x^2+2}{x-2}=\dfrac{27}{2}$; b) $\left(x^2-6x\right)^2-2(x-3)^2=17$;

c) $\dfrac{z^2-z}{z^2-z+1}-\dfrac{z^2-z+2}{z^2-z-2}=1$; d) $\dfrac{16}{x^2+2x-8}-\dfrac{15}{x^2+2x-3}=3$;

e) $\dfrac{6}{(x+1)(x+2)}+\dfrac{8}{(x-1)(x+4)}=1$; f) $\dfrac{u^2}{2-u^2}+\dfrac{u}{2-u}=-\dfrac{22}{7}$;

g) $(x+1)^2(x+2)+(x-1)^2(x-2)=-12$;

h) $x^3+x^2(a+1)-x-a-1=0$; i) $\left(x^2-3x\right)^2-2\left(x^2-3x\right)=8$;

j) $\dfrac{1}{x^2}+\dfrac{1}{(x+2)^2}=2$; k) $\dfrac{x^2+1}{x}+\dfrac{x}{x^2+1}=2.9$;

l) $\dfrac{x^2+2x+3}{x^2}+\dfrac{2x^2}{x^2+2x+3}=\dfrac{19}{3}$; m) $x^3-\dfrac{11}{2x^3-5}=7$;

n) $\dfrac{1}{x^2(x^2-4x+4)}+\dfrac{2}{x(x-2)}=\dfrac{5}{4}$.

V. 53. Find $m \in \mathbf{R}$ such that the point $A\left(2m + 6, 16 - m^2\right)$ is in the second quadrant.

V. 54. Solve the inequalities:

a) $x^2 + 7x + 3 > -3$; b) $-x^2 + x - 2 \geq -4x + 4$; c) $3x^2 + x - 2 \leq -7x + 9$;

d) $4x^2 - 3x + 1 \geq -x^2 + 3$; e) $\dfrac{x^2 - 3}{4} - \dfrac{x-1}{3} > \dfrac{2x-6}{3} - x$;

f) $\dfrac{x^2 + 2x + 13}{x^2 - 3x + 2} \geq 2$; g) $\dfrac{2x^2 + 5(x+1) - 10}{3x^2 + x - 2} \leq 1$; h) $\dfrac{x-2}{x+2} \geq \dfrac{2(x+1) - 5}{x+3}$;

i) $\dfrac{3x - 3}{x^2 - 6x + 5} > 2$; j) $\dfrac{3x+1}{x-2} - \dfrac{2x+1}{x+2} \leq \dfrac{1}{2}$; k) $\dfrac{x}{x-1} > \dfrac{x+1}{x}$;

l) $\dfrac{x+2}{x+3} \geq \dfrac{2x+1}{x-1}$; m) $-1 < \dfrac{x^2 + 3x + 2}{x^2 - 4x + 3} \leq 2$; n) $\dfrac{(x-1)(x-2)(x-3)}{(x+1)(x+2)(x+3)} > 1$;

o) $\dfrac{6}{(x+2)(x-3)} - \dfrac{1}{x+2} \geq 2$; p) $\dfrac{1}{x-1} \geq \dfrac{4}{x+2} + \dfrac{6}{x-2}$;

q) $\dfrac{\sqrt{1+x^3} + x - 2}{x-1} \geq x + 1$; r) $\dfrac{\left|x^2 + 3x\right| + 3}{x^2 + |x-3|} \leq 1$; s) $\dfrac{\sqrt{4 + 3x - x^2}}{x-3} \geq 0$.

V. 55. Solve the systems of inequalities:

a) $\begin{cases} x^2 - 3x > 4 \\ -5x^2 + 6x - 1 < 0 \end{cases}$; b) $\begin{cases} 2x - 7 \leq 0 \\ x^2 - 7x + 12 \leq 0 \\ -x^2 + 2x + 3 < 0 \end{cases}$; c) $\begin{cases} 3(x-1) \geq 2(x-2) \\ 2x^2 + 4x > 0 \end{cases}$;

d) $\begin{cases} -4x + 3 \geq x - 2 \\ (x+3)(x+2) \geq (x-3)(x-2); \\ x^2 - 3x + 2 \geq 0 \end{cases}$ e) $\begin{cases} \dfrac{|2x-1| + |1-2x|}{4 - x^2} \leq 1 \\ \dfrac{|2x-1| + |1-2x|}{4 - x^2} \geq -1 \end{cases}$;

f) $\left| \dfrac{x^2 - 6x + 8}{x^2 - 9} \right| \leq 1$; g) $\left| \dfrac{x^2 - 5x + 4}{x^2 - 4} \right| \leq 1$.

V. 56. Find the domain of the functions:

a) $f(x) = \sqrt{9 - x^2} + \sqrt[4]{x^2 - 1}$; b) $f(x) = \sqrt{9 - x^2} + \sqrt[3]{x^2 - 1}$;

c) $f(x) = \sqrt[3]{9 - x^2} + \sqrt{x^2 - 1}$; d) $f(x) = \sqrt{\dfrac{-x^2 + 5x - 4}{x^2 - 4}}$;

e) $f(x) = \sqrt{\dfrac{-x^2 + 6x - 8}{\sqrt{x-3}}}$;   f) $f(x) = \dfrac{\sqrt{1-x^2}}{x}$.

V. 57. For the function $f : \mathbf{R} \to \mathbf{R}, f(x) = x^2 - 4x + 3$ calculate:

a) $f([0, 3])$; b) $f([\,0, 4))$; c) $f([1, 5])$; d) $f((-\infty, 0])$;
e) $f([3, +\infty))$; f) $f((-\infty, 3])$; g) $f((-1,+\infty))$.

V. 58. Draw the graph of the function

$$f : \mathbf{R} \to \mathbf{R}, \ f(x) = \begin{cases} x^2 - 2x & \text{if } x \le 2 \\ x^2 - 8x + 12 & \text{if } x > 2 \end{cases},$$

Find range of f, $f((\infty, -1)$, $f([-1,1.2])$, $f((3, 7))$.

V. 59. Find the range for the functions:

a) $f : \mathbf{R} \to \mathbf{R}, \ f(x) = x^2 - x + 1$;   b) $f : \mathbf{R} \to \mathbf{R}, \ f(x) = x^2 + 5x - 6$;

c) $f : \mathbf{R} \to \mathbf{R}, f(x) = |x^2 - 4| - 5$;   d) $f : \mathbf{R} \to \mathbf{R}, \ f(x) = \dfrac{x^2 - x - 3}{x^2 + x + 2}$;

e) $f : \mathbf{R} \to \mathbf{R}, \ f(x) = \dfrac{x^2 - 4x + 3}{x^2 - 2x + 3}$;   f) $f : \mathbf{R} \to \mathbf{R}, \ f(x) = \dfrac{-x^2 - 2x - 3}{x^2 + 2x + 1}$.

V. 60. Consider the function $f : \mathbf{R} \to \mathbf{R}, \ f(x) = \dfrac{3x^2 + ax + b}{x^2 + 1}$. Find the real numbers a and b such that the range of the function be the interval $[-3, 5]$.

V. 61. The roots of the equation $x^2 - 2(m - 2)x - 2m + 3 = 0$

are x_1 and x_2. Find the value of $m \in \mathbf{R}$ such that $\dfrac{x_1}{x_2^2} + \dfrac{x_2}{x_1^2} \ge \dfrac{1}{x_1} + \dfrac{1}{x_2}$.

V. 62. The roots of the equation $mx^2 - 2(m + 1)x + 8 = 0$ are x_1 and x_2.

Find the value of $m \in \mathbf{R}$ such that $\dfrac{x_1^2 + x_2^2}{x_1 x_2} > x_1 + x_2$.

V. 63. Consider the equations $x^2 - (m + 2)x + m^2 + 1 = 0$ with the roots x_1

and x_2 and $y^2 - (2m - 1)y + m^2 + 2m = 0$ with the roots y_1 and y_2.
Find the value of m such that $(x_1 + x_2)(y_1 + y_2) \ge x_1 x_2 + y_1 y_2$.

V. 64. The roots of the equation

$$4(m+1)x^2 - 2(m+1)x + m - m^2 = 0 \text{ are } x_1 \text{ and } x_2. \text{ Find the value of}$$

$m \in \mathbf{R} \setminus \{-1\}$ such that $-2 \le \dfrac{1}{x_1} + \dfrac{1}{x_2} + \dfrac{1}{2x_1x_2} \le 0$.

V. 65. The equation $x^2 - 2mx + m + 5 = 0$ has the roots x_1 and x_2. Find the value of $m \in \mathbf{R}$, such that

a) one of the roots is the double of the other;

b) $\dfrac{x_1^2 + x_2^2}{x_1 + x_2} \ge x_1 x_2$.

V. 66. Consider the family of quadratic functions

$f_m(x) = x^2 + 2(m-1)x + m - 1$. Find the value of $m \in \mathbf{R}$ such that the vertices of the parabolas lie above the x-axis.

V. 67. Consider the family of quadratic functions

$f_m(x) = x^2 - 2(m+2)x + m + 2$, $m \in \mathbf{R}$.

a) Show that the vertices of the parabolas lie on a parabola.
b) For what value of m, do the vertices of the parabolas lie under the x-axis?

V. 68. For the family of quadratic functions

$f_m(x) = x^2 - 2(m-1)x + m - 2$, $m \in \mathbf{R}$ show that the vertices of the parabolas lie on a parabola.

V. 69. Consider the family of quadratic functions

$f_m(x) = mx^2 + 2(m+1)x + m + 2$, $m \in \mathbf{R} \setminus \{0\}$
a) Show that the vertices of the parabolas lie on the equation of the line $y = x + 1$.
b) Let A and B be the x intercept and V and F are the vertices and the feet of the

perpendicular from V onto x-axis, show that for any $m \in \mathbf{R} \setminus \{0\}$ AB = 2 FV.
c) Show that all parabolas from the family pass through a fix point and find the coordinates of this point.

V. 70. Consider the family of quadratic functions

$f_m(x) = (m+2)x^2 + (m+1)x - 3$, $m \in \mathbf{R} \setminus \{-2\}$.
a) For $m = -1$ find the intervals where the function decreases or increases.

133

b) Find $m \in \mathbf{R} \setminus \{-2\}$ such that the vertices of the parabolas lie on the first bisector.

c) Find $m \in \mathbf{R} \setminus \{-2\}$ such that $f_m(x) > 0$ for any $x \in \mathbf{R}$.

V. 71. Find the parameter $m \neq -1$ such that the graph of the parabola

$f(x) = (m+1)x^2 - (2m+3)x + m - 1$ intersect the x-axis in two distinct points.

V. 72. Prove that for any $m \neq 0$, the graph of the function

$f(x) = mx^2 - 2(m-3)x + m - 6$ intersects the x-axis in two distinct points.

V. 73. Find the numerical value of $m, n \in \mathbf{R}$, such that the functions

$f(x) = x^2 - 2x - n + 1$ and $g(x) = -x^2 - 2mx - 6$ have the same vertex.

V. 74. Show that the graphs of the functions $f(x) = 2x^2 - 2x + 3$ and

$g(x) = x^2 - 2x + 4$ have two common points.

V. 75. Let $f(x) = x^2 - 2x - n + 1$ and $g(x) = -mx^2 - 2mx - 6$.
Find the numerical value of $m, n \in \mathbf{R}$, such that the combined function
$h(x) = f(x) + g(x)$ passes through the points $(1, 3)$ and $(-1, 5)$.

V. 76. Consider the quadratic family of functions
$f_m(x) = (m-1)x^2 + 2(m+2)x + m + 1, \ m \neq 1$.
a) Find the parameter m if;
   1) the graph of the functions is below the x-axis;
   2) the graph of the parabolas f_m is tangent to x-axis;
   3) the graph of the parabolas f_m lies above the line $y = -3$;
   4) the vertices of the parabolas lie on the line $x + y = 3$.

b) Show that the vertices of the parabolas f_m lie on a line for any $m \in \mathbf{R}$.
c) Show that the family of functions passes through a fix point for any $m \in \mathbf{R}$.

V. 77. Consider the family of functions
$f_m(x) = x^2 - 2(m-1)x + m^2 - 3m, \ m \in \mathbf{R}$.

a) Find the value of m such that:

1) the vertices of the parabolas f_m lie below x-axis;

2) the graph of the parabolas f_m be tangent to x-axis;

3) the vertices of the parabolas lie on the line $2x + 3y = 1$;

4) the vertices of the parabolas lie on the parabola $y = x^2 - 3x + 1$;

b) Show that the vertices of this family of functions f_m lie on a line for $m \in \mathbf{R}$.

V. 78. Find the numeric value of $m \neq 1$, such that

$(m-1)x^2 + 2mx + m + 2 > 0$, for any $x \in \mathbf{R}$.

V. 79. Find the value of m, such that $(m-2)x^2 + 2(2m-3)x + m - 2 < 0$, for any $x \in \mathbf{R}$.

V. 80. For what value of m, does the inequality

$(m-1)x^2 + 2(m+2)x + m + 2 > 0$ have no solution?

V. 81. For what value of $m \in \mathbf{R}$, is the range of the function

$f(x) = \sqrt{(m-1)x^2 + (m+2)x + m + 3}$ given by $y \in [2, \infty)$?

V. 82. Let $f : \mathbf{R} \to \mathbf{R}, f(x) = x^2 - 6x + 9 + m$ be a quadratic function, where $m \in \mathbf{R}$. Find the value of m such that $f(x) \geq 0$ for any $x \in \mathbf{R}$ and for $m = 0$ find the value of x such that $(f \circ f)(x) = 0$.

V. 83. Find the value of $m \in \mathbf{R}$ such that

a) $\dfrac{x^2 + mx + m^2 - m}{x^2 + 4mx + 5m^2 + 3} > 0$;   b) $\dfrac{x^2 + (m+1)x + m + 2}{x^2 + x + m} \geq 0$;

c) $\dfrac{x^2 - 4x + 5}{mx^2 + 2(m+1)x + m - 3} < 0$ are true for any $x \in \mathbf{R}$.

V. 84. Let x_1 and x_2 be the roots of the equation

$2x^2 - 2(m+1)x + m^2 - 2m + 3 = 0$, $m \in \mathbf{R}$. Show that:

a) $(x_1 - 2)^2 + (x_2 - 2)^2 = 2$;   b) $2 - \sqrt{3} \leq \dfrac{x_1}{x_2} \leq 2 + \sqrt{3}$;

135

c) Find a relation between x_1 and x_2 which does not depend on m.

V. 85. Let x_1 and x_2 be the roots of the equation $f_m(x) = 0$, where $f_m(x) = 2x^2 - 2(m+3)x + m^2 - 4m + 24$, $m \in \mathbf{R}$. Show that:

a) $(x_1 - 5)^2 + (x_2 - 5)^2 = 5$; b) $x_1, x_2 \in \left[5 - \sqrt{5},\ 5 + \sqrt{5}\right]$;

c) $-\sqrt{10} \le x_1 - x_2 \le \sqrt{10}$; d) $10 - \sqrt{10} \le x_1 + x_2 \le 10 + \sqrt{10}$;

e) $x_1^2 + x_2^2 \le 55 + 10\sqrt{10}$; f) $0.5 \le \dfrac{x_1}{x_2} \le 2$;

g) Determine the parameter $m \in \mathbf{R}$ such that $f_m(x) \ge 0$ for any $x \in \mathbf{R}$;

h) Find the smallest possible value of $f_m(x)$;

i) Find a relation between x_1 and x_2 which does not depend on m;

j) Find $m \in \mathbf{R}$ such that the equation $f_m(x) = 0$ have a root between 5 and 6.

V. 86. Solve the systems of equations:

a) $\begin{cases} x^2 - xy - y = 5 \\ x - 2y = 1 \end{cases}$; b) $\begin{cases} 2x - y = 3 \\ 3x^2 - y^2 - 2xy = 7 \end{cases}$; c) $\begin{cases} x + 2y = 7 \\ x^2 + 2y^2 - xy = 1 \end{cases}$;

d) $\begin{cases} \dfrac{x^2 - xy - 6}{x + y} = x - y \\ x + 3y = 10 \end{cases}$; e) $\begin{cases} 2x - 3y = -8 \\ 2x^2 - xy + 5y^2 = 24 \end{cases}$; f) $\begin{cases} \dfrac{x^2 - xy - 1}{x - y} = 1 \\ 2x + y = 5 \end{cases}$.

V. 87. Find the real solutions of the system of equations:

a) $\begin{cases} y^2 - xy = -12 \\ x^2 - xy = 28 \end{cases}$; b) $\begin{cases} x^4 - y^4 = 15 \\ x^3 y - xy^3 = 6 \end{cases}$; c) $\begin{cases} (x - y)xy = 30 \\ (x + y)xy = 120 \end{cases}$;

d) $\begin{cases} x^2 + y^2 + 6x + 2y = 0 \\ x + y + 8 = 0 \end{cases}$; e) $\begin{cases} \dfrac{1}{x + y} + \dfrac{1}{x - y} = \dfrac{4}{3} \\ (x + y)^2 + (x - y)^2 = 10 \end{cases}$; f) $\begin{cases} x^2 + y^4 = 5 \\ x^4 + y^2 = 17 \end{cases}$;

g) $\begin{cases} \dfrac{4}{2x - y + 1} - \dfrac{5}{x - y - 1} = \dfrac{13}{2} \\ \dfrac{3}{2x - y + 1} + \dfrac{1}{x - y - 1} = \dfrac{5}{2} \end{cases}$; h) $\begin{cases} \dfrac{3}{x^2 + y^2 - 1} + \dfrac{2y}{x} = 5 \\ x^2 + y^2 + \dfrac{3x}{y} = 5 \end{cases}$;

i) $\begin{cases} x^3 + 3xy^2 = 14 \\ 3x^2 y + y^3 = -13 \end{cases}$; j) $\begin{cases} \dfrac{x}{y} + \dfrac{x^2}{y^2} + \dfrac{x^3}{y^3} = 39 \\ x + y = 4 \end{cases}$.

V. 88. Solve the *homogeneous* system of equations:

a) $\begin{cases} x^2 - 3y^2 = 1 \\ x^2 + xy = 6 \end{cases}$;　b) $\begin{cases} 2y^2 - x^2 = -1 \\ xy = 6 \end{cases}$;　c) $\begin{cases} x^2 - 2xy + 3y^2 = 17 \\ 3x^2 - 2xy + y^2 = 11 \end{cases}$;

d) $\begin{cases} 2x^2 - 3xy - 5y^2 = -6 \\ x^2 - 6y^2 = -5 \end{cases}$;　e) $\begin{cases} 2y^2 - 7x^2 = -5 \\ xy = -1 \end{cases}$;　f) $\begin{cases} 2x^2 - 3xy + 7y^2 = 6 \\ 3x^2 - xy + y^2 = 3 \end{cases}$.

V. 89. Solve the *symmetric* system of equations:

a) $\begin{cases} x^2 + y^2 = 5 \\ xy = 2 \end{cases}$;　b) $\begin{cases} x^2 + y^2 = 13 \\ x + y + xy = 11 \end{cases}$;　c) $\begin{cases} 2x^2 + 2y^2 - x^2 y^2 = 0 \\ xy = x + y \end{cases}$;

d) $\begin{cases} xy(x+1)(y+1) = 24 \\ (x-1)(y-1) = 0 \end{cases}$;　e) $\begin{cases} xy + x + y = 11 \\ x^2 y + xy^2 = 30 \end{cases}$;　f) $\begin{cases} x^3 + y^3 = 26 \\ x^2 y + xy^2 = -6 \end{cases}$;

g) $\begin{cases} 5x^2 - 6xy + 5y^2 = 29 \\ 7x^2 - 8xy + 7y^2 = 43 \end{cases}$;　h) $\begin{cases} x^3 + y^3 = 19 \\ (xy + 7)(x + y) = 1 \end{cases}$;　i) $\begin{cases} \dfrac{x^2 + y^2}{x + y} = \dfrac{10}{3} \\ \dfrac{1}{x} + \dfrac{1}{y} = \dfrac{3}{4} \end{cases}$;

j) $\begin{cases} x^{-2} + y^{-2} = 13 \\ x^{-1} + y^{-1} = 5 \end{cases}$;　k) $\begin{cases} \dfrac{x}{y} + \dfrac{y}{x} = \dfrac{17}{4} \\ x + y = 5 \end{cases}$;　l) $\begin{cases} x + xy + y = 17 \\ x^2 + x^2 y^2 + y^2 = 129 \end{cases}$;

m) $\begin{cases} x^3 + y^3 = -9 \\ x^3 y^3 = 8 \end{cases}$;　n) $\begin{cases} x^2 + y^4 = 20 \\ x^4 + y^2 = 20 \end{cases}$;　o) $\begin{cases} x^4 + 5x^2 y^2 + y^4 = 37 \\ x^3 y + xy^3 = 10 \end{cases}$;

p) $\begin{cases} x^2 + xy + y^2 = 13 \\ x^4 + x^2 y^2 + y^4 = 91 \end{cases}$;　q) $\begin{cases} x^6 + y^6 = 793 \\ x^4 - x^2 y^2 + y^4 = 61 \end{cases}$.

V. 90. Solve the system of equations:

a) $\begin{cases} (x - 0.8)^2 + (y - 0.7)^2 = 1 \\ x + y = 2.9 \end{cases}$;　b) $\begin{cases} x - y = 2 \\ x^3 - y^3 = 26 \end{cases}$;　c) $\begin{cases} x^3 + y^3 = 9 \\ 3xy^2 - 2x^2 y = 8 \end{cases}$;

d) $\begin{cases} \dfrac{x}{y} - \dfrac{y}{x} = -\dfrac{16}{15} \\ x^2 - y^2 = -16 \end{cases}$;　e) $\begin{cases} \left(\dfrac{x}{y}\right)^2 + \left(\dfrac{y}{x}\right)^2 = \dfrac{257}{16} \\ (xy)^2 + xy = 20 \end{cases}$;　f) $\begin{cases} \dfrac{x + y}{x - y} - \dfrac{x - y}{x + y} = 4\dfrac{4}{5} \\ xy = 6 \end{cases}$;

g) $\begin{cases} (x - y)(x^2 - y^2) = 3a^3 \\ (x + y)(x^2 + y^2) = 15a^3 \end{cases}$, $a \ne 0$;　h) $\begin{cases} \left(\dfrac{x}{y}\right)^2 + \left(\dfrac{x}{y}\right)^3 = 36 \\ x^2 y^2 + xy = 12 \end{cases}$;

i) $\begin{cases} (x-y)(x-2y)(x-3y) = -105 \\ (y-x)(y-2x)(y-3x) = 60 \end{cases}$; j) $\begin{cases} \dfrac{x^3}{y} + xy = 30 \\ \dfrac{y^3}{x} + xy = \dfrac{10}{3} \end{cases}$.

V. 91. Solve the system of equations:

a) $\begin{cases} x+y = -3 \\ x+z = -2 \\ xy+xz+yz = 2 \end{cases}$; b) $\begin{cases} x+y+2z = 1 \\ x+2y+z = 1 \\ (x+y)^3 + 2(x+2y)^3 - 3(z+1)^3 = 0 \end{cases}$;

c) $\begin{cases} xy+yz = 18 \\ yz+zx = 20; \\ zx+xy = 14 \end{cases}$ d) $\begin{cases} x+y+z = 9 \\ x(y+z) = 14; \\ y(x+z) = 18 \end{cases}$ e) $\begin{cases} x+y+z = 9 \\ \dfrac{1}{x} + \dfrac{1}{y} + \dfrac{1}{z} = 1; \\ xyz = 27 \end{cases}$

f) $\begin{cases} \dfrac{1}{x} + \dfrac{1}{y} + \dfrac{1}{z} = 1 \\ \dfrac{1}{xy} + \dfrac{1}{yz} + \dfrac{1}{zx} = \dfrac{1}{3}; \\ xyz = 27 \end{cases}$ g) $\begin{cases} \dfrac{3}{xy} + \dfrac{12}{yz} = 2 \\ \dfrac{6}{yz} + \dfrac{2}{xz} = 1; \\ \dfrac{4}{xz} + \dfrac{3}{xy} = 2 \end{cases}$ h) $\begin{cases} \dfrac{x}{y} + \dfrac{y}{z} + \dfrac{z}{x} = 3 \\ \dfrac{y}{x} + \dfrac{z}{y} + \dfrac{x}{z} = 3; \\ x+y+z = 3a \end{cases}$

i) $\begin{cases} x+y+z = 6 \\ x^2+y^2+z^2 = 14; \\ xz+yz = (xy+1)^2 \end{cases}$ j) $\begin{cases} (x+y)^2 - z^2 = 9 \\ (y+z)^2 - x^2 = 45; \\ (z+x)^2 - y^2 = 27 \end{cases}$ k) $\begin{cases} x+yz = 2 \\ y+zx = 2; \\ z+xy = 2 \end{cases}$

l) $\begin{cases} x-y+z = 9 \\ x^2+y^2+z^2 = 29; \\ x^3 - y^3 + z^3 = 99 \end{cases}$ m) $\begin{cases} (y+z)(x+y+z) = 2 \\ (z+x)(x+y+z) = 6 \\ (x+y)(x+y+z) = 10 \end{cases}$.

V. 92. Prove that $\sqrt{\sqrt[3]{x} + \sqrt[3]{y} + \sqrt[3]{z}} > \sqrt[3]{\sqrt{x} + \sqrt{y} + \sqrt{z}}$, for any x, y, z positive integers.

QUADRATIC ROOT WORD PROBLEMS

Numbers

V. 93. The product of a number and that number decreased by 4 is 45. Find the number.

V. 94. The product of two consecutive even integers is 528. Find the integers.

V. 95. The difference between a number and its square is -56. Find the number.

V. 96. Adding 18 to the square of a number, the result is nine times the original number. Find the number.

V. 97. The sum of two numbers is 15. The sum of their squares is 113. What are the numbers?

V. 98. The sum of the squares of two consecutive integers is 421. What are the numbers?

V. 99. Two numbers differ by 8 and the sum of their squares is 274. Find the numbers.

V. 100. If the square of a positive number is decreased by 8, then the result is seven times the original number. Find the number.

V. 101. The sum of a number and its reciprocal is $\dfrac{181}{90}$. Find the number.

V. 102. Double of a number minus 15 times its reciprocal is equal to 29. Find the number.

V. 103. The sum of the squares of three consecutive even integers is 116. Find the integers.

Shapes

V. 104. A right triangle ABC ($\angle A = 90°$) has sides 2 cm and 4 cm respectively less than its hypotenuse. Find the length of each side of the triangle.

V. 105. The hypotenuse of a right triangle is 15 cm. The sum of the other two sides is 21 cm. Find the lengths of the other two sides of the triangle.

V. 106. A right triangle has a height of 8 cm more than twice the length of the base. If the area of the triangle is 96 cm², find the dimensions of the triangle.

V. 107. Double of the hypotenuse of a right angled triangle is 20 cm more than twice the length of the shortest side. The other side is 5 cm longer than the shortest side. Find the length of each side of the triangle.

V. 108. A rectangle has length 4 cm greater than its width. Find its length given that its area is 96 cm².

V. 109. A rectangular courtyard has perimeter 54 m and area 180 m². What are its dimensions?

V. 110. A farmer plants 1200 trees. The number of trees in each row is 20 more than four times the number of rows. If equal numbers of trees were planted in each row, how many rows did the farmer plant?

V. 111. A rectangular skating rink measuring 30 m by 20 m is to be doubled in area by adding a strip at one end and a strip of the same width along one side. Find the width of the strips.

V. 112. A rectangle has length 4 cm greater than its width. If it has an area of 140 cm², find the dimensions of the rectangle.

V. 113. A garden, which measures 60 m by 40 m is to be doubled in area by extending each side an equal amount. By how much should each side be extended?

QUADRATIC MAX/MIN WORD PROBLEMS

V. 114. A manufacturer produces CD's at a cost of $2 apiece. The CD's are sold for $5 apiece. At that price, consumers have bought 4,000 CD's a month. The manufacturer intends to raise the unitary price and he estimates that for each $1 increase, he will sell 400 CD less every month.
a) Express the manufacturer's monthly profit as a function of price at which CD's are sold.
b) What is the price to maximize the profit? What is the maximum profit?

V. 115. A manufacturer has determined that if his company sells x items per day then their profit is given by $P(x) = -x^2 + 650x - 100000$. How many items must be sold in order to make a profit?

V. 116. A ferryboat carries 300 cars daily from a place to another place charging $8 per car. The captain realized that if he dropped the charge by $0.25, he would have 10 cars more on board of his ferryboat and he would make $2,200 in fares every day. What fare should they charge to maximize the revenue?

V. 117. An apple orchard has an average yield of 30 bushels per tree only if there are at most 40 trees per acre. For more than 40 trees per acre, the average yield decreases by half a bushel per tree for every tree over 40. Find the number of trees which maximize the yield per acre.

V. 118. An object is thrown in the air. The height (in metres) of the object is given by the equation $h(t) = 25t - 5t^2$, where t is time in seconds. What is the maximum height reached by the object?

V. 119. A stone is thrown into the air. Its height above the ground is given by the function $F(t) = -4t^2 + 25t + 2$, where t is the time in seconds from when the stone is released and the height is in meters.
a) How high above the ground was the stone released?
b) How high above the ground is the stone at time $t = 3$ seconds?
c) At what times was the stone's height above the ground 38 m?
d) At what time was the maximum stone's height above the ground?

V. 120. A fence is 100 feet long and it forms three sides of a rectangular garden. One side is occupied by a building. Find the necessary dimensions of the fence to enclose the largest area.

V. 121. On the segment $AB = a$, consider the point M and the equilateral triangles with the sides AM and MB. Find the minimum of the total area of the triangles.

V. 122. Let AD be the median of $\triangle ABC$. Find a point $M \in (AD)$ such that $S = MA^2 + MB^2 + MC^2$ be minimum.

V. 123. Consider a circle $\mathbf{C}$ of radius r. On a diameter AB of this circle, consider a point M. Construct the circles $\mathbf{C'}, \mathbf{C''}$ with diameter AM, respectively MB. Find the position of the point M such that the area between the circles $\mathbf{C'}, \mathbf{C''}$ and the circle of radius r be maximized.

V. 124. On the segment $AB = 2a$, consider a point M and the circles of diameters AM and MB. Find the minimum of the area of the two circles and the length of AM.

V. 125. The area of a rectangle is 100 m². Find the dimensions of the rectangle such that the perimeter be minimized.

V. 126. In $\triangle ABC$, $AB = 8$ cm and $AC + BC = 10$ cm. Find the altitude of the triangle such that its area be maximized.

V. 127. On the side $AB = a$ of the quadrilateral $ABCD$ consider the point P such that $BP = b$ and on the sides AD and CD the points M and N, respectively, such that $MN \parallel CP$. Find the maximum area of the quadrilateral $MNCP$.

DISTANCE, VELOCITY, TIME

V. 128. A jet flew from Town A to Town B, a distance of 4000 km. On the return trip, the speed was increased by 300 km/h. If the total trip took 13 h, what was the speed from Town A to Town B?

V. 129. Vlad drove from B to C, a distance of 720 km. On the return trip he increased this speed by 10km/h. If the total trip took 17 h, what was his speed from C to B?

V. 130. A jet flew from A to B, a distance of 4800 km. On the return trip the speed was decreased by 200 km/h. If the difference in the times of the flights was 2 h, what was the speed from B to A?

Integer Part Function (Floor Function)

$f : \mathbf{R} \to \mathbf{Z}$, $f(x) = [x]$, $[x] \le x < [x] + 1$.

V. 131. Solve the equations:

a) $[2x - 1] = 4$; b) $\left[\dfrac{x-2}{3}\right] = 5$; c) $\left[\dfrac{x-2}{4}\right] = 4 - x$.

V. 132. Solve the equations in the set of $\mathbf{R}$:

a) $\left[\dfrac{2x+1}{3}\right] = \dfrac{x-4}{2}$; b) $\left[\dfrac{x+1}{2}\right] = \dfrac{x-1}{3}$; c) $3[x-1] = 2x + 4$;

d) $4[x] - 5 = 3x$; e) $\left[3x - \dfrac{2}{5}\right] = 2x + 1$; f) $\left[2x - \dfrac{2}{3}\right] = 4x - 3$;

g) $\left[\dfrac{5x+2}{3}\right] = x - \dfrac{1}{2}$; h) $\left[5x - \dfrac{1}{4}\right] = 6x - 1$; i) $\left[\dfrac{x+1}{3}\right] = \dfrac{x-1}{2}$.

V. 133. Solve in $\mathbf{R}$ the equation $\left[\dfrac{4x+1}{4x-5}\right] = \dfrac{2x+5}{3}$.

V. 134. Solve the equations:

a) $\left[x + \dfrac{2}{3}\right] + \left[x + \dfrac{3}{5}\right] = 1$; b) $\left[x + \dfrac{1}{2}\right] + \left[x + \dfrac{1}{3}\right] = 1$.

V. 135. Prove that $[x] + \left[x + \dfrac{1}{2}\right] = [2x]$. Solve the equations:

a) $\left[\dfrac{2x+3}{3}\right] + \left[\dfrac{4x+9}{6}\right] = 5x + 6$;

b) $\left[\dfrac{3x+2}{12}\right] + \left[\dfrac{3x-4}{12}\right] + \left[\dfrac{3x-7}{6}\right] = \dfrac{3x+1}{2}$;

c) $\left[\dfrac{2x+3}{8}\right] + \left[\dfrac{2x+7}{8}\right] + \left[\dfrac{2x+5}{4}\right] + \left[\dfrac{2x+4}{2}\right] = \dfrac{x+3}{2}$.

V. 136. If n and k are positive integers such that $2^k \le n < 2^{k+1}$, prove that

$$\left[\dfrac{n+1}{2}\right] + \left[\dfrac{n+2}{2^2}\right] + \left[\dfrac{n+2^2}{2^3}\right] + \ldots + \left[\dfrac{n+2^k}{2^{k+1}}\right] = n.$$

V. 137. Calculate:

a) $\left[\sqrt{n^2+n+1}\right]$, $n \in \mathbf{N}^*$; b) $\left[\sqrt{9n^2+6n}\right]$, $n \in \mathbf{N}^*$.

V. 138. Find the value of the sum $S = \left[\displaystyle\sum_{k=2}^{2013} \frac{1}{k^2}\right]$.

V. 139. Solve the system $\begin{cases}[x]+3\{y\}=3.9 \\ \{x\}+3[y]=3.4\end{cases}$,

where $[x]$ and $\{x\}$ are the known notations .

V. 140. Find the real root(s) of the equation

$25^{[x]}+5^x = 6\cdot 5^{[x]}$, where $[x]$ is the integer part (floor) of x.

Chapter VI

COMPLEX NUMBERS

$z = a + bi$, where $a \in \mathbf{R}$, $b \in \mathbf{R}$, $i^2 = -1$ is the algebraic representation of a complex number.

The set of the complex number is $\mathbf{C} = \{a + bi \mid a \in \mathbf{R},\ b \in \mathbf{R},\ i^2 = -1\}$,

where a is the real part of the complex number z or $a = \operatorname{Re}(z)$,

$\qquad$ b is the imaginary part of the complex number z or $b = \operatorname{Im}(z)$.

$\bar{z} = a - bi$ is the *conjugate* of the complex number z.

$|z| = \sqrt{a^2 + b^2}$ is the *modulus* (*absolute value*) of the complex number z.

If $z = a + bi$ is a complex number, the point $M(x, y)$ is called the geometric imagine of z, therefore the distance from the origin to M is $|z|$.

Properties:

a) $z + \bar{z} = 2a$; b) $z \cdot \bar{z} = a^2 + b^2$; c) $\overline{\displaystyle\sum_{k=1}^{n} z_k} = \displaystyle\sum_{k=1}^{n} \bar{z}_k$; d) $\overline{\displaystyle\prod_{k=1}^{n} z_k} = \displaystyle\prod_{k=1}^{n} \bar{z}_k$;

e) $\overline{z^n} = (\bar{z})^n$; f) $\overline{\left(\dfrac{z_1}{z_2}\right)} = \dfrac{\bar{z}_1}{\bar{z}_2}$; g) $z \in \mathbf{R} \Leftrightarrow \bar{z} = z$.

VI. 1. Write the following complex numbers in standard algebraic form:

a) $z_1 = (3 + 3i) + (2 - i)$; b) $z_2 = (3 + i) - (2 + 4i)$; c) $z_3 = (3 + 2i)(1 + 4i)$;

d) $z_4 = (3 + i) \div (2 + i)$; e) $z_5 = (1 - i) \div (1 + 2i)$; f) $z_6 = \dfrac{1 + i}{3 - 4i} + \dfrac{4}{4 + 3i}$.

VI. 2. Compute:

a) $(1 + i)(1 - i)^2 (-1 - i)^3 (-1 + i)^4$; b) $(1 + i) + (1 - i)^2 + (-1 - i)^3 + (-1 + i)^4$;

c) $i + i^3 + i^5 + i^7$; d) $\dfrac{1}{i} + \dfrac{1}{i^3} + \dfrac{1}{i^5} + \dfrac{1}{i^7}$; e) $\dfrac{i + i^2 + i^3 + i^4}{1 + i}$;

f) $i^{2010} + i^{2011} + i^{2012} + i^{2013}$; g) $(1 + i)^{2013} - (1 - i)^{2013}$.

VI. 3. Calculate $\left(1-i\right)\left(1-i^2\right)\left(1-i^3\right)\cdot...\cdot\left(1-i^{2013}\right)$.

VI. 4. Let $z_1 = 4-2i$, $z_2 = 5+3i$ be two complex numbers,

calculate: **a)** $z_1 + z_2$; **b)** $z_1 - z_2$; **c)** $z_1 \cdot z_2$; **d)** $\dfrac{1}{z_1}$; **e)** $2\cdot z_1 + 4\cdot z_2$.

VI. 5. Is the number $1+i$ a root of the equation $z^4 + 4 = 0$?

VI. 6. Consider the complex numbers:

a) $z_1 = -4+3i$; **b)** $z_2 = 3-2i$; **c)** $z_3 = -2+i$; **d)** $z_4 = 1-2i$.
Compute $\operatorname{Re}(z)$, $\operatorname{Im}(z)$, and $|z|$ in each case.

VI. 7. If $z \in \mathbf{C}$ show that:

a) $z\cdot\bar{z} \in \mathbf{R}$, $z+\bar{z} \in \mathbf{R}$, $\dfrac{z}{\bar{z}}+\dfrac{\bar{z}}{z} \in \mathbf{R}$;

b) $z-\bar{z} \in \mathbf{C}-\mathbf{R}$, $\dfrac{z}{\bar{z}}-\dfrac{\bar{z}}{z} \in \mathbf{C}-\mathbf{R}$.

VI. 8. If $z_1 = 1+2i$, $z_2 = -3$, $z_3 = 5+5i$, and $z_4 = -8-6i$,

then calculate: **a)** $\dfrac{z_3 - \bar{z}_3}{z_1\bar{z}_1}$; **b)** $\dfrac{z_1 z_3}{z_2 z_4}$; **c)** $\dfrac{z_1\cdot z_2}{z_3 + z_4}$.

VI. 9. If $z = -2-4i$ then calculate: **a)** $\bar{z}$; **b)** $|z|$.

VI. 10. Write each complex expression in standard form:

a) $\dfrac{1+i}{2-5i}$; **b)** $\dfrac{-3+i}{2+3i}$; **c)** $\dfrac{-3+2i}{2+i}$; **d)** $\dfrac{(2+i)^3 - (2-i)^3}{(2+i)^3 + (2-i)^3}$;

e) $\dfrac{(1+i)^4 + (1-i)^4}{(1+2i)^4 + (1-2i)^4}$; **f)** $\left(-2(1+i)^3 + \dfrac{17+31i}{3+4i}\right)\dfrac{-1+i}{6i} - 1$;

g) $\dfrac{i^{13} - i^{10}}{1+i^1} + i^{14}$; **h)** $\dfrac{3}{(1+i)^4} + \dfrac{31-17i}{4-28i} - \dfrac{1}{i}$; **i)** $\dfrac{1+i}{3-i} + \dfrac{i}{3+i}$;

j) $\dfrac{(1-i)(1-2i)(1-3i)(1-4i)}{(1+i)(1+2i)(1+3i)(1+4i)}$.

VI. 11. Find the real numbers x and y such that:

a) $(1+3i)x + (2-i)y = 3+2i$; b) $(x+i)x + (1+3yi)y = 2+4i$;

c) $3\sqrt{x^2 - 2y} + (1-i)x^2 = 2(1+2i)y - 12i$; d) $4x+5+(y-2)i = -19+5i$;

e) $\overline{3x-5i} + 2x + y = \overline{3+i} + 2yi$; f) $\dfrac{x-3+(y-3)i}{x+2+(y+4)i} = i$;

g) $\dfrac{x-2+(y-1)i}{y-3} = -1+3i$; h) $\dfrac{2+5i}{x-y} + \dfrac{-1+3i}{x+y} = \dfrac{7x-12i}{x^2-y^2}$.

VI. 12. Find the real numbers a and b such that:

a) $\dfrac{11+i}{1-11i} = a+bi$; b) $\dfrac{5+2i}{2-5i} = a+bi$; c) $\dfrac{5-3i}{2+3i} = a+bi$;

d) $\dfrac{2-i}{2+i} + \dfrac{2+i}{2-i} = a+bi$; e) $\dfrac{(1+i)^3}{(1-i)^5} = a+bi$.

VI. 13. Calculate $|z|$ for the complex numbers:

a) $z = \dfrac{3+4i}{4-3i}$; b) $z = \dfrac{1-i}{8+6i}$; c) $z = \left(\sqrt{2}+i\right)^{20}$;

d) $z = \left(\sqrt{5+\sqrt{5}} + i\sqrt{5-\sqrt{5}}\right)^4$; e) $z = \left(\sqrt{3+\sqrt{2}} + i\sqrt{3-\sqrt{2}}\right)^4$;

f) $z = \left(\sqrt{1+\sqrt{2}} + i\sqrt{-1+\sqrt{2}}\right)^6$.

VI. 14. Show that the number $z = \left(1+i\sqrt{3}\right)^3$ is an integer number.

VI. 15. The complex number $z = 2+i$ is a root of the equation $z^2 - mz + 5 = 0$. Find m.

VI. 16. If $z_1 = 1+2i$ and $z_2 = 2-i$, show that $z_1\bar{z}_2 + z_2\bar{z}_1 \in \mathbf{R}$.

VI. 17. Find the complex number z such that $\dfrac{\bar{z}+5+7i}{z} = 6$.

VI. 18. If $z \cdot \bar{z} = 5$ and $(z-1)(\bar{z}-1) = 10$, find the number z.

VI. 19. Solve the equation $z^4 - 10z^2 + 41 = 0$ and compute the expressions

$$E_1 = \frac{z_1 + z_2 + z_3 + z_4}{z_1 z_2 z_3 z_4} \quad \text{and} \quad E_2 = \frac{z_1 z_2 z_3 + z_1 z_2 z_4 + z_2 z_3 z_4}{z_1 z_2 z_3 z_4}.$$

VI. 20. Solve in **C** the equations:

a) $9x^2 + 6x + 10 = 0$; b) $x^2 + 5x + 7 = 0$;

c) $2x^2 - 3x + 2 = 0$; d) $-x^2 + 3x - 8 = 0$.

VI. 21. Find the real numbers m and n such that each of the equations have the indicated root:

a) $x^2 + mx + n = 0$, $x_1 = 3 + i$; b) $x^2 + mx + n = 0$, $x_1 = 4 - i$;

c) $x^2 + mx + n = 0$, $x_1 = \dfrac{1 - 2i}{2 + i}$; d) $x^2 + mx + n = 0$, $x_1 = \dfrac{2 + i}{3 - 4i}$.

VI. 22. Prove that the complex numbers $1 + 3i$ and $1 - 3i$ are the solutions of the equation $x^3 - 4x^2 + 14x - 20 = 0$.

VI. 23. Prove that the complex numbers $-\dfrac{1}{2} + i\dfrac{\sqrt{3}}{2}$ and $-\dfrac{1}{2} - i\dfrac{\sqrt{3}}{2}$ are roots of the equation $x^4 + x^2 + 1 = 0$.

VI. 24. If x_1 and x_2 are the roots of the equation $x^2 + x + 1 = 0$, then calculate:

a) $\left(x_1^3 + x_1^2 - 1\right)^{2002} + \left(x_2^3 + x_2^2 - 1\right)^{2002}$;

b) $\left(x_1^4 + x_1^3\right)^{2002} + \left(x_2^4 + x_2^3\right)^{2002}$;

c) $\left(x_1^4 + x_1^3 + x_1^2 + 1\right)^{2013} + \left(x_2^4 + x_2^3 + x_2^2 + 1\right)^{2013}$.

VI. 25. If ε is one of the roots of the unity, $x^3 - 1 = 0$ and $\varepsilon \neq 1$, calculate the following complex numbers:

a) $z_1 = \varepsilon^4 + \varepsilon^2 + 1$; b) $z_2 = \dfrac{1 + \varepsilon}{(1 - \varepsilon)^2} + \dfrac{1 - \varepsilon}{(1 + \varepsilon)^2}$;

c) $z_3 = \dfrac{\varepsilon^3 - 1}{1 - 2\varepsilon^2}$; d) $z_4 = (a + b + c)(a + b\varepsilon + c\varepsilon^2)(a + b\varepsilon^2 + c\varepsilon)$;

e) $z_5 = (1 + \varepsilon)(1 + \varepsilon^2)(1 + \varepsilon^3)(1 + \varepsilon^4)(1 + \varepsilon^5)(1 + \varepsilon^6)$;

f) $z_6 = (1 - \varepsilon)(1 - \varepsilon^2)(1 - \varepsilon^4)(1 - \varepsilon^5)$.

VI. 26. Prove the identities:

a) $\left(\dfrac{-\sqrt{3}+i}{2}\right)^5 - \left(\dfrac{\sqrt{3}+i}{2}\right)^5 = \sqrt{3}$; b) $\left(\dfrac{-1+i\sqrt{3}}{2}\right)^5 + \left(\dfrac{-1-i\sqrt{3}}{2}\right)^5 = -1$;

c) $\left(\dfrac{-1+i\sqrt{3}}{2}\right)^6 + \left(\dfrac{-1-i\sqrt{3}}{2}\right)^6 = 2$; d) $\left(\dfrac{\sqrt{3}-i}{2}\right)^6 + \left(\dfrac{\sqrt{3}+i}{2}\right)^6 = -2$.

VI. 27. Calculate $z^{13} + \dfrac{1}{z^{13}}$ if z is a root of the equation $z + \dfrac{1}{z} = 1$.

VI. 28. Calculate the complex number:

$$z = \left(\dfrac{1+i}{1-i}\right)^{2010} + \left(\dfrac{1+i}{1-i}\right)^{2011} + \left(\dfrac{1+i}{1-i}\right)^{2012} + \left(\dfrac{1+i}{1-i}\right)^{2013}.$$

VI. 29. If x_1 and x_2 are the roots of the equation $x^2 + x + 1 = 0$ then calculate the value of the expression

$$E = \dfrac{(x_1 + 1)^{2014}}{x_1^2 + 1} + \dfrac{(x_2 + 1)^{2014}}{x_2^2 + 1}.$$

VI. 30. Let z_1, z_2 be two complex numbers such that $|z_1| = |z_2| = 1$.

Show that $\dfrac{z_1 + z_2}{1 + z_1 z_2}$ is a real number.

VI. 31. Let z be a complex number such that $|z| = 1$. Show that $\dfrac{1 + z^{2n}}{z^n}$ is a real number.

VI. 32. Let $z \in \mathbf{C}$ be a complex number such that $z \neq \bar{z}$ and

$\dfrac{1 - z + z^2}{1 + z + z^2} \in \mathbf{R}$. Prove that $|z| = 1$.

VI. 33. Let $z_1 = x_1 + iy_1$ and $z_2 = x_2 + iy_2$ be two complex numbers

such that $\bar{z}_1 = \dfrac{1 - z_2}{1 + z_2}$. Then:

149

a) $y_1 y_2 > 0$;

b) If $z_1 + z_2 \in \mathbf{R}$ find a relation between y_1 and y_2.

VI. 34. Show that the number $z = (3+2i)^{4n} + (2+3i)^{4n}$ is a real number for any n integer positive number.

Prove that in general $z = (a+b\ )^{4n} + (b+\bar{a}\ )^{4n} \in \mathbf{R}$, where $a, b \in \mathbf{R}$.

VI. 35. Show that the number $z = (1+i)^m (1-i)^n$, $m, n \in \mathbf{R}$ is a real number if and only if $m-n$ is multiple of 4.

VI. 36. Let n be an integer positive number and

$z_n = (1+i)^n + (1-i)^n$. Prove that for any n:

a) z_n is a real number;

b) $z_{n+2} = 2 z_{n+1} - 2 z_n$;

c) $z_n = 2\left(\sqrt{2}\right)^n \cos \dfrac{n\pi}{4}$.

VI. 37. Represent the geometric image of the following complex numbers:

a) $z_1 = -2 + 3i$; b) $z_2 = 3 - 2i$; c) $z_3 = -2$; d) $z_4 = 2i$;

e) $z_5 = (2 - 2i)(3 - i)$; f) $z_6 = \dfrac{3 + 2i}{1 - i}$.

VI. 38. The geometric image of the complex numbers

$z_1 = 1 + 3i$ and $z_2 = 3 - i$ in the complex plane is represented by the points A and B, respectively. Prove that $\triangle OAB$ is a right triangle.

VI. 39. Show that the geometric images of the complex numbers

$z_1 = -1 - 2i$, $z_2 = 1 + 4i$, and $z_3 = 2 + 7i$ are three collinear points.

VI. 40. Let $z_1 = 1 + i$, $z_2 = 4 - 2i$, and $z_3 = -2 + 4i$ be three complex numbers. Show that the geometric imagine in the complex plane is an isosceles triangle.

VI. 41. The points A, B, C, D are the geometric image in the complex plane of the numbers $z_1 = 1 - 2i$, $z_2 = 1 + 2i$, $z_3 = -2 + i$, and $z_4 = -2 - i$. Show that they lie on a circle and find the equation of this circle, moreover A, B, C, D is an isosceles trapezoid.

VI. 42. Solve for the complex variable z and express z in the form $a + bi$:

a) $|z| = |z + 1 - i| = |\bar{z} + 2 - 4i|$; b) $|z + i| = |z + 1 - i| = |z - 3 - 2i|$;

c) $\left|\dfrac{z + 2i}{z + 4i}\right| = 1$ and $\left|\dfrac{z + 2i}{z - 1}\right| = \dfrac{1}{\sqrt{2}}$; d) $\left|\dfrac{z + 1}{z}\right| = 2$ and $\left|\dfrac{iz + 1}{1 + i}\right| = 1$;

e) $\left|\dfrac{z - 1}{z - i}\right| = \dfrac{1}{2}$ and $\left|\dfrac{z + 1}{z}\right| = \sqrt{2}$.

VI. 43. Find the set of points in the complex plane such that:

a) $|z - i| = 2$; b) $|z - 1| + |z + 1| = 4$; c) $|z - 3| = |z + 3|$; d) $|z + i| < 2$;

e) $|z - 2| - |z + 2| < 2$; f) $1 < |2z + 3 - 2i| \leq 2$; g) $|i - z| > 4$;

h) $|z + i - 2| \leq 2$.

VI. 44. Solve the equations:

a) $x^3 - 1 = 0$; b) $x^3 - 8 = 0$; c) $27x^3 + 1 = 0$; d) $64x^3 - 27 = 0$;

e) $x^4 - 16 = 0$; f) $x^4 - 81 = 0$; g) $x^6 - x^3 - 2 = 0$; h) $x^6 + 3x^3 + 2 = 0$.

VI. 45. Let $z \in \mathbf{C}$ be a complex number. Show that:

a) $\left| iz \right| + \left| \left(1 - i\sqrt{3}\right) z \right| = 3|z|$; b) if $|z| < 1$ then $|2 - i\bar{z}| < 3$;

c) if $|z| < \dfrac{1}{2}$ then $\left| (1 + 2i)z + 2i|z| \right| < 3$;

d) if $|z| < \dfrac{1}{3}$ then $\left| \left(\sqrt{2} - i\right) z^3 - iz \right| < \dfrac{3}{4}$;

e) if $|z| < \dfrac{1}{2}$ then $\left| (1 + i)z^3 + i|z| \right| < 3$;

f) if $\left| z^3 + \dfrac{1}{z^3} \right| \leq 2$ then $\left| z + \dfrac{1}{z} \right| \leq 2$.

VI. 46. If $\varepsilon \neq 1$ is one of the roots of the equation $x^3 - 1 = 0$, prove that

a) $(1 + \varepsilon)^{2013} + \left(1 + \varepsilon^2\right)^{2013} = -2$;

151

b) $(1+\varepsilon)^n + (1+\varepsilon^2)^n + (\varepsilon+\varepsilon^2)^n = \begin{cases} -3 \ \text{if} \ n = 3k \\ 0 \ \text{if} \ n \neq 3k \end{cases}, \ k \in \mathbf{N};$

c) $|z-1|^2 + |z-\varepsilon|^2 + |z-\varepsilon^2|^2 = 3(1+|z|^2), \ \forall z \in \mathbf{C}.$

VI. 47. Prove that the number

$z = \left(\sqrt[4]{a} + i \cdot \sqrt[4]{b}\right)\left(\sqrt[4]{a} + i^2 \cdot \sqrt[4]{b}\right)\ldots\left(\sqrt[4]{a} + i^{4k} \cdot \sqrt[4]{b}\right)$ is real, where a, b are positive real numbers.

VI. 48. Find $a \in \mathbf{R}$ such that:

a) $\left|\dfrac{2+4i}{1+\dot{a}}\right| = 2$; **b)** $\dfrac{a+2i}{3+\dot{a}} \in \mathbf{R}$.

VI. 49. Find the sets of the points in the complex plane such that:

a) $A = \left\{(x,y) \in \mathbf{R} \times \mathbf{R} \mid \text{Re}\left(\dfrac{z-2}{z-6}\right) = \text{Im}\left(\dfrac{z-2}{z-6}\right), \ z = x + iy\right\};$

b) $B = \left\{(x,y) \in \mathbf{R} \times \mathbf{R} \mid \text{Re}\left(\dfrac{z-2}{z-1}\right) = 0, \ z = x + iy\right\};$

c) $C = \left\{(x,y) \in \mathbf{R} \times \mathbf{R} \mid \dfrac{i+\overline{z}}{i-\overline{z}} \in \mathbf{R}, \ z = x + iy\right\};$

d) $D = \left\{(x,y) \in \mathbf{R} \times \mathbf{R} \mid \left|\dfrac{z^2-i}{z^2-2i}\right| = 1, \ z = x+iy, \ z^2 \neq 2i\right\};$

e) $E = \left\{(x,y) \in \mathbf{R} \times \mathbf{R} \mid \text{Im}\left(\dfrac{z+1+i}{2-iz}\right) = 0, \ z = x + iy\right\}.$

VI. 50. Find the set of the points in the complex plane such that

$\log_{\frac{1}{2}} \dfrac{|z-3|+4}{3|z-3|-2} < 1.$

VI. 51. Determine m $\mathbf{R}$ such that the equation

$(m+2)x^2 + (m-1)x + m = 0$ doesn't have real roots.

VI. 52. Let x_1 and x_2 be the roots of the equation $2x^2 - 2mx + m = 0$.
a) Calculate, in terms of the parameter $m \in \mathbf{R}$, the expressions

$x_1^2 + x_2^2, \ x_1^3 + x_2^3$, and $\dfrac{1}{x_1} + \dfrac{1}{x_2};$

b) Find m such that the equation does not have real roots;

c) Find m when $|x_1| + |x_2| = 4$.

VI. 53. If $z = 1 + 2i$ is a root of the equation $z^2 - (1 - 2i)z + 2a - i = 0$ find a **C**.

VI. 54. If $z = 1 + i$ is a root of the equation $z^3 + z^2 + (a - 1)z + b = 0$ find a, b **R**.

VI. 55. Factor each polynomial over the set of complex numbers:

a) $x^4 - x$; **b)** $x^3 + 8$; **c)** $x^4 + x^2$; **d)** $x^5 + 27x^2$.

VI. 56. Find $z \in \mathbf{C}$ such that:

a) $z + 4\bar{z} = 20 + 9i$; **b)** $|z + i| - z + 1 = 4\sqrt{2} + 2i$; **c)** $|z| + z = 2 - 4i$;

d) $z^2 + 2z + 2\bar{z} + 4 = 0$; **e)** $|z| + |z + i| = 2|\bar{z}|$; **f)** $|z| - iz = 4 - 2i$;

g) $|z| + i(4z + \bar{z} + 10) = 0$; **h)** $\bar{z}^2 - z^2 - |z| = -5 + 48i$;

i) $|(1 + 3i)z| - 2(1 - 2i)z = -5i$.

VI. 57. Find the complex numbers z such that:

a) $z^2 - \bar{z} = 0$; **b)** $z^2 - i\bar{z} = 0$; **c)** $z^2 + |z| = 0$;

d) $z^2 + 3|z| = 0$; **e)** $z^2 + z|z| + |z|^2 = 0$.

VI. 58. Let $z_1 = \dfrac{1 - i\sqrt{3}}{a + (a + 1)i}$ and $z_2 = \dfrac{b + i}{b - 1 - 2i}$ be two complex numbers, where $a \in \mathbf{R}$, $b \in \mathbf{C}$.

a) Find a such that $z_1 \in \mathbf{R}$.

b) Find the points in the complex plane such that z_2 is a real number.

VI. 59. Find $z = x + iy$, $x, y \in \mathbf{R}$ where:

a) $z^2 = -3 - 4i$; **b)** $z^2 = 1 - 4\sqrt{3}i$; **c)** $z^2 = 3 + 4i$;

d) $z^2 = 7 + 24i$.

VI. 60. Solve for the complex variable z:

a) $z^2 - 4iz - 3 = 0$; b) $z^2 - (1-i)z + 2 + i = 0$;

c) $(2-i)z^2 - (4-12i)z + 14 - 12i = 0$; d) $z^2 - 5iz - 7 + i = 0$;

e) $z^2 + 4iz + 5 = 0$; f) $z^2 - (5-2i)z + 6 - 4i = 0$;

g) $iz^2 + (1-2i)z - 1 + 3i = 0$; h) $z^2 - (5-2i)z + 5 - 5i = 0$;

i) $z^2 - (2+i)z - 1 + 7i = 0$; j) $z^2 - (5+2i)z + 9 + 7i = 0$;

k) $z^2 - (8+3i)z + 13 + 13i = 0$.

VI. 61. Find $z = x + iy$, x, y **R** from the equations:

a) $(1+i)z + (5-3i)\bar{z} = 20 - 4i$; b) $(2+2i)z - 3\,\mathrm{Re}(z) = -18 + 20i$;

c) $(1+2i)(z+i) + (3-4i)(1+zi) = 1 + 7i$;

d) $(-2+i)(i-z) + (3-4i)(i+z) = 1 + 7i$;

e) $(1+3i)z - 11i = (3-2i)z + 13$; f) $2z + i \cdot \bar{z} = 9 + 3i$.

VI. 62. Solve for the complex variable z and express z in the form $z = a + bi$, $a, b \in$ **R** :

a) $iz^4 + (4+2i)z^2 + (1-i)^6 = 0$; b) $z^4 - 3(1-2i)z^2 - 8 - 6i = 0$;

c) $z^4 - 2z^3 + 4z^2 - 2z + 3 = 0$; d) $z^4 - (5-2i)z^2 + 8 + 10i = 0$;

e) $\left(\dfrac{5z-2i}{2z-5i}\right)^3 + \left(\dfrac{5z-2i}{2z-5i}\right)^2 + \dfrac{5z-2i}{2z-5i} + 1 = 0$.

VI. 63. Find $z = a + bi$, $a, b \in$ **R** such that:

a) $z^3 = -11 + 2i$; b) $z^3 = -9 - 46i$; c) $z^3 = -8i$;

d) $z^3 = -6 + 6i\sqrt{3}$; e) $z^4 = -8 - 8i\sqrt{3}$; f) $z^6 = -64$.

VI. 64. Simplify the fractions:

a) $\dfrac{z^2 - 8(1-i)z + 63 - 16i}{z^2 - 6z + 25}$; b) $\dfrac{z^2 - iz - 1 - i}{z^2 - 2z + 2}$;

c) $\dfrac{z^3 - (3+i)z^2 + (2+3i)z - 2i}{z^3 - (3-i)z^2 + (2-3i)z - 2i}$.

VI. 65. Solve the system of equations, where $x, y \in \mathbf{C}$:

a) $\begin{cases} (1+3i)x - (1-2i)y + 1 - i = 0 \\ 2(2+i)x - (2+i)y - (7-4i) = 0 \end{cases}$; b) $\begin{cases} (2+3i)x + (2-3i)y = 4 - 2i \\ (1+i)x + (1-i)y = 2 \end{cases}$;

c) $\begin{cases} 4x + (3-2i)y = 5 - i \\ (3+2i)x + 4y = 5 + i \end{cases}$; d) $\begin{cases} (2+i)x + (2-i)y = 6 + 3i \\ (1+i)x + (1-i)y = 3 + i \end{cases}$.

VI. 66. Solve the systems of equations, where $z_1, z_2 \in \mathbf{C}$:

a) $\begin{cases} z_1 + 2z_2 = 1 + i \\ 3iz_1 - z_2 = 3 + 2i \end{cases}$; b) $\begin{cases} 4z_1 + 5iz_2 = 14 + 4i \\ 3z_1 + 2iz_2 = 7 + 3i \end{cases}$; c) $\begin{cases} 2z_1 - 2z_2 = 9 - 6i \\ iz_1 - 3z_2 = 4 - 3i \end{cases}$;

d) $\begin{cases} z_1 + 2iz_2 = 4 + 9i \\ z_1 + z_2 = 5 + 2i \end{cases}$; e) $\begin{cases} iz_1 - 5z_2 = -6 - 15i \\ 2z_1 - 3iz_2 = 9 - i \end{cases}$.

VI. 67. Solve in $\mathbf{C}$ the systems of equations:

a) $\begin{cases} x + y = 3 \\ xy = 4 \end{cases}$; b) $\begin{cases} x + y = 2i \\ x^2 + y^2 = -1 \end{cases}$; c) $\begin{cases} x^2 - y^2 = -3 \\ 2(x^2 + y^2) + xy = -4 \end{cases}$;

d) $\begin{cases} 2x + y = 2 \\ x^2 + y^2 + xy + 3x = 0 \end{cases}$.

VI. 68. Consider the function

$f : \mathbf{C} \to \mathbf{C}$, $f(z) = z^2 + 3z - \bar{z}$.

a) Compute: $f(-1)$, $f(i)$, $f(2-i)$, $f(\bar{z})$, $f(|z|)$;

b) Verify the eguality $f(z) - f(\bar{z}) = 4(\bar{z} - z)$.

VI. 69. Consider the function $f : \mathbf{C} \setminus \{1\}$ $\mathbf{C}$, $f(z) = \dfrac{z^2 - z - 2}{z - 1}$. Find the points in the plane such that $f(z)$ **R**.

VI. 70. Consider the complex function f having the property

$f(x) + f(\varepsilon x) = 4 - \varepsilon x$, $\forall x \in \mathbf{C}$, where ε is a complex root of the equation $x^3 = 1$. Prove that $f(x) + f(\varepsilon^2 x) = 4 - x$, $\forall x \in \mathbf{C}$ and find $f(x)$.

VI. 71. Consider the complex functions $f, g : \mathbf{C} \to \mathbf{C}$ having the properties $f(x)f(ix) = x^2$ and $g(x) + g(\varepsilon x) = x$, $\forall x \in \mathbf{C}$, where ε is a complex root of the equation $x^3 = 1$. Prove that $f(-x) = -f(x)$, $\forall x \in \mathbf{C}$ and find $g(x)$.

VI. 72. Let ε be one of the roots of the equation $z^n - 1 = 0$, $n \in \mathbf{N}$, $n \geq 2$ and the function $f : \mathbf{R} \to \mathbf{R}$, such that

(1) $f(x+a) + \varepsilon \cdot f(x-a) = b$, $\forall x \in \mathbf{R}$, where $a \in \mathbf{R}^*$, $b \in \mathbf{C}$. Show that f is a periodic function.

VI. 73. If α is one of the roots of the equation $z^n - 1 = 0$, $n \in \mathbf{N}$ and $n \geq 2$, then calculate $E = \alpha + 2\alpha^2 + 3\alpha^3 + ... + n\alpha^n$;

VI. 74. Let $z = \cos t + i \sin t$ be a complex number. Prove that

$$\cos nt = \frac{z^{2n} + 1}{2z^n} \text{ and } \sin nt = \frac{z^{2n} - 1}{2iz^n}, \ n \in N^\bullet.$$

Calculate the product $A = \sin 10° \sin 50° \sin 70°$.

VI. 75. Prove that if $z_1, z_2, ..., z_n \in \mathbf{C} \setminus \{0\}$ and $|z_1| = |z_2| = ... = |z_n|$ then the

expression $\left(1 + \frac{z_2}{z_1}\right) \cdot \left(1 + \frac{z_3}{z_2}\right) \cdot ... \cdot \left(1 + \frac{z_n}{z_{n-1}}\right) \cdot \left(1 + \frac{z_1}{z_n}\right)$ is a real number.

VI. 76. Solve the equation $(2 - i)z^3 - 3 - i = 0$.

VI. 77. Solve the equation $\left(\dfrac{1 + iz}{1 - iz}\right)^n = \dfrac{1 + ia}{1 - ia}$, $n \in \mathbf{N}$, $a \in \mathbf{R}$ and show that all of its roots are real.

VI. 78. Solve the equation

$\left(\dfrac{x - i}{x + i}\right)^{n-1} + \left(\dfrac{x - i}{x + i}\right)^{n-2} + ... + \left(\dfrac{x - i}{x + i}\right)^1 + 1 = 0$, $n \in \mathbf{N}$.

VI. 79. Solve the equation

$\left(z + i\sqrt{1 - z^2}\right)^n + \left(z - i\sqrt{1 - z^2}\right)^n = 0$, where $n \in \mathbf{N}$.

VI. 80. Consider the complex numbers z_1, z_2, z_3 such that $z_1 \neq z_2 \neq z_3$, then the geometric imagines of the numbers $A(z_1)$, $B(z_2)$, $C(z_3)$ are collinear if and only if $\dfrac{z_3 - z_1}{z_2 - z_1} \in \mathbf{R}^*$.

VI. 81. Consider the function $f : \mathbf{C} \to \mathbf{C}, f(z) = az + b$, where $a, b \in \mathbf{R}, a \neq 2$ and the complex numbers z_1, z_2, z_3. Show that if the geometric images of the numbers z_1, z_2, z_3 are collinear, then the geometric imagines of the numbers $f(z_1)$ $f(z_2)$ and $f(z_3)$ are also collinear.

VI. 82. Consider the function $f : \mathbf{C} \to \mathbf{C}$, $f(z) = 3z + |z|$. Prove that f is a bijection and find its inverse.

Chapter VII

EXPONENTIAL AND LOGARITHMIC FUNCTIONS

$f : \mathbf{R} \to (0, \infty)$, $f(x) = a^x$, $a > 0$ and $a \neq 1$.

f decreases for $a \in (0, 1)$ and f increases for $a \in (1, \infty)$,

$a^0 = 1$, $a^m \cdot a^n = a^{m+n}$,

$\left(a^x\right)^y = a^{xy}$,

$\dfrac{a^m}{a^n} = a^{m-n}$

$\left(\dfrac{a}{b}\right)^m = \dfrac{a^m}{b^m}$

$a^{-m} = \dfrac{1}{a^m}$ $m, n \in \mathbf{R}$.

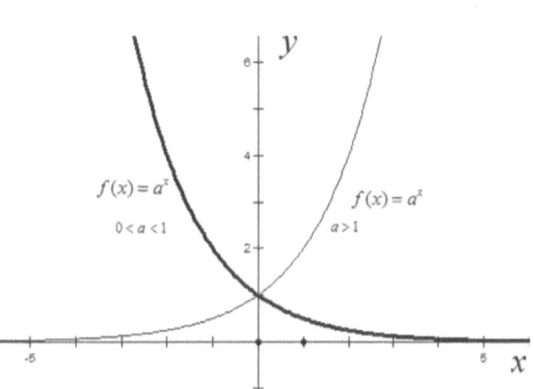

LOGARITHMIC FUNCTION

If $a^x = y$, the unique solution of this equation is $x = \log_a y$, therefore
$a^{\log_a y} = y$.

$f : (0, \infty) \to \mathbf{R}$, $f(x) = \log_a x$, $a > 0$ and $a \neq 1$,

f decreases for $a \in (0, 1)$ and f increases for $a \in (1, \infty)$,

$f^{-1} : \mathbf{R} \to (0, \infty)$, $f^{-1}(x) = a^x$.

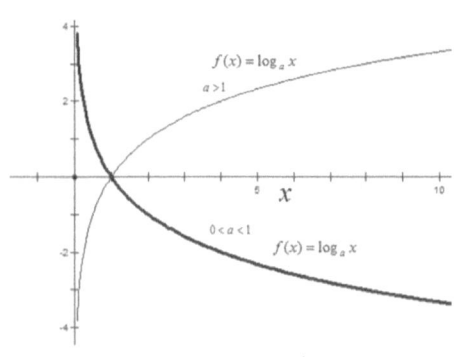

Laws of the Logarithmic function:

a) $\log_a 1 = 0$, $\log_a a = 1$,

b) $\log_a xy = \log_a x + \log_a y$,

c) $\log_a \dfrac{x}{y} = \log_a x - \log_a y$,

d) $\log_a x^p = p \log_a x$, $p \in \mathbf{R}$

and $\log_a \sqrt[n]{x} = \dfrac{1}{n} \log_a x$, $n \in \mathbf{N}^*$,

$\log_a a_1 \cdot a_2 \cdot ... \cdot a_n = \log_a a_1 + \log_a a_2 + ... + \log_a a_n$, $a_1, a_2, ..., a_n > 0$,

$$\log_a \prod_{k=1}^{n} a_k = \sum_{k=1}^{n} \log_a a_k$$

e) $\log_A B = \dfrac{\log_a B}{\log_a A}$, $A, B > 0$, $A \neq 1$.

Also $\log_{10} x = \log x$, $\log_e x = \ln x$, where e = 2.718281828459045... which is

the value approached by (1 + 1/n)n as n becomes bigger and bigger.

VII. 1. Specify the domains of the functions:

a) $f_1(x) = \sqrt{(0.5)^x - 16}$; b) $f_2(x) = \sqrt{(0.2)^x - 25}$;

c) $f_3(x) = \sqrt[4]{5^x - 0.04}$; d) $f_4(x) = \sqrt[3]{3^{-3x} - \sqrt{3}}$;

e) $f_5(x) = \sqrt{a^x - a^{-x}}$; $a \in \mathbf{R}$; f) $f_6(x) = \sqrt[5]{4^x - 0.05}$.

VII. 2. Find the domains and ranges of the functions:

a) $f(x) = 2^{\sqrt{1-x^2}}$; b) $f_1(x) = 2^{-\sqrt{x^2-1}}$; c) $f_2(x) = \sqrt{3^x - 27}$;

d) $f_3(x) = 3^{-|x|}$; e) $f_4(x) = 4^x - 2^{x+3} + 17$.

VII. 3. Draw the graph of the following exponential functions:

a) $f : \mathbf{R} \to (0, \infty)$, $f(x) = 3^x$;

b) $f : \mathbf{R} \to [2, \infty)$, $f(x) = 2^x + 2^{-x}$;

c) $f : \mathbf{R} \to \mathbf{R}$, $f(x) = \dfrac{3^x - 3^{-x}}{2}$;

d) $f : \mathbf{R} \to (-1.5, \infty)$, $f(x) = 2 \cdot 3^{x-3} - 1.5$;

e) $f : \mathbf{R} \to (-\infty, 2)$, $f(x) = -3 \cdot 2^x + 2$;

f) $f : \mathbf{R} \to (-\infty, -3)$, $f(x) = -2 \cdot 3^{-x-3} - 4$.

VII. 4. Find the domain of each function.

a) $f_1(x) = \log_{10}(2 + 3x)$;   b) $f_2(x) = \log_2(1 - 3x)$;

c) $f_3(x) = \log_3(x^2 - 4)$;   d) $f_4(x) = \ln(x - x^2)$;

e) $f_5(x) = \ln x + \ln(3 - x)$;   f) $f_6(x) = \sqrt{x - 4} + 3\ln(10 - x)$.

VII. 5. Compare the domains of the functions

a) $f(x) = \ln x^2$; b) $g(x) = 2\ln x$.

VII. 6. Find the domain of the functions:

a) $f(x) = \log|x + 1|$; b) $f_1(x) = \log_{x+1}(9 - x^2)$; c) $f_2(x) = \log_2 \sqrt[3]{x^2 - 5x - 6}$;

d) $f_3(x) = \log_{x+1} \log_3(10 - x^2)$; e) $f_4(x) = \log_3\left(\log_2\left(\dfrac{6 + \sqrt[4]{x}}{4} \right) - \dfrac{\log_2 x}{4} \right)$;

f) $f_5(x) = \sqrt{\log_{\frac{1}{2}} \dfrac{x - 1}{x + 5}}$; g) $f_6(x) = \log_{\frac{1}{2}}\left(6^{|x|} - 1 \right)$; h) $f_7(x) = \sqrt{\log_{\frac{1}{2}} \dfrac{x^2 - 1}{\sqrt{x^2 + 3x}}}$.

VII. 7. Graph $y = 2^x$ and $y = -2^{x-3} + 4$ on the same system of axes.

VII. 8. Graph $y = -3 \cdot \left(\dfrac{1}{2} \right)^{x-3} + 2$ and state the basic (parent) function. Describe each transformation applied and what the shape, domain, range, asymptotes, and y-intercept are.

VII. 9. For the function represented by the equation $y = 2^x$, state the equation of the graph that has been reflected in the x-axis, given a vertical stretch by a factor of 2, translated 3 units left and 6 units down.

VII. 10. Graph $y = \log_{\frac{1}{3}} x$ and $y = 2\log_{\frac{1}{3}}(x - 1) - 3$ on the same system of axis.

VII. 11. Graph $y = 8\log_{\frac{1}{2}} \sqrt[4]{x + 2} - 1$, state the basic function and find a simpler form of the equation. Describe each transformation applied, as well as the shape, domain, range, asymptotes, and y-intercept.

VII. 12. For the function represented by the equation $y = \log_2 x$ state the equation of the graph that has been reflected in the x-axis and y-axis, a vertically stretched by a factor of 2, translated 3 units to the left and 5 units up.

160

VII. 13. Draw the graph of the following logarithmic functions and on the same graph, as well as the parent function specifying the transformations applied:

a) $f:(1,\infty)\to \mathbf{R}$, $f(x)=\dfrac{1}{3}\log_2(x-1)-4$;

b) $f:\left(\dfrac{1}{3},\infty\right)\to(-2,\infty)$, $f(x)=\left|\log_2(3x-1)\right|-2$;

c) $f:\mathbf{R}\setminus\{0\}\to \mathbf{R}$, $f(x)=\log_2|x|+1$;

d) $f:(-\infty,5)\to \mathbf{R}$, $f(x)=-3\log_3\left(1-\dfrac{x}{5}\right)-4$;

e) $f:(-2,\infty)\to \mathbf{R}$, $f(x)=\log_3\sqrt{x+2}-3$;

f) $f:(2,\infty)\to \mathbf{R}$, $f(x)=-5\log_2\dfrac{x-2}{3}-4$.

VII. 14. Use a calculator to find the value of each of the following, correct to four decimal places.

a) $\log_5 29$; b) $\log_7 125$; c) $\log_6 5.24$; d) $\log_4 4.17$.

VII. 15. Evaluate without using a calculator:

a) $e^{\ln 3}$; b) $\ln e^4$; c) $2\ln e$; d) $e^{5\ln 2}$; e) $\ln\sqrt{e}$; f) $\ln 2+2\ln 3-\ln 18$;

g) $\log_3\sqrt[4]{\dfrac{1}{27}}$; h) $\log_{0.5}\sqrt[5]{\dfrac{1}{16}}$; i) $\log_{\frac{1}{3}}\dfrac{\sqrt[3]{81}}{27}$.

VII. 16. Evaluate without using a calculator:

a) $\log_{0.10}\dfrac{\sqrt[4]{1000}}{10}$; b) $\log_2\sqrt[3]{64^5}$; c) $\log_{0.2}\sqrt[3]{625^{-1}}$; d) $\log_3\dfrac{1}{27}+\log_2\sqrt[5]{32}$;

e) $\log_2\log_2\sqrt{2\sqrt{2\sqrt{2}}}$; f) $\log_2\left(\sqrt{7}+\sqrt{3}\right)+\log_2\left(\sqrt{7}-\sqrt{3}\right)$.

VII. 17. Express as a single logarithm.

a) $\dfrac{2}{3}\mathrm{h}\ \sqrt{x}-\dfrac{1}{2}\ln(x-1)+\dfrac{1}{3}\ln(x+1)$;

b) $2\ln x-\dfrac{1}{2}\ln(x^2-1)+3\ln(x^2+1)$.

VII. 18. Compute:

a) $\log_5 625$; b) $\log_3\dfrac{1}{2187}$; c) $\log_{\frac{1}{2}}\left[\log_{\sqrt{6}}\left(\log_3 729\right)\right]$;

d) $\log_2\left[\log_3\left(\log_{\frac{1}{2}}\dfrac{1}{512}\right)\right]$; e) $\log_2\log_2\sqrt{\sqrt[4]{2}}$; f) $\log_3\dfrac{81}{\sqrt{3}}-2\cdot\log_3 27$;

g) $\dfrac{\log 20+\log 3-\log 6}{\log_{125}4+\log_{125}10-\log_{125}5}$; h) $\dfrac{\log_3 6+\log_3 2-\log_3 4}{2\log_3 4-\log_3 2-\log_3 32}$;

i) $\log_{\frac{49}{4}}\left(\log_4 6^{\log_6 128}\right)$; j) $\log_{\frac{2}{3}}\left(\log_5 3^{\log_3\sqrt[3]{25}}\right)$.

VII. 19. Simplify the expressions:

a) $\dfrac{\log_6 x^3+\log_6 x^4}{\log_5 x^4+\log_5 x^3}$; b) $\left(\log_a b+\log_b a+2\right)\left(\log_a b-\log_{ab} b\right)\log_b a-1$;

c) $\dfrac{\log_4\sqrt{a}+\log_4\sqrt{a^3}+\log_4\sqrt{a^5}}{\log_4\sqrt[3]{a^2}+\log_4\sqrt[3]{a^4}+\log_4\sqrt[3]{a^6}}$; d) $\dfrac{1-\log_a^3 2}{\left(\log_a 2+\log_2 a+1\right)\log_a\dfrac{a}{2}}$;

e) $\dfrac{\log_a\sqrt[3]{a^2-1}\cdot\log_{a^{-2}}\sqrt{a^2-1}}{\log_{a^4\cdot}\left(a^2-1\right)\log_{\sqrt[6]{a}}\sqrt[4]{a^2-1}}$, $a>1$; $a\neq\sqrt{2}$;

f) $\left(\sqrt{\log_a\sqrt[4]{ab}+\log_b\sqrt[4]{ab}}+\sqrt{\log_a\sqrt[4]{\dfrac{b}{a}}+\log_b\sqrt[4]{\dfrac{a}{b}}}\right)\cdot\sqrt{\log_a b}$.

VII. 20. Simplify the expressions:

a) $\left(b^{\frac{\log_{100}a}{\log_{10}a}}\cdot a^{\frac{\log_{100}b}{\log_{10}b}}\right)^{2\log_{ab}(a+b)}$; b) $\left(x^{1+\frac{1}{2\log_4 x}}+8^{\frac{1}{3\log_x\sqrt{2}}}+1\right)^{\frac{1}{2}}$;

c) $\left(9^{\frac{2}{\log_5 9}}+3^{\frac{3}{\log_6 9}}\right)\left[\left(\sqrt{5}\right)^{\frac{2}{\log_{25}5}}-125^{\log_{25}6}\right]$;

d) $\dfrac{\left(3^{\frac{9}{\log_2 27}}+5^{\log_{25}49}\right)\left(2\cdot 81^{\frac{2}{\log_2 9}}-8^{\log_4 9}\right)}{\left(3+5^{\log_{25}16}\right)\cdot 5^{\log_5 3}}$;

e) $\dfrac{\log_a b\cdot\log_{ab}b^2}{b^{2\log_b(\log_a b)}-1}\cdot\dfrac{\log_a b+\log_a\left(b^{\frac{1}{3}\log_a b^3}\right)}{\log_a b^2-\log_{ab}b^2}$;

f) $\log_2 2x^2+\log_2 x\cdot x^{\log_x(\log_2 x+1)}+8\log_4^2 x+2^{3\log_2(\log_2 x)}$.

162

VII. 21. Show that each of the following statements is true.

a) $\dfrac{1}{\log_5 a} + \dfrac{1}{\log_3 a} = \dfrac{1}{\log_{15} a}$;   b) $\dfrac{1}{\log_8 a} - \dfrac{1}{\log_4 a} = \dfrac{1}{\log_2 a}$;

c) $\dfrac{k}{\log_6 a} = \dfrac{1}{\log_{6^k} a}$, $k \in \mathbf{R}$;

d) $\dfrac{1}{\log_6 a} + \dfrac{1}{\log_{10} a} + \dfrac{1}{\log_{15} a} = \dfrac{1}{\log_4 a} + \dfrac{1}{\log_9 a} + \dfrac{1}{\log_{25} a}$;

e) $\dfrac{1}{\log_a b} + \dfrac{1}{\log_{a^2} b} + \dfrac{1}{\log_{a^3} b} + \dfrac{1}{\log_{a^4} b} = 10 \log_b a$;

f) $\dfrac{1}{\log_{x^n y^m} a} = \dfrac{n}{\log_x a} + \dfrac{m}{\log_y a}$, $a, x, y > 1$, $m, n \in \mathbf{R} \setminus \{0\}$;

g) $\dfrac{1}{\displaystyle\sum_{k=1}^{n} \log_x a_k} = \log_{a_1 a_2 \dots a_n} x$.

VII. 22. Prove that:

a) $\log_3 2 \cdot \log_4 3 \cdot \log_5 4 \cdot \log_6 5 \cdot \log_7 6 \cdot \log_8 7 = \dfrac{1}{3}$;

b) $(\log_c a)^2 - (\log_c b)^2 = \log_c (ab) \log_c \left(\dfrac{a}{b}\right)$, where $a, b > 0$, $c \neq 1$, $c > 0$;

c) $\log_{ab} c = \dfrac{(\log_a c)(\log_b c)}{\log_a c + \log_b c}$;   d) $\dfrac{\log_a x}{\log_{ab} x} = 1 + \log_a b$;

e) $(\log_a b)(\log_b a) = 1$, where $a > 1$, $b > 1$;

f) $\log_a \dfrac{1}{x} = \log_{\frac{1}{a}} x$, where $a > 0$, $a \neq 1$, $x > 0$;

g) $\left(\dfrac{a^2}{bc}\right)^{\log \frac{b}{c}} \cdot \left(\dfrac{b^2}{ac}\right)^{\log \frac{c}{a}} \cdot \left(\dfrac{c^2}{ab}\right)^{\log \frac{a}{b}} = 1$.

VII. 23. If :
a) $a = \log_{12} 27$ calculate $\log_6 16$ in terms of a;
b) $a = \log_3 12$ calculate $\log_3 18$ in terms of a;
c) $a = \log_{30} 3$, $b = \log_{30} 5$, calculate $\log_{30} 16$ in terms of a and b;
d) $a = \log 2$, $b = \log 3$, calculate $\log_6 18$ in terms of a and b;
e) $a = \log_3 7$, $b = \log_7 5$, $c = \log_5 4$, calculate $\log_3 12$ in terms of a, b, and c.

VII. 24. If:

a) $\log_{10} 5 = a$, $\log_{10} 3 = b$, calculate $\log_{30} 8$ in terms of a and b;

b) $a = \log_{60} 3$, $b = \log_{60} 5$, calculate $\log_3 4 + \log_{12} 2$ in terms of a and b;

c) $a = \log_{14} 42$, $b = \log_{14} 36$, calculate $\log_{49} 8$ in terms of a and b;

d) $\log_7 b = \dfrac{1}{1 - \log_7 a}$, $\log_7 c = \dfrac{1}{1 - \log_7 b}$, calculate $\log_7 a$ in terms of $\log_7 c$.

VII. 25. If $\log_a\left(\dfrac{a}{3}\right) = \log_{\frac{b}{3}} b$, where $a \in (0, \infty) \backslash \{1\}$ and $b \in (0, \infty) \backslash \{3\}$, prove that $ab = 3$.

VII. 26. a) If $a^{\log_b (ab)} = b^{\log_c (bc)} = c^{\log_a (ac)}$, where $a, b, c \in (0, \infty) \backslash \{1\}$, prove that $a = b = c$;

b) If $a, b, c, A \in (0, \infty) \backslash \{1\}$, such that $abc = 1$, prove that

$$\left(\frac{1}{\log_a x}\right)^3 + \left(\frac{1}{\log_b x}\right)^3 + \left(\frac{1}{\log_c x}\right)^3 = \frac{3}{\log_a x + \log_b x + \log_c x}, \text{ for}$$

$x \in (0, \infty) \backslash \{1\}$;

b) If $a, b, c \in (0, \infty) \backslash \{1\}$ and x, y, z are nonzero real numbers such that

$(ab)^x = (bc)^y = (ac)^z = A^2$, prove that $\dfrac{1}{\log_a A} + \dfrac{1}{\log_b A} + \dfrac{1}{\log_c A} = \dfrac{1}{x} + \dfrac{1}{y} + \dfrac{1}{z}$.

VII. 27. Let a, b, and c be positive non zero real numbers.

a) If $2^a = 3$ and $2^b = 5$, calculate $\log_3 10$;

b) If $2^a = 35$, $5^b = 14$, and $7^c = 10$, $abc - 2 = a + b + c$;

c) If $x = \log_a bc$, $y = \log_b ac$, and $z = \log_c ab$, $xyz - 2 = x + y + z$.

VII. 28. Calculate:

a) $E = \left(a^{\frac{1}{\log_2 a}} \cdot a^{\frac{1}{\log_4 a}} \cdot a^{\frac{1}{\log_8 a}} \cdot \ldots \cdot a^{\frac{1}{\log_{512} a}} \right)^{\frac{1}{9}}$, $a > 0$, $a \neq 1$;

b) $F = 3^{\log_{3^{-1}} 2^{-1}} + \log_{\sqrt{2}} \dfrac{4}{\sqrt{3} + \sqrt{2}} + \log_{2^{-1}} \dfrac{1}{5 + 2\sqrt{6}}$.

VII. 29. Find a simple form for the expression

$$E = \left(\sqrt{3}\right)^{\frac{1}{\log_{\sqrt{2}}3}} + \left(\sqrt{3}\right)^{\frac{1}{\log_{\sqrt{3}}3}} + \left(\sqrt{3}\right)^{\frac{1}{\log_{\sqrt{4}}3}} + \ldots + \left(\sqrt{3}\right)^{\frac{1}{\log_{\sqrt{n}}3}}.$$

VII. 30. If $2\log_a(2x+3y) = \log_a(3x) + 2\log_a(2\sqrt{2y})$, calculate the ratio $\dfrac{x}{y}$.

VII. 31.

a) If $\log_a b = p^3$, and $\log_b a = \dfrac{4}{p^2}$, show that $p = \dfrac{1}{4}$;

b) If $a^2 + b^2 = 7ab$, where $a > 0$, $b > 0$, show that

$$\log\left(\frac{a+b}{3}\right) = \frac{1}{2}(\log a + \log b);$$

c) If $a^3 - b^3 = 3a^2 b + 5ab^2$, where $a > 0$, $b > 0$, show that

$$\log\left(\frac{a-b}{2}\right) = \frac{1}{3}(\log a + 2\log b);$$

d) If $\log_3(2x+y) - 1 = \dfrac{\log_3 x + \log_3 y}{2}$, show that $4x^2 + y^2 = 5xy$, where $x > 0$, $y > 0$.

VII. 32. For the non-zero positive real numbers a and b, $a \ b$, prove the inequality $\log_2(a+b) > 1 + \dfrac{1}{2}(\log_2 a + \log_2 b)$.

VII. 33. Prove the inequalities:

a) $\log_{a_3}\sqrt[n]{a_1 a_2} + \log_{a_4}\sqrt[n]{a_2 a_3} + \ldots + \log_{a_1}\sqrt[n]{a_{n-1}a_n} + \log_{a_2}\sqrt[n]{a_n a_1} \geq 2$, where $a_1, a_2, \ldots, a_n \in (0, \infty) \setminus \{1\}$;

b) $\dfrac{1}{\log_3 n} + \dfrac{1}{\log_5 n} + \ldots + \dfrac{1}{\log_{2n-1} n} < n$, $n \in \mathbf{N}$, $n > 1$.

VII. 34. Solve for x and state the restrictions:

a) $6^{x-2} = 3$; b) $2^{x-4} = 5$; c) $3^{x-5} = 5^{x-2}$; d) $e^x = 3$; e) $e^{3x-4} = 10$;

f) $e^{-4x} = \dfrac{1}{2}$; g) $e^{5x+4} = 10$; h) $e^{e^x} = 7$; i) $7^{\sqrt{x-5}} - 5 = 0$.

VII. 35. Find x in the following exponential equations:

a) $2^{3x+1}3^{5x-1} = 6^4$; b) $2^{2x+1}3^{2x}5^{2x-1} = 360$; c) $\left(\dfrac{2}{3}\right)^x \left(\dfrac{9}{8}\right)^x = \dfrac{64}{27}$;

d) $\left(\dfrac{5}{9}\right)^{2x-3} = (1.8)^{3x-2}$; e) $0.6^x \left(\dfrac{9}{25}\right)^{x^2-3} = \left(\dfrac{125}{27}\right)$.

VII. 36. Find x in the following exponential equations:

a) $\left(3+2\sqrt{2}\right)^x = \left(1+\sqrt{2}\right)^2$; b) $\sqrt[4]{16^{5\sqrt{x}}} = 2^{x\left(\sqrt{x}-4\right)}$; c) $\dfrac{3^x}{3^{\frac{1}{x}-1}} = 3^{\frac{1}{x}}$;

d) $2^{\frac{1}{\sqrt{x}-1}} \cdot \left(\dfrac{1}{2}\right)^{\frac{1}{\sqrt{x}+1}} = 2^{\frac{2\sqrt{x}}{x+\sqrt{x}}}$; e) $9^{\sqrt{x-5}} - 27 = 6 \cdot 3^{\sqrt{x-5}}$;

f) $2.5^{\frac{4+\sqrt{9-x}}{\sqrt{9-x}}} \cdot 0.4^{1-\sqrt{9-x}} = 5^{10} \cdot 0.1^5$; g) $\sqrt[3]{5} \cdot 0.2^{\frac{1}{2x}} - 0.04^{1-x} = 0$.

VII. 37. Find x in the following exponential equations:

a) $\dfrac{2^{-\frac{2}{3}} \cdot 4^{\frac{4}{3}}}{\sqrt[3]{64}} = \dfrac{\left(\sqrt{2}\right)^{\frac{3}{2}} \cdot 2^x}{\sqrt[4]{32}}$; b) $\dfrac{\sqrt[3]{25} \cdot 25^{-\frac{1}{4}}}{\sqrt[6]{5^4} \cdot 5^{-0.5}} = \dfrac{\left(\sqrt[4]{25}\right)^{-2} \cdot 5^x}{\left(\sqrt[4]{625}\right)^{0.5}}$;

c) $\dfrac{2^{\frac{1}{6}} \cdot 2^{2x-2}}{\left(\sqrt[3]{2}\right)^{-2}} = \dfrac{2^{\frac{5}{3}} \cdot 4^{-\frac{2}{3}}}{\sqrt[6]{32}}$; d) $\dfrac{9^{\frac{3}{4}} \cdot \sqrt[4]{9^x}}{3^{-3} \cdot \sqrt{27}} = \dfrac{\left(\sqrt[5]{81}\right)^{\frac{5}{2}} \cdot 3^2}{\left(\sqrt[6]{9}\right)^3}$.

VII. 38. Solve the equations:

a) $3^{x+1} + 3^x = 36$; b) $2^{x+1} - 5 \cdot 2^{x-2} - 12 = 0$; c) $3 \cdot 5^x = 5^x + 2 \cdot 5^3$;

d) $2^{x+2} - 2^{x-1} = 56$; e) $5^x + 5^{x-1} + 5^{x-2} = 155$; f) $6^{2x+2} + 6^{2x-1} = 1302$;

g) $6^{2x+1} - 6^{2x-1} = 210$; h) $3^{6x+12} + 9^{3x+6} + 27^{2x+4} = 3^{4x+27}$.

VII. 39. Solve the equations:

a) $4 \cdot 3^x - 5 \cdot 3^{x-1} + 3^{x-2} = 66$; b) $3^{x-1} + 3^{x-2} - 4 \cdot 3^{x-4} = 32$;

c) $2^{2x-1} + 2^{2x-2} + 2^{2x-3} = 224$; d) $2^{x+3} + 2^{x+2} + 3 \cdot 2^{x+1} + 2^x = 38$;

e) $11 \cdot 2^{x+4} - 9 \cdot 2^{x+3} - 7 \cdot 2^{x+2} - 5 \cdot 2^{x+1} - 3 \cdot 2^x = 504$.

VII. 40. Solve the equations:

a) $\dfrac{5-2^x}{2^x-5} = \dfrac{2}{2^x-5} + 1$; b) $\dfrac{2 \cdot 5^x - 3}{5^x - 5} = \dfrac{7}{5^x - 5} + 4$; c) $\dfrac{5^x - 5^{-x}}{5^x + 5^{-x}} = \dfrac{3}{5}$;

d) $\dfrac{3 \cdot 2^x - 7}{2^x - 8} = \dfrac{17}{2^x - 8} + 2$; e) $\dfrac{41}{3^x - 9} = \dfrac{5 \cdot 3^x - 4}{3^x - 9} - 5$.

VII. 41. Solve the equations:

a) $\left(\dfrac{3}{4}\right)^{x-1} \sqrt[x]{\dfrac{4}{3}} = \dfrac{\sqrt{3}}{2}$; b) $\left(\dfrac{4}{3}\right)^x \sqrt{\dfrac{3}{4}} = \dfrac{2}{\left(\sqrt[4]{3}\right)^{3x-1}}$; c) $5^{x^2-1} \cdot 2^{2x^2} = 4 \cdot 3^{x^2-1}$;

d) $\sqrt{5} \cdot 0.04^{\frac{1}{4x}} - 0.04^{1-x} = 0$; e) $\sqrt{2^x \cdot \sqrt[3]{16^{\frac{x}{2}} \cdot 0.125^{\frac{1}{x}}}} = 4 \cdot \sqrt[3]{2}$;

f) $0.6^{\frac{x}{2}} \cdot \left(\dfrac{3}{5}\right)^{12-x^2} = \left(\dfrac{27}{125}\right)^{\frac{3}{2}}.$

VII. 42. Solve the equations:

a) $2 \cdot 2^{2x} - 3 \cdot 2^x + 1 = 0$; **b)** $2^{x-2}\left(2^x + 10\right) = 36$; **c)** $8^x - 4 = 2^{x+1}(2^x - 1)$;

d) $3^x + 78 = 9^{x+1}$; **e)** $3 \cdot 5^{2x-1} - 2 \cdot 5^{x-1} = 5^{-1}$; **f)** $5^{2x-1} + 2 \cdot 5^{x-1} = 5^x + 2$.

VII. 43. Solve the equations:

a) $9^{1+x^2} - 9^{1-x^2} = 80$; **b)** $4^x - 3 \cdot 2^{x+1} + 8 = 0$; **c)** $6^{x+1} = 36^x - 16$;

d) $9^{x^2-1} - 36 \cdot 3^{x^2-3} + 3 = 0$; **e)** $4^{\sqrt{x+1}} - 10 \cdot 2^{\sqrt{x+1}} + 16 = 0.$

VII. 44. Solve the equations:

a) $2^{2x+1} + 2^{2+x} = 160$; **b)** $2 \cdot 3^x + 3^{1-x} = 7$; **c)** $2^{2x} - 3 \cdot 2^{x+1} + 8 = 0$;

d) $3^{4x+4} - 4 \cdot 3^{2x+3} + 27 = 0$; **e)** $2^{2-x^2} \cdot 2^{7x} = 8^{-2}$.

VII. 45. Solve the equations:

a) $2^x + 8 \cdot 2^{x-2} + 8^{\frac{x+2}{3}} = 7 \cdot 16^{\frac{2x-1}{4}}$; **b)** $64^{\frac{x-\sqrt{x-1}}{3}} - 3 \cdot 8^{\frac{x-\sqrt{x-1}}{3}} - 4 = 0$;

c) $5^x + 25^{\frac{x+1}{2}} + 2 \cdot 5^{x+1} = 16 \cdot 125^{\frac{2x-1}{3}}$; **d)** $3^x + 2 \cdot 3^{x-1} + 27^{\frac{x+2}{3}} = 11 \cdot 3^{2x-1} - 1.$

VII. 46. Solve the equations:

a) $2^{x+1} + 2^x + 8 \cdot 2^{x-1} = 6^{x+1} + 6^x$; **b)** $3^{x+1} + 5^{x+2} = 3^{x+5} - 3 \cdot 5^{x+3}$;

c) $2^{x+3} + 3 \cdot 2^{x+1} = 7^x + 7^{x-1}$;

d) $14 \cdot 7^{2x-1} - 7^{2x} + 7^{2x+1} = 5 \cdot 5^{2x-1} + 2 \cdot 5^{2x} + 5^{2x+1}.$

VII. 47. Solve the equations:

a) $3 \cdot 4^x + \dfrac{1}{3} \cdot 9^{2+x} = 6 \cdot 4^{x+1} - \dfrac{1}{2} \cdot 9^{1+x}$; **b)** $27^x - 13 \cdot 9^x + 13 \cdot 3^{x+1} - 27 = 0$;

c) $3^{3x} + 3^{-3x} + 3^{x+1} + 3^{1-x} = 8$; **d)** $3^{2x+1} + 4 \cdot 3^{1-2x} + 4 \cdot 3^{x+\frac{1}{2}} - 8 \cdot 3^{\frac{1}{2}-x} = 17$;

e) $\dfrac{5}{3^{2x} + 3^{x+1} - 3} + \dfrac{3}{3^{2x} - 3^x - 3} = \dfrac{4}{3^x}.$

VII. 48. Solve the equations:

a) $\sqrt[5]{3}^x + \sqrt[10]{3}^{x-10} = 84$; **b)** $3^{x+3} + 9 = \sqrt{9^{4x-2}} + 3^{4x}$;

c) $3^2 \cdot 9^{-1+\sqrt{x+5}} - 27 = 2 \cdot 3^{1+\sqrt{x+5}}$; **d)** $2^{x+\sqrt{x^2-4}} - 5 \cdot \sqrt{2}^{x-2+\sqrt{x^2-4}} = 6$;

e) $\left(1+\sqrt{3}\right)^x + 2^{x-1}\left(2+\sqrt{3}\right)^x = 4$.

VII. 49. Solve the equations:

a) $\left(2+\sqrt{3}\right)^x + \left(2-\sqrt{3}\right)^x = 14$; b) $\left(5+2\sqrt{6}\right)^x + \left(5-2\sqrt{6}\right)^x = 10$;

c) $\left(\sqrt{7+4\sqrt{3}}\right)^x + \left(\sqrt{7-4\sqrt{3}}\right)^x = 14$; d) $\left(4+\sqrt{15}\right)^x + \left(4-\sqrt{15}\right)^x = 62$.

VII. 50. Solve the equations:

a) $\left(6+\sqrt{35}\right)^x - \left(6-\sqrt{35}\right)^x = 2\sqrt{35}$;

b) $\left(2+\sqrt{3}\right)^{x^2-2x+1} + \left(2-\sqrt{3}\right)^{x^2-2x-1} = 4\cdot\left(2+\sqrt{3}\right)$;

c) $\left(49+10\sqrt{24}\right)^x + \left(49-10\sqrt{24}\right)^x = 98$;

d) $\left(\dfrac{3+\sqrt{5}}{2}\right)^x - \left(\dfrac{3-\sqrt{5}}{2}\right)^x = \sqrt{5}$; i) $\left(3+\sqrt{5}\right)^x + \left(3-\sqrt{5}\right)^x = 3\cdot 2^x$.

VII. 51. Solve the equations:

a) $\left(x^2-x-1\right)^{x^2-1} = 1$; b) $|x-2|^{x^2+4x} = (x-2)^5$;

c) $|x-3|^{x^2+3x} = (x-3)^4$; d) $\left(x^2-5x+6\right)^{x^2+3x} = 1$;

e) $\left(x^2-5x+6\right)^{x^2-3x}(x+6)^{3x-x^2} = 1$.

VII. 52. Solve the equations:

a) $2^{|x|} - \left|2^{x-1}-1\right| = 2^{x-1}+1$;   b) $\left|3^{x-1}-1\right| + \left|3^{x-1}-3\right| = 2$;

c) $\left|2^{x-1}-1\right| + \left|2^{x-1}-2\right| = 1$.

VII. 53. Solve the exponential equations:

a) $6\cdot 9^x - 13\cdot 6^x + 6\cdot 4^x = 0$;   b) $3^{2x+4} + 45\cdot 6^x - 9\cdot 2^{2x+2} = 0$;

c) $7\cdot 4^{x^2} - 9\cdot 14^{x^2} + 2\cdot 49^{x^2} = 0$;   d) $5^{2x-1} + 2^{2x} + 2^{2x+2} = 5^{2x}$;

e) $6\cdot 9^{\frac{1}{x+1}} + 6\cdot 4^{\frac{1}{x+1}} - 13\cdot 6^{\frac{1}{x+1}} = 0$ $x \neq -1$; f) $27^x + 12^x = 2\cdot 8^x$;

g) $2^{2x} - 6^{x-1} - 2\cdot 3^{2x} = 0$; h) $2\cdot 3^{4x} = 6^{2x} + 3\cdot 4^{2x}$;

i) $3^{3x} - 3\cdot 18^x - 12^x + 3\cdot 2^{3x} = 0$.

VII. 54. Solve the equations:

a) $3\cdot 5^{x-2008} + 5^{x-2009} + 2\cdot 5^{x-2012} = 2002$;

b) $9\cdot 6^{x-2000} + 6^{x-2001} + 3\cdot 6^{x-2002} + 4\cdot 6^{x-2003} = 2002$.

VII. 55. Prove that the number $\log_5 10$ is not a rational number.

VII. 56. Find the solution for each equation, correct to six decimal places.
a) $\ln x = 6$; b) $\ln(2x - 3) = 1$; c) $\ln(x + 5) = 3$; d) $\ln(e^{2-x}) = 8$;
e) $\ln x = \ln 4 + \ln 5$; f) $\ln(\ln x) = 2$.

VII. 57. Solve the equations:
a) $2 \log_x 9 = 2$; b) $\log_7(2x + 1) = 1$; c) $\log_{x+1}\left(x^2 + 3x\right) = 2$;
d) $\log_3(10 - x) = 2$; e) $\log_6(x + 3) - \log_6(2x - 5) = 1$; f) $\log(3x - 2) = 2$;
g) $\log_6\left(x^2 + x + 1\right) = \log_6(2x + 3)$; h) $\log_2(4 - x) + \log_2(2 - x) = 3$;
i) $\log_5(2x - 1) + \log_5(2x - 5) = 1$.

VII. 58. Solve the equations:
a) $2\log_5(x - 3) - \log_5(3x - 9) = 2\log_5 2$;
b) $\log_5(3x + 19) + \log_5(x - 17) = 3 + 3\log_5 2$;
c) $\log 2\sqrt{x - 23} - 0.5\log\sqrt{x - 1} = 2\log\sqrt{2}$;
d) $\log_{2x+1}\left(x^2 + x + 2\right) = \log_{2x+1}(x + 3)$;
e) $\log_{x+0.5}(5x + 3) = \log_{x+0.5}(3x + 9)$;
f) $\log_{\frac{1}{2}}(x - 1) - \log_{\frac{1}{2}}(x + 1) - \log_{\frac{1}{\sqrt{2}}}(7 - x) = 2$.

VII. 59. Solve the equations:
a) $\log_x\left(x^2 - 3x\right) = 1$; b) $x\log_{x+2} 7 \cdot \log_{\frac{1}{\sqrt[3]{7}}}(x + 2) = \dfrac{x - 4}{x}$;

c) $3\log_8(x - 2) = \log_2\sqrt{2x - 1}$; d) $\log_3\left(x^3 + 1\right) - 0.5\log_3\left(x^2 + 2x + 1\right) = \log_3 7$;
e) $\log_{10} x + \log_{\sqrt{10}} x + \log_{\sqrt[3]{10}} x + \ldots + \log_{\sqrt[10]{10}} x = 5.5$;

f) $\log_4 x + \log_{16} x = \log_{\frac{1}{4}}\sqrt{3}$; g) $\log_3\left[\log_{\frac{1}{2}}(x + 1)\right] = 1$;

h) $\log_4(\log_2 x) + \log_2(\log_4 x) = 2$.

VII. 60. Solve the equations:
a) $(\log_2 x)^2 + \log_2(4x) = 4$; b) $\log_2\left(\log_3\left(x^2 + 1\right)\right) = 1$; c) $\log_3\left(\log_2\left(x^2 + 1\right)\right) = 1$;
d) $\log_5^2 x - \log_5 x - 2 = 0$; e) $\log_5^2 x + \log_5 7 \cdot \log_7 x = 2$;
f) $\log\sqrt{1 + x} + 3\log\sqrt{1 - x} = \log\sqrt{1 - x^2} - 2$;
g) $(3\log_a x - 2)\log_x^2 a = \log_{\sqrt{a}} x - 3$, $(a > 0, \ a \neq 1)$;

h) $\log_5^4(x-2)^2 - \log_5^2(x-2)^3 = 7$.

VII. 61. Solve the equations:

a) $\log_6(2+x) + 2\log_6\sqrt{3+x} = 1$;   b) $3\log_2(2+x) + \log_2(x-4)^3 = 0$;

c) $\log_4 x + \log_2 x = \log_{\frac{1}{3}}\dfrac{1}{27}$;   d) $3\log_8(x-2) = 2\log_4\sqrt{2x-1}$;

e) $\dfrac{\log_a(x-1)}{\log_a(x^2-4)} = \dfrac{1}{2}$;   f) $\dfrac{\log_5(\sqrt{2x-3}+1)}{\log_5(\sqrt{2x-3}+7)} = \dfrac{1}{2}$.

VII. 62. Solve the equations :

a) $\log_x\sqrt{5} + \log_x 5x = (\log_x\sqrt{5})^2 + 2.25$;   b) $(\log_{3^{-1}} 9x)^2 + \log_3\dfrac{x^2}{27} = 8$;

c) $\sqrt{\log_a x} + \sqrt{\log_x a} = \dfrac{10}{3}$;   d) $\dfrac{1+2\log_9 2}{\log_9 x} = 1 + \log_{\sqrt{x}} 3 \cdot \log_9(12-x)$;

e) $\dfrac{2-4\log_{12} 2}{\log_{12}(2x-5)} = 1 + \dfrac{\log_1(15-2x)}{\log_1(2x-5)}$;   f) $\dfrac{\log_{4\sqrt{x}} 2}{\log_{2x} 2} + \log_{2\sqrt[3]{x}} 2 \cdot \log_{0.5}(2 \cdot \sqrt[3]{x}) = 0$;

g) $\log_{2x+1}(x^3 + 7x + 16) \cdot \log_{x+1}(2x+1) = 3$.

VII. 63. Solve the equations:

a) $\log_6(2 \cdot 3^x + 1) = \log_6(3^x + 10)$;

b) $3\log_8(4^{x^2-1} - 1) + 2 = 3\log_8(2^{x^2+2} - 7)$;

c) $\log_2(9^{x-2} + 7) = 2 + \log_2(3^{x-2} + 1)$;

d) $\log_3(\sqrt[3]{9}) + \log_8(9^{x+1} - 1) = 1 + \log_8(3^{x+1} + 1)$;

e) $\log(3^x + x - 5) = x\log 30 - x$;

f) $\log 2 + \log(4^{x-2} + 9) = 1 + \log(2^{x-2} + 1)$;

g) $\log_2(4^x + 1) = x + \log_2(2^{x+3} - 6)$.

VII. 64. Solve the equations:

a) $\sqrt{\log_x\sqrt{4x}} + \log_x 4 = 0$;   b) $\sqrt{\log_x\sqrt{3x}} \cdot \log_3 x = -1$;

c) $\sqrt{\log_2(2x^2)} \cdot \log_4(16x) = 3 \cdot \log_4 x$;

d) $2 \cdot \sqrt[3]{2\log_{16}^2 x} - \sqrt[3]{\log_2 x} - 6 = 0$;   e) $\sqrt{(2x)^{\log\sqrt{2x}}} = 10$;

f) $\sqrt{\log_5 x} + \sqrt[3]{\log_5 x} = 2$.

VII. 65. Solve the equations:

a) $\log_x\left(x \cdot \sqrt[3]{4}\right) + 3\log_x\left(x \cdot \sqrt[3]{2}\right) + \log_x^2\left(\sqrt[3]{4}\right) = 13$;

b) $\log_x^2 6 + \log_{\frac{1}{6}}^2 \dfrac{1}{x} + \log_{\frac{1}{\sqrt{x}}} \dfrac{1}{6} + \log_{\sqrt{6}} x + \dfrac{3}{4} = 0$;

c) $\log_x \sqrt{5} + \log_x 5x = \left(\log_x \sqrt{5}\right)^2 + 2.25$;

d) $2 - \log_{a^2}(1+x) = 3\log_a \sqrt{x-1} - 2\log_{a^2}\left(x^2 - 1\right)$, $\quad a \in (0,\infty)\setminus\{1\}$;

e) $\log_x 3 + \log_3 x = \log_{\sqrt{x}} 3 + \log_3 \sqrt{x} + 0.5$;

f) $\log_a x + \log_{a^2} x + \log_{a^3} x = 1$, $a \in (0,\infty)\setminus\{1\}$;

g) $2\log_x \sqrt[4]{2} - \left(\log_x \sqrt{2}\right)^2 = \log_2 8 - \log_x(2x)$.

VII. 66. Solve the equations:

a) $\log_{3x} \dfrac{3}{x} + \log_3^2 x = 1$; b) $3\log_x 4 + 2\log_{4x} 4 + 3\log_{16x} 4 = 0$;

c) $5\log_{\frac{x}{9}} x + 2\log_9 x^3 + 8\log_{9x^2} x^2 = 5$;

d) $30\log_{9x} \sqrt[3]{x} + 7\log_{81x} x^3 - 3\log_{\frac{x}{3}} x^2 = 0$;

e) $3\log_x a^2 + 3\log_{a^2 x} a - \dfrac{14\log_{ax} a}{\log_a x} = 0$, $a \in (0,\infty)\setminus\{1\}$;

f) $\log_{\frac{2}{x^2}} x + \log_{\frac{2}{x}} x^2 - 4\log_{2x} x^2 = 0$;

g) $\dfrac{\log_3 x}{\log_9 3x} = \dfrac{\log_{27} 9x}{\log_{81} 27x}$; h) $\dfrac{\log_a x}{\log_{a^2} ax} = \dfrac{\log_{a^3} a^2 x}{\log_{a^4} a^3 x}$, $a \in (0,\infty)\setminus\{1\}$.

VII. 67. Solve the equations:

a) $x^{2+\log_3 x} = 3^8$;

b) $\log_2 3 + 2\log_4 x = x^{\frac{\log_9 16}{\log_3 x}}$;

c) $\sqrt{x^{\log_{10}\sqrt{x}}} = 10$;

d) $\left(\sqrt[4]{x}\right)^{\log x + 7} = 10^{\log x + 1}$;

e) $x^{2\log^3 x - 1.5\log x} = \sqrt{10}$;

f) $x^{\log_2 \frac{x}{98}} \cdot 14^{\log_2 7} = 1$;

g) $4^{\log_3 x^2} - 4^{\log_3 x + 1} + 4^{\log_3 x + 2} = 28$.

VII. 68. Solve in the set of $\mathbf{R}$ the equations:

a) $343^x + 512^x + 729^x = 3 \cdot 504^x$;

b) $9^x + 36^x + 2 \cdot 32^x = 4 \cdot 24^x$.

VII. 69. Solve the systems of exponential equations:

a) $\begin{cases} x^{x+y} = y^{x-y} \\ x^2 y = 1 \end{cases}$; b) $\begin{cases} y = 2^x \\ y^x = 16 \end{cases}$; c) $\begin{cases} 2^x \cdot 3^y = 18 \\ 2^y \cdot 3^x = 12 \end{cases}$;

d) $\begin{cases} \left(\dfrac{2}{3}\right)^{x-y} - \left(\dfrac{3}{2}\right)^{x-y} = -\dfrac{65}{36}; \\ x^2 + y^2 = 10 \end{cases}$ e) $\begin{cases} 16^{x+2y} = 4 \cdot 8^{x-y+1} \\ 4 \cdot 32^{x+3y+0.2} = 4^{2x+4y}; \end{cases}$

f) $\begin{cases} x^{y+1} = 27 \\ x^{2y-5} = \dfrac{1}{3}; \end{cases}$ g) $\begin{cases} 3^x + 3^y = 12 \\ 3^{x+y} = 27 \end{cases}$; h) $\begin{cases} 4^{x+y} - 9^{x-y} = -65 \\ 3 \cdot 2^{x+y} + 2 \cdot 3^{x-y-1} = 18. \end{cases}$

VII. 70. Solve the systems of exponential equations:

a) $\begin{cases} x^{\sqrt[4]{x+\sqrt{y}}} = y^{\frac{8}{3}} \\ y^{\sqrt[4]{x+\sqrt{y}}} = x^{\frac{2}{3}} \end{cases}$; b) $\begin{cases} 4^{\sqrt[3]{x}} 3^{\sqrt{y}} = 144 \\ 4^{2\sqrt[3]{x}} + 3^{2\sqrt{y}} = 337 \end{cases}$; c) $\begin{cases} 3^{\sqrt[3]{x}} \cdot 5^{\sqrt[4]{y-2}} = 45 \\ 9^{\sqrt[3]{x}} + 25^{\sqrt[4]{y-2}} = 106 \end{cases}$;

d) $\begin{cases} z^x = x \\ z^y = y; \\ y^y = x \end{cases}$ e) $\begin{cases} 2^x \cdot 3^y + 2^z = 124 \\ 3^y \cdot 2^z + 2^x = 436; \\ 2^z \cdot 2^x + 3^y = 91 \end{cases}$ f) $\begin{cases} 2^{x+y+1} = 2^{2y} + 1 \\ 2^{y+z+1} = 2^{2z} + 1. \\ 2^{z+x+1} = 2^{2x} + 1 \end{cases}$

VII. 71. Solve the systems of logarithmic equations:

a) $\begin{cases} \log_x(x+y) = 2 \\ \log_x(4x-2y) = 2 \end{cases}$; b) $\begin{cases} \log_x y + \log_y x = 2 \\ x^2 - 2y = 8 \end{cases}$;

c) $\begin{cases} \log_3 x + 2\log_3 y = 5 \\ 3\log_3 x - \log_9 y^2 = 1 \end{cases}$; d) $\begin{cases} 2^x + 2^y = 20 \\ \log_2 x + \log_2 y = 3; \end{cases}$

e) $\begin{cases} \log_3 x + \log_2 y = 2 \\ 3^x - 2^y = 23 \end{cases}$; f) $\begin{cases} x^{\log_3 y} + 2y^{\log_3 x} = 27 \\ \log_3 x - \log_3 y = -1 \end{cases}$;

g) $\begin{cases} \log_3 \dfrac{\sqrt[4]{x^3}}{\sqrt[3]{y}} + \log_3 \dfrac{\sqrt[4]{y^3}}{\sqrt[3]{x}} = \dfrac{5}{3} \\ x^2 + y^2 = 738 \end{cases}$; h) $\begin{cases} 9^{\sqrt[4]{xy^2}} - 27 \cdot 3^{\sqrt{y}} = 0 \\ \dfrac{1}{4}\log_4 x + \dfrac{1}{2}\log_4 y = \log_4\left(4 - \sqrt[4]{x}\right). \end{cases}$

VII. 72. Solve the systems of logarithmic equations:

a) $\begin{cases} x + y = 7 \\ \log_3 x + \log_3 y = 1 + 2\log_3 2 \end{cases}$; b) $\begin{cases} \log_3 x + 2\log_3 y = 5 \\ 3\log_3 x - \log_3 y = 1 \end{cases}$;

c) $\begin{cases} x^2 + y^2 = 34 \\ \log_5 x + \log_5 y = \log_5 15 \end{cases}$; d) $\begin{cases} xy = 27 \\ x^{\log_3 y} = 9 \end{cases}$; e) $\begin{cases} \log_{10}\left(x^2 + y^2\right) = 2 \\ \log_2 x + 2 = \log_2 3 + \log_2 y \end{cases}$;

f) $\begin{cases} \log_{\wp} x \cdot \left(\dfrac{1}{\log_x 2} + \log_2 y \right) = \log_2 x \\ \log_2 x \cdot \left[\log_3 (x + y) \right] = 3\log_3 x \end{cases}$;

g) $\begin{cases} \left(\log_a x\right)^2 + \left(\log_a y\right)^2 = 2m, \\ x^{\log_a x} + y^{\log_a y} = 2 \cdot \dfrac{p^2 + q^2}{p^2 - q^2} \, a^m \end{cases}$, $(p \neq q)$;

h) $\begin{cases} x^{\log y} \cdot y^{\log z} \cdot z^{\log x} = 10^{-1} \\ x^{\log y \log z} \cdot y^{\log x \log z} \cdot z^{\log x \log y} = 1. \\ xyz = 1 \end{cases}$

VII. 73. Solve the system of equations

$\begin{cases} \log_{10}\left(x^3 - x^2\right) = \log_5 y^2 \\ \log_{10}\left(y^3 - y^2\right) = \log_5 z^2 \\ \log_{10}\left(z^3 - z^2\right) = \log_5 x^2 \end{cases}$,

where $x, y, z > 1$.

VII. 74. Solve the exponential inequations:

a) $6^x + 1 \geq 2^x + 3^x$;   b) $4^x - 6 \cdot 2^x + 8 \leq 0$;   c) $0.25^{\sqrt{x}} + 2 > 3 \cdot 0.5^{\sqrt{x}}$;

d) $4^x \leq 3 \cdot 2^{\sqrt{x} + x} + 4^{1 + \sqrt{x}}$;   e) $0.008^x + 5^{1 - 3x} + 0.2^{3 + 3x} < 30.04$;

f) $8 \cdot 3^{\sqrt{x} + \sqrt[4]{x}} + 9^{1 + \sqrt[4]{x}} \geq 9^{\sqrt{x}}$;   g) $3^{2|x-1|} + 3 < 4 \cdot 3^{|x-1|}$;

h) $\sqrt{2^3 + 2^{1 + \sqrt{3-x}} - 2^{\sqrt{12-4x}}} + 2^{1 + \sqrt{3-x}} > 5$;

i) $x^2 + (2 \cdot 3^x - 5)x + 9^x - 5 \cdot 3^x + 4 > 0$;

j) $\dfrac{1}{3} \leq \dfrac{3^x + 1}{3^{2x} + 3} \leq \dfrac{1}{2}$;   k) $\dfrac{3 \cdot 2^{x-1}}{3^x - 2^x} \geq 1 + \left(\dfrac{2}{3} \right)^x$;   l) $\sqrt[4]{x^{3x}} < \sqrt{x^{-x^2 - x + 1}}$.

VII. 75. Solve the logarithmic inequations:

a) $\log_3\left(x^2 + 5x + 4\right) < 0$;   b) $\log_3\left[-\log_{0.5}\left(2 - \log_4 x\right) - 1 \right] \leq 1$;

c) $\log_2\left(2^x - 2\right) \cdot \log_{0.5}\left(2^{x+1} - 4\right) < -2$;

d) $\dfrac{1}{\log_{0.5} \sqrt{x+3}} \leq \dfrac{1}{\log_{0.5} \sqrt{x+1}}$;   e) $\dfrac{\ln\left(2x^2 - 3x + 1\right)}{x^2 - 3x} \geq 0$.

173

VII. 76. Solve the logarithmic inequalities:

a) $\log_{\frac{x}{2}} 8 + 3\log_{\frac{x}{4}} 2 < \dfrac{2\log_2 x^2}{\log_2 x^2 - 4}$;

b) $\log_3 \log_4 \dfrac{2x-1}{x-1} - \log_{\frac{1}{3}} \log_{\frac{1}{4}} \dfrac{x-1}{2x-1} < 0$;

c) $\log_{0.5} \log_8 \dfrac{x^2 - 2x}{x-3} \le 0$;   i) $\log_2 \log_{0.5}\left(2^x - 4^x\right) \le 1$;

d) $\left(\log_2 x\right)^4 - \left(\log_{0.5} \dfrac{x^3}{8}\right)^2 + 45 < 4\left(\log_{0.5} x\right)^2 + 9\log_2 x^2$.

VII. 77. Prove the inequality:

$(ab)^{\sqrt{\log_a c \cdot \log_b c}} + (bc)^{\sqrt{\log_b a \cdot \log_c a}} + (ac)^{\sqrt{\log_a b \cdot \log_c b}} \ge a^2 + b^2 + c^2$, for any
$a, \ b, \ c > 1$.

VII. 78. Solve the equations:

a) $3^x + 4^x = 5^x$;   b) $\sqrt{6^x} + \sqrt{8^x} = \sqrt{10^x}$;

c) $3 \cdot 5^{2x-1} - 7 \cdot 2^{4x-3} = 19$,   d) $\sqrt{3}^x + 1 = 2^x$;

e) $5 \cdot 2^{x-5}\left(3^x - 1\right) = 3^x$;   f) $3 + \sqrt[3]{5^x} = 2^x$.

VII. 79. Solve the equations:

a) $\log_3\left(2^x + 1\right) = \log_2\left(3^x - 1\right)$; b) $\log_5\left(4^x + 1\right) = \log_4\left(5^x - 1\right)$;

c) $e^x - e^{-x} = 2\ln\left(x + \sqrt{x^2 + 1}\right)$.

VII. 80. Solve the equations:

a) $2^{x + \frac{1}{x}} + 2^{\sqrt{x} + \frac{1}{\sqrt{x}}} = 8$;   b) $\left(2^{\log_5 x} + 3\right)^{\log_5 2} = x - 3$;

c) $2^{1 - 2\sin^2 x} = 2 + \log_2\left(1 - \sin^2 x\right)$.

VII. 81. Solve the equation $\log_x\left(\sqrt{x} + \sqrt[4]{x}\right) \le \dfrac{1}{4}\log_3 12$.

VII. 82. Consider the function $f : [1; \infty) \to \mathbf{R}$,

$f(x) = \sqrt{x + 3 - 4\sqrt{x-1}} + \sqrt{x + 8 - 6\sqrt{x-1}}$.
Solve the equation $f(\log x) = 1$.

VII. 83. Prove that the function $f: \mathbf{R} \to \mathbf{R}$, $f(x) = 2^x - 2^{-x} + 1$ is invertible and find its inverse.

VII. 84. If $a, b, c \in (0, 1)$, $a < b < c$, then compare the numbers
$A = \log_b a + \log_c b + \log_a c$ and $B = \log_a b + \log_b c + \log_c a$.

VII. 85. Consider the function $f: A \times A \to \mathbf{R}$, where $A = (0, \infty) \setminus \{1\}$,
$f(x, y) = \log_x y + \log_y x$.

a) Prove that $f(a, a^3)$ does not depend of a, $a \in (0, \infty) \setminus \{1\}$;

b) Prove that $f\left(x, \dfrac{1}{y}\right) = f\left(\dfrac{1}{x}, y\right)$, for any $x, y \in (0, \infty) \setminus \{1\}$;

c) Solve the equation $f(a^x, b) - f(a, b^x) = \dfrac{15}{4}(\log_b a - \log_a b)$, $a, b \in (0, \infty) \setminus \{1\}$, $a \neq b$.

Chapter VIII

ARITHMETIC AND GEOMETRIC SEQUENCES

$a_1, a_2, a_3, ..., a_n$; a_n is the nth term, d is the ratio or common difference.

$a_n = a_{n-1} + d$, $\qquad a_n = a_1 + (n-1)d$,

$$a_k = \frac{a_{k-1} + a_{k+1}}{2} , \quad k = 2,3,4,..,n-1,$$

$$a_1 + a_n = a_i + a_{n-i+1} , \quad i = 1,2,3,..,n,$$

$$S_n = a_1 + a_2 + ... + a_n = \sum_{k=1}^{n} a_k , \quad S_n = \frac{(a_1 + a_n)n}{2} = \frac{[2a_1 + (n-1)d]n}{2} .$$

VIII. 1. The general term of a sequence is given. Write the first five terms.

a) $t_n = 2n$; b) $t_n = 10 - n$; c) $t_n = 3n$; d) $t_n = 2^n$; e) $t_n = 1 + 2n$; f) $t_n = 11n$;

g) $a_n = 3n + 1$; h) $a_n = n^2 - 2$; i) $a_n = \dfrac{(-1)^n}{n!}$; j) $y_n = (-1)^{n-1}\dfrac{1}{n} + (-1)^n n$;

k) $z_n = \dfrac{n}{n^2 + 1}$; l) $c_n = \left(1 + \dfrac{1}{n}\right)^n$; m) $a_n = \dfrac{1 + (-1)^n + (-1)^{n+1}}{n}$.

VIII. 2. Find the formula for the general term for each of the sequences:

a) $1, -\dfrac{1}{2}, \dfrac{1}{3}, -\dfrac{1}{4}, ...$; b) $\dfrac{1}{2}, \dfrac{1}{5}, \dfrac{1}{10}, \dfrac{1}{17}, \dfrac{1}{26}, ...$; c) $\dfrac{2 \cdot 4}{1 \cdot 3}, \dfrac{4 \cdot 6}{3 \cdot 5}, \dfrac{6 \cdot 8}{5 \cdot 7}, ...$;

d) $\dfrac{2}{5}, \dfrac{6}{25}, \dfrac{18}{125}, ...$; e) $b_1 = 1$, $b_{n+1} = b_n - \dfrac{1}{n(n+1)}$; f) $a_1 = 1$, $a_{n+1} = (n+1)a_n$;

g) $x_1 = \dfrac{2}{3}$, $x_{n+1} = x_n + \dfrac{1}{(n+2)(n+3)}$;

h) $x_1 = a$, $a \in \mathbf{R}^*$, $\dfrac{x_1 + 2x_2 + 3x_3 + ... + nx_n}{x_1 + x_2 + x_3 + ... + x_n} = \dfrac{2n+1}{3}$, $n \geq 2$.

VIII. 3. Describe each sequence and write an expression for the general

term, t_n :
a) $1, 3, 5, 7, 9, ...$; b) $6, 12, 18, 24, 30, ...$;
c) $4, 9, 14, 19, 24, ...$; d) $10, 100, 1000, 10000, ...$;
e) $5, 50, 500, 5000, ...$; f) $7, 13, 19, 25, 31, ...$

VIII. 4. What is the recursive formula for the sequence
349, 321, 293, 265, ...?

VIII. 5. The 8th term of an arithmetic sequence is 3 and the 100th term is
49. What is the recursive formula for the sequence?

VIII. 6. Two arithmetic sequences 6, 12, 18, 24, 30,... and 4, 9, 14, 19,
24,... share common terms. The common terms form a new sequence.
Find the general formula for the new sequence.

VIII. 7. How many terms are there in the sequence
−2, 1, 6, 13, 22, ... , 166?

VIII. 8. A sequence has the form $x_n = an^2 + bn$. The 2nd term is −2, and
the 5th term is 10. Find x_{13}.

VIII. 9. A sequence has the form $x_n = an^2 + bn + c$. The 3rd term of
the sequence is 5, the 6th term is 47, and the 8th term is 95. Find the constant
numbers a, b, and c.

VIII. 10. Find the indicated terms in the sequences:

a) $t_n = 12 + 2n$, t_{10} and t_{12}; b) $t_n = 6n + 7$, t_2 and t_5;

c) $t_n = n^2 - 4$, t_4 and t_9; d) $t_n = (-2)^n$, t_2 and t_5;

e) $t_n = \dfrac{n+1}{2n+1}$, t_3 and t_6; f) $t_n = 1 + \dfrac{1}{3^n}$, t_3 and t_4.

VIII. 11. Find the general term of the sequence

$$\frac{\sqrt{7}}{2}, \frac{\sqrt{12}}{7}, \frac{\sqrt{17}}{14}, \frac{\sqrt{2}}{23}, \frac{\sqrt{27}}{34} \dots$$

VIII. 12. In a sequence:

a) $a_1 = 1$, $a_2 = 6$, $a_3 = 15$, $a_4 = 28$. Find the formula for the general term;

b) $S_n = 3n^2 - 6n + 8$, for any n a positive integer. Find the formula for the
general term. Is this sequence an arithmetic sequence?

VIII. 13. In a sequence, the kth is given by the formula $t_k = 2^k - 2k + 1$. Determine the sum of the first n terms.

VIII. 14. Find the first three terms of a series for which:

a) $S_n = 3n$; b) $S_n = 3n^2 - n$; c) $S_n = 2n^2 - 3n$;

d) $S_n = 5n - 3$; e) $S_n = n^2 + 2n$.

VIII. 15. For a given series, $S_5 = 44$. Find t_6 if S_6 is:
a) 40; b) 60; c) 76; d) 84.

VIII. 16. The sum of the first n terms of a series is given.

a) $S_n = n^2 - n$; b) $S_n = 3n^2 - 5n$; c) $S_n = 2^n - 1$;

d) $S_n = 2n^2 - 3n$; e) $S_n = 2(3^n - 1)$; f) $S_n = n^2 + 2n$.

Find: i) S_{n-1}; ii) t_n.

VIII. 17. Let $(a_n)_{n \geq 1}$ be a sequence, such that

$a_1 + a_2 + ... + a_n = n^2 + 3n, \forall n \in \mathbf{N}, \ n \geq 1$. Then:

a) is 2012 one of the terms of $(a_n)_{n \geq 1}$?

b) if $(a_n)_{n \geq 1}$ is an arithmetic sequence, find a_1 and the ratio.

VIII. 18. Given: $3 + 7 + 11 + 15 + ...$

a) Find: i) t_{20} and t_n; ii) S_{20} and S_n;
b) How many terms are less than 500?

VIII. 19. For the arithmetic sequence $-8, -3, 2, 7,...$, find:

a) t_{17} and t_{33};
b) what term is 242?

VIII. 20. Find the missing terms in the arithmetic sequence:
a) _, _, 8, 2, _;   b) _, 9, 16, _, _;   c) 12, _, 22, _, _;
d) 3, _, _, 24, _;   e) _, -4, _, _, -19;   f) 15, _, _, _, -21.

VIII. 21. Find the missing terms for the arithmetic sequence

$a_1, a_2, a_3, 2, 6, 10,...$

178

VIII. 22. Consider the sequence (a_n), $n > 1$, with the nth term given by the formula $a_n = 3n^2 + 2$.
Which of these numbers are terms of this sequence?
a) 200;   b) 256;   c) 434;   d) 4109.

VIII. 23. Consider the sequence (a_n), where $a_0 = 2$, $a_1 = 3$, for any positive integer k, such that $a_{k+1} = 3a_k - 2a_{k-1}$. Find the formula for the general term.

VIII. 24. Consider the sequence of real numbers (a_n), where $a_1 = 1$, $a_2 = 4$ and $a_{k+1} = 2a_k - a_{k-1} + 2$ for any k, a positive integer, find the formula for general term.

VIII. 25. Let $(a_n)_{n\geq1}$ be a sequence, such that $a_1 = 1$ and

$$a_{n+1} = a_n + \frac{1}{2}\left(\sqrt{8a_n + 1} + 1\right), \text{ for any } n \in \mathbf{N}, n \geq 1.$$

Compute a_2, a_3, a_4 and find the general term of the sequence.

VIII. 26. Let $(a_n)_{n\geq1}$ be a sequence, such that $a_1 = 0$ and

$a_{n+1} = a_n + \sqrt{4a_n + 1} + 1$, for any $n \in \mathbf{N}$, $n \geq 1$. Find the formula for general

term and show that $\sqrt{4a_1 + 1} + \sqrt{4a_2 + 1} + ... + \sqrt{4a_n + 1} = n^2$.

VIII. 27. Which of the expressions S_n is the sum of the first n terms of:

a) $1 + 4 + 7 + ...?$ i) $S_n = \dfrac{n^2 - 7n}{2n}$;   iv) $S_n = \dfrac{3n^2 + n}{2}$;

b) $2 + 5 + 8 + ...?$ ii) $S_n = \dfrac{3n^2 - n}{2}$;   v) $S_n = 2n^2 + 5n$;

c) $7 + 11 + 15 + ...?$ iii) $S_n = 3n^2 + 2n$;   vi) $S_n = \dfrac{7n^2 + 3n}{2}$.
d) $5 + 12 + 19 + ...?$

VIII. 28. In a circle, one cord splits the circle in two regions, the second cord intersecting the first cord splits the circle in four regions, the third cord intersecting the first two cords and not passing through the existing point of intersection splits the circle in seven regions and so on. Write the first five

179

terms of the sequence representing the number of sections created and find a general formula for this sequence.

VIII. 29. In an arithmetic sequence, the third term is 11 and the 8th term is 46. Find the general formula for the sequence.

VIII. 30. The third term of an arithmetic sequence is 14 and the 9th term is -1. Find the twentieth term.

VIII. 31. A sequences is given by $a_1 = 1$ and $a_{n+1} = a_n + n^2 - n + 1$. Find a_{100}.

VIII. 32. Find the first five terms of the arithmetic sequence (a_n) and write the formula for the nth term:

a) $a_1 = 2, d = -3$; b) $a_{10} = 20, d = 2$;

c) $a_5 = 27, a_{27} = 60$; d) $a_3 = 14, a_6 = 29$.

VIII. 33. How many terms are in each sequence?

a) 2, 6, 10, ..., 122; b) $-9, -4, 1, ..., 171$;

c) 4, 15, 26, ..., 444; d) 17, 12, 7, ..., -83;

e) 5, 12, 19, ..., 222; f) 4, 12, 20, ..., 796.

VIII. 34. Calculate the sum of the first 20 terms of an arithmetic sequence with $a_3 = 8$ and $a_8 = 143$.

VIII. 35. Find the sum of the 15 terms of an arithmetic sequence if the middle term is 92.

VIII. 36. For the arithmetic series $5 + 11 + 17 + 23 + ...$, find:
a) the 30th term;
b) the sum of the first 30 terms.

VIII. 37. For an arithmetic sequence $(a_n)_{n \in N}$ find a_{10} and S_{12} if $a_5 = 55$ and $a_{15} = 5$.

VIII. 38. Calculate the sum $S_n = 1 + 5 + 9 + ...+101$.

VIII. 39. Find the sum of all positive multiples of 7 that are less than 100.

VIII. 40. Find $x \in \mathbf{R}$ if the numbers:

a) $5 + 3x$, 8, $1 + 2x$; b) $x - 2$, x^2, $x + 6$ are consecutive terms in an arithmetic sequence.

VIII. 41. Find $a \in \mathbf{R}$ if the numbers:

a) 2^a, $4^a + 1$, 2^{a+2}; b) $\log a$, $\dfrac{3}{2}$, $\log a^2$ are the consecutive terms in an arithmetic sequence.

VIII. 42. If $x + 2y + 1$, $2x - 4y - 2$, and $2x - 5y - 7$ are consecutive terms in an arithmetic sequence, find the relation between x and y.

VIII. 43. The equations $x^2 - 8x + m = 0$ and $x^2 - 16x + n = 0$ have the roots x_1, x_2 and x_3, x_4, repectively. If the roots are in an increasing arithmetic sequence in this order, find the numbers m and n.

VIII. 44. Find the 50th term of the arithmetic sequence
(a_n) if $a_{48} = \sqrt{18 - 8\sqrt{2}}$ and $a_{49} = \sqrt{27 - 10\sqrt{2}}$.

VIII. 45. For the arithmetic series $44 + 41 + 38 + 35 + \ldots$, find the sum of the first:
a) 15 terms; b) 30 terms; c) 60 terms.

VIII. 46. If the sum of the first and the fifth term of an arithmetic series is equal to 8, and the product of the third term and the fifth term is equal 80, find the sum of the first twenty terms of the series.

VIII. 47. The sum of the first 5 terms of an arithmetic series is 85 and the sum of the first 6 terms is 123. Find the series.

VIII. 48. For an arithmetic sequence, $(a_n)_{n>1}$ $a_1 + a_2 + a_3 = 14$ and $a_2 + a_3 + a_4 = 28$, find a_1, d, and S_{10}.

VIII. 49. For an arithmetic sequence, the fifth term is 16 and the sum of the first 20 terms is 650. Find a_1 and d.

VIII. 50. Find a_1 and d for an arithmetic sequence if:

a) $\begin{cases} a_2 - a_6 + a_4 = -2 \\ a_7 - a_5 = 10 \end{cases}$;
b) $\begin{cases} a_1 a_2 a_3 = 312 \\ a_1 + a_2 + a_3 = 24 \end{cases}$;

c) $\begin{cases} a_2 + a_3 + a_4 = a_7 + 2 \\ a_1 + a_2 + a_4 = a_6 + 1 \end{cases}$;
d) $\begin{cases} a_2 a_3 = a_1 a_7 \\ a_1 a_5 + a_3 = 2a_6 - a_3 - 2 \end{cases}$.

VIII. 51. In an arithmetic sequence $a_1, a_2, a_3, ..., a_{10}$, $S_{10} = 140$ and $a_2 a_9 = 147$. Find the sequence.

VIII. 52. Find a_1 and d for an arithmetic sequence when:

a) $a_1^2 + a_2^2 + a_3^2 = 56$ and $a_1^3 + a_2^3 + a_3^3 = 288$;

b) $a_1 + a_2 + a_3 + a_4 = 36$ and $a_1^2 + a_2^2 + a_3^2 + a_4^2 = 504$.

VIII. 53. The sum of the terms of an arithmetic sequence is 1,050, the first term is 5 and the last term is 100. Find the common difference.

VIII. 54. In an arithmetic series, $a_1 = 6$ and $S_9 = 108$. Find the common difference and the sum of the first 20 terms.

VIII. 55. Insert three numbers between 7 and 23 such that all five numbers are in an arithmetic sequence.

VIII. 56. Find a_1 and d for an arithmetic sequence knowing:

a) $S_n = 5n^2 + 6n$; b) $S_n = 3n^2 + 4n$.

VIII. 57. In an arithmetic sequence the sum of three consecutive terms is 6, and the sum of the squares of these terms is 62. Find the numbers.

VIII. 58. Let a_n, a_{n+1}, a_{n+2} be three consecutive terms of an arithmetic sequence. Solve the equation

$$(a_{n+1} - a_{n+2})x^2 + (a_{n+2} - a_n)x + (a_n - a_{n+1}) = 0 .$$

VIII. 59. Let $a, b, c > 1$ be real numbers such that $2\log_a b = 1 + \log_a c$.

Show that the numbers $1 - \log_{abc}(bc)$, $1 - \log_{abc}(ac)$, $1 - \log_{abc}(ab)$ are in an arithmetic sequence.

VIII. 60. In an arithmetic sequence (a_n), $a_1 = \dfrac{\alpha - \beta}{2}$ and $d = \dfrac{\alpha + \beta}{2}$. Find n when $a_n = 5\alpha + 4\beta$, $(\alpha, \beta \in \mathbf{R})$.

VIII. 61. Find the formula for the nth term for an arithmetic sequence (a_n), if $a_1 = \alpha$ and $\dfrac{S_m}{S_n} = \dfrac{m^2}{n^2}$, where $m, n \in \mathbf{N}^*$, $\alpha \in \mathbf{R}$.

VIII. 62. In an arithmetic sequence with $m + n$ terms, $m, n \in \mathbf{N}^*$, the sum of the first m terms is equal to the sum of the next n terms. Prove that $S_{m+n} = 0$.

VIII. 63. Prove that for any $a, b \in \mathbf{R}$, the numbers $(a+b)^2$, $a^2 + b^2$, $(a-b)^2$ form an arithmetic sequence.

VIII. 64. Prove that if the numbers

$\dfrac{1}{a+2b}$, $\dfrac{1}{3c+a}$, $\dfrac{1}{3c+2b}$, $a \neq -2b \neq -3c$ form an arithmetic sequence, then the

numbers a^2, $4b^2$, $9c^2$ are also in an arithmetic sequence.

VIII. 65. If a, b, c and m_a, m_b, m_c are the sides and the medians of the $\triangle ABC$, respectively, and a^2, b^2, c^2 form an arithmetic sequence, then m_a^2, m_b^2, m_c^2 also form an arithmetic sequence.

VIII. 66. If the numbers a, b, c are positive numbers and form an arithmetic sequence, show that the numbers $\dfrac{1}{\sqrt{b} + \sqrt{c}}$, $\dfrac{1}{\sqrt{c} + \sqrt{a}}$, and $\dfrac{1}{\sqrt{a} + \sqrt{b}}$ form an arithmetic sequence.

VIII. 67. Show that the elements of the set $\{a, b, c\}$ form an arithmetic sequence if and only if $2(a^3 + b^3 + c^3) + 21abc = 3(a + b + c)(ab + bc + ca)$.

VIII. 68. If the numbers $\dfrac{1}{1-x}$, $\dfrac{m}{x(1-x)}$, and $\dfrac{2x - m}{x(1-x)}$,

$x \in \mathbf{R} \setminus \{0, 1\}$ are positive numbers and form an arithmetic sequence, find $m \in \mathbf{R}$ such that $x = 3$.

183

VIII. 69. Let a, b, c, d be non zero real numbers such that the numbers

$$a\sqrt[3]{\frac{a^2}{bcd}}\ ,\ b\sqrt[3]{\frac{b^2}{acd}}\ ,\ c\sqrt[3]{\frac{c^2}{abd}}\ ,\ d\sqrt[3]{\frac{d^2}{abc}}\quad \text{form an arithmetic sequence. Prove that}$$

the numbers a^2, b^2, c^2, d^2 form an arithmetic sequence.

VIII. 70. Solve the equation $\dfrac{1}{x}+\dfrac{2}{x}+...+\dfrac{x-1}{x}=10$, where x is a non-zero positive integer number.

VIII. 71. Solve the equations:

a) $3+7+11+...+x=210$;

b) $(2x+3)+(2x+7)+(2x+1)+...+(2x+39)=210$;

c) $(2x^2+1)+(2x^2+4)+(2x^2+7)+...+(2x^2+37)=999$.

VIII. 72. In an arithmetic sequence $a_1,a_2,a_3,...,a_{2n+1}$ with positive terms and a non zero ratio r, prove that

$$\prod_{k=1}^{n}\frac{a_{2k-1}}{a_{2k}}\le\sqrt{\frac{a_1}{a_{2n+1}}}\ .$$

VIII. 73. In an arithmetic sequence $a_1,a_2,a_3,...,a_n$ where d is the ratio, prove the identities:

a) $\dfrac{1}{\sqrt{a_1}+\sqrt{a_2}}+\dfrac{1}{\sqrt{a_2}+\sqrt{a_3}}+...+\dfrac{1}{\sqrt{a_{n-1}}+\sqrt{a_n}}=\dfrac{n-1}{\sqrt{a_1}+\sqrt{a_n}}$;

b) $a_1^k-\binom{n}{1}\cdot a_2^k+\binom{n}{2}\cdot a_3^k+...+(-1)^n\binom{n}{n}\cdot a_{n+1}^k=0$, $k\ge1$;

c) $\left(1-\dfrac{d^2}{a_2^2}\right)\cdot\left(1-\dfrac{d^2}{a_3^2}\right)\cdot...\cdot\left(1-\dfrac{d^2}{a_{n-1}^2}\right)=\dfrac{a_1a_n}{a_2a_{n-1}}$.

VIII. 74. In an arithmetic sequence $a_1,a_2,a_3,...,a_n$, $n\in\mathbf{N}^*$, where $d\ne0$ is the ratio, calculate:

$$S_1=\frac{1}{a_1a_2}+\frac{1}{a_2a_3}+...+\frac{1}{a_{n-1}a_n}\ ;$$

$$S_2=\frac{a_1+a_2}{a_1^2a_2^2}+\frac{a_2+a_3}{a_2^2a_3^2}+...+\frac{a_{n-1}+a_n}{a_{n-1}^2a_n^2}\ ;$$

$$S_3=\frac{a_2^3-d^3}{a_2^3+d^3}+\frac{a_3^3-d^3}{a_3^3+d^3}+...+\frac{a_{n-1}^3-d^3}{a_{n-1}^3+d^3}\ ;$$

$$S_4 = \frac{1}{\sqrt{a_1} + \sqrt{a_2}} + \frac{1}{\sqrt{a_2} + \sqrt{a_3}} + \dots + \frac{1}{\sqrt{a_{n-1}} + \sqrt{a_n}} .$$

VIII. 75. Calculate the sums

$$A_n = a_1 a_2 a_3 \dots a_k + a_2 a_3 a_4 \dots a_{k+1} + \dots + a_{n+1} a_{n+2} a_{n+3} \dots a_{n+k}$$

and $B_n = \dfrac{1}{a_1 a_2 \dots a_k} + \dfrac{1}{a_2 a_3 \dots a_{k+1}} + \dots + \dfrac{1}{a_n a_{n+1} \dots a_{n+k-1}}$, where

$(a_n)_{n>1}$ is an arithmetic sequence.

VIII. 76. Let $a_1, a_2, \dots, a_n$ be a sequence in an arithmetic sequence and

$$S_{n+1} = \sum_{i=1}^{n+1} a_i . \text{ Prove that:}$$

a) $3S_n - 3S_{2n} + S_{3n} = 0$;

b) $(p - q)pqS_k + (k - q)kqS_p = (k - p)kpS_q , \quad \forall k, p, q \in \mathbf{N}^*$;

c) $\displaystyle\sum_{k=0}^{n} C_n^k a_{k+1} = \dfrac{2^n S_{n+1}}{n+1}$.

VIII. 77. Consider the arithmetic sequence $(a_n)_{n\geq 1}$ and the sequence

$(b_n)_{n\geq 1}$ such that $b_n = 2^{a_n}$. Show that the sequence $(b_n)_{n\geq 1}$ is a geometric sequence and

$$a_1^2 - a_2^2 + a_3^2 - a_4^2 + \dots + a_n^2 - a_{n+1}^2 = \frac{n+1}{2n} \cdot \left(a_1^2 - a_{n+1}^2 \right) .$$

GEOMETRIC SEQUENCES

$b_1, b_2, b_3, \dots, b_n$; r b_n is the nth term, r is the ratio.

$$b_n = b_1 r^{n-1}$$

$$b_k^2 = b_{k-1} b_{k+1} , \quad k = 2, 3, 4, \dots, n-1 ,$$

$$b_1 b_n = b_i b_{n-i+1} , \quad i = 1, 2, 3, \dots, n ,$$

$$S_n = b_1 + b_2 + \dots + b_n = \sum_{k=1}^{n} b_k$$

$$S_n = \frac{b_1 \left(1 - r^n\right)}{1 - r} , \quad r \neq 1 ,$$

185

If $|r| < 1$, $S = \dfrac{b_1}{1-r}$, where $S = b_1 + b_2 + ... = \displaystyle\sum_{k=1}^{\infty} b_k$.

VIII. 78. Let $b_n = b_1 r^{n-1}$ be a geometric sequence. Show that the sequence $q_n = b_{n+1} - b_n$, is also a geometric sequence with the same ratio.

VIII. 79. Find the first five terms for the geometric sequence (b_n), knowing that:

a) $b_1 = 2$, $r = 3$; b) $b_2 = -6$, $r = -2$;

c) $b_2 = \sqrt{6}$, $b_5 = 9\sqrt{2}$; d) $b_3 = 20$, $b_9 = 1280$.

VIII. 80. If 9th term in a geometric sequence is 45 927 and the common ratio is 3, what is the first term?

VIII. 81. If 8th term of a geometric sequence is 256 and the 5th term is 32, what is the common ratio?

VIII. 82. Find the general term of the sequence
6, 42, 294, 2058, 14406,...

VIII. 83. The 3rd term of a geometric sequence is 36, and the 6th term is 4.5. What is the recursive formula for the sequence?

VIII. 84. a) Consider the sequences $(a_n)_{n \geq 1}$ and $(b_n)_{n \geq 1}$ where $a_n = \left(\dfrac{1}{2}\right)^n$ and $b_n = 9^n$. Show that these sequences are geometric sequences and find their ratios.

b) In the sequence $(x_n)_{n \geq 1}$, $x_0 = 2$ and $x_{n+1} = \dfrac{6x_n}{x_n + 3}$, $\forall n \in \mathbf{N}$. Prove that the sequence $y_n = \dfrac{x_n}{3 - x_n}$, $\forall n \in \mathbf{N}$ is a geometric sequence.

VIII. 85. Consider the sequences $(x_n)_{n \geq 0}$ and $(y_n)_{n \geq 0}$ such that $x_0 = 4$,

$x_{n+1} = \dfrac{3x_n + 1}{x_n + 3}$, $\forall n \in \mathbf{N}$ and $y_n = \dfrac{x_n + 1}{x_n - 1}$, $\forall n \in \mathbf{N}$. Prove that $(y_n)_{n \geq 0}$ is a geometric sequence and find the general formula for $(x_n)_{n \geq 0}$ and $(y_n)_{n \geq 0}$.

VIII. 86. Consider the sequences $(a_n)_{n \geq 0}$ and $(b_n)_{n \geq 0}$. In the system of coordinates xOy consider the points $A_n(\log_2 a_n, \log_3 b_n)$, $n \geq 1$. Write the equation of the line which passes through the points A_1 and A_2 and prove that the points $A_n(\log_2 a_n, \log_3 b_n)$ lie on the line $A_1 A_2$ for any $n \geq 1$.

VIII. 87. Find the terms b_2, b_3, b_5, b_6 of the following geometric sequence $\sqrt{2}$, b_2, b_3, $3\sqrt{6}$, b_5, b_6, $27\sqrt{2}$,.....

VIII. 88. The numbers $\sqrt{3}$, x, y, $18\sqrt{2}$ are consecutive terms in a geometric sequence. Find x and y.

VIII. 89. Show that the numbers 1, $\log_3 9$, $\sqrt[3]{64}$ are consecutive terms in a geometric sequence.

VIII. 90. How many terms are in the geometric sequence 2, 6, 18, 54, ..., 118 098?

VIII. 91. For the geometric sequence $(b_n)_{n \geq 1}$, $b_1 = -1$, $r = 3$, calculate b_5 and S_{10}.

VIII. 92. In a geometric sequence $(b_n)_{n \geq 1}$, $b_2 + b_3 = 14$ and $b_3 + b_4 = 28$. Find b_1, r, S_{10}.

VIII. 93. Insert two numbers between 5 and 135 such that the four numbers form a geometric sequence.

VIII. 94. What number must be added to each term of the sequence 1, 9, 33 to make it into a geometric sequence?

VIII. 95. Find b_1 and r for a geometric sequence if:

a) $\begin{cases} b_5 - b_1 = 80 \\ b_4 - b_2 = 24 \end{cases}$;   b) $\begin{cases} b_5 + b_2 - b_4 = 114 \\ b_4 + b_1 - b_3 = 38 \end{cases}$.

VIII. 96. In a geometric sequence of positive numbers, each term after the second is the sum of the two preceding terms. Find the common ratio.

VIII. 97. If in a geometric sequence of positive terms any term is equal to the sum of the next two terms, find the ratio.

VIII. 98. The positive numbers x, y, z form a geometric sequence. Prove that the numbers $\log x$, $\log y$, $\log z$ form an arithmetic sequence.

VIII. 99. Prove the identity

$$\left(1 + x + x^2 + \ldots + x^{n-1}\right)\left(1 + x + x^2 + \ldots + x^{n+1}\right) = \left(1 + x + x^2 + \ldots + x^n\right)^2 - x^n.$$

VIII. 100. The equations $x^2 - 9x + m = 0$ and $x^2 - 36x + n = 0$ have the roots x_1, x_2 and x_3, x_4, repectively. If the roots are in an increasing geometric sequence in this order, find the numbers m and n.

VIII. 101. The sum of the first three terms of a geometric sequence is 21 and their product is 64. Find the numbers.

VIII. 102. The sum of the first and fifth terms of a geometric series is equal to twice the sum of the second and sixth terms. What is the ratio of the geometric series?

VIII. 103. The product of the first three terms of a geometric sequence is $3\sqrt{3}$ and the sum of their cubes is $28 + 3\sqrt{3}$. Find these first three terms.

VIII. 104. Let a, b, c be three real numbers in an arithmetic sequence with ratio r. In each case find the numbers a, b, c ; if:
a) $a + 4, b + 2, c + 6$ are a geometric sequence and $a + b + c = 30$;
b) $a, b + 2, c + 12$ are a geometric sequence with ratio $r + 1$.

VIII. 105. a) The numbers a, b, c, and d are in a geometric sequence, if the numbers $a + 1, b + 6, c + 6$, and $d - 4$ are an arithmetic sequence, find the numbers;
b) The numbers a, b, c, and d are in an arithmetic sequence, if the numbers $a + 1, b - 1, c - 1$, and $d + 3$ are a geometric sequence, find the numbers.

VIII. 106. The numbers a, b, and c are in a geometric sequence, $a + b + c = 21$, and $a^2 + b^2 + c^2 = 189$. Find the numbers.

188

VIII. 107. The sum of three numbers in a geometric sequence is 13. If 1 is added to the first and second, and it is subtracted from the third, they will be in a geometric sequence. Find these numbers.

VIII. 108. The sum of the first three terms of a geometric series is 21 times the first term. The sum of the first and the third term is 17. What is the second term?

VIII. 109. The sum of an infinite geometric series is equal to three times its initial term. What is the ratio of the series?

VIII. 110. Determine the sum of the following infinite geometric series with first term b_1 and ratio r:

a) $b_1 = 7, r = -\dfrac{3}{4}$; b) $b_1 = -2, r = \dfrac{1}{3}$;

c) $b_1 = 3.865, r = 0.227$; d) $b_1 = -4, r = -\dfrac{1}{2}$;

e) $b_1 = 35, r = \dfrac{12}{17}$; f) $b_1 = 5, r = 0.33$.

VIII. 111. In an arithmetic sequence, $a_5 + a_6 = 44$ and the terms a_1, a_4, a_{16} form a geometric sequence. Find a_1 and the common difference for the arithmetic sequence.

VIII. 112. Three numbers form a geometric sequence. Increasing the second number by 8, the numbers form an arithmetic sequence and if, in the arithmetic sequence, the last term is increased by 64, the numbers form a geometric sequence. Find the numbers.

VIII. 113. In a geometric sequence, find four successive terms such that the second term be smaller than the first by 54, and the third term be larger than the forth by 6.

VIII. 114. The first term of an arithmetic sequence is 1 and the sum of the first nine terms is equal to 369. The first and ninth terms of the arithmetic sequence coincide with the first and ninth terms of a geometric sequence, respectively. Find the ratio of the geometric sequence.

VIII. 115. The sum of four consecutive terms of a geometric sequence is 60, and the fourth term is 4 times the second term. Find the fourth term.

VIII. 116. The sum of the first five terms in a geometric sequence is 62. At the same time, the fifth, eighth, and eleventh term of this sequence are the first, the second, and the tenth in an arithmetic sequence, respectively. Find the first term of the geometric sequence.

VIII. 117. The sum of the first three terms in a geometric sequence is 65 and the sum of their logarithms in base 15 is 3. Find the first term and ratio.

VIII. 118. In a geometric sequence of nine positive terms, the third and fifth terms are the smallest and largest solution of the equation
$\frac{1}{2}\left(1+\frac{1}{2}\log_2(3x-2)\right) = \log_4\left(1+\sqrt{10x-1}\right)$. Find the sum of the geometric sequence.

VIII. 119. Let (a_n) and (b_n), $n \geq 1$ be an arithmetic and a geometric sequence, such that $a_1 + a_2 + ... + a_{10} = 155$ and $b_1 + b_2 = 9$. Find these sequences, if a_1 is the ratio of the geometric sequence and b_1 is the ratio of the arithmetic sequence.

VIII. 120. If the sum of the first and the fifth term of an geometric sequence with positive terms is 164 and the product of the third term and the fifth term is 2916, find the 10th term of the sequence.

VIII. 121. The numbers a, b, and c are simultaneously the 5th, 17th, and 37th terms of an arithmetic and a geometric sequence, prove that $a^{b-c}b^{c-a}c^{a-b} = 1$.

VIII. 122. The numbers a, b, and c form an geometric sequence. Prove that:

a) $a^2b^2c^2\left(\frac{1}{a^3}+\frac{1}{b^3}+\frac{1}{c^3}\right) = a^3 + b^3 + c^3$;

b) $\log_a \frac{a}{b^2} = \log_{\frac{c}{b^2}} c$.

VIII. 123. Let a, b, and c be three real numbers. If $(ab + bc + ca)^3 = abc(a+b+c)^3$, then the numbers a, b, and c form an geometric sequence.

VIII. 124. The real positive numbers $b_1, b_2, b_3, ..., b_{2n+2}$, form a decreasing geometric sequence.

Prove that $\dfrac{b_1}{b_2} \cdot \dfrac{b_3}{b_4} \ldots \dfrac{b_{2n-1}}{b_{2n}} < \sqrt{\dfrac{b_1}{b_{2n+2}}}$.

VIII. 125. In a sequence of real numbers (a_n), $a_n = 2^{1-n} \cdot 3^n$ for any n, a positive integer.

a) Prove that the sequence $(a_n)_{n \in \mathbb{N}^*}$ is a geometric sequence;

b) Find $n \in \mathbb{N}^*$ such that $a_1 + a_2 + \ldots + a_n = \dfrac{633}{16}$.

VIII. 126. Consider the sequences of real numbers (a_n) and (b_n), where $a_n = 2n - 1$ and $b_n = 2^{a_n}$ for any n, a positive integer.

a) Prove that $(a_n)_{n \in \mathbb{N}^*}$ is a arithmetic sequence and $(b_n)_{n \in \mathbb{N}^*}$ is a geometric sequence;

b) Find $n \in \mathbb{N}^*$ such that $a_1 + a_2 + \ldots + a_n = 2014^2$.

VIII. 127. Evaluate the sums:

a) $S = 1 + 2 \cdot 3 + 3 \cdot 3^2 + 4 \cdot 3^3 + 5 \cdot 3^4 + \ldots + 100 \cdot 3^9$;

b) $S_n = 3 + 3\ + 333 + \ldots + \underbrace{333\ldots3}_{n \text{ times}}$.

VIII. 128. Calculate the sums:

a) $S_1 = q + 2q^2 + 3q^3 + \ldots + nq^n$;

b) $S_2 = q + 4q^2 + 9q^3 + \ldots + n^2 q^n$, $q \in \mathbb{R}$;

c) $S_3 = \dfrac{1}{b_1} + \dfrac{1}{b_2} + \ldots + \dfrac{1}{b_n}$, where $b_1 = 3$, $b_6 = 96$, and $(b_n)_{n \in \mathbb{N}^*}$ is a geometric sequence.

VIII. 129. If $a \in \mathbb{R}$, $a \neq 0$, calculate the numbers:

$S_1 = \left(a + \dfrac{1}{a} \right)^2 + \left(a^2 + \dfrac{1}{a^2} \right)^2 + \ldots + \left(a^n + \dfrac{1}{a^n} \right)^2$ and

$S_2 = \left(a - \dfrac{1}{a} \right)^2 + \left(a^2 - \dfrac{1}{a^2} \right)^2 + \ldots + \left(a^n - \dfrac{1}{a^n} \right)^2$.

VIII. 130. Let $b_1, b_2, b_3, ..., b_n$ be a geometric sequence and $r \neq 1$ its ratio. Calculate:

a) $A_n = \displaystyle\sum_{k=2}^{n} \frac{\sqrt{b_{k-1}}}{\sqrt{b_k} - \sqrt{b_{k-1}}}$;   b) $B_n = \displaystyle\sum_{k=1}^{n} \frac{b_k b_{k+1}}{b_k + b_{k+1}}$.

VIII. 131. Let $b_1, b_2, b_3, ..., b_n$ be a geometric sequence and r be its ratio.

If $A = b_1 + b_2 + ... + b_n$ and $B = \dfrac{1}{b_1} + \dfrac{1}{b_2} + ... + \dfrac{1}{b_n}$, calculate $C = b_1 b_2 ... b_n$.

VIII. 132. In a geometric sequence $b_1, b_2, b_3, ..., b_n$, $b_7 = \left(x^4 + 1\right)^4$ and $b_{11} = \left(x^4 + 1\right)^6$, $x \in \mathbf{R}$. Calculate the sum $S = \displaystyle\sum_{k=1}^{n} \log_{b_1} b_k$.

VIII. 133. Consider the geometric sequence $(b_n)_{n \geq 1}$. Show that

$$\sum_{k=1}^{2n} (-1)^{k-1} b_k^2 \geq \sum_{k=2n+1}^{4n} (-1)^{k-1} b_k^2 , \; n \geq 1.$$

VIII. 134. Let $b_1, b_2, b_3, ..., b_n$ be n positive numbers in a geometric sequence. Prove that $b_1 b_2 ... b_n = \sqrt{\left(b_1 b_n\right)^n}$.

VIII. 135. Let $b_1, b_2, b_3, ..., b_n$ be in a geometric sequence, $b_i < b_{i+1}$ for any $i = 1, 2, ..., n - 1$, and p a non-zero positive integer. Prove that the ratio of the sums

$$S_1 = \frac{1}{b_2^p + b_1^p} + \frac{1}{b_3^p + b_2^p} + ... + \frac{1}{b_{n+1}^p + b_n^p} \text{ and}$$

$$S_2 = \frac{1}{b_2^p - b_1^p} + \frac{1}{b_3^p - b_2^p} + ... + \frac{1}{b_{n+1}^p - b_n^p} \text{ does not depend on } n.$$

VIII. 136. Prove that in a geometric sequence $\dfrac{S_{2n} - S_n}{S_{3n} - S_{2n}} = \dfrac{S_n}{S_{2n} - S_n}$.

VIII. 137. Let $b_1, b_2, ,, ..., b_n$ be a sequence of distinct positive numbers. Prove that this sequence is a geometric sequence if and only if

$$\frac{b_1}{b_2} \sum_{k=1}^{n-1} \frac{b_n^2}{b_k b_{k+1}} = \frac{b_n^2 - b_1^2}{b_2^2 - b_1^2}, \; \forall n \geq 2.$$

VIII. 138. Let x_1, x_2, $,...,x_n$ be a sequence of distinct positive numbers,

where $x_1 = a$ and $2\sum_{k=1}^{n}\sqrt{x_k} = (n+1)\sqrt{x_n}$, $\forall n \geq 2$. Calculate $\sum_{k=1}^{n} x_k$.

VIII. 139. Consider the numbers $x_0, x_1,..., x_n ,...$ such that

$x_n = 3x_{n-1} - \dfrac{1}{p^2} x_{n-1}{}^3$, where $|x_0| \leq 2p$ and p is a positive

number. Calculate x_n in terms of n and x_0.

VIII. 140. Consider the sequence $(x_n)_{n\in N}$, such that $x_0 = a$ and $x_1 = b$

where a, b are real numbers and $x_n = \dfrac{5}{7}\cdot x_{n-1} + \dfrac{2}{7}\cdot x_{n-2}$, $\forall n \geq 2$. Find x_n in terms of a, b and n.

Some considerations on **Fibonacci numbers**

Let $\qquad F_{n+1} = F_n + F_{n-1}$, $n \in N$ $\qquad$ (1)
be a recurrent sequence. Suppose that

$F_0 = 0$, $F_1 = 1$, $F_2 = 1$, $F_3 = 2$, $F_4 = 3$, $F_5 = 5$
Writing the terms of this sequence, we obtain
$\qquad$ 0, 1, 1, 2, 3, 5, 8, ... $\qquad$ (2)

called the Fibonacci sequence and F_n is called the nth Fibonacci number. Fibonacci was one of the greatest European mathematicians of the Middle Ages. His full name was Leonardo of Pisa or Leonardo Pisano in Italian, he was born in Pisa, Italy, the city of the famous Leaning Tower, about 1175 AD.

The Lucas numbers are defined by the equations $L_1 = 1$,

$\qquad L_n = F_{n+1} + F_{n-1}$ $\quad$ (3)

and satisfy the same recurrence $L_{n+1} = L_n + L_{n-1}$ where the first few numbers are
$\qquad$ 1, 3, 4, 7, 11, 18, 29, 47, 76,123, ...

The French mathematician Edouard Lucas (1842-1891), was the first who called the series of numbers 0, 1, 1, 2, 3, 5, 8, 13 the Fibonacci Numbers.

Assuming that the sequence F_n has the form $F_n = \lambda^n$, where λ is a real

parameter and substituting F_n in (1)

$$\lambda^{n+1} = \lambda^n + \lambda^{n-1} \qquad (4)$$

or equivalent $\lambda^{n-1}\left(\lambda^2 - \lambda - 1\right) = 0$

Since $F_n \neq 0$, $n \in \mathbb{N}^*$, the last equality becomes $\lambda^2 - \lambda - 1 = 0$ and it has the roots

$$\lambda_1 = \frac{1 + \sqrt{5}}{2} \quad \text{and} \quad \lambda_2 = \frac{1 - \sqrt{5}}{2} \qquad (5)$$

Thus the sequences of general term $a_n = \left(\dfrac{1 + \sqrt{5}}{2}\right)^n$, $b_n = \left(\dfrac{1 - \sqrt{5}}{2}\right)^n$

verify the equality (1). So we conclude that the equation (1) can have more solutions. In general there are an infinite number of sequences verifying (1). It is easy to observe that the general solution for (1) has the form

$$F_n = c_1 \left(\frac{1 + \sqrt{5}}{2}\right)^n + c_2 \left(\frac{1 - \sqrt{5}}{2}\right)^n \qquad (6)$$

where c_1, c_2 are arbitrary real numbers, verifying (1), as well. Also you can prove that any sequence of the form (6) satisfies (1).

For $n = 0$ and $n = 1$ in (6), we obtain the linear system

$$\begin{cases} c_1 + c_2 = 0 \\ c_1 \left(\dfrac{1 + \sqrt{5}}{2}\right) + c_2 \left(\dfrac{1 - \sqrt{5}}{2}\right) = 1 \end{cases}$$

having the solutions $c_1 = \dfrac{1}{\sqrt{5}}$, $c_2 = -\dfrac{1}{\sqrt{5}}$

Finally, the general term of the Fibonacci sequence has the form

$$F_n = \frac{1}{\sqrt{5}} \left(\frac{1 + \sqrt{5}}{2}\right)^n - \frac{1}{\sqrt{5}} \left(\frac{1 - \sqrt{5}}{2}\right)^n, \quad n \in \mathbb{N}. \qquad (7)$$

Some proprieties of the Fibonacci sequence:

1. $F_1 + F_2 + ... + F_n = F_{n+2} - 1$ (8)

Proof: $F_1 = F_3 - F_2$

             $F_2 = F_4 - F_3$

           ....

            $F_{n-1} = F_{n+1} - F_n$

            $F_n = F_{n+2} - F_{n+1}$

Summing all equalities we get $F_1 + F_2 + ... + F_n = F_{n+2} - F_2$,

since $F_2 = 1$ so (8) is proven.

2. $F_1 + F_3 + ... + F_{2n-1} = F_{2n}$.

3. $F_2 + F_4 + ... + F_{2n} = F_{2n+1} - 1$.

  (2. and 3. can be proven in a similar manner).

4. $F_1^2 + F_2^2 + ... + F_n^2 = F_n F_{n+1}$ (9)

Proof: It is easy to observe that

$$F_n F_{n+1} - F_{n-1} F_n = F_n(F_{n+1} - F_{n-1}) = F_n^2 , \quad n \in N .$$

From this we have successively the equalities:

$$F_1^2 = F_1 F_2$$

$$F_2^2 = F_2 F_3 - F_1 F_2$$

$$F_3^2 = F_3 F_4 - F_2 F_3$$

$$.....................$$

$$F_n^2 = F_n F_{n+1} - F_{n-1} F_n$$

Summing all these equalities we get (9).

5. Prove that $F_{n+m} = F_{n-1} F_m + F_n F_{m+1}$ (10)

where F_n is the nth term of the Fibonacci sequences.

Proof: Using (1) and (9), (10) is easily proven.

To prove (10) using mathematical induction is an option. Proceed by using mathematical induction with respect to $m \in N$. For $m = 1$,

equality (10) becomes $F_{n+1} = F_{n-1} F_1 + F_n F_2$, which is obvious, so (10) is true for $m = 1$. (For example when $m = 2$ the formula (10) is true

$F_{n+2} = F_{n-1} F_2 + F_n F_3 = F_{n-1} + 2F_n = F_{n+1} + F_n$).

Assuming that (10) is true for $m = k$ and $m = k + 1$,
prove that (10) is true for $m = k + 2$.

Therefore the following equalities hold $F_{n+k} = F_{n-1} F_k - F_n F_{k+1}$

$$F_{n+k+1} = F_{n-1} F_{k+1} - F_n F_{k+2} ,$$

$$F_{n+k+2} = F_{n-1} F_{k+2} - F_n F_{k+3}$$

By adding these equalities term by term, we obtain (10)
for $m = k + 2$.

6. Prove that $F_{2n} = F_n L_n$ (sometimes called *double angle formula*).

Proof: By taking $m = n$ in (10) we obtain

$$F_{2n} = F_{n-1} F_n + F_n F_{n+1} = F_n(F_{n-1} + F_{n+1}) = F_n L_n ,$$

7. Prove that $F_n^2 = F_{n-1} F_{n+1} + (-1)^{n+1} , \quad n > 1$ (11)

Proof: By mathematical induction.

For $n=2$, (11) becomes, $F_2^2 = F_1 F_3 - 1$, which is true. Thus (11) is true for $n=2$. Assuming that (11) is true for n, prove that it is true for $n+1$, as well. So

$F_n^2 = F_{n-1} F_{n+1} + (-1)^{n+1}$ is true. Adding on both sides the last equality of the

number $F_n F_{n+1}$ we obtain, $F_n^2 + F_n F_{n+1} = F_{n-1} F_{n+1} + F_n F_{n+1} + (-1)^{n+1}$ or

$F_n(F_n + F_{n+1}) = F_{n+1}(F_{n-1} + F_n) + (-1)^{n+1}$.

Next, using (1), we have $F_n F_{n+2} = +F_{n+1}^2 + (-1)^{n+1}$, and finally

$F_{n+1}^2 = F_n F_{n+2} + (-1)^{n+2}$. Hence, (11) is true for $n+1$.

8. Prove that $2^{n+1} - 7F_{n-1}^2 = (F_{n-1} + 2F_n)^2$. (use mathematical induction).

Chapter IX

POLYNOMIALS AND ALGEBRAIC EQUATIONS

$$P(x) = a_n x^n + a_{n-1} x^{n-1} + a_{n-2} x^{n-2} + ... + a_1 x + a_0 ,$$

$a_n, a_{n-1}, a_{n-2}, + ... + a_1, a_0$ are the coefficients. If $a_n \neq 0$ then the degree of the polynomial is n.

Vieta's formulas
$$\begin{cases} x_1 + x_2 + .. + x_n = (-1)^1 \dfrac{a_{n-1}}{a_n} \\ x_1 x_2 + x_1 x_3 + .. + x_{n-1} x_n = (-1)^2 \dfrac{a_{n-2}}{a_n} \\ .... \\ x_1 x_2 \cdot .. \cdot x_n = (-1)^n \dfrac{a_0}{a_n} \end{cases} .$$

Bézout's theorem (*remainder theorem*): The remainder of a polynomial when divided by $x - a$ is $r = P(a)$.

Factor theorem: If $P(x)$ is a polynomial, then $P(a) = 0$ if and only if $x - a$ is a factor of $P(x)$.

Fundamental theorem of algebra (d'Alembert-Gauss). Every polynomial with complex coefficients of the degree at last one has at least a complex root. An equivalent form of this theorem is: Every polynomial of degree n, $n > 0$ can be factored into n linear factors, that is, $P(x) = (x - x_1)(x - x_2)...(x - x_n)$.

IX. 1. Divide $P(x) = x^4 + x^3 + 2x^2 + 1$ by $x + 1$ using synthetic division (*Horner's algorithm*).

IX. 2. Using synthetic division (*Horner's Algorithm*), find the quotient q and the remainder r when dividing the polynomial f by the polynomial g:

a) $f(x) = x^3 + 3x^2 - 7x - 6$, $g(x) = x - 2$;

b) $f(x) = 2x^4 + 3x^3 - 5x^2 - 3x + 1$, $g(x) = x - 3$;

c) $f(x) = 3x^4 + 6x^3 - 8x^2 + 3x - 3$, $g(x) = x - 1$;

d) $f(x) = x^4 - x^3 + 5x^2 - 6x + 3$, $g(x) = x + 1$;

e) $f(x) = 4x^3 + 6x^2 - 7x + 8$, $g(x) = 2x + 1$;

f) $f(x) = x^5 - x^4 + 3x^2 + 3x - 2$, $g(x) = x + 2$;

g) $f(x) = x^5 - 2x^4 + 3x^2 - 5$, $g(x) = x - 1$;

h) $f(x) = x^8 - 4x^6 + 6x^4 - 2x^2 + 2$, $g(x) = x + 1$.

IX. 3. Using long division, find the quotient q and the remainder r when dividing the polynomial f to g:

a) $f(x) = 2x^3 + 2x^2 - 6x + 1$, $g(x) = x^2 - x + 2$;

b) $f(x) = x^4 + 2x^3 + 3x^2 + 7x + 1$, $g(x) = x + 3$;

c) $f(x) = x^4 + 6x^3 + 8x^2 + 1$, $g(x) = x^3 - x + 1$;

d) $f(x) = x^4 + x^3 + 7x^2 - 6x + 8$, $g(x) = x^2 - x + 1$;

e) $f(x) = x^5 - x^4 + 2x^3 - 3x^2 + x - 4$, $g(x) = x^2 - x + 3$;

f) $f(x) = x^5 - x^4 + 3x^2 - 4$, $g(x) = x^2 - x$;

g) $f(x) = x^8 - 3x^6 + 5x^4 - 2x^2 + 1$, $g(x) = x^3 - 4x + 1$.

IX. 4. Is $x - 1$ a factor of the polynomial
$f(x) = x^3 + 2x^2 - 6x + 3$?

IX. 5. Determine the remainder when $x^3 + 2x^2 - 6x + 1$ is divided by $x + 2$.

IX. 6. Consider the polynomial $P(x) = x^3 - 2x^2 - 5x + 6$.
a) Calculate $P(1)$;
b) Find the quotient and the remainder when P is divided by $x - 1$;
c) Solve the equation $P(x) = 0$.

IX. 7. Show that both $x - 2$ and $x - 3$ are factors of the polynomial $2x^3 - 11x^2 + 17x - 6$ and find another factor of this polynomial.

IX. 8. Use a polynomial to find three consecutive integers with a product of -1320.

IX. 9. The dimensions of a rectangular solid are $x - 1$, $x - 2$, and $x + 3$. The volume of the solid is 42 cm³. Find the dimensions of the solid.

IX. 10. The radius of a sphere is r. The volume, $V(r)$, of a larger sphere is $V(r) = \frac{4}{3}\pi\left(r^3 + 12r^2 + 48r + 64\right)$. What is the radius of the larger sphere?

IX. 11. Consider the polynomial function
$$f(x) = x^4 - 5x^3 + 4x^2 + 5x - 6.$$
a) Calculate $f(1)$ and $f(-1)$;

b) Calculate the quotient and remainder when f is divided by $(x-1)(x+1)$;

c) Solve the equation $f(x) + 1 = 0$.

IX. 12. When a linear polynomial $P(x)$ is divided by $x+1$ the remainder is 2 and when divided by $x-1$ the remainder is -3. Find the polynomial.

IX. 13. Using the Factor Theorem, prove that:

a) $x^3 - 6x^2 + 3x + 10$ is divisible by $x^2 - x - 2$;

b) $(x+a)^3 + (x+b)^3 + (a-b)^3$ is divisible by $x+a$.

IX. 14. Find k if:

a) $x-2$ is a factor of $x^3 + kx^2 + 3x - 10$;

b) $x+4$ is a factor of $x^3 + 3x^2 - kx + 12$;

c) $x-3$ is a factor of $x^3 + 3x^2 + 2x + k$.

IX. 15. Find the numeric value of m such that the polynomial function $f(x) = x^3 - 3x^2 + mx - 6$ is divisible by $x-2$.

IX. 16. Find the real number m, such that the polynomial function $f(x) = 2x^4 - mx^3 - 5x^2 - 8x + 7$ divided by $x-1$, yield the remainder is -3.

IX. 17. When the polynomial $f(x) = x^3 - mx^2 - 7mx + 1$ is divided by $x-2$, the remainder is -9. Find the value of m.

IX. 18. Find $a \in \mathbf{R}$ such that the polynomial function $P(x) = x^4 + (a+2)x^3 + 3a^2x^2 - 5x - 14$ be divisible by $x+2$.

IX. 19. The polynomial $P(x) = x^6 - mx^4 + (m^2 + 4)x^2 - 2$, $m \in \mathbf{R}$, yields a remainder of 5 when divided by $x - 1$. Find the value of m.

IX. 20. Suppose that we have the factorization
$6x^3 + ax^2 - 21x - 10 = (mx - 5)(3x^2 + nx + 2)$, where a, m, n are real numbers. What are the values of a, m, n?

IX. 21. Find the numeric value(s) of $m \in \mathbf{R}$ such that the polynomial function $f(x) = x^3 - 4x^2 + 9x - 6$ is divisible by $x - m$.

IX. 22. Let $f(x) = x^3 + ax^2 + bx - 6$ be a polynomial. If $f(x)$ is divisible by $x + 1$ yields a remainder of -2 and when divided by $x + 2$ the remainder is -19, what are the values of a and b?

IX. 23. Find a quadratic polynomial function $f(x) = ax^2 + bx + c$, $a \neq 0$, such that:
a) $f(-1) = 0$, $f(1) = 2$, $f(2) = 9$;
b) $f(1) = f(2) = 0$, $f(-1) = 18$.

IX. 24. Find the value of a and b if the polynomial
$f(x) = x^5 + x^4 - 9x^3 + ax^2 + bx + c$ is divisible by $(x + 3)(x^2 - 4)$.

IX. 25. Consider the polynomial function $f(x) = x^3 + ax + b$, a and b real parameters.
a) Find a and b such that f be divisible by $x + 1$ and $f(1) = 4$;
b) For $a = 1$ and $b = 2$, solve the equation $f(x) = 0, x \in \mathbf{C}$;
c) For $a = 1$ and $b = 2$, solve the inequation $f(x) < 0$, $x \in \mathbf{R}$.

IX. 26. The divisions $(2x^3 + 4x^2 - kx + 5) \div (x + 3)$ and
$(6y^3 - 3y^2 + 2y + 7) \div (2y - 1)$ yields the same remainder. Determine the value of k.

IX. 27. If the polynomial $4x^3 + ax^2 + bx + 11$ is divided by $x + 2$, the remainder is -7 and if the polynomial is divided by $x - 1$, the remainder is 14. Find the values of a and b.

IX. 28. If the polynomial $x^4 + ax^3 + x^2 + bx + 3$ is divided by $x-1$, the remainder is 3 and if the polynomial is divided by $x-2$, the remainder is 5. Find the values of a and b.

IX. 29. Find the values of k when $f(x) = x^3 + 6x^2 + kx - 4$ gives the same remainder when divided by either $x-1$ or $x+2$.

IX. 30. When $f(x) = x^5 - 2x^4 - mx^3 - x^2 + nx - 2$ is divided by $x+1$ the remainder is -7 and divided by $x-2$ the remainder is 32. Find the values of m and n.

IX. 31. When $10x^3 + mx^2 - x + 10$ is divided by $5x - 3$, the quotient is $2x^2 + nx - 2$ and the remainder is 4. Find the values of m and n.

IX. 32. A polynomial $f(x)$ divided by $x+2$, yields remainder 3, find the remainder when $g(x) = f(x) + x^3 + 11$ is divided by $x+2$.

IX. 33. Let $P(x) = ax^3 + bx^2 + cx + d$ be a polynomial. Find a, b, c, and d such that $P(-1) = -1$, $P(0) = 0$, $P(1) = 1$, and $P(2) = 0$.

IX. 34. Consider the polynomial function
$$f(x) = 1 + (x + a)^1 + (x + a)^2 + \ldots + (x + a)^n.$$
Calculate the sum of coefficients of f.

IX. 35. What is the degree of the polynomial
$$f(x) = (m^2 + 3m + 2)x^4 + (m^3 + 2m^2 - m - 2)x^3 + (m^2 + 4m + 3)x^2 + (m^2 - 1)x + 3,$$
$m \in \mathbf{R}$?

IX. 36. Show that if $a, b \in \mathbf{R}$, $a \neq b$ the remainder of dividing a polynomial $P(x)$ of degree at least two by $(x - a)(x - b)$ is
$$r(x) = \frac{P(b) - P(a)}{b - a} x + \frac{bP(a) - aP(b)}{b - a}.$$

IX. 37. If a polynomial of degree of at least three is divided by $x-1$ the remainder is zero, divided by $x+1$ the remainder is one, and if the polynomial is divided by $x-2$, the remainder is -3. Find the remainder if the polynomial is divided by $(x-1)(x+1)(x-2)$.

IX. 38. The polynomial $f(x) = x^5 - mx^2 - mx + 1$ is divisible by the polynomial $g(x) = x^2 + 2x + 1$. Find the value of $m \in \mathbf{R}$.

IX. 39. The polynomial $f(x) = x^5 - 2x^4 + 3x^3 - 7x^2 + 8x + a$ is divisible by the polynomial $g(x) = x^3 - 3x^2 + 3x - 1$.
Find the numeric value of $a \in \mathbf{R}$.

IX. 40. Find a, b, $c \in \mathbf{R}$ such that the polynomial
$f(x) = (x+1)^6 + a(x+1)^3 + bx + c$ be divisible by $(x-1)^3$.

IX. 41. Let $P(x)$ be a polynomial of degree at least two. If $P(x)$ divided by $x+1$ the remainder is 2 while $(x+1) P(x) + x P(x+3) = 1$. Find the remainder when $P(x)$ is divided by $x^2 - x - 2$.

IX. 42. Let $P(x) = x^3 - 5x^2 + 6x$. Find the zeros of the polynomial and sketch the graph.

IX. 43. Determine the equation of the polynomial with zeros ± 1, -2, and y-intercept of -6.

IX. 44. A polynomial has the roots -3, -1, 1, and 2 and the graph passes through the point $(0, -12)$. Find the polynomial.

IX. 45. Find the equation of the quartic function that satisfies the following conditions:
a) the graph of the function crosses the x-axis at $x = -2$;
b) the function is tangent to the x-axis at 2;
c) the graph of the function passes through the points $(1, -18)$ and $(3, -10)$.

IX. 46. Determine the cubic function that passes through the points $(1, 12)$, $(0, -4)$, $(2, 0)$ and $(-1, 0)$.

IX. 47. Sketch the graph of each polynomial function:

a) $f(x) = -0.5(x+1)(x-2)(x-4)$;

b) $f(x) = 0.4(x+1)(x-2)^2(x-3)$;

c) $f(x) = x(x+1)^2(x-2)^3$.

IX. 48. Let $f(x) = 0.05(x-4)(x-1)(x+3)^2$ be a polynomial function.

a) Describe the end behavior and draw the graph;

b) State the interval(s) on which $f(x) \geq 0$;

c) State the interval(s) on which $f(x) < 0$.

IX. 49. Find a possible polynomial equation for the graph:

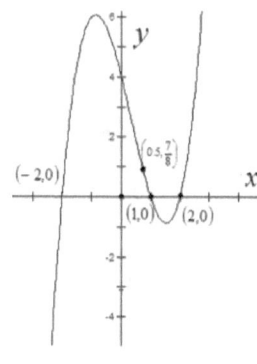

IX. 50. Given the graph of the function $y = f(x)$, sketch each transformation:

a) $y = 2f(x-2)$;   b) $y = f(-x) + 3$;

c) $y = -2f\left(\frac{1}{3}x\right)$;   d) $y = \frac{1}{3}f\left(\frac{1}{2}x\right) - 1$.

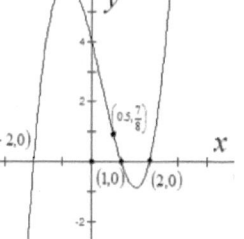

IX. 51. Find the equation of a polynomial function with x-intercepts at –3, 1, 4 and degree 3, passing through the point (2, –20).

IX. 52. a) Find the family of polynomial functions that has degree 4 and have x-intercepts –3, 0, 2, 4;

b) Find the polynomial of this family that passes through the point (1, –24).

IX. 53. a) Factor $f(x) = x^4 + 2x^3 - x^2 - 2x$, given that $f(-2) = 0$;

b) Sketch the graph of $f(x)$;

203

c) Using the graph of f, find x such that $f(x) \geq 0$.

IX. 54. The polynomial function $f(x) = ax^4 - 30x^2 + 20x + b$ has a root $x = 2$ and the graph passes through the point $(4, 160)$.
a) Find a and b;

b) Factor $f(x)$;

c) Find y-intercept and draw the graph of $f(x)$.

IX. 55. Consider the polynomial $f(x) = ax^3 + bx^2 + cx - 12$. Find the real numbers a, b, c such that the graph of f intersects the x- axis at the points $(-3, 0)$, $(-1, 0)$, and $(2, 0)$.

IX. 56. Prove that each polynomial function:

a) $f(x) = x^{n+1} - (n+1)x + n$, $n \in \mathbf{N}^*$ is divisible by $(x-1)^2$;

b) $f(x) = nx^{n+1} - (n+1)x^n + 1$. $n \in \mathbf{N}^*$ is divisible by $(x-1)^2$;

c) $f(x) = (2n-1)x^{2n} - 2nx^{2n-1} + 1$. $n \in \mathbf{N}^*$ is divisible by $(x-1)^2$;

d) $f(x) = x^{n+2} - x^{n-1} - 3x + 3$. $n \in \mathbf{N}^*$ is divisible by $(x-1)^2$.

IX. 57. Find $m \in \mathbf{R}$ such that the polynomial function
$f(x) = x(x+1)^{2n+1} + (m-1)x^n$ be divisible by $x^2 + x + 1$.

IX. 58. Prove that each of the polynomials:

a) $f(x) = (x^2 + x - 1)^{4n+1} - x$, $n \in \mathbf{N}^*$ is divisible by $x^2 - 1$;

b) $f(x) = (x^2 + x + 1)^{8n+1} - x$, $n \in \mathbf{N}^*$ is divisible by $x^2 + 1$;

c) $f(x) = (x^2 - x + 1)^{2n+2} + (x^4 + x - 1)^{2n+4} - x^2 - 1$, $n \in \mathbf{N}^*$ is divisible by $x^2 + 1$;

d) $f(x) = (2x^2 + x + 2)^{4n+1} - x$, $n \in \mathbf{N}^*$ is divisible by $x^2 + 1$;

e) $f(x) = x^{6n-1} + x + 1$, $n \in \mathbf{N}^*$ is divisible by $x^2 + x + 1$;

f) $f(x) = (x+1)^{6n+1} + (x+1)^{6n+5} + (x^2 + 1)^{6n+3}$, $n \in \mathbf{N}^*$ is divisible by $x^2 + x + 1$;

g) $f(x) = (x-1)^{2n+1} - x^{n+2}$, $n \in \mathbf{N}^*$ is divisible by $x^2 - x + 1$.

IX. 59. Prove that the polynomial f is divisible by g for any $n \in \mathbf{N}^*$:

a) $f(x) = (3x+1)^{6n+1} + 3x^2$, $\quad g(x) = 3x^2 + 3x + 1$;

b) $f(x) = (2x+1)^{3n+2} + 2x + 2$, $\quad g(x) = 4x^2 + 6x + 3$.

IX. 60. If $f(x)$ is a polynomial by degree 5 such that $f(x) - 2$ is divisible by $(x+1)^3$ and $f(x)$ is divisible by x^3, find $f(x)$.

IX. 61. Find a polynomial of degree five, knowing that $f(x) + 1$ is divisible by $(x-1)^3$ and $f(x) - 1$ is divisible by $(x+1)^3$.

IX. 62. Find the roots of the polynomial

$$f(x) = ax^3 + bx^2 + cx + d, \quad a,b,c,d \in \mathbf{R}, \text{ if } b + d = a + c.$$

IX. 63. Consider the polynomial $f(x) = ax^3 + bx^2 + cx + d$, $a,b,c,d \in \mathbf{R}$. Prove that if $bc = ad$ then $x_1 + x_2 = 0$ and vice versa.

IX. 64. Solve each equation with the indicated condition and in some cases find the parameter m as well.

a) $2x^3 - x^2 - 7x + m = 0$, $\qquad\qquad x_1 + x_2 = 1$;

b) $x^4 + 3x^3 - mx^2 - 3x + 10 = 0$, $\qquad x_1 + x_2 = -3$;

c) $x^4 - mx^3 + 12x - 5 = 0$, $\qquad\qquad x_1 + x_2 = 2$;

d) $x^3 + 5x^2 + 2x - 8 = 0$, $\qquad\qquad x_1 = 2x_2$;

e) $x^4 - 10x^3 + 35x^2 - 50x + 24 = 0$, $\qquad x_1 = 3x_2$;

f) $x^3 - 8x^2 + (3m+2)x - 2m = 0$, $\qquad 3x_1 = x_2 + x_3$;

g) $x^4 - 10x^3 + 5mx^2 - 50x + 24 = 0$, $\qquad x_1 - x_4 = x_3 - x_2$;

h) $x^4 - 4x^3 + (m+3)x^2 + 18x + 14 = 0$, $\qquad x_1 + x_2 = x_3 + x_4$;

i) $x^3 - 9x^2 + 6x + 56 = 0$, $\qquad\qquad x_1 - x_2 = 3$;

j) $x^3 + (1 - \sqrt{3})x^2 - (2 + \sqrt{3})x + 2\sqrt{3} = 0$, $\qquad x_1^2 = x_2^2 + x_3^2$;

k) $6x^4 - (m+1)x^3 - 75x^2 - 10x + 24 = 0$, $\qquad \dfrac{1}{x_1 x_2} + \dfrac{1}{x_3 x_4} = 1$;

205

l) $x^4 + 2x^3 - 5x^2 - 10x + m + 2 = 0$, $\qquad\qquad x_1^2 + x_2^2 = x_3^2 + x_4^2$;

m) $3x^3 + 8x^2 + 13x + 6 = 0$, $\qquad\qquad x_1 = \dfrac{1}{x_2} + \dfrac{1}{x_3}$.

IX. 65. The polynomial $f(x) = x^3 - ax + b$ has the distinct roots x_1, x_2, x_3.
Compute $(x_1 - x_2)^2 (x_2 - x_3)^2 (x_3 - x_1)^2$ in terms of a and b.

IX. 66. If one of the roots of the polynomial $f(x) = x^3 - x^2 + 5x + 3$ is $-1 + \sqrt{2}$, find the sum of the other two roots.

IX. 67. Consider the polynomial $f(x) = x^3 + 3x^2 + 3x + 5$ with the roots $x_1, x_2, x_3 \in \mathbf{R}$ and $g(x) = x^2 - 2x + 5$ with the roots $y_1, y_2 \in \mathbf{R}$
a) Calculate $S - S'$ where $S = x_1 + x_2 + x_3$ and $S' = y_1 + y_2$;
b) Calculate the value of the expressions $E = \dfrac{1}{x_1^2} + \dfrac{1}{x_2^2} + \dfrac{1}{x_3^2}$,
$F = x_1^5 + x_2^5 + x_3^5$, and $G = f(y_1) + f(y_2)$.

IX. 68. For the equation $x^3 + x + m = 0$, find $m \in \mathbf{R}$ such that $x_1^5 + x_2^5 + x_3^5 = 1$, where x_1, x_2, x_3 are the roots of the equation. Find the value of the expression

$$E = \frac{x_1^2 + x_2^2}{x_3^2} + \frac{x_2^2 + x_3^2}{x_1^2} + \frac{x_1^2 + x_3^2}{x_2^2} \quad \text{when } m = 2.$$

IX. 69. Consider the polynomial function
$f(x) = x^3 - mx^2 - 3x + m + 2, \ m \in \mathbf{R}$.
a) Calculate $S(m) = x_1^3 + x_2^3 + x_3^3$, where x_1, x_2, x_3 are the roots of the polynomial;
b) Find $m \in \mathbf{R}$ such that $S(m) \le (x_1 x_2 x_3)^2$;
c) Find $m \in \mathbf{R}$ such that f be divisible by $x - 2$.

IX. 70. For the polynomial $f(x) = x^3 - x^2 + mx - 1$, $m \in \mathbf{R}$, consider $S_n = x_1^n + x_2^n + x_3^n$, $n \in \mathbf{N}^*$ where $x_1, x_2, x_3 \in \mathbf{C}$ are the roots of the polynomial f. Show that $S_n - S_{n-1} + mS_{n-2} - S_{n-3} = 0$ and find $m \in \mathbf{R}$ such that $S_3 = 1$.

IX. 71. If $x_1, x_2, x_3 \in \mathbf{C}$ are the roots of the equation $x^3 + mx + n = 0$, $m, n \in \mathbf{R}$, show that:

a) $3\left(x_1^2 + x_2^2 + x_3^2\right)\left(x_1^5 + x_2^5 + x_3^5\right) = 5\left(x_1^3 + x_2^3 + x_3^3\right)\left(x_1^4 + x_2^4 + x_3^4\right)$;

b) $12\left(x_1^7 + x_2^7 + x_3^7\right) = 7\left(x_1^3 + x_2^3 + x_3^3\right)\left(x_1^2 + x_2^2 + x_3^2\right)^2$.

IX. 72. Let $x^3 + x^2 + mx - 1 = 0$, $m \in \mathbf{R}$ be an equation.

a) Show that the equation has a root on the interval $[-1, 1]$;
b) Find m if the equation has a double root;
c) Find $m \in \mathbf{R}$ such that the roots of the equation fulfill the condition

$$\frac{1}{x_1^2} + \frac{1}{x_2^2} + \frac{1}{x_3^2} = \frac{4}{x_1 x_2 x_3}.$$

IX. 73. Show that the equation

$x^4 + 2(a+1)x^3 + \left(2a^2 + 4a + 3\right)x^2 + 5x + 9 = 0$ cannot have all real roots.

IX. 74. The polynomial $f(x) = x^6 - 3x^4 + 5x^2 + 4x + b$ has the roots $x_1 = -8$, $x_2 = 1$, $x_3 = 2$, $x_4 = 3$, $x_5 = 5$. Find x_6 and b.

IX. 75. Consider the polynomial $f(x) = x^3 + x^2 + ax + b$. Find the numeric value for a and b such that $f(x-1)$ divided by $x+1$ yields remainder is -4 and the roots of the equation $f(x) = 0$ fulfill the condition $x_1^3 + x_2^3 + x_3^3 = 8$.

IX. 76. Consider the polynomial $f(x) = x^3 - 6x^2 + ax + b$. Find the numeric value for a and b such that $f(x-1)$ divided by $x-5$ yields remainder 21 and the roots of the equation $f(x) = 0$ fulfill the condition $x_1^3 + x_2^3 + x_3^3 = 36$.

IX. 77. Find parameter $m \in \mathbf{R}$ and solve the equations with the roots in an arithmetic sequence:

a) $x^3 - 12x^2 + 3(2m+3)x - 28 = 0$;

b) $x^4 - 4x^3 - 34x^2 + 2(5m+8)x + 105 = 0$;

c) $x^3 - 2(2m-1)x^2 + 3(3m-5)x + 4m + 2 = 0$;

d) $x^4 + (m-1)x^3 - 3(2m+1)x^2 - (7m+1)x + 40 = 0$.

IX. 78. Let $f(x) = x^3 + ax^2 + bx + c$, $a, b, c \in \mathbf{R}$ be a polynomial function. Find he condition between the coefficients a, b, c such that the roots of $f(x) = 0$ are in a geometric sequence.

IX. 79. Find the parameter $m \in \mathbf{R}$ and solve the equations with the roots in a geometric sequence:

a) $x^4 - 15x^3 + 70x^2 - 120x + m = 0$;

b) $2x^4 - 15x^3 + 35x^2 - 30x + m = 0$;

c) $x^3 - (m+5)x^2 + 2(2m+3)x - 8 = 0$;

d) $x^4 + 5(3m-5)x^3 - 30x^2 - 40x + 64 = 0$.

IX. 80. Consider the polynomial $f(x) = ax^3 + bx^2 + cx + d$, $a,b,c,d \in \mathbf{R}$. Find a,b,c,d such that $f(1) + f(2) + \ldots + f(n) = n^4$, for any n, a natural number.

IX. 81. Consider the polynomial $P(x) = x^3 - 6x + 5$. Find the value of P for indicated value of x.

a) $\sqrt[3]{7+\sqrt{41}} + \sqrt[3]{7-\sqrt{41}}$;   b) $\sqrt[3]{4-2\sqrt{2}} + \sqrt[3]{4+2\sqrt{2}}$;

c) $\sqrt[3]{6-2\sqrt{7}} + \sqrt[3]{6+2\sqrt{7}}$.

IX. 82. Prove that each polynomial function has the indicated root:

a) $f(x) = x^3 - 3x - 2\sqrt{5}$, $\sqrt[3]{\sqrt{5}+2} + \sqrt[3]{\sqrt{5}-2}$;

b) $f(x) = x^3 + 3x - 2$, $\sqrt[3]{\sqrt{2}+1} - \sqrt[3]{\sqrt{2}-1}$;

c) $f(x) = x^3 - 3x - 2\sqrt{3}$, $\sqrt[3]{\sqrt{3}-\sqrt{2}} + \sqrt[3]{\sqrt{3}+\sqrt{2}}$.

IX. 83. Find a polynomial with real coefficients such that $x = -1 + \sqrt{2} + \sqrt{3}$ is one of the roots.

IX. 84. Show that the number $x = \sqrt[5]{27} - \sqrt[5]{9}$ is a root of the equation $x^5 + 15x^3 - 225x - 18 = 0$.

IX. 85. Find roots of the polynomials in the set C (*complex numbers*), knowing that each of them has a real root.

a) $f(z) = z^3 - (4+i)z^2 + (7+3iz) - 2i - 6$;

b) $f(x) = x^3 - 4x^2 + (6-i)x - 3 + i$.

IX. 86. Consider the polynomial $f(x) = x^3 + ax^2 + 4x + b$, where $a, b \in$ **R**. Find the numeric value for a and b such that $x_1 = \dfrac{1 - i\sqrt{3}}{2}$ be a root of the equation $f(x) = 0$.

IX. 87. The equation $x^3 - 3x^2 + ax + b = 0$ has rational coefficients and one of the roots is $-1 + i\sqrt{3}$; determine the value of a and b.

IX. 88. One of the roots of the polynomial function $f(x) = x^4 - mx^3 + 13x^2 - 14x + n$, $m, n \in$ **R** is $x_1 = 1 - i$. Find m and n and solve the equation $f(x) = 0$.

IX. 89. A *quartic* polynomial with real coefficients has four zeroes, two of them are $-1 + i\sqrt{3}$ and $-1 + i$. What are the other two roots? Find the polynomial.

IX. 90. Consider the polynomial $f(x) = 2x^3 + mx^2 + 4x + 4$. Find $m \in$ **R** such that f have a double root.

IX. 91. Consider the polynomial $f(x) = x^4 + ax^2 + bx + c$. Find $a, b, c \in$ **R** and the roots such that f has a triple root and -1 is another one.

IX. 92. Consider the polynomial $f(x) = x^4 + ax^3 + bx + c$. If $f(0) = f(1)$

and $x_1 = \dfrac{1 + i\sqrt{3}}{2}$ is a root, find the other three roots.

IX. 93. Consider the polynomial function $f(x) = x^4 - 5x^3 + ax^2 + bx + c$,

$a, b, c \in \mathbf{R}$. If $f(0) = f(1)$ and $x_1 = 2 - i$ is a root, find the other three roots and a, b, c.

IX. 94. Solve each equation with the indicated root.

a) $x^3 - 3x^2 - 3x + 1 = 0$, $x_1 = 2 - \sqrt{3}$;

b) $x^3 - 3x^2 - 5x + 7 = 0$, $x_1 = 1 - 2\sqrt{2}$;

c) $x^4 - 2x^3 - 2x - 1 = 0$, $x_1 = 1 - \sqrt{2}$;

d) $x^3 - 7x^2 + 16x - 10 = 0$, $x_1 = 3 + i$;

e) $x^4 - 4x^3 + 11x^2 + 8x - 26 = 0$, $x_1 = 2 + 3i$;

f) $x^4 - 7x^3 + 19x^2 + mx + n = 0$, $m, n \in \mathbf{R}$, $x_1 = 2 + i$;

g) $x^4 + 2x^3 + mx^2 - 2x + n = 0$, $m, n \in \mathbf{R}$, $x_1 = -1 + i$;

h) $x^5 + x^4 - 16x^3 - 16x^2 + 4x + 4 = 0$, $x_1 = -\sqrt{3} - \sqrt{5}$;

i) $x^6 - 23x^4 + 76x^2 - 48 = 0$, $x_1 = -\sqrt{3} - \sqrt{7}$.

IX. 95. Solve the reciprocal equations:

a) $2x^3 + 3x^2 + 3x + 2 = 0$;

b) $2x^3 + 7x^2 + 7x + 2 = 0$;

c) $2x^4 + x^3 + x^2 + x + 2 = 0$;

d) $2x^4 - 5x^3 + 4x^2 - 5x + 2 = 0$;

e) $x^5 + 2x^4 + 3x^3 + 3x^2 + 2x + 1 = 0$;

f) $2x^5 + 4x^4 + 2x^3 + 2x^2 + 4x + 2 = 0$.

IX. 96. Consider the polynomial $P(x) = x^{2n} - x^n + x^3 + 1$. Find the remainder of $P(x)$ when divided by $(x - 1)^2$.

IX. 97. Find the remainder when the polynomial

$f(x) = x^n + x^{n-1} + \ldots + x + 1$, $n \geq 3$ is divided by the polynomial

$g(x) = x^2(x - 1)$.

210

IX. 98. Consider the polynomial function $f : \mathbf{R} \to \mathbf{R}$,

$$f(x) = (m-1)x^4 - (m+3)x^3 - 7x^2 - (m+6)x + m - 4, \quad m \in \mathbf{R}.$$

a) Show that for any $m \in \mathbf{R}$, $f(x)$ is divisible by $x^2 + x + 1$;

b) Find $m \in \mathbf{R}$ such that the polynomial $f(x)$ have a double root;

c) Find $m \in \mathbf{R}$ such that $\dfrac{1}{x_1} + \dfrac{1}{x_2} + \dfrac{1}{x_3} + \dfrac{1}{x_4} \geq \dfrac{5}{4}$, where x_1, x_2, x_3, and x_4

are the roots of the polynomial f.

d) Find $m, \beta \in \mathbf{R}$ such that $x_0 = -1 + i \tan \beta$ be a root of f and solve the equation $f(x) = 0$.

IX. 99. Given a polynomial $f(x)$ such that $f(x^2 + 1) = x^4 + 3x^2 + 1$, determine $f(x^2 - 1)$.

IX. 100. Let a be one of the roots of the equation $x^3 - x - 5 = 0$. Find an equation of degree at least three such that one of the roots is

a) $y = a^4 - 5a + 3$; b) $y = 2a^2 - a + 2$; c) $y = a^4 - a + 6$;

d) $y = a^5 - a^3 + 1$; e) $y = a^6 - a + 1$.

IX. 101. Let x_1, x_2, x_3 be the roots of the equation $x^3 - 2x - 1 = 0$.

a) If $P(x) = x^5 - x^4 - x - 4$, calculate $P(x_1) + P(x_2) + P(x_3)$;

b) If $P(x) = x^5 - 2x^4 + 6x + 1$, calculate $P(x_1) + P(x_2) + P(x_3)$.

IX. 102. The roots of the equation $x^n = 1$, $n > 1$, $n \in \mathbf{N}$ are $x_1, x_2, ..., x_{n-1}, 1$. Prove that:

a) $(1 - x_1)(1 - x_2)...(1 - x_{n-1}) = n$;

b) $\dfrac{1}{1 - x_1} + \dfrac{1}{1 - x_2} + ... + \dfrac{1}{1 - x_{n-1}} = \dfrac{n-1}{2}$;

c) $(1 + x_1)(1 + x_2)...(1 + x_{n-1}) = \dfrac{1 + (-1)^{n+1}}{2}$;

d) $(a + b)(a + bx_1)(a + bx_2)...(a + bx_{n-1}) = a^n + (-1)^{n+1} b^n$.

IX. 103. The polynomial $P(x) = x^n - 4x + 5$ has the roots $x_1, x_2, ..., x_n$,

where $n \in \mathbf{N}^*$. Calculate $\displaystyle\sum_{k=1}^{n} \frac{x_k^2 + 1}{x_k - 1}$.

IX. 104. If $P_n(x+1) - P_n(x) = x^n$ where $P_n(x) \in \mathbf{Q}[x]$ and

$P_n(0) = 0, (\forall) n \in \mathbf{N}$. Prove that:

a) $(\forall) n \geq 1$, $P_n(1) = 0$;

b) $(\forall) k \in \mathbf{N}$, $P_n(k) = 1^n + 2^n + 3^n + ... + (k-1)^n$.

IX. 105. Prove that all of the real roots of the equation

$x^{2n+1} - x^{2n} + x^{2n-1} + 2nx^n - n^2 = 0$ are positive numbers, for any $n \in \mathbf{N}^*$.

IX. 106. Consider the polynomial $f = x^5 + 2ax^3 + (2b+1)x + 2c + 1$ where
$a, b, c \in \mathbf{Z}$. Prove that f does not have any integer root.

IX. 107. Let f be a polynomial of degree n, with real coefficients, such

that $\displaystyle\sum_{k=1}^{n} f(x^k) = \left(\sum_{k=1}^{n} x^k\right) \cdot f(x)$. Find f.

IX. 108. If the roots of the polynomial

$P(x) = x^n + a_1 x^{n-1} + ... + a_{n-1}x + (-1)^n$ have the same module as $P(-1)$ is a
real number.

IX. 109. Let ε be one of the roots of the equation $x^n = 1$, where n is a

positive integer. Prove that for any polynomial $f(x) = a_0 + a_1 x + ... + a_n x^n$,

where $a_0, a_1, a_2, ..., a_n$ are real numbers, the sum $\displaystyle\sum_{k=1}^{n} f\left(\frac{1}{\varepsilon^k}\right)$ is a real number.

Chapter X

PERMUTATIONS AND COMBINATIONS.
NEWTON'S BINOMIAL THEOREM

Permutations of n objects taken k at a time is denoted variously

by $P(n,k)$, A_n^k, etc. $P(n,k) = \dfrac{n!}{(n-k)!} = n \cdot (n-1) \cdot ... \cdot (n-k+1)$

The number of ways of a *combinations* of n objects taken k at a time is denoted variously by $\binom{n}{k}$, $_nC_k$, C_n^k, etc. We shall use $\binom{n}{k}$,

$\binom{n}{k} = \dfrac{n!}{k! \cdot (n-k)!} = \dfrac{n \cdot (n-1) \cdot ... \cdot (n-k+1)}{1 \cdot 2 \cdot 3 \cdot ... \cdot k}$.

Binomial formula

$(a+b)^n = \binom{n}{0}a^n + \binom{n}{1}a^{n-1}b + \binom{n}{2}a^{n-2}b^2 + ... + \binom{n}{n}b^n$

$(a+b)^n = \sum\limits_{k=0}^{n} \binom{n}{k}a^k b^{n-k}$,

$t_{k+1} = \binom{n}{k}a^{n-k}b^n$ ($k = 0, 1, 2, ..., n$) is the *general term* of the binomial formula.

X. 1. How many numbers can be formed using the set $\{1, 3, 5, 7, 9\}$?

X. 2. In how many ways can 6 students standing in a line be arranged?

X. 3. Using the digits 0, 1, 2, 3, 4, 5 find:
a) the total number of numbers which can be formed;
b) how many numbers begin with the digit one?
c) how many numbers end with the digit one?
d) how many numbers begin with 50?
e) how many numbers do not begin with 50?

X. 4. Find the number of arrangements possible using the letters of the words SUCCESS, PARALLELIPIPED, and MISSISSIPPI.

X. 5. A resort manager must assign 15 guards to three buildings: 6 in the first building, 5 in the second building, and 4 in the third building. In how many ways can this be done?

X. 6. In a math contest, how many ways can 14 students rank in the top three?

X. 7. In how many ways can 5 of 7 books be arranged on a bookshelf? In how many ways can these 7 books be arranged on the shelf?

X. 8. How many different numbers can be formed with the digits 0, 1, 2, 3, 4 if in each number any digit can be written at most once?

X. 9. A college promotion committee consists of 7 members. A favorable majority decision is reached if at least 4 members vote favorably. In how many ways can the committee reach a majority decision in favor of a problem?

X. 10. A club has 20 members. How many different 4-member committees are possible?

X. 11. A team consists of 8 boys and 4 girls. In how many ways can 6 people be chosen under each of the following restrictions?
a) only one girl is included,
b) at least one girl is included,
c) one girl, Alexandra, is included.

X. 12. In how many ways can a team be selected consisting of 2 teachers and 5 students if there are 4 teachers and 10 students?

X. 13. In a class there are 25 students. Determine in how many ways a 5-person committee can be selected to organize a trip
a) with no restrictions
b) with Alex on the committee.

X. 14. How many lines can be drawn through 6 points if any 3 of them are not collinear?

X. 15. What is the maximum number of points of intersection between 3 lines?

X. 16. How many 5-digit numbers can be formed such that in each number the digits are in an increasing order from left to right? How about if they are in a decreasing order from left to right?

X. 17. From a group of 15 kids; consisting of 9 boys and 6 girls, a team of 7 kids has to be formed with at least 3 girls. Determine how many ways this group can be formed.

X. 18. At a party there are 9 boys and 7 girls. In how many ways can 4 pairs consisting of one boy and one girl be formed?

X. 19. In a class there are 25 students, 15 boys, and 10 girls. Determine how many ways a team can be selected with
a) 4 boys and 6 girls
b) 8 students with at least 3 girls in each team.

X. 20. Prove the identity
$$\binom{15}{5}\cdot\binom{10}{3}+\binom{15}{4}\cdot\binom{10}{4}+\binom{15}{3}\cdot\binom{10}{5}+\binom{15}{2}\cdot\binom{10}{6}+\binom{15}{1}\cdot\binom{10}{7}+\binom{15}{0}\cdot\binom{10}{8}=$$
$$=\binom{25}{8}-\binom{15}{6}\cdot\binom{10}{2}-\binom{15}{7}\cdot\binom{10}{1}-\binom{15}{8}\cdot\binom{10}{0}.$$

X. 21. Compute $\binom{9}{n}$ if $\binom{18}{n}=\binom{18}{n+4}$.

X. 22. Solve the equations:
a) $\binom{x+9}{x+4}=5\dfrac{(x+7)!}{(x+4)!}$; **b)** $P(x,2)=42$; **c)** $\dfrac{(x-1)!}{(x-3)!}=72$;

d) $3\binom{2x}{x+1}=5\binom{2x-1}{x}$; **e)** $13\binom{2x}{x+1}=7\binom{2x+1}{x-1}$; **f)** $11\binom{2x}{x}=6\binom{2x+1}{x+1}$;

g) $\binom{x}{2}+\binom{x+1}{2}=x+2$; **h)** $\binom{x}{1}+2\binom{x}{2}+\binom{x}{3}=460$;

i) $\binom{x+3}{x+1}=\binom{x+1}{x-1}+\binom{x+1}{x}+\binom{x}{x-2}$; **j)** $\binom{x+3}{x+1}-5\cdot\binom{3x}{2}+19x^2=6$,

k) $6\cdot\binom{x+1}{1}+6\cdot\binom{x+3}{3}=13\cdot\binom{x+2}{2}$; **l)** $\binom{n}{n-1}+\binom{n}{n-2}+\binom{n}{n-3}+...+\binom{n}{n-10}=1024$;

m) $\binom{n-10}{n-11}+\binom{n-9}{n-11}+\binom{n-8}{n-11}+...+\binom{n-1}{n-11}=65$.

X. 23. Prove the identities:
a) $P(n,k)=P(n-1,k)+kP(n-1,k-1)$; **b)** $k\binom{n}{k}=n\binom{n-1}{k-1}$; **c)** $\binom{n}{k}=\binom{n-1}{k}+\binom{n-1}{k-1}$;

d) $\binom{n}{k}=\binom{n-2}{k}+2\binom{n-2}{k-1}+\binom{n-2}{k-2}$; **e)** $\binom{n}{k}=\binom{n-3}{k}+3\binom{n-3}{k-1}+3\binom{n-3}{k-2}+\binom{n-3}{k-3}$;

f) $\binom{n}{k}-\binom{n-2}{k}-\binom{n-2}{k-2}=\dfrac{2(n-k)}{n-1}\binom{n-1}{k-1}$; **g)** $(k-1)k\binom{n}{k}=(n-1)n\binom{n-2}{k-2}$.

X. 24. Calculate and simplify the expressions:
a) $P(5,3)+P(5,4)$; **b)** $\binom{9}{6}$; **c)** $\dfrac{\binom{6}{3}\cdot3!}{P(6,2)}$; **d)** $\dfrac{P(10,6)+10!}{7!\cdot\binom{9}{6}}$;

e) $\dfrac{P(n+k,n-k+2)+P(n+k,n-k+1)}{P(n+k,n-k)}$; **f)** $\dfrac{(n-k)!\,P(n-1,k-1)}{(k-1)!}$;

g) $\dfrac{\binom{n}{k}}{\binom{n}{k}+\binom{n}{k+1}}$; h) $\dfrac{\binom{n}{k}-\binom{n-2}{k}-\binom{n-2}{k-2}}{\binom{n-1}{k-1}}$; i) $\dfrac{\binom{n+1}{k}}{\binom{n+2}{k}}-\dfrac{\binom{n}{k}}{\binom{n+1}{k}}$.

X. 25. Prove that the number $\binom{2n+1}{1}\cdot\binom{2n+1}{2}\cdot\ldots\cdot\binom{2n+1}{2n}$ is a perfect square.

X. 26. Solve the inequations:

a) $\binom{2n}{7}\ge\binom{2n}{5}$; b) $\binom{17}{n}\le\binom{17}{n-2}$; c) $7\binom{n}{4}>3\binom{1+n}{5}$; d) $\binom{19}{n-1}<\binom{19}{n}$;

e) $5\binom{n}{3}>3\binom{2+n}{4}$; f) $\binom{15}{n}<\binom{15}{n-2}$; g) $\binom{n+8}{n+3}\le 5P(n+8,3)$;

h) $11\binom{2n}{n}\ge 6\binom{2n+1}{n+1}$; i) $(n+2)\cdot\binom{n+1}{n-1}\le 6\cdot\binom{n}{n-1}+9n$.

X. 27. Solve the system of equations:

a) $\begin{cases} P(x,y+1)=2P(x,y-1) \\ 21\binom{x}{y+1}=\binom{x}{y-1} \end{cases}$;

b) $\begin{cases} P(2y,3x)=8P(2y,3x-1) \\ 9\binom{2y}{3x}=8\binom{2y}{3x-1} \end{cases}$;

c) $\begin{cases} P(2x,y-2)=8P(2x,y-3) \\ 3\binom{2x}{y-2}=8\binom{2x}{y-3} \end{cases}$;

d) $\begin{cases} P(n,k)=7P(n,k-1) \\ 6\binom{n}{k}=5\binom{n}{k+1} \end{cases}$;

e) $\begin{cases} P(x,2)+P(y,2)=20 \\ P(x,3)+P(y,3)=61 \end{cases}$;

f) $\begin{cases} 4\binom{x}{y+1}=3\binom{x}{y} \\ 3\binom{x}{y}=4\binom{x}{y-1} \end{cases}$;

g) $\begin{cases} 5P(x,y)=2P(x+1,y) \\ 3\binom{x}{y}=2\binom{x}{y-1} \end{cases}$.

X. 28. Prove that:

a) $\dfrac{\binom{2}{1}+\binom{4}{2}+\binom{6}{3}+\ldots+\binom{2n}{n}}{\binom{1}{1}+\binom{3}{2}+\binom{5}{3}+\ldots+\binom{2n-1}{n}}=2$; b) $\dfrac{\binom{2}{1}\cdot\binom{4}{2}\cdot\binom{6}{3}\cdot\ldots\cdot\binom{2n}{n}}{\binom{1}{1}\cdot\binom{3}{2}\cdot\binom{5}{3}\cdot\ldots\cdot\binom{2n-1}{n}}=2^n$.

X. 29. Calculate the sums:

a) $S_1 = 1!\cdot 1+2!\cdot 2+\ldots+n!\cdot n$; b) $S_2 = \dfrac{1}{2!}+\dfrac{2}{3!}+\ldots+\dfrac{n}{(n+1)!}$;

c) $S_3 = \dfrac{1}{3\cdot 1!}+\dfrac{1}{4\cdot 2!}+\dfrac{1}{5\cdot 3!}+\ldots+\dfrac{1}{n\cdot(n-2)!}$, $n>2$;

d) $S_4 = \displaystyle\sum_{k=2}^{n}\dfrac{k^2-2}{k!}$; e) $S_5 = \displaystyle\sum_{k=1}^{n}\dfrac{(k-1)!}{(k+1)!}$; f) $S_6 = \displaystyle\sum_{k=2}^{n}\dfrac{k^2+k+1}{(k+1)!k(k+1)}$;

g) $S_7 = \dfrac{3}{1!+2!+3!}+\dfrac{4}{2!+3!+4!}+\dfrac{5}{3!+4!+5!}+\ldots+\dfrac{n+1}{(n-1)!+n!+(n+1)!}$.

X. 30. Prove the identity $\binom{k}{k}+\binom{k+1}{k}+\ldots+\binom{n+k-1}{k}=\binom{n+k}{k+1}$.

X. 31. Calculate the sums:

a) $S_1 = 1\cdot2\cdot3+2\cdot3\cdot4+\ldots+n\cdot(n+1)\cdot(n+2)$;

b) $S_2 = 1\cdot2\cdot3\cdot\ldots\cdot k+2\cdot3\cdot4\cdot\ldots\cdot k\cdot(k+1)+\ldots+n\cdot(n+1)\cdot\ldots\cdot(n+k-1)$.

X. 32. Calculate the sums: a) $\sum_{k=2}^{n}\binom{k}{2}$; b) $\sum_{k=2}^{n}\binom{k}{2}^2$.

X. 33. Find the positive integers a, b, c such that

$$\sum_{k=1}^{n}k(bk+a)(ak+c)= n(n+1)(n+2)(n+3)$$

is true for any positive integer n.

X. 34. Prove that $\binom{n}{k}^p = \binom{n-1}{k-1}^p + \frac{n^p-k^p}{(n-k)^p}\binom{n-1}{k}^p,(\forall)k,n,p\in N^*,k<n$.

X. 35. Show that $\sqrt{\binom{n+1}{k}\cdot\binom{n-1}{k}}\le\binom{n}{k}\le\dfrac{\binom{n+1}{k}+\binom{n-1}{k}}{2}$, $n, k \in N, n\ge1$.

X. 36. Prove that for any integer positive n, the number $\binom{2n}{n}$ is divisible by $n+1$.

X. 37. Prove the identity $\displaystyle\sum_{k=0}^{n}\frac{(k+1)\binom{n}{k+1}}{\binom{n}{k}} = \frac{n(n+1)}{2}$.

X. 38. Prove that the fraction $\dfrac{\binom{n+1}{n-1}!}{\prod_{k=1}^{n}k!}$ is a natural number for any-non zero positive integer n.

X. 39. Prove the identities:

a) $\dfrac{1}{\binom{2n}{k}} = \dfrac{2n+1}{2(n+1)}\left(\dfrac{1}{\binom{2n+1}{k}}+\dfrac{1}{\binom{2n+1}{k+1}}\right)$; b) $\displaystyle\sum_{k=0}^{2n}(-1)^k\cdot k!(2n-k)!=\dfrac{(2n+1)!}{n+1}$.

217

X. 40. Expand:

a) $(1+2x)^4$; b) $\left(x+\sqrt{x^2-1}\right)^6 + \left(x-\sqrt{x^2-1}\right)^6$.

X. 41. Find the fifth term of the expansion $\left(\sqrt[3]{x} + \dfrac{1}{\sqrt[3]{x}}\right)^9$.

X. 42. Write down the expansion by binomial theorem $\left(\dfrac{x}{2}+2y\right)^5$, and collect the terms.

X. 43. a) Find, in ascending powers of x, the first three terms in the expansion of $(3x-2)^8$.

b) Find the term independent of x from the expansion of $\left(x^2 - \dfrac{2}{x^2}\right)^4$.

X. 44. Find the term independent of x from the expansion of $\left(x+\dfrac{2}{\sqrt[5]{x}}\right)^{2010}$.

X. 45. a) Find the value of the expression $\left(\sqrt{2}+1\right)^6 + \left(\sqrt{2}-1\right)^6$.

b) If the coefficients of x^2 and x^3 in the expansion of $(3+ax)^9$ are the same, find the value of a.

X. 46. Find the exact value of $\left(\sqrt{3}+1\right)^5 - \left(\sqrt{3}-1\right)^5$. Hence show, without calculators, that the value of $\left(\sqrt{3}+1\right)^5$ lies between 152 and 153.

X. 47. Using the formula $b_1 \dfrac{r^n-1}{r-1}$ for the sum of the first n terms of a geometric progression (r ratio), calculate the sum of the series $1+(1+x)+(1+x)^2+...+(1+x)^{20}$. Using binomial expansion theorem verify that the coefficients of x, in the given expression and in your answer, are equal.

X. 48. In the expansion of $(1+x)^n$ the first terms are $1+3+4+...$. Calculate the values of n and x, and the value of the fourth term of the expansion.

X. 49. Find the coefficient of x^{-16} in the expansion of $\left(x^2 - \dfrac{1}{x}\right)^{25}$.

X. 50. The expansion of $(1 + ax)^n$, where $n > 0$, by the binomial theorem is
$1 + 20x + 45a^2x^2 + bx^3 + \dots$
Calculate n, a, and b.

X. 51. The first three terms of a binomial expansion are

$1 - \dfrac{16}{\sqrt{3}}x + \dfrac{112}{3}x^2$. Find the fourth term.

X. 52. In the binomial expansion of $\left(1 + \dfrac{1}{5}\right)^n$, the second and third terms are equal. Calculate the value of n.

X. 53. In the expansion $\left(1 + \sqrt[3]{x}\right)^n$ the binomial coefficient of the sixth and eleventh terms are equal. Find the value of n.

X. 54. In the expansion $\left(\dfrac{\sqrt{a}}{\sqrt[5]{x}} + \dfrac{\sqrt[3]{x}}{\sqrt{a}}\right)^n$ the binomial coefficient of the third term is 190. Find the value of n.

X. 55. In the expansion of $(k + x)^8$, where k is a positive number, the coefficients of x^2 and x^3 are equal. Find the value of k.

X. 56. In the expansion of $\left(x^3 + \dfrac{k}{x}\right)^8$, where k is a real number, the coefficient of x^8 is 70. Find the value of k.

X. 57. Find the positive integer n, $n > 1$ such that the fourth term of the

expansion $\left(\dfrac{\sqrt[5]{x^4}}{\sqrt[n]{x^{n-1}}} + x \cdot \sqrt[n+1]{x^{n-1}}\right)^8$ is $56 \cdot x^{7.9}$.

X. 58. Find the term in the expansion $\left(x + \dfrac{1}{x^2}\right)^{12}$ which does not contain x.

X. 59. Find the term in the expansion $\left(x - \dfrac{1}{x^2}\right)^{3n}$ which does not contain x.

X. 60. Find the sixth term in the expansion $\left(x^{\frac{1}{2}} + y^{\frac{1}{3}}\right)^{n}$ if the binomial coefficient of the third term from the end is 45.

X. 61. Find which term of the binomial expansion $\left(\sqrt{x} + \dfrac{1}{\sqrt[4]{x^3}}\right)^{n}$, contains $x^{\frac{13}{2}}$ if the ninth term has the largest coefficient.

X. 62. In the binomial expansion $\left(x \cdot \sqrt[5]{\dfrac{x}{3}} - \dfrac{y}{\sqrt[7]{x^3}}\right)^{n}$ the sum of the binomial coefficients of the odd terms is $2{,}048$. Find the term which contains x^3.

X. 63. Let $a_k, a_{k+1}, a_{k+2}, a_{k+3}$, $1 \le k \le n - 3$ ($n \in \mathbf{N}$) be four consecutive binomial coefficients in the expansion of $(x + y)^n$. Prove that

$$\frac{a_k}{a_k + a_{k+1}} + \frac{a_{k+2}}{a_{k+2} + a_{k+3}} = \frac{2a_{k+1}}{a_{k+1} + a_{k+2}}.$$

X. 64. In the expansion $\left(x^2 + \dfrac{1}{\sqrt[3]{x}}\right)^{n}$

a) the sum of the second, the third, and the fourth binomial coefficients is 41.
b) the sum of the first three binomial coefficients is 56.
Find the value of n in each case.

X. 65. Find the coefficient of the expansion $\left(\sqrt[3]{\dfrac{x}{\sqrt{y}}} + \sqrt{\dfrac{y}{\sqrt[3]{x}}} \right)^{21}$, such that x and y have equal exponents.

X. 66. Find the positive number x, if the third term of the expansion $\left(x + x^{\log x} \right)^{5}$ is 10^{6}, where the base of log is 10.

X. 67. Find the positive integer m such that the 10th term of the expansion $(5 + m)^{m}$ be the largest.

X. 68. Find n and x in the expansion $\left(\sqrt{2^{x}} + \sqrt{2^{1-x}} \right)^{n}$ knowing that the sum of the first three coefficients is 22 and the sum between the third and the fifth term is 135.

X. 69. For what value of x, the third term of the expansion $\left(\dfrac{1}{\sqrt[7]{x^{2}}} + x^{\log \sqrt{x}} \right)^{9}$ is 36 000, where the base of log is 10?

X. 70. Find n and x in the expansion $\left(\sqrt{3^{x+1}} + \dfrac{1}{\sqrt{3^{x}}} \right)^{n}$ if the sum of the first three binomial coefficients is 22 and the sum of the third an fifth terms is 420.

X. 71. In the expansion $\left(\dfrac{2}{\sqrt[x]{4}} + \dfrac{\sqrt[x-4]{4}}{4} \right)^{6}$, the fourth term is 10. Find the real number x.

X. 72. In the expansion $\left(\sqrt{x^{\frac{1}{1+\log x}}} + \sqrt[12]{x} \right)^{6}$, the fourth term is 200, where the base of log is 10. Find the real number x.

X. 73. In the binomial expansion $\left(\sqrt{x} + \dfrac{1}{2 \cdot \sqrt[4]{x}} \right)^{n}$ the second, third, and fourth binomial coefficients form an arithmetic sequence. Find the rational terms in the expansion.

X. 74. The sixth term of the expansion $\left[\sqrt{2^{\log\left(10-3^x\right)}} + \sqrt[5]{2^{(x-2)\log 3}} \right]^n$ is 21

and the coefficients of the second, third, and fourth terms are in an arithmetic sequence. Find the numerical value of x and n.

X. 75. In the expansion $\left(x \cdot \sqrt[3]{x} - \dfrac{2}{\sqrt[5]{x^2}} \right)^n$ the sum of the binomial

coefficients is 256. Find the term containing x^2.

X. 76. In the expansion $\left(\sqrt[5]{x} - \dfrac{1}{\sqrt[5]{x^3}} \right)^n$ the sum of the coefficients of odd

terms is 256. Find the term containing $\dfrac{1}{x^3}$.

X. 77. Find the rational term(s) of the expansions:

a) $\left(\sqrt[3]{3} + \sqrt{2} \right)^5$; b) $\left(\sqrt[3]{4} + \sqrt[5]{2} \right)^{20}$; c) $\left(\sqrt{3} + \sqrt[14]{5} \right)^{200}$; d) $\left(\sqrt[5]{2} - \sqrt[7]{3} \right)^{24}$.

X. 78. Consider the binomial expansion $\left(\sqrt[3]{x} + \dfrac{1}{2 \cdot \sqrt[5]{x^2}} \right)^{100}$. Determine in each case:

a) the 4th binomial coefficient of the expression;
b) the term containing x^{26};
c) the sum of the coefficients of the expansion;
d) the middle term of the expansion;
e) the number of rational terms of the expansion when $x = 3$.

X. 79. Find the sum of the coefficients of the expansions:

a) $\left(9x^3 - 8y \right)^{20}$; b) $\left(7x^4 - 5y \right)^{22}$.

X. 80. Find the largest term in the binomial expansion:

a) $\left(\dfrac{1}{2} + \dfrac{1}{2} \right)^{100}$; b) $\left(\dfrac{9}{10} + \dfrac{1}{10} \right)^{100}$; c) $\left(\sqrt{2} + \sqrt{5} \right)^{20}$.

X. 81. Find the coefficient of x^2 in the expansion $(2 + x)(1 - ax)^5$.

X. 82. Solve for x the following equations:

a) $\displaystyle\sum_{k=0}^{9} \binom{9}{k}\cdot(-1)^k x^{9-k} = 1$; b) $\displaystyle\sum_{k=0}^{13} \binom{13}{k}\cdot(-1)^k x^{13-k} 2^{2k} = 0$;

c) $\displaystyle\sum_{k=0}^{13} \binom{10}{k}\cdot 2^{-k} \sin^{10-k} x = 0$.

X. 83. Simplify: a) $\displaystyle\sum_{k=0}^{n} \binom{n}{k}\sin^{2k} x \cos^{2n-2k} x$;

b) $\displaystyle\sum_{k=0}^{n} \binom{n}{k}(-1)^k \cosh^{2k} x \cdot \sinh^{2n-2k} x$.

X. 84. Prove that for any positive integer n, $\left(2+\sqrt{2}\right)^n + \left(2-\sqrt{2}\right)^n$ is an integer positive number.

X. 85. Prove the identities:

a) $1 - \binom{n}{2} + \binom{n}{4} - \binom{n}{6} + \ldots = 2^{\frac{n}{2}} \cos\dfrac{n\pi}{4}$;

b) $\binom{n}{1} - \binom{n}{3} + \binom{n}{5} - \binom{n}{7} + \ldots = 2^{\frac{n}{2}} \sin\dfrac{n\pi}{4}$;

c) $\left[1 - \binom{n}{2} + \binom{n}{4} - \binom{n}{6} + \ldots\right]^2 + \left[\binom{n}{1} - \binom{n}{3} + \binom{n}{5} - \binom{n}{7} + \ldots\right]^2 = 2^n$;

d) $\left[1 - 3\binom{n}{2} + 9\binom{n}{4} - 27\binom{n}{6} + \ldots\right]^2 + 3\left[\binom{n}{1} - 3\binom{n}{3} + 9\binom{n}{5} - 27\binom{n}{7} + \ldots\right]^2 = 2^{2n}$.

X. 86. Prove the identities:

a) $\binom{n}{0} + \binom{n}{3} + \binom{n}{6} + \ldots = \dfrac{1}{3}\left(2^n + 2\cos\dfrac{n\pi}{3}\right)$;

b) $1 - 3\binom{n}{2} + 9\binom{n}{4} - 27\binom{n}{6} + \ldots = (-1)^n 2^n \cos\dfrac{2n\pi}{3}$;

c) $\binom{n}{1} - 3\binom{n}{3} + 9\binom{n}{5} - \ldots = \dfrac{(-1)^{n+1} 2^{n+1}}{\sqrt{3}} \sin\dfrac{2n\pi}{3}$.

X. 87. Prove the identity $\binom{n}{1} - \dfrac{1}{3}\binom{n}{3} + \dfrac{1}{9}\binom{n}{5} - \dfrac{1}{27}\binom{n}{7} + \ldots = \dfrac{2^n}{3^{\frac{n-1}{2}}} \sin\dfrac{n\pi}{6}$.

X. 88. Prove the identity $\displaystyle\sum_{k=0}^{3n} \binom{6n}{2k}(-3)^k = 2^{6n}$.

223

X. 89. Prove the identities:

a) $\binom{n}{0}+\binom{n}{4}+\binom{n}{8}+\ldots=\dfrac{1}{2}\left(2^{n-1}+2^{\frac{n}{2}}\cos\dfrac{n\pi}{4}\right)$;

b) $\binom{n}{1}+\binom{n}{5}+\binom{n}{9}+\ldots=\dfrac{1}{2}\left(2^{n-1}+2^{\frac{n}{2}}\sin\dfrac{n\pi}{4}\right)$;

c) $\binom{n}{2}+\binom{n}{6}+\binom{n}{10}+\ldots=\dfrac{1}{2}\left(2^{n-1}-2^{\frac{n}{2}}\cos\dfrac{n\pi}{4}\right)$;

d) $\binom{n}{3}+\binom{n}{7}+\binom{n}{1}+\ldots=\dfrac{1}{2}\left(2^{n-1}-2^{\frac{n}{2}}\sin\dfrac{n\pi}{4}\right)$.

X. 90. Calculate the following sums:

a) $\binom{n}{0}+\binom{n}{1}+\binom{n}{2}+\ldots+\binom{n}{n}$;

b) $\binom{n}{1}+2\binom{n}{2}+3\binom{n}{3}+\ldots+n\binom{n}{n}$;

c) $\binom{n}{0}+2\binom{n}{1}+3\binom{n}{2}+\ldots+(n+1)\binom{n}{n}$;

d) $\binom{n}{2}+2\binom{n}{3}+3\binom{n}{4}+\ldots+(n-1)\binom{n}{n}$;

e) $\binom{n}{0}+3\binom{n}{1}+5\binom{n}{2}+\ldots+(2n+1)\binom{n}{n}$;

f) $k\binom{n}{0}+(k+1)\binom{n}{1}+(k+2)\binom{n}{2}+\ldots+(k+n)\binom{n}{n}$;

g) $\binom{n}{0}-2\binom{n}{1}+3\binom{n}{2}+\ldots+(-1)^{n}(n+1)\binom{n}{n}$;

h) $3\binom{n}{1}+7\binom{n}{2}+11\binom{n}{3}+\ldots+(4n-1)\binom{n}{n}$;

i) $\binom{n}{1}-2\binom{n}{2}+3\binom{n}{3}-\ldots+(-1)^{n-1}n\binom{n}{n}$;

j) $\dfrac{\binom{n}{0}}{1}+\dfrac{\binom{n}{1}}{2}+\dfrac{\binom{n}{2}}{3}+\ldots+\dfrac{\binom{n}{n}}{n+1}$;

k) $\dfrac{\binom{n}{0}}{2}+\dfrac{\binom{n}{1}}{3}+\dfrac{\binom{n}{2}}{4}+\ldots+\dfrac{\binom{n}{n}}{n+2}$;

l) $\dfrac{\binom{n}{0}}{1\cdot 2}+\dfrac{\binom{n}{1}}{2\cdot 3}+\dfrac{\binom{n}{2}}{3\cdot 4}+\ldots+\dfrac{\binom{n}{n}}{(n+1)(n+2)}$;

m) $\dfrac{\binom{n}{0}}{1}-\dfrac{\binom{n}{1}}{2}+\dfrac{\binom{n}{2}}{3}-\ldots+(-1)^{n}\dfrac{\binom{n}{n}}{n+1}$.

X. 91. Calculate the following sums:

a) $\binom{n}{0}-\binom{n}{1}+\binom{n}{2}-\ldots+(-1)\binom{n}{m}$;

b) $\binom{n}{k}+\binom{n+1}{k}+\binom{n+2}{k}+\ldots+\binom{n+m}{k}$.

X. 92. Calculate the following sums:

a) $\binom{n}{0}^2+\binom{n}{1}^2+\binom{n}{2}^2+\ldots+\binom{n}{n}^2$;

b) $\binom{n}{0}^2-\binom{n}{1}^2+\binom{n}{2}^2+\ldots+(-1)^n\cdot\binom{n}{n}^2$;

c) $\binom{2n}{0}-\binom{2n-1}{1}+\binom{2n-2}{2}-\ldots+(-1)^n\binom{n}{n}$;

d) $\binom{2n}{n}+2\binom{2n-1}{n}+4\binom{2n-2}{n}+\ldots+2^n\binom{n}{n}$;

e) $\binom{n}{1}^2+2\binom{n}{2}^2+3\binom{n}{3}^2+\ldots+n\binom{n}{n}^2$.

X. 93. Prove that $\displaystyle\sum_{k=0}^{p}\binom{p}{k}P(m,k)P(n-p,m-k)=P(n,m)$.

X. 94. Prove that the equality $\binom{3n}{n}=\displaystyle\sum_{k=0}^{n}\binom{2n}{k}\binom{n}{k}$ holds for any $n\geq 1$.

X. 95. Prove the identity $\left(\displaystyle\sum_{k=0}^{n}2^k\binom{2n}{2k}\right)^2-2\cdot\left(\displaystyle\sum_{k=0}^{n-1}2^k\binom{2n}{2k+1}\right)^2=1$.

X. 96. Prove the identity $\dfrac{\binom{n}{1}}{\binom{n}{0}}+\dfrac{2\binom{n}{2}}{\binom{n}{1}}+\dfrac{3\binom{n}{3}}{\binom{n}{2}}+\ldots+\dfrac{n\binom{n}{n}}{\binom{n}{n-1}}=\dfrac{n(n+1)}{2}$.

X. 97. Prove the identity

$\binom{n}{0}(n-1)^n+\binom{n}{1}(n-1)^{n-1}+\binom{n}{2}(n-1)^{n-2}+\ldots+\binom{n}{n}(n-1)^{n-n}=n^n$, $n\in\mathbf{N}$.

X. 98. Calculate the sums

$S_1=1+\binom{n}{1}\cos x+\binom{n}{2}\cos 2x+\ldots+\binom{n}{n}\cos nx$ and

$S_2=\binom{n}{1}\sin x+\binom{n}{2}\sin 2x+\ldots+\binom{n}{n}\sin nx$.

Chapter I

NATURAL AND INTEGER NUMBERS

I. 1. a) >; b) <; c) =; d) <; e) =; f) <; g) =; h) =; i) =; j) =.

I. 2. a) 4; b) 52; c) 15; d) 28; e) 112; f) 42; g) 89; h) 33; i) 589; j) 33.

I. 3.

x	18	59	76
y	22	12	15
$x+y$	40	71	91

a	b	c	$a+b$	$a+c$	$a+b+c$
32	14	19	46	51	65
14	15	13	29	27	39

I. 4. a) 1; b) 5; c) 5; d) 5; e) 2; f) $b = 40$; g) $x = 1$; h) $b = 52$; i) $a = 50$; j) $a = 40$.

I. 5. a)

$43 + 18 = 61$	$55 - 25 - 8 = 22$
$35 + 29 = 64$	$84 - 56 - 8 = 20$
$83 - 27 = 56$	$38 + 29 + 17 = 84$
$84 - 36 = 48$	$46 + 9 - 16 = 39$

b) $2 + 13 + 7 + 18 = (13 + 7) + (2 + 18) = 40$.

c)

$n = 51$	$a = 36$
$d = 4$	$a = 24$
$x = 11$	$x = 11$

d) 71; e) 120; f) 75; g) 0, 5, 10, <u>15</u>, <u>20</u>, <u>25</u>, <u>30</u>, <u>35</u>, <u>40</u>, <u>45</u>, 50;

0,3,6, <u>9</u> , <u>12</u>, <u>15</u>, <u>18</u>, <u>21</u> , <u>24</u>, <u>27</u>, 30; 0,10,20, <u>30</u>, <u>40</u>, <u>50</u>, <u>60</u>, <u>70</u>, <u>80</u>, <u>90</u>, 100;

h) i) 13,806. ii) 582. iii) 571. iv) 197. i) 7 times. j) $a = 42$; $a = 6$; $a = 9$.

I. 6.

a	b	c	d	c−b	3a−b	3c−d+1	a+3b−c	3b+2c−d+2
7	21	63	189	42	0	1	7	2
10	16	16	48	0	14	1	42	34
7	15	15	30	0	6	16	37	47
5	15	45	135	30	0	1	5	2
4	19	36	144	17	−7	−35	25	−13

I. 7. a) 0; b) 0; c) 217; d) 3; e) 1; f) 0; g) 0; h) 0; i) 0; j) 0.

I. 8. a) −5; b) 1; c) 1; d) −8; e) 8; f) −8; g) −3.

I. 9. a) −49; b) −65; c) 189; d) 51; e) 1; f) -3; g) 16.

I. 10. a) 1; b) 72; c) -1; d) 0; e) 4; f) 8; g) -16; h) -10.

I. 11. a) -12; b) −63; c) −58; d) 17.

I. 12.

a	b	c	a−b	b−c	(a−b)+2c	(2a−2b)+(b−c)
26	15	14	11	1	39	23
26	25	24	1	1	49	3
125	100	100	25	0	225	50
45	40	30	5	10	65	20
60	16	6	30	10	42	70
8	7	1	1	6	3	8

I. 13. a) 1; b) −46; c) 40; d) -39; e) 65. I. 14. a) 6; b) −12; c) 6;
d) -31; e) 35; f) −7. I. 15. a) −144; b) 48; c) −192; d) −36; e) 7;
f) −5; g) −3; h) 4. I. 16. a) -30; b) 9; c) 23; d) −1; e) −4; f) 1;
g) −1; h) −6; i) 73; j) -17; I. 17. a) 60; b) 120; c) −32; d) 39;
e) -130; f) −72; g) 14. I. 18. a) $(80 + 9 − 9) \div (−40) = -2$;
b) $50 \cdot (59 − 59) = 0$; c) $22 \div (17 − 6) = 2$; d) $3 \cdot (18 + 11) = 87$;
e) $7 \cdot 17 = 119$; f) $10 + 53 = 63$; g) 1; h) -218; i) 5; j) $12 − 7 = 5$;
k) $−3 − 4 + 7 = 0$; l) $−12 \div (−6) − 7 = 2 − 7 = −5$.

I. 19. a) 5; b) $−3 \cdot [72 − 65 − (34 − 45)] = -3 \cdot 18 = −54$; c) 92;
d) $−9 \cdot [−5 − (12 − 9)] \cdot (−4) = −9 \cdot (−8) \cdot (−4) = -288$.

I. 20. a) -11; b) 1; c) -3; d) -124; e) 30; f) 18;

I. 21.

a	b	c	d	a²c−3a²b+a+b	a²+3ab²−abc
6	18	54	162	24	36
3	9	27	81	12	9
11	33	99	297	44	121
9	27	81	243	36	81
17	51	153	459	68	289

I. 22. a) −8; b) −5; c) 225; d) −1; e) 1; f) 17; g) 1; h) 1; i) 0; j) −144 ;

k) 190; l) $-29 \cdot 10^{23}$; m) 2^{81}; n) 3^8; o) 0; p) -1.

I. 23. a) -2^{12}; b) 3^{12}; c) 2^5; d) -1. **I. 24.** a) 1; b) 3 ;

c) -2^4 ; d) 3 . **I. 25.** a) −4998; b) −1; c) 1.

I. 26. a) 3^6; b) −3; c) 1; d) 1; e) 7; f) -2^{111}.

I. 27. a) $8 \cdot \left\{10 + 1 \cdot \left[2^4 \cdot 2^4 \div 2^4 - 5 \cdot (27 - 25)\right] \div 3 + 2\right\} =$

$= 8 \cdot \left[10 + (16 - 10) \div 3 + 2\right] = 8 \cdot (10 + 6 \div 3 + 2) = 8 \cdot (10 + 2 + 2) = 112;$

b) $\left(5^2 + 3^2\right) \div 2 = (25 + 9) \div 2 = 34 \div 2 = 17;$

c) $\left(2 \cdot 3^2\right) + 1 = 18 + 1 = 19;$ d) $\left(3^2 + 2^4 + 5^9\right) \div \left(3^2 + 2^4 + 5^9\right) = 1$.

I. 28. a) 0; b) $(-1)^{n+1}$; c) $(-1)^n + 2$; d) −50.

I. 29. a) i) $1 - 8 + 9 = 2$; ii) $(25 + 5 + 1) \div (27 + 4) = 31 \div 31 = 1;$

b) $\left(1000 \div 125 + 2^6 \div 2^4 - 27\right) \cdot 2 + 1 + 1 = (8 + 4 - 27) \cdot 2 + 2 = -28;$

c) $968 - \left\{182 - 5 \cdot \left[8 \cdot 5 \div (9 \cdot 5 \div 15 + 5) + 2 \cdot (13 - 12)\right]\right\} \cdot 5 =$

$= 968 - \left\{182 - 5 \cdot \left[40 \div (45 \div 15 + 5) + 2\right]\right\} \cdot 5 =$

$= 968 - \left\{182 - 5 \cdot \left[40 \div (3 + 5) + 2\right]\right\} \cdot 5 = 968 - \left[182 - 5 \cdot (5 + 2)\right] \cdot 5 =$

$= 968 - (182 - 5 \cdot 7) \cdot 5 = 968 - 147 \cdot 5 = 968 - 735 = 233$;

d) i) $4^2 (15 - 14) = 4^2 = 16$, ii) $11^2 (20 - 18 - 1) = 11^2 = 121$,

iii) $11^2 (30 - 28) = 11^2 \cdot 2 = 121 \cdot 2 = 242$, iv) $2 + 2 = 4$,

v) $3^2 + 1 = 9 + 1 = 10$; vi) 24 ; e) 100; f) 326 . **I. 30.** a) $-9x^5$; b) $12m^6$;

c) $-8x^7$; d) $40a^4b^4$; e) $14x^6y^3$; f) $72x^4y$; g) $-12a^3b^4c$; h) $10m^4n^6$;

i) $-20a^8b^7$. **I. 31.** c). **I. 32.** a) $8a$; b) $3x^2y$; c) $-5mn$; d) $4m^2-3$;

e) $-2x^3+4x^3y$; f) $-rs^2+6rs$; g) $-7a^2b+2ab$ h) x^2yz; i) $-2x^2y+5xy^2$.

I. 33.

a	b	c	d	6a³c÷a–18a³b÷a+b	3a³c²÷b–a³cd÷b+b÷a
5	15	45	135	15	3
3	9	27	81	9	3
1	3	9	27	3	3
9	27	81	243	27	3
17	51	153	459	51	3

I. 35. $7 \mid (5x+y) \Rightarrow 7 \mid [7x-(5x+y)] \Rightarrow 7 \mid (2x-y)$, $x, y \in \mathbf{Z}$.
I. 36. a) $a=3$; b) $a=4$; c) $a=3$; d) $a=7$; e) $a=9$; f) $a=14$;
g) $a=77$; h) $a=38$; i) $a=40$; j) $x=17$. **I. 37.** a) $x=7$; b) $y=9$; c) $z=5$;
d) $x=-4$; e) $b=-3$; f) $x=10$; g) $x=39$; h) $x=-5$; i) $y=0$; j) $z=19$; k) $b=0$;
l) $x=15$; m) $x=14$; n) $x=-11$.
I. 38. a) $x=12$; b) $y=-4$; c) $a=-4$; d) $x=-16$; e) $y=5$; f) $z=3$; g) $x=-8$;
h) $t=-11$; i) $x=-1$.
I. 39. a) $x=3$; b) $y=2$; c) $z=-6$; d) $t=-18$; e) $x=-7$; f) $y=-4$; g) $x=-4$;
h) $x=-41$; i) $y=-45$.
I. 40. a) $x=-8$; b) $x=3$; c) $y=-3$; d) $x=-5$; e) $t=2$; f) $x=-7$; g) $a=-5$;
h) $x=-3$; i) $y=-15$; j) $x=-1$; k) $a=12$; l) $t=-4$; m) $y=3$; n) $x=-5$;
o) $z=12$; p) $x=-7$; q) $x=0$; r) $x=1$; s) $x=3$; t) $x=6$; u) $x=-2$; v) $x=-17$; w) y
$=4$; x) $x=-2$; y) $z=5$; z) $a=-4$.
I. 41. a) $x=7$; b) $x=12$; c) $y=20$; d) $a=17$; e) $x=12$; f) $x=13$; g) $x=-9$;
h) $y=4$; i) $x=7$; j) $x=-5$; k) $y=9$; l) $a=5$; m) $x=-6$; n) $y=3$; o) $t=3$;
p) $x=-12$; q) $x=9$; r) $x=3$; s) $x=-2$; t) $x=-3$; u) $x=1$; v) $x=1$; w) $x=1$; x)
$x=-1$.
I. 42. a) $x>3$ or $x \in (3, \infty)$; b) $x \le 7$ or $x \in (-\infty, 7]$; c) $x<4$ or $x \in (-\infty, 4)$;
d) $x \le -7$ or $x \in (-\infty, -7]$; e) $x<4$ or $x \in (-\infty, 4)$; f) $x \le -8$ or $x \in (-\infty, -8]$;
g) $x>-5$ or $x \in (-5, \infty)$; h) $x \le 5$ or $x \in (-\infty, 5]$; i) $x \le 3$ or $x \in (-\infty, 3]$;
j) $x<18$ or $x \in (-\infty, 18)$; k) $x \le 9$ or $x \in (-\infty, 9]$.

I. **43**. **a)** $x > 5$ or $x \in (5, \infty)$; **b)** $x > 9$ or $x \in (9, \infty)$; **c)** $x > -5$ or $x \in (-5, \infty)$;
d) $x > 1$ or $x \in (1, \infty)$; **e)** **f)** $x \leq 3$ or $x \in (-\infty, 3]$; **f)** $x \leq 6$ or $x \in (-\infty, 6]$;
g) $x < 1$ or $x \in (-\infty, 1)$; **h)** $x > -2$ or $x \in (-2, \infty)$; **i)** $x \leq -1$ or $x \in (-\infty, -1]$.
I. **44**. **a)** $x < -2$ or $x \in (-\infty, -2)$; **b)** $x < -$ or $x \in (-\infty, -9)$;
c) $x \geq -14$ or $x \in [-14, \infty)$; **d)** $y \geq 11$ or $y \in [11, \infty)$; **e)** **f)** $a \leq -3$ or $a \in (-\infty, -3]$;
f) $y > 13$ or $y \in (13, \infty)$.
I. **45**. **a)** $x \in (-3, 4]$; **b)** $x \in [2, 4)$; **c)** $x \in [-4, -1)$; **d)** $x \in (1, 6]$; **e)** $x \in (-1, 4]$;
f) $x \in [-2, 4)$; **g)** $x \in (-2, 2]$; **h)** $x \in [-1, 3]$; **i)** $x \in [1, 7]$; **j)** $x \in [-1, 1$

TEST I. 1
1) 1,4,7,10,13,16 and 3,7,11,15,19,23; **2) a)** 865 030;
b) 7 005 806; **3) a)** 2000; **b)** 35 000; **c)** 2886; **d)** 57; **e)** 79; **f)** 83; **4)** 92; **5)** 47; **6)**
a) \$35.19; **b)** \$11.73; **7)** \$5.78; **8)** \$5.78;
9) Matthew will walk 400m, 1600m, 4800 m and Alex 425m, 1700m, 5100 m,
10) \$7.33.

TEST I. 2
1) $a = 31$; **2)** $a = 80$; **3)** $a = 12$; **4)** $a = 56$; **5)** $a = 19$; **6)** $a = 2$; **7)** $a = 14$;
8) $a = 64$; **9)** $a = 41$; **10)** $a = 36$.

TEST I. 3
1) $100+(-1+101)+(-2+102)+(-3+103)+(-4+104)-5=500-a$
$100+100+100+100+100-5=500-a$. Therefore $a = 5$.
2) Let x be my age. $(4x+24) \div 2-2x = 2x+12-2x=12$.
Therefore my age is 12.
3) $2^7 - 2^6 - 2^5 - 2^4 = 2^4(2^3 - 2^2 - 2 - 1) = 2^4$
4) If you subtract 6 from one of the numbers it would be the same as if you
subtract 1 from each of the six numbers, so that means their average is 1 less
than the 12. Therefore the answer is 11.
5) The perimeter of the hexagon can't be 5 because if one side is 1, for example,
the minimum perimeter would be 6, and 77.
6) $1^2 + 2^2 + 3^2+...+18^2 + 19^2 + 20^2 = 2870$.
So then $1^2 + 2^2 + 3^2+...+18^2 + 19^2 = 2870 - 20^2 = 2870 - 400 = 2470$.

7) $[(2+8)+(4+6)+10] \div [10 + (8+2)+(6+4)] = 1$.
8) A cube has 12 edges because there are 4 top edges, 4 bottom edges, and 4
vertical edges.
9) The amount he spent is $13 \times 7¢=91¢$. The amount he has left is
\$2.50-91¢=\$1.59. With 14¢ for a gumball he can buy 11 gumballs because \$1.59
$\div 14¢=11.35$ so he will have 5¢ left over.
10) $270 \div 30=9$. Therefore it will take her 9 days to read the whole book.

TEST I. 4

1) 273; 2) i) $x = 153$, ii) $x = 4$, iii) $x = 9$, iv) $x = 40$; v) $x = 9$;
3) 6 ; 4) $a = 36$; 5) 250 ; 6) 70 ; 7) $a = 190$, $b = 38$; 8) $a = 212$, $b = 53$;
9) $a = 503$, $b = 10$; 10) \$30.

TEST I. 5

1) a) 6; b) 9; c) 0.8, d) 9; 2) $a = 13$; 3) $x = 185.24$; 4) 7;
5) $a = 45$, $b = 70$, $c = 25$; 6) 10 pigs and 30 hens; 7) a = 165 ;
8) 424.5; 9) $a = 30$, $c = 15$, $b = 75$; 10) 2950.

TEST I. 6

1) a) $a = 6$; b) $a = 9$; c) $a = 8$; d) $a = 9$; 2) $a = 12$; 3) $x = 3$;
4) 7; 5) $c = 55$; 6) 153,846; 7) 4,885; 8) $a = 135$, $b = 89$; 9) 793;
10) $x = 30$, $y = 156$, $z = 26$.

TEST I. 7

1) a) 2^9; b) 3^{10}; c) a^{15}; d) 7^{14}; e) 5^5; f) 2^2; g) 3^8;
h) 5^{12}; 2) a) 2^8; b) 3^2; c) 2^{12}; d) 35^8; 3) a) $2^{10} > 8^3$;
b) $32^{20} = 16^{25}$; c) $2^{15} < 3^{30}$; d) $3^{100} > 2^{150}$; e) $5^{30} < 3^{45}$;
4) a) 5; b) 2^5; c) 1; d) 32; e) 1; f) 4; g) 1;
5) a) $x = 4$, b) $x = 2^3$; c) $x = 3^2$; d) $x = 6$;
6) a) $x = 4$; b) $x = 66$; c) $x = 65$; 7) a > b;
8) $a = 6^{2n+2}$, $b = 3^{2n+4} \cdot 2^{4n+4}$, $a < b$;
9) a) $n = 45$; b) $31 \cdot 3^n \cdot 5^{2n}$; 10) $x = 4$.

TEST I. 8

1) a) $1 - 8 + 9 = 2$; b) $(25 + 5 + 1) \div (27 + 4) = 31 \div 31 = 1$;
2) $\left(1000 \div 125 + 2^6 \div 2^4 - 27\right) \cdot 2 + 1 + 1 = (8 + 4 - 27) \cdot 2 + 2 = -28$;
3) $968 - \{182 - 5 \cdot [8 \cdot 5 \div (9 \cdot 5 \div 15 + 5) + 2 \cdot (13 - 12)]\} \cdot 5 =$
$= 968 - \{182 - 5 \cdot [40 \div (45 \div 15 + 5) + 2]\} \cdot 5 =$
$= 968 - \{182 - 5 \cdot [40 \div (3 + 5) + 2]\} \cdot 5 = 968 - [182 - 5 \cdot (5 + 2)] \cdot 5 =$
$= 968 - (182 - 5 \cdot 7) \cdot 5 = 968 - 147 \cdot 5 = 968 - 735 = 233$
4) a) $4^2 (15 - 14) = 4^2 = 16$, b) $11^2 (20 - 18 - 1) = 11^2 = 121$,
c) $11^2 (30 - 28) = 11^2 \cdot 2 = 121 \cdot 2 = 242$, d) 4, e) $3^2 + 1 = 10$;

f) $2^7 - 2^6 - 2^5 - 2^4 + 2^3 = 2^3\left(2^4 - 2^3 - 2^2 - 2 + 1\right) = 2^3 \cdot 3 = 24;$

5) $\{[(49-14)+32\div 4]\div(27+16)+9\}\cdot 11 - [(4+9)\cdot 5-5\ ]=$
$= [(35+8)\div 43+9]\cdot 11 - (13\cdot 5-55) = (43\div 43+9)\cdot 11 - (65-55)=$
$= 10\cdot 1\ -10 = 110-10 = 100;$

6) $\{[(625-25)\div 25+(32+64)\cdot 4]-81\}-1 =$

$= [(600\div 25+96\cdot 4)-81]-1 = [(24+384)-81]-1 = 326;$

7) $15\cdot(3x-14+9)+(2x+25\cdot 2-4\cdot 9)\cdot 5 = 5\cdot 8^2 \div 2$

$15\cdot(3x-5)+(2x+50-36)\cdot 5 = 5\cdot 64\div 2,$

$15\cdot(3x-5)+(2x+14)\cdot 5 = 160, 3\cdot(3x-5)+(2x+14) = 32$
$9x-15+2x+14 = 32, 9x+2x = 32-14+15\ \ x = 3;$

8) $\left[\left(5x-20+2^{10}\div 2^9-3^{10}\div 3^9\right)\cdot 5-2^{10}\div 2^9\right]\div 3 = 1,$

$[(5x-20+2-3)\cdot 5-2]\div 3 = 1, [(5x-21)\cdot 5-2]\div 3 = 1, (25x-105-2)\div 3 = 1,$
$25x-107 = 3, 25x = 3+107, 25x = 110,\ \ x = 4.4;$

9) a) 5220; b) 49; c) 280; 10) $a = 1,\ \ b = 2^n,\ \ c = 2^{2n}, n = 1.$

TEST I. 9

1) a, $a+1$, $a+2$, $3(a+1)$; 2) a) 10; b) 50; 3) $a = 6254$, $b = 10$;
4) $(x-2)^{2x-4} = 8^{16}$, $x = 10$; 5) A $= 6ab$; 6) $x = 0, y = 2^{1950}$, $(2\cdot 3)^3$;
7) 891, 781, 561, 451, 341, 231, 121; 8) $E = 0$; 9) $a = 1$;
10) The last digit is 4.

TEST I. 10

1) 520; 2) 7; 3) $a = 22$, $b = 112$, $c = 280$;

4) $(7\cdot 8+4\cdot 1\ )-50 = 14$; 5) 39; 6) 89; 7) 45; 8) 25;
9) 396.5; 10) 410.

TEST I. 11

1) a) 10; b) 25; c) 1; d) 8; e) 2; f) $6\frac{1}{9}$; g) 23; h) 0.15;

2) a) $x = 6.98$; b) $x = 20.47$; c) $x = 0.4$; d) $x < 4.2$;

e) $x = 16.86$; f) $x \leq 20.5$; g) $x = 4$ .3) a) 10^2; b) $\left(3^7\right)^2$;

c) 1000^2; d) $\left(2^4\right)^2$; 4) a) $a = 4,\ \ b = -4$;

b) $x = 28$; c) $-20{,}200$; 5) a) 11^2; b) $\left(2^5\right)^2$; c) 2010^2; d) $\left(2^8\right)^2$.

6) a) $10a + a + 10b + b = 1 \cdot (a+b)$;

b) 18, 27, 36, 45, 54, 63, 72, 81; 7) $x + y = 9$; 8) a) $\underbrace{100...00}_{37\ zeros}455$; b) 15;

c) $\overline{...5} - \overline{...5}$; d) odd –odd; 9) $N = 6^{n-1} \cdot 720$, the remainder is 0;

10) $(\overline{a1b} + \overline{a2b} + ... + \overline{a9b}) \div (\overline{a5b}) = 9$.

TEST I. 12

2) $a = 5$; 3) $a = 3$; 4) $N = 10^n \cdot x + 2x = 3x + 10^n \cdot x - x =$

$= 3x + x(10^n - 1) = 3x + 3x \cdot 3x = 3x(3x+1)$, $N = \underbrace{\overline{333...3}}_{n\ times} \cdot \underbrace{\overline{333...4}}_{n\ times}$;

5) From (1) $a + b + c = d$, $\overline{abc} + \overline{bca} + \overline{cab} = 111(a+b+c) = 111d = \overline{ddd}$;

6) $a = 3^2$; 7) $a = 3111x$; 8) $\overline{abc} = 187$; 9) $\overline{abc2} = 1000a + 100b + 10c + 2$,

$3 \cdot \overline{abc} = 300a + 30b + 3c$, $\overline{4abc} = 4000 + 100a + 10\ b + c$, $1200a + 120b + 12c = 5232$, $100a + 10b + c = 436$,

$\overline{abc} = 436$, $a = 4$; $b = 3$; $c = 6$;

10) $20 \cdot a + 2 \cdot b = 2(10a + b) = 2 \cdot \overline{ab} \Rightarrow \overline{abb}^{\overline{ab}} = (\overline{ab}^2)^{\overline{ab}} \Rightarrow \overline{abb} = \overline{ab}^2 \Rightarrow$

$10\overline{ab} + b = \overline{ab}^2$. Therefore b is a divisor of

$\overline{ab} \Rightarrow b = 0$ and $a = 1, \overline{ab} = 10$.

TEST I. 13

1) A $= 2002^2$; 2) $A = (3^{n+2} \cdot 2^{2n+2})^2$; 3) $n = 2^{2010}(2^3 - 2^2 - 2^1 - 2^0) = (2^{1005})^2$;

4) $2^{6n+10} + 2^{6n+10} + 2^{6n+9} = 5 \cdot 2^{3n+24} \Rightarrow 2^{6n+9}(2 + 2 + 1) = 5 \cdot 2^{3n+24}$,

$n = 5$, the last digit is 2; 5) $(2^0 + 2^1 + 2^2)(2^0 + 2^3 + 2^6 + ... + 2^{96})$;

6) $(3^1 + 3^2 + 3^3 + 3^4 + 3^5)(3^0 + 3^6 + ... + 3^{2005})$; 7) N $= \overline{...7}$;

8) The last digit is 2; 9) $a = 2^{2n} \cdot 3^{2n+2}(4 \cdot 9 + 2 \cdot 3 - 1) = 6^{2n} \cdot 369$;

10) $x = 2^n, y = 3^{n-1}, z = 3$; $xyz = 6^n$.

TEST I. 14

1) a) 123400 ; b) $a = 1$, $b = 0$, $c = 1$; 2) a) $9^{60} = 27^{40}$,

b) $7^{1275} > 5^{1275}$; 3) $a = 2^{102}$, $b = 2^{101} + 2^{99} \cdot x$, $x = 4$;

4) a) $x = (2^{1500})^2 \cdot 5^2$, $y = 5^2 \cdot (3^{1000})^2$; b) $x = (2^{2999} \cdot 5)^2 \cdot 10$; c) $x < y$;

5) $x = 121(a+b)$; 6) $x = 110(a+b)$; 7) $x = 211(a+b)$;

233

8) $x \cdot 1000 + y \cdot 100 + z \cdot 10 + 0 + x \cdot 100 + y \cdot 10 + z = 2002,\ 100x + 10y + z = 182,$
$z = 2$ and $y = 8$;

9) $\overline{abc} \cdot 10^{2007} + \ldots + \overline{abc} \cdot 10^3 + \overline{abc} = \left(10^{2007} + \ldots + \cdot 10^3 + 10^0\right)\overline{abc} =$
$= \left(10^3 + 10^0\right)\left(10^{2004} + \ldots + \cdot 10^3 + 10^0\right)\overline{abc}$; 10) $x = 222(a+b)$.

TEST I. 15
1) $x = \underbrace{1000000...00}_{2006\ times} + 9\underbrace{00...00}_{2002\ times} - 2 = 1000\underbrace{900...00}_{2003\ times} - 2 = 10008\ \underbrace{9...9}_{2001\ times}\ 8$.

Then the sum of the digits is $9 \times 2001 + 16 + 1 = 18.026$.

2) Let $a - 1$, a, $a + 1$ be the three consecutive numbers. We have
$(a-1)^2 + a^2 + (a+1)^2 = 2\underbrace{99...99}_{n-1\ times}82\underbrace{00...00}_{n-2\ times}29$. Expanding and

collecting similar terms we get $3a^2 = 2\underbrace{99...99}_{n-1\ times}82\underbrace{00...00}_{n-2\ times}27$ and

$a^2 = \underbrace{99...99}_{n-1\ times}94\underbrace{00...00}_{n-2\ times}09$.

On the other hand

$a^2 = 9 \cdot 10^{2n} + \ldots + 9 \cdot 10^{n+2} + 9 \cdot 10^{n+1} + 4 \cdot 10^n + 9 = \left(10^{n+1} - 3\right)^2$.
Finally, $a = 10^{n+1} - 3$.

4) $x = 4$.

5) $x \cdot 1000 + y \cdot 100 + z \cdot 10 + 0 + x \cdot 100 + y \cdot 10 + z = 2002$
$x(1000 + 100) + y(100 + 10) + z(10 + 1) = 2002,\ 1100x + 110y + 11z = 2002,$
$100x + 10y + z = 182,\ z = 2,\ y = 8,\ x = 1$.

6) a) $x = 3$, b) $x = 4.4$. 7) $\overline{abc} = 148$. 8) Denoting $x = \underbrace{111...11}_{n\ times}$, we have

$N = x \cdot 10^{n-1} + 2x = x \cdot \left(10^{n-1} - 1\right) + 3x = 3x \cdot x + 3x = 3x(3x+1) =$
$= \underbrace{333...33}_{n\ times} \cdot \underbrace{333...34}_{n\ times}$. 9) $n = 1$. 10) The last digit is 5.

Chapter II

RATIONAL NUMBERS AND FRACTIONS

II. 1. a) 0.75 ; **b)** 0.6; **c)** 0.73; **d)** $0.8\bar{3}$; **e)** $0.\bar{6}$; **f)** 0.35; **g)** $0.9\overline{714285}$;

h) $0.8\bar{6}$; **i)** 0.12. **II. 2. a)** $\dfrac{7}{10}$; **b)** $\dfrac{231}{100}$; **c)** $\dfrac{1}{100}$; **d)** $\dfrac{13}{1000}$; **e)** $\dfrac{34}{99}$; **f)** $\dfrac{7}{10}$;

g) $\dfrac{7}{3}$; **h)** $\dfrac{142}{99}$; **i)** $\dfrac{14}{15}$; **j)** $\dfrac{17}{30}$; **k)** $\dfrac{32}{15}$; **l)** $\dfrac{17}{12}$. **II. 3. a)** $\dfrac{1}{3}$; **b)** $-\dfrac{7}{11}$; **c)** $\dfrac{11}{909}$;

d) $-\dfrac{5}{7}$; **e)** $\dfrac{1}{2}$; **f)** $\dfrac{1}{7}$; **g)** $\dfrac{2}{5}$; **h)** $\dfrac{1}{4}$; **i)** $\dfrac{1}{2}$; **j)** $\dfrac{108}{107}$.

II. 4. a) $-\dfrac{3}{4} = -0.75$; **b)** $-\dfrac{3}{4} < -\dfrac{2}{4}$; **c)** $-\dfrac{2}{5} < \dfrac{1}{7}$; **d)** $0.47 = \dfrac{47}{100}$;

e) $1.\bar{4} = 1\dfrac{4}{9}$; **f)** $-0,35 < -0.34$; **g)** $-\dfrac{3}{8} < -\dfrac{1}{3}$; **h)** $-\dfrac{2}{3} < -\dfrac{7}{8}$; **i)** $\dfrac{2}{3} < \dfrac{3}{4}$.

II. 5. a) $-2, \ -\dfrac{4}{3}, \ -\dfrac{1}{3}, \ 0, \ \dfrac{2}{3}, \ \dfrac{4}{3}, \ 2$; **b)** $-1\dfrac{1}{5}, -\dfrac{4}{5}, -\dfrac{2}{5}, 0, \dfrac{2}{5}, 1\dfrac{1}{5}, 1\dfrac{3}{5}$;

c) $-2\dfrac{1}{4}, -\dfrac{1}{4}, \dfrac{1}{4}, \dfrac{1}{2}, \dfrac{3}{4}$; **d)** $\dfrac{2}{9}, \dfrac{1}{4}, \dfrac{5}{7}, \dfrac{3}{3}, \dfrac{9}{5}, \dfrac{8}{2}$; **e)** $\dfrac{4}{7}, \dfrac{3}{5}, 2\dfrac{5}{7}, \dfrac{4}{1}, 1\dfrac{9}{3}, 5\dfrac{3}{2}$.

II. 6. a) $-\dfrac{1}{4}$; **b)** $\dfrac{1}{3}$; **c)** $-\dfrac{5}{7}$; **d)** $\dfrac{1}{7}$; **e)** 0; **f)** 1.6; **g)** 1.21; **h)** 1.26; **i)** -0.4;

j) 0.16; **k)** -0.7; **l)** -3.22; **m)** 5.25; **n)** -1.19; **o)** 6.481.

II. 7. a) $\dfrac{3}{14}$; **b)** $\dfrac{1}{14}$; **c)** $-\dfrac{3}{14}$; **d)** $-\dfrac{11}{10}$; **e)** $-\dfrac{4}{15}$; **f)** $-\dfrac{14}{15}$; **g)** $\dfrac{1}{10}$; **h)** $-\dfrac{1}{4}$;

i) $-\dfrac{11}{12}$; **j)** $-\dfrac{11}{12}$; **k)** $\dfrac{5}{6}$; **l)** -11.68 ; **m)** 1.96; **n)** 0.92; **o)** -0.1; **p)** -0.01.

II. 8. a) $\dfrac{3}{7}$; **b)** 1; **c)** $-\dfrac{11}{14}$; **d)** $\dfrac{14}{45}$; **e)** $\dfrac{5}{24}$; **f)** $\dfrac{4}{3}$; **g)** $-\dfrac{23}{20}$; **h)** $-\dfrac{3}{4}$; **i)** $\dfrac{11}{12}$; **j)** $\dfrac{3}{32}$.

II. 9. a) $\dfrac{4}{15}$; **b)** $\dfrac{1}{35}$; **c)** $-3\dfrac{3}{20}$; **d)** $\dfrac{37}{72}$; **e)** $\dfrac{11}{12}$; **f)** $\dfrac{23}{36}$; **g)** $-\dfrac{13}{144}$; **h)** $\dfrac{4}{3}$;

i) $-8\dfrac{23}{30}$; **j)** 3; **k)** $-\dfrac{103}{20}$; **l)** $1\dfrac{7}{12}$; **m)** $-\dfrac{13}{24}$.

II. 10. a) 2; **b)** -4 ; **c)** 1; **d)** -0.3 ; **e)** 3.75; **f)** 1.5; **g)** $\dfrac{1}{3}$; **h)** 0.21 ; **i)** $-\dfrac{1}{15}$.

II. 11. a) $\dfrac{2}{3}$; **b)** $\dfrac{12}{125}$; **c)** $\dfrac{16}{49}$; **d)** $\dfrac{2}{5}$; **e)** $4\dfrac{1}{2}$; **f)** -1; **g)** 12; **h)** -7; **i)** -70;

j) -5.1; k) -2.6; l) -32; m) $\dfrac{8}{3}$; n) $\dfrac{1}{2}$. **II. 12. a)** $\dfrac{17}{21}$; b) $\dfrac{3}{10}$; c) $-\dfrac{7}{8}$;

d) $\dfrac{7}{10}$; e) $1\dfrac{2}{5}$; f) $-\dfrac{3}{2}$; g) $3\dfrac{1}{10}$; h) $\dfrac{1}{4}$; i) $\dfrac{1}{12}$; j) 0; k) $-\dfrac{4}{5}$; l) $\dfrac{2}{27}$;

m) $-\dfrac{23}{140}$; n) $\dfrac{5}{12}$; o) $\dfrac{1}{12}$; p) $-\dfrac{2}{63}$; q) $\dfrac{5}{6}$; r) $\dfrac{4}{5}$; s) $-11\dfrac{2}{5}$; t) -1.

II. 13. a) $\dfrac{5}{6}$; b) -6; c) 1; d) $-5\dfrac{1}{17}$; e) -6; f) 4; g) 2; h) $\dfrac{5}{8}$.

II. 14. a) $2\dfrac{8}{1}$; b) $-\dfrac{17}{36}$; c) $-\dfrac{5}{6}$; d) $\dfrac{4}{5}$; e) $\dfrac{5}{6}$; f) $1\dfrac{4}{7}$; g) $\dfrac{9}{4}$; h) $103\dfrac{1}{3}$;

i) $3\dfrac{3}{25}$. **II. 15. a)** 0; b) 5; c) -20; d) 1; e) 3; f) $1\dfrac{15}{16}$; g) 30; h) 1; i) 1; j) 24.

II. 16. a) $\dfrac{\left(\dfrac{43}{10}\cdot\dfrac{12}{43}-\dfrac{1}{5}\right)\cdot 5}{\dfrac{31}{12}-\dfrac{1}{4}\cdot\dfrac{28}{3}\cdot\dfrac{5}{7}}\cdot\dfrac{1}{10}+\dfrac{76}{50}+\dfrac{362}{25}=\dfrac{\left(\dfrac{6}{5}-\dfrac{1}{5}\right)\cdot 5}{\dfrac{31}{12}-\dfrac{20}{12}}\cdot\dfrac{1}{10}+8=14$;

b) $25\dfrac{1}{2}$; c) $7\dfrac{1}{5}$; d) $1\dfrac{1}{16}$; e) 1; f) 2; g) 9; h) $\dfrac{54}{55}$; i) $A=\dfrac{75}{2}$; $B=16\dfrac{1}{2}$; $C=3$;

$x=\dfrac{7500}{11}$.

II. 17. a) $\dfrac{27}{125}$; b) $-\dfrac{8}{27}$; c) $\dfrac{9}{25}$; d) $\dfrac{2401}{81}$; e) 1; f) $-\dfrac{3}{22}$; g) 1; h) $\dfrac{23}{11}$; i) $-\dfrac{1331}{1000}$;

j) $-\dfrac{16}{81}$; k) $-\dfrac{169}{100}$; l) $\dfrac{343}{1000}$; m) $-\dfrac{512}{125}$; n) $-\dfrac{2197}{1728}$.

II. 18. a) $\dfrac{1}{9}$; b) $\dfrac{1}{125}$; c) $\dfrac{1}{16}$; d) $-\dfrac{1}{216}$; e) $-\dfrac{1}{5}$; f) -4; g) $-\dfrac{5}{2}$; h) $\dfrac{16}{9}$;

i) $-\dfrac{27}{343}$; j) $-\dfrac{3}{2}$; k) -125; l) $\dfrac{9}{4}$; m) $-\dfrac{729}{125}$.

II. 19. a) 2^5, 2^4, 2^3, 2^2, 2^1, 2^0, $\dfrac{1}{2^1}$, $\dfrac{1}{2^2}$, $\dfrac{1}{2^3}$, $\dfrac{1}{2^4}$, $\dfrac{1}{2^5}$;

b) $\dfrac{1}{5^4}$, $\dfrac{1}{5^3}$, $\dfrac{1}{5^2}$, $\dfrac{1}{5^1}$, 5^0, 5^1, 5^2, 5^3, 5^4;

c) $\dfrac{1}{10^4}$, $\dfrac{1}{10^3}$, $\dfrac{1}{10^2}$, $\dfrac{1}{10^1}$, 10^0, 10^1, 10^2, 10^3, 10^4.

II. 20. a) 10^{-2}; b) 7^{-3}; c) 10^{-4}; d) 3^{-3}; e) 2^{-4}; f) 11^{-2}; g) 3^{-6}; h) 6^{-3}; i) 5^{-5}.

II. 21. a) 2^9; b) 7^6; c) -8^8; d) -3^5; e) 4^3; f) $\dfrac{1}{7}$; g) $-\dfrac{1}{3^5}$; h) $\left(\dfrac{2}{5}\right)^8$;

i) $\left(-\dfrac{2}{5}\right)^5$; j) $\left(\dfrac{3}{7}\right)^4$; k) 2^3 ; l) -3 ; m) $\left(-\dfrac{2}{3}\right)^5$; n) -3 ; o) $\left(\dfrac{3}{7}\right)^3$; p) $\left(\dfrac{5}{1}\right)^6$;

q) $\left(\dfrac{2}{3}\right)^4$; r) 2.3 ; s) 1.

II. 22. a) 5^4 ; b) 2^6 ; c) $\dfrac{1}{2^8}$; d) $\dfrac{1}{2^6}$; e) $\left(\dfrac{2}{5}\right)^4$; f) $-\dfrac{5}{6}$; g) $\left(\dfrac{3}{4}\right)^3$; h) $-\dfrac{2}{3}$;

i) $\dfrac{5}{3}$; j) $-\dfrac{4}{5}$; k) $\left(\dfrac{19}{10}\right)^3$; l) $\left(\dfrac{4}{9}\right)^2$; m) $\left(\dfrac{10}{38}\right)^8$; n) $\left(\dfrac{13}{10}\right)^7$; o) $\left(\dfrac{8}{10}\right)^4$.

II. 23. a) -6^{16}; b) 4^8 ; c) 3^{12} ; d) -2^{12} ; e) $\left(\dfrac{2}{5}\right)^6$; f) $\left(-\dfrac{3}{4}\right)^6$; g) $\left(-\dfrac{3}{4}\right)^6$;

h) $\left(\dfrac{2}{3}\right)^8$; i) $\left(\dfrac{3}{4}\right)^{30}$; j) $\left(\dfrac{1}{2}\right)^4$; k) $\left(\dfrac{9}{5}\right)^6$; l) $\left(\dfrac{1}{3}\right)^8$; m) $\left(-\dfrac{1}{2}\right)^{81}$; n) $\left(\dfrac{3}{2}\right)^6$;

o) 2^6 ; p) -2^6 ; q) $\left(\dfrac{5}{2}\right)^{12}$; r) $\left(\dfrac{3}{4}\right)^4$; s) $\left(\dfrac{5}{3}\right)^6$; t) $\left(\dfrac{2}{3}\right)^{24}$; u) 1; v) 1.

II. 24. a) $-\dfrac{3}{2}$; b) $\left(\dfrac{2}{5}\right)^3$; c) 0.3; d) -2.

II. 25. a) 3^4 ; b) -2^{10} ; c) $-\dfrac{5^2}{2^3}$; d) 3^3 ; e) $-\dfrac{1}{2^{11}}$; f) -3^9 ; g) 3^{18} ; h) $\dfrac{1}{5}$;

i) 2^{32} ; j) $\dfrac{1}{3^6}$; k) 6^3 ; l) 3^{-7} ; m) $-\dfrac{1}{2^7}$; n) -1 ; o) $-\dfrac{2}{9}$; p) $\left(\dfrac{2}{3}\right)^9$; q) 6^5.

II. 26. a) $\dfrac{25}{49}$; b) $\dfrac{1}{1000}$; c) $\dfrac{1}{2}$; d) $\dfrac{1}{2}$; e) 9 ; f) 6^5 ; g) 3^{11} ; h) 2^{20} ;

i) 5^8 ; j) $\dfrac{1}{2^3}$; k) $\dfrac{1}{6^8}$.

II. 27. a) 5; b) $-\dfrac{21}{2^4}$; c) $-\dfrac{1}{3^4}$; d) $\dfrac{19}{2^2}$; e) 7; f) 2; g) $-2\dfrac{1}{4}$; h) 0; i) $\dfrac{5}{3}$;

j) $\dfrac{16}{121}$; k) $-3\dfrac{3}{8}$; l) $2\dfrac{17}{30}$; m) $1\dfrac{1}{4}$; n) $-\dfrac{9}{4}$; o) $1\dfrac{1}{9}$.

II. 28. a) $\dfrac{3}{7}$; b) $\left(\dfrac{3}{2}\right)^{10}$; c) 1; d) 1; e) $-\dfrac{2}{27}$; f) 24; g) $3\dfrac{1}{2}$; h) 20; i) -24.

II. 29. a) $1\dfrac{1}{2}$; b) $1\dfrac{1}{3}$; c) $-\dfrac{2}{5}$; d) $3\dfrac{2}{3}$; e) $4\dfrac{1}{2}$; f) $\dfrac{1}{12}$; g) $\dfrac{1}{15}$; h) $x=\dfrac{1}{4}$;

i) $-\dfrac{5}{12}$; j) -15 ; k) 10; l) $1\dfrac{1}{7}$; m) -6 ; n) -20 ; o) -7 ; p) 3; q) -3 ; r) 2;

s) $\dfrac{3}{5}$. **II. 30.** a) $\dfrac{1}{14}$; b) $-\dfrac{5}{6}$; c) $8\dfrac{18}{49}$; d) $10\dfrac{2}{7}$; e) $1\dfrac{1}{8}$; f) -1 ; g) $-\dfrac{5}{6}$;

h) $\dfrac{15}{28}$; i) $-\dfrac{30}{7}$; j) -2 ; k) 3; l) -9 ; m) -39 ; n) -2 ; o) $-\dfrac{2}{5}$.

II. 31. a) -1 ; b) -8 ; c) 22; d) 30; e) $-\dfrac{1}{3}$; f) $\dfrac{1}{8}$; g) 12; h) 40; i) 5; j) $\dfrac{2}{3}$;

k) $1\dfrac{1}{2}$; l) $\dfrac{10}{11}$; m) -50 ; n) $1\dfrac{1}{8}$; o) $11\dfrac{1}{2}$; p) -32 ; q) 13; r) impossible; s) 12.

II. 32. a) $x = -3$; b) $x = -\dfrac{1}{8}$; c) $x = 2\dfrac{2}{5}$; d) $x = -5$; e) $x = 5\dfrac{1}{2}$; f) $x = \dfrac{4}{5}$;

g) $x = \dfrac{2}{3}$; h) $x = -5\dfrac{1}{3}$.

II. 33. a) $x = 1$; b) $x = -1$; c) $x = 0.5$; d) $x = 2$; e) $x = 2$; f) $x = 0.5$;

g) $x = 5$; h) $a = 0$; i) $z = -2$.

II. 34. a) $x = 10$; b) $x = 10$; c) $x = 5\dfrac{2}{11}$; d) $x = 6\dfrac{1}{5}$; e) $x = -2\dfrac{1}{10}$;

f) $x = 3900$; g) $x = 2\dfrac{4}{13}$; h) $x = 1\dfrac{182}{807}$; i) $x = 26$; j) $x = 2010$.

II. 35. a) $4\dfrac{1}{5}$; b) $\dfrac{5}{7}$; c) $-\dfrac{7}{15}$; d) $1\dfrac{11}{30}$; e) $-\dfrac{3}{4}$; f) $\dfrac{5}{3}$; g) $-\dfrac{8}{27}$; h) $-1\dfrac{7}{9}$;

i) $2\dfrac{1}{23}$; j) $2\dfrac{17}{20}$; k) $\dfrac{5}{6}$; l) -18 ; m) $2\dfrac{29}{67}$; n) -126 ; o) $\dfrac{15}{14}$; p) $\dfrac{8}{11}$; q) 10;

r) $\dfrac{13}{50}$; s) $\dfrac{10}{49}$; t) 8; u) $-\dfrac{5}{61}$; v) $3\dfrac{3}{2}$; w) -7 ; x) $2\dfrac{19}{24}$; y) $8\dfrac{2}{11}$; z) $\dfrac{50}{177}$.

II. 36. a) -1 ; b) $-3\dfrac{4}{7}$; c) 5; d) $-\dfrac{17}{24}$; e) $\dfrac{44}{3}$; f) $\dfrac{28}{27}$; g) $3\dfrac{14}{29}$; h) -48 ;

i) $2\dfrac{23}{90}$; j) $-5\dfrac{51}{219}$.

II. 37. a) $a = 1$; b) $x = 2\dfrac{33}{71}$; c) $a = 70$; d) $x = \dfrac{4}{25}$; e) $x = 30$; f) $x = -\dfrac{20}{23}$;

g) $x = 45$; h) $x = 105$; i) $x = -60$; j) $x = 4\dfrac{1}{2}$.

II. 38. a) $x = 1$; b) $x = 1$; c) $x = -2$; d) $x = 4$; e) $x = 2$.

II. 39. a) $x_1 = -\dfrac{1}{3}$, $x_2 = \dfrac{3}{4}$; b) $x_1 = -\dfrac{21}{4}$, $x_2 = -\dfrac{3}{2}$; c) $x_1 = -\dfrac{24}{35}$, $x_2 = \dfrac{4}{3}$;

d) $x_1 = \dfrac{91}{6}$, $x_2 = \dfrac{9}{14}$; e) $x_1 = \dfrac{49}{8}$, $x_2 = -\dfrac{36}{25}$.

II. 40. a) -4; b) $-\dfrac{1}{4}$; c) $\dfrac{2}{9}$; d) $-\dfrac{30}{13}$; e) $\dfrac{31}{7}$; f) $-\dfrac{18}{23}$; g) $\dfrac{5}{2}$; h) $-\dfrac{10}{19}$;

i) $-\dfrac{1}{4}$; j) $\dfrac{33}{5}$; k) 0; l) $-\dfrac{6}{19}$.

II. 41. a) -5; b) $\dfrac{125}{216}$; c) 4; d) 516; e) $-\dfrac{5^4}{3^6}$; f) $\dfrac{b}{c} = -\dfrac{1}{20}$, $\dfrac{c}{b} = -20$.

II. 42. a) $a = 4\dfrac{3}{5}$; b) $a = \dfrac{5}{2}$; c) $m = -\dfrac{3}{2}$; d) $m = \dfrac{3}{14}$; e) $m = -\dfrac{15}{34}$.

II. 43. a) $x = -1$, $m \in \mathbf{R} \setminus \left\{\dfrac{1}{2}\right\}$; b) $x = \dfrac{a-3}{a-2}$, $a \in \mathbf{R} \setminus \{2\}$;

c) $x = \dfrac{12}{a-4}$, $a \in \mathbf{R} \setminus \{4\}$; d) $x = \dfrac{9}{2-2a}$, $m \in \mathbf{R} \setminus \{1\}$;

e) $x = \dfrac{3a-4}{4a}$, $a \in \mathbf{R} \setminus \{0\}$; f) $x = \dfrac{3a-2}{a+5}$, $a \in \mathbf{R} \setminus \{-5\}$;

g) $x = \dfrac{1}{a-5}$, $a \in \mathbf{R} \setminus \{5\}$; h) $x = \dfrac{4a-3}{a-8}$, $a \in \mathbf{R} \setminus \{8\}$; i) $x = \dfrac{6m}{m-1}$, $a \in \mathbf{R} \setminus \{1\}$.

II. 44. a) 1; b) Denote $n! = 1 \cdot 2 \cdot 3 \cdot \ldots \cdot n$, then

$$\left(\dfrac{1}{1!} - \dfrac{1}{2!}\right) + \left(\dfrac{1}{2!} - \dfrac{1}{3!}\right) + \ldots + \left(\dfrac{1}{x!} - \dfrac{1}{(x+1)!}\right) = \dfrac{2010! - 1}{2010!}, x = 2009;$$

c) $\left(\dfrac{2x}{x+1} - 1\right) + \left(\dfrac{2x+1}{x+2} - 1\right) + \ldots + \left(\dfrac{2x+2013}{x+2014} - 1\right) = 0$

$\left(\dfrac{x-1}{x+1}\right) + \left(\dfrac{x-1}{x+2}\right) + \left(\dfrac{x-1}{x+3}\right) + \ldots + \left(\dfrac{x-1}{x+2014}\right) = 0$

$(x-1)\left(\dfrac{1}{x+1} + \dfrac{1}{x+2} + \ldots + \dfrac{1}{x+2014}\right) = 0$.

Only $x = 1$ is the solution of the equation; d) 1. **II. 45.** 8-foot and 4-foot.
II. 46. 24-foot and 12-foot. **II. 47.** 25 \$140 suits and 15 \$210 suits.
II. 48. Let $a - 2$, $a - 1$, a, $a + 1$, and $a + 2$ be the five positive consecutive

numbers. $a(a-1)(a-2) = a(a+1)(a+2) - 216$, $a = 6$.
The numbers are 4, 5, 6, 7, and 8.

II. 49. The first gear has 39 teeth and the second gear has 57 teeth.

I. 50. Let x and y be the numbers. $x = y - 27.8$ and $x + y = 66.7$.
Therefore $x = 19.45$ and $y = 47.25$.

II. 51. Let m and n be the two numbers. Then $m = 3n$ and $m + n = \dfrac{34}{9}$.

Therefore $m = \dfrac{17}{6}$ and $n = \dfrac{17}{18}$. **II. 52.** $5.8a = a - 4$, $a = \dfrac{5}{6}$.

II. 53. The length is 418 yards and the width is 202 yards.

II. 54. $x^2 + 83 = (x + 3)(x + 5)$, $x = \dfrac{17}{2}$.

II. 55. The length is 20 feet and the width is 11 feet.

II. 56. When the angles of the triangle are $2x$, $2x$, and $5x$, $x = 20$. When the angles of the triangle are $5x$, $5x$, and $2x$, $x = 15$.

II. 57. $\dfrac{3}{4}$. **II. 58.** Let x be the length and y be the width. $x + y = 4\dfrac{1}{2}$ and

$x = 2y + 1.2$. Then $x = \dfrac{17}{5}$ and $y = \dfrac{11}{10}$.

II. 59. Let A, B, and C be the three angles, then $\angle A = 72°$. $B = 3C$ and $B + C = 108$. Therefore $\angle B = 81°$ and $\angle C = 27°$.

II. 60. Let x and y be the price for 1 kg of apples and 1 kg of oranges,

respectively. $3x + 2y = \dfrac{101}{20}$ and $x = y + \dfrac{3}{10}$, $x = \dfrac{89}{100}$ and $y = \dfrac{119}{100}$. 1 kg of

apples costs \$0.89 and 1 kg of oranges costs \$1.19.

II. 61. Let x and y be the math and physics textbooks, respectively, then

$10x + 15y = 2000$ and $x + y = 163$. Therefore $x = 89$ and $y = 74$.

II. 62. Let x and y be the two numbers. $x + y = 150$ and $x = 4y + 15$. Then $x = 123$ and $y = 27$.

II. 63. Let a and b be the length and width, respectively. $a - b = 13$ and $a + b = 63$. The length is 38 cm and the width is 25 cm.

II. 64. a) 1.7 and $\dfrac{10}{17}$; b) $\dfrac{70}{51}$ and $\dfrac{51}{70}$; c) $\dfrac{6}{5}$ and $\dfrac{5}{6}$; d) $\dfrac{10}{9}$ and $\dfrac{9}{10}$.

II. 65. a) $-\dfrac{9}{5}$; b) $\dfrac{65}{3}$. **II. 66.** a) 2π; b) 8π. **II. 67.** a) $\dfrac{3}{4}$; b) $\dfrac{3}{4}$.

II. 68. a) $\dfrac{2}{3}$; b) $\dfrac{12}{7}$; c) $\dfrac{5}{9}$; d) $\dfrac{1}{3}$; e) $\dfrac{1}{2}$; f) $\dfrac{3}{5}a^2$; g) $\dfrac{1}{6}$; h) $\dfrac{3}{8}$; i) $\dfrac{1}{3}$; j) $\dfrac{75}{2}$;

k) $\dfrac{3}{8}$; l) $\dfrac{5}{6}$. **II. 69.** a) $-\dfrac{43}{15}$; b) $-\dfrac{14}{53}$; c) $\dfrac{25}{158}$; d) $-\dfrac{2}{23}$; e) $-1\dfrac{2}{69}$; f) $\dfrac{5}{9}$;

g) 4; h) $-\dfrac{73}{4}$; i) $\dfrac{a}{x} = \dfrac{13}{6}$, $\dfrac{b}{x} = \dfrac{13}{21}$, $\dfrac{c}{x} = \dfrac{13}{12}$; j) $-\dfrac{4}{3}$ or $-\dfrac{15}{14}$; k) $a = 14$, $b = 4$,

$c = 12$, $d = 8$, $e = 10$. **II. 70.** a) $x = 20$; b) $x = \dfrac{117}{40}$; c) $x = 8\dfrac{4}{5}$;

d) $a = 8$; e) $z = \dfrac{19}{2}$; f) $z = \dfrac{10}{63}$; g) $a = 2\dfrac{1}{2}$; h) $n = 2\dfrac{1}{2}$; i) $x = 7$; j) $x = 7\dfrac{1}{7}$;

k) $x = 2$; l) $x = 52$; m) $x = 27\dfrac{1}{2}$; n) $x = \dfrac{91}{1200}$.

II. 71. a) $x = 15, y = 35$; **b)** $x = \dfrac{91}{6}$, $y = \dfrac{49}{6}$; **c)** $x = -\dfrac{20}{7}$, $y = -\dfrac{90}{7}$;

d) $x = -25\dfrac{32}{43}$, $y = -10\dfrac{21}{43}$; **e)** $x = -16\dfrac{7}{13}$, $y = -14\dfrac{5}{26}$; **f)** $x = \dfrac{1}{7}$, $y = 1\dfrac{13}{21}$.

II. 72. a) 2; **b)** 30; **c)** 22; **d)** 30; **e)** $\dfrac{20}{29}$; **f)** $-\dfrac{15}{2}$;

g) $-\dfrac{3}{14}$; **h)** 13; **i)** 4; **j)** Impossible; **k)** $\dfrac{17}{4}$; **l)** 48; **m)** $\dfrac{26}{15}$.

II. 73. a) $\dfrac{3}{8}$; **b)** $\dfrac{22}{15}$; **c)** $\dfrac{1}{2}$; **d)** $-\dfrac{4}{7}$; **e)** $\dfrac{18}{35}$; **f)** $-\dfrac{1}{9}$; **g)** $-\dfrac{1}{45}$; **h)** -1;

i) $0.\overline{6}$. **II. 74. a)** $\dfrac{3}{4}$; **b)** $\dfrac{1}{4}$; **c)** $\dfrac{3}{25}$; **d)** $\dfrac{1}{250}$; **e)** $\dfrac{1}{2}$; **f)** $\dfrac{7}{1000}$; **g)** $\dfrac{23}{1000}$;

h) $\dfrac{2}{125}$. **II. 75. a)** 0.05; **b)** 0.03; **c)** 0.34; **d)** 0.17; **e)** 0.0033; **f)** 0.0415;

g) 0.0025; **h)** 0.0145. **II. 76. a)** 100; **b)** 6; **c)** 16.5; **d)** 0.54; **e)** 2.25;

f) 78; **g)** 99; **h)** 40.6; **i)** $10.8\overline{3}$.

II. 77. a) $16.\overline{6}\%$; **b)** 80%; **c)** 75%; **d)** 64%; **e)** 48%; **f)** $32.\overline{432}\%$;

g) 5000%; **h)** 500%; **i)** 80%; **j)** $66.\overline{6}\%$; **k)** 64%.

II. 78. a) $\dfrac{20}{100} \cdot x = 15$, $x = 75$; **b)** $x = 26.\overline{6}$; **c)** 22.5; **d)** $x = 106.\overline{6}$;

e) $x = 200$; **f)** $x = 600$; **g)** $x = 500$; **h)** $63.\overline{3}$; **i)** $x = 72$.
II. 79. $x = 18, y = 24$, and $z = 30$. **II. 80.** $x = 4, y = 6, z = 10$, and $t = 14$.
II. 81. $x = 6, y = 4$, and $z = 8$. **II. 82.** $x = 8, y = 12$, and $z = 15$.

II. 83. (2, 3, 4); **b)** $\left(\dfrac{1}{18}, \dfrac{1}{9}, \dfrac{1}{6}\right)$. **II. 84.** $x = 20, y = 12$, and $z = 24$.

II. 85. $x = 5, y = 3$, and $z = 4$. **II. 86.** $x = \dfrac{8}{9}$, $y = \dfrac{4}{3}$, and $z = \dfrac{16}{9}$.

II. 87. $x = 6, y = 8$, and $z = 12$. **II. 88.** $a = 4, b = 6, c = 8$, and $d = 10$.

II. 89. a) $\dfrac{2}{3}$; **b)** $\dfrac{2}{3}$; **c)** $\dfrac{4}{9}$. **II. 90. a)** $\dfrac{2}{3}$; **b)** $1\dfrac{2}{11}$; **c)** $1\dfrac{2}{5}$.

II. 92. a) $a = 10^n \cdot 2 + 1 = \underbrace{2\,00...001}_{n-1\ zeros}$, the sum of the digits is three which means

241

a is divisible by three. $b = 10^n \cdot 5 + 1 = \overline{5\underbrace{00...00}_{n-1\ zeros}1}$, $c = 10^n \cdot 8 + 7 = \overline{8\underbrace{00...00}_{n-1\ zeros}7}$,

$d = 10^{n+1} \cdot 25 - 1 = \overline{24\underbrace{99...99}_{n+1\ zeros}9}$.

b) $\dfrac{7d+c}{b-a} = \dfrac{10^{n+1} \cdot 175 - 7 + 10^n \cdot 8 + 7}{10^n \cdot 5 + 1 - 10^n \cdot 2 - 1} = \dfrac{10^n(174+8)}{10^n(5-2)} = \dfrac{183}{3} = 61$.

c) No, c is not a perfect square because the last digit is seven and a perfect

square does not have the last digit seven.

II. 93. a) 13^{101}; b) -6. **II. 94.** The number is 111.

II. 95. $n = a^4 + 6a^3 + 11a^2 + 6a + 1 = \left(a^2 + 3a + 1\right)^2$.

II. 96. $\overline{\underbrace{100...0200...01}_{2n+3\ times}} = 1 \cdot 10^{2n+2} + 2 \cdot 10^{n+1} + 1 = \left(10^{n+1} + 1\right)^2$.

II. 97. $\overline{aaa} - \overline{bbb} = 100a + 10a + a - (100b\_10b + b) = 111(a-b)$.

II. 98. Let $\overline{abcd} = 5x$. We obtain $5 + 10 + 15 + ... + 5x = 5000x$.

Then $5(1 + 2 + 3 + ... + x) = 5000x \Rightarrow 5 \cdot \dfrac{x(x+1)}{2} = 5000x$.

Therefore we get $x = 1999$ or $abcd = 9995$.

II. 99. $3^{3n+1} + 2 \cdot 3^{n+1} = (x-1)^3 + x^3 + (x+1)^3 \Rightarrow 3(3^{3n} + 2 \cdot 3^n) = 3(x^3 + 2x)$.

For $x = 3^n$, the given number can be written as the sum of three cubes.

II. 100. We have

$A^2 = \left(3 \cdot \underbrace{1\,...1}_{n\ times} + 1\right)^2 = 9 \cdot \underbrace{11\,...1^2}_{n\ times} + 6 \cdot \underbrace{11\,...1}_{n\ times} + 1 = \underbrace{11\,...1}_{n\ times}\left(9 \cdot \underbrace{11\,...1}_{n\ times} + 6\right) + 1 =$

$= \underbrace{11\,...1}_{n\ times}\left(\underbrace{99...9}_{n\ times} + 6\right) + 1 = \underbrace{11\,...1}_{n\ times}\left(10^n + 5\right) + 1 = \underbrace{11\,...1}_{n\ times} \cdot 10^n + 5 \cdot \underbrace{11\,...1}_{n\ times} + 1 =$

$= \underbrace{11\,...1}_{n\ times} \cdot 10^n + \underbrace{55...5}_{n-1\ times}6 = \underbrace{11\,...1}_{n\ times}\underbrace{55...6}_{n-1\ times}$.

TEST II. 1

1) The coins he might have in his pocket is $\dfrac{1}{20} = \dfrac{5}{100} = 0.05$ dollars; 5 cents.

2) a) $\dfrac{3}{5} = \dfrac{6}{10} = 06$; b) $\dfrac{3}{2} = \dfrac{15}{10} = 1.5$; c) $2\dfrac{1}{5} = 2\dfrac{2}{10} = 2.2$; d) $5\dfrac{4}{5} = 5\dfrac{8}{10} = 5.8$.

3) a) $\dfrac{1}{4} = \dfrac{25}{100} = 0.25$; b) $\dfrac{3}{10} = \dfrac{30}{100} = 0.3$; c) $\dfrac{2}{50} = \dfrac{4}{100} = 0.04$;

d) $\dfrac{2}{25} = \dfrac{8}{100} = 0.8$.

4) a) $7\dfrac{1}{2} = 7\dfrac{50}{100} = 7.5$; b) $6\dfrac{2}{5} = 6\dfrac{40}{100} = 6.4$; c) $3\dfrac{11}{25} = 3\dfrac{44}{100} = 3.44$;

d) $11\dfrac{1}{100} = 11.01$.

5) a) $\dfrac{75}{100} = \dfrac{3}{4}$; b) $0.08 < 0.8$; c) $0.6 = \dfrac{60}{100}$; d) $\dfrac{16}{10} > \dfrac{13}{10}$.

6) a) $0.9 = \dfrac{9}{10} = \dfrac{90}{100}$; b) $0.40 = \dfrac{4}{10} = \dfrac{40}{100}$;

c) $0.75 = \dfrac{75}{100} = \dfrac{15}{20}$; d) $0.5 = \dfrac{5}{10} = \dfrac{50}{100}$.

7) a) $77 \div 8 = \dfrac{77}{8} = 9\dfrac{5}{8}$; b) $84 \div 9 = \dfrac{84}{9} = 9\dfrac{3}{9}$; c) $45 \div 7 = \dfrac{45}{7} = 6\dfrac{3}{7}$;

d) $79 \div 6 = \dfrac{79}{6} = 13\dfrac{1}{6}$. 8) $7 \div 5 = \dfrac{7}{5} = \dfrac{140}{100} = 1.4$ of a pizza.

9) The area of the yard is $28.54 \times 31 = 884.74$ m^2 .
10) $0.94 \text{ m} \times 6 = 5.64 \text{ m}$.

TEST II. 2

a) $1\dfrac{1}{5}$; b) $\dfrac{24}{125}$; c) $\dfrac{4}{51}$; d) 12 ; e) 12 ; f) 7 ; g) 1.3 ; h) $1\dfrac{4}{5}$; i) -12 ; j) $\dfrac{2}{3}$.

TEST II. 3

1) $x = 4$; 2) $x = -11$; 3) $x = 2$; 4) $x = 1/5$; 5) $x = -2$; 6) $x = -1$;
7) $x = -6$; 8) $x = -7/19$; 9) $x = 7/2$; 10) $x = 1$.

TEST II. 4

1) $x = 2$, $y = 14$, $\dfrac{x + y}{2} = 8$; 2) $x = 9\dfrac{1}{8}$; 3) $x = 63$, $y = 70$, $z = 84$;

4) a) 1, b) 2; 5) $a = 1$; 6) a) $x = 7$; b) $x = 6$; c) $x = 2$; 7) a) 6.1; b) 30; c) 1;

8) $x = \dfrac{29}{9}$; 9) 1,536; 10) $a = 10$, $b = \dfrac{81}{10}$, $c = \dfrac{439}{560}$.

TEST II. 5

1) $x = \dfrac{1.5 \cdot 0.5}{0.2} = \dfrac{0.75}{0.2} = \dfrac{7.5}{2} = 3.75$. 2) a) $x = 6$; b) $x = 20/9$;

c) $x = 0.75$; d) $x = 1.2$.

3) We have $\dfrac{x}{0.2} = \dfrac{y}{0.05} = \dfrac{x+y}{0.2+0.05} = \dfrac{10}{0.25} = 40$. From $\dfrac{x}{0.2} = 40 \Rightarrow x = 8$,

and $\dfrac{y}{0.05} = 40 \Rightarrow y = 2$. 4) We have $\dfrac{x}{3} = \dfrac{y}{4} = \dfrac{z}{5} = \dfrac{x+y+z}{3+4+5} = \dfrac{24}{12} = 2$.

Therefore $x = 6$, $y = 8$, $z = 10$. 5) $x = 20$, $z = 24$, $y = 8$. 6) $x = 25$, $y = 10$, $z = 20$.

7) $a = 6$, $b = 8$, $c = 10$.

8) Let be $\dfrac{x}{12} = \dfrac{y}{15} = \dfrac{z}{30} = k$. Then $x = 12k$, $y = 15k$, $z = 30k$, and

$xyz = (12k)(15k)(30k) = 5400k^3$, so $5400k^3 = 1600 \Rightarrow k^3 = \left(\dfrac{2}{3}\right)^3 \Rightarrow k = \dfrac{2}{3}$.

Hence $x = 12 \cdot \dfrac{2}{3} = 8$, $y = 15 \cdot \dfrac{2}{3} = 10$, $z = 30 \cdot \dfrac{2}{3} = 20$.

9) $x = 33$, y $= 42$, $z = 51$.

TEST II. 6

1) a) 50; b) 2; 2) a) 32 ; b) 1; 3) $a = -1$, $b = -0.3$,

$x = -2\dfrac{7}{30}$; 4) $\dfrac{1}{256}$; 5) $\dfrac{1}{7} \cdot \left[\dfrac{1}{5} \cdot \left(\dfrac{1}{3} \cdot (x-2) - 4\right) + 6\right] + 8 = 3$,

$\dfrac{1}{7} \cdot \left[\dfrac{1}{5} \cdot \left(\dfrac{1}{3} \cdot (x-2) - 4\right) + 6\right] = -5$, $\dfrac{1}{5} \cdot \left(\dfrac{1}{3} \cdot (x-2) - 4\right) = -41$,

$\dfrac{1}{3} \cdot (x-2) - 4 = -205$, $x = -601$. 6) $\dfrac{2}{669}$. 7) a)

$\left(\dfrac{1}{\sqrt{2}} - \dfrac{1}{1}\right) + \left(\dfrac{1}{\sqrt{3}} - \dfrac{1}{\sqrt{2}}\right) + \dots + \left(\dfrac{1}{\sqrt{2010}} - \dfrac{1}{\sqrt{2009}}\right) = -1 + \dfrac{1}{\sqrt{2010}}$;

b) 2012^{2012} . 8) $x\left(\dfrac{1}{2} + \dfrac{1}{3} + \dots + \dfrac{1}{n}\right) = \dfrac{1}{2} + \dfrac{1}{3} + \dots + \dfrac{1}{n} \Rightarrow x = 1$.

9) AB = BD =6, BC = BD − CD = 6 − 4 = 2 and AM = AB + $\dfrac{BC}{2}$ = 6 + 1 = 7.

10) AB + CD + DF = AG − (BC + FG) = 57 − 24 = 33cm,

AB = 11cm, BF = BC + CD + DF = 12 + 11 + 11 = 34cm,

CG = CD + DF + FG = 11+11+12 = 34cm. If E is the midpoint of CG, then AE =

AC + $\dfrac{CG}{2}$ = 23 + 17 = 40cm.

TEST II. 7

1) $x = 2$, $y = 84$, $\dfrac{x+y}{2} = 43$; **2)** $x = 9\dfrac{1}{8}$; **3)** $x = 63$, $y = 70$, $z = 84$;

4) a) 1, **b)** 2; **5)** $a = 1$; **6) a)** $x = 7$; **b)** $x = 6$; **c)** $x = \dfrac{42}{19}$; **7) a)** 1; **b)** 50; **c)** $\dfrac{1}{100}$;

8) $x = -\dfrac{1}{3}$; **9)** 48, 32; **10)** $a = 10$, $b = \dfrac{81}{10}$, $c = \dfrac{439}{560}$.

TEST II. 8

1) a) 1. **b)** 50. **c)** 1. **2)** $x = 1$. **3)** $\dfrac{1}{3}$. **4)** $x = 5$. **5)** $x = 5\dfrac{1}{3}$. **6)** 1.

7) a) 1; **b)** 4. **8)** $a = 1$. **9) a)** 0; **b)** $x = 4$; **c)** $x = 1, 2, 3$. **10)** 7.5.

TEST II. 9

1) a) Mary is 10 years old, **b)** When Mary is 20 years old her mother is 40 years old. **2) a)** 850; **b)** $a = 300$.

3) b) $S = \dfrac{24}{25}$, **d)** $n = 100$. **4) a)** $\dfrac{a}{b} = -52$;

b) $n = \left(7 + 7^2 + 7^3\right)\left(1 + 7^3 + \ldots + 7^{2007}\right)$. **5)** $N = 2^{6n} \cdot 64$.

6) $\dfrac{a}{4} = \dfrac{b}{6} = \dfrac{c}{8} = \dfrac{a+c}{12}$, $\dfrac{a+c}{12} = \dfrac{b}{6}$, $a + c = 2b$. $a = \dfrac{504}{13}$, $b = \dfrac{756}{13}$,

$c = \dfrac{1008}{13}$. **7)** $\dfrac{49}{145}$. **8)** $a = 15$; $b = 24$; $c = 36$.

9) a) 1400; **b)** 36; **c)** $k = 280$. **10)** 55.

TEST II. 10

1) a) $\dfrac{10+x}{99} + \dfrac{20+x}{99} + \ldots + \dfrac{90+x}{99} = \dfrac{26x}{11}$, $\dfrac{450+9x}{99} = \dfrac{26x}{11}$, $x = 2$;

b) $x = 1$. **2)** $a = 6^{2n+2} \cdot 16$. **3)** $S_1 = 6^{n+1} \cdot 5$, $S_2 = 15^n \cdot 25$, $S_3 = 231^n \cdot 2000$,

$n \in N$, $S_4 = 2^{3n} \cdot 3^{4n} \cdot 1118$.

4) $\dfrac{\left(3^n + 4\right) \cdot 3^n + 17 \cdot \left(3^n + 4\right)}{2 \cdot \left(3^n + 4\right)} = \dfrac{3^n + 17}{2}$, $3^n + 17$ is an even number.

5) $A = \dfrac{3^{2n} \cdot 2^{2n} \cdot 5^{n+1} \cdot 7 - 2^{2n+1} \cdot 3^{2n+1} \cdot 5^n}{3^n \cdot 2^n \cdot 5^{n-1} \cdot 19 + 2^{n+1} \cdot 5^n \cdot 3^n} =$

$$= \frac{3^{2n} \cdot 2^{2n} \cdot 5^n (5 \cdot 7 - 2 \cdot 3)}{3^n \cdot 2^n \cdot 5^{n-1} (19 + 2 \cdot 5)} = 6^n \cdot 5. \ \ 6) \ \mathbf{a)} \ \frac{7^x \cdot 2}{6^x} \ ; \ \mathbf{b)} \ \frac{11^x \cdot 10}{13^x} \ .$$

7) a) $N = 666$, $a = 1$, $b = 2$, $c = 3$. **b)** $\dfrac{111}{185}(a + b + c)$;

8) $6 \cdot \overline{ac} = \overline{abc}$, $6(10a + c) = 100a + 10b + c$, $\overline{abc} = 108$.

9) $a + b + c = 3$, 111, 102, 120, 210, 201.

9) $(a,b) = \{(1,6); \ (2,5); \ (3,4)\}$.

11) $a + \dfrac{\overline{bcd}}{999} + b + \dfrac{\overline{cda}}{999} + c + \dfrac{\overline{dab}}{999} + d + \dfrac{\overline{abc}}{999} = 10 \Rightarrow$

$3d + \dfrac{111(a + b + c + d)}{999} = 10 \Rightarrow d = 3$.

Chapter III

ALGEBRAIC EXPRESSIONS. COORDINATE GEOMETRY. SLOPES. LINEAR FUNCTIONS

III. 1. a) $\dfrac{xb^2 - 7a^2b}{2}$; b) $5nm^2 - 12an$; c) $0.18ax$; d) $\dfrac{13x - 10x^2}{4}$;

e) $\dfrac{-5bxy - 6x^2y}{4}$; f) $\dfrac{48x}{5}$; g) $\dfrac{320b - 9a}{20}$; h) $-ax^3$; i) $\dfrac{79by}{10}$;

j) $\dfrac{13a^2b + 21ab^2}{6}$; k) $x^2 - \dfrac{5}{6}y^2z$.

III. 2. a) $\dfrac{3ab^2}{5}$; b) $4(m-n)^2$; c) $4(a+b)^3$; d) $8 - x^2$; e) $-\left(5x^3 + x^2 + 5\right)$;

f) $11b + 6c - 6a$; g) $-\left(x^2 + 1\right)$; h) $a + b + c$; i) $\dfrac{8ab}{7} - \dfrac{bc}{10} - \dfrac{7ac}{15}$;

j) $\dfrac{7a^2x}{6} - \dfrac{ax^2}{12} + \dfrac{25x^2}{24}$.

III. 3. a) $9a^6$; b) $14x^{10}$; c) $6x^6$; d) $-x^9$; e) $-6a^6b^3$;

f) $5a^2b^5$; g) $3x^4y^3z^4$; h) 0.

III. 4. a) $x^2 + 2x + 1$; b) $x^2 - 2x + 1$; c) $x^2 - 4x + 4$; d) $x^2 + 6x + 9$;

e) $4x^2 + 4x + 1$; f) $9x^2 - 6x + 1$; g) $4x^2 - 8x + 4$; h) $16x^2 - 16x + 4$;

i) $25x^2 + 30x + 9$; j) $16x^2 - 24x + 9$; k) $9x^2 - 6x + 1$; l) $x^2 + 6x + 9$;

m) $36x^2 - 12x + 1$; n) $0.25x^2 + 4x + 16$; o) $a^4 + 2a^2 + 1$; p) $25u^2 - 20u + 4$;

q) $a^2 + 14a + 49$; r) $4b^2 - 12b + 9$; s) $9x^2 + 6x + 1$; t) $4x^4 - 4x^3 + x^2$;

u) $x^2 + 10x^3 + 25x^4$; v) $a^4 - 4a^2x + 4x^2$; x) $t^4 - 10t^2y^3 + 25y^6$.

III. 5. a) $\dfrac{x^2}{4} - \dfrac{xy}{3} + \dfrac{y^2}{9}$; b) $x^2 - \dfrac{2x}{3} + \dfrac{1}{9}$; c) $\dfrac{16}{9} - \dfrac{2b}{3} + \dfrac{b^2}{16}$;

d) $\dfrac{9a^2}{4} - \dfrac{3ab}{5} + \dfrac{4b^2}{25}$; e) $4r^2 - \dfrac{16r}{3} + \dfrac{16}{9}$; f) $\dfrac{4a^2}{25} + \dfrac{a}{2} + \dfrac{25}{16}$;

g) $\dfrac{9a^2}{64} + \dfrac{a}{2} + \dfrac{4}{9}$; h) $\dfrac{a^4}{36} + \dfrac{2a^3}{9} + \dfrac{4a^2}{9}$; i) $\dfrac{9x^8y^2}{4} + 2x^7y^3 + \dfrac{4x^6y^4}{9}$;

j) $\dfrac{4a^{10}b^2}{49} + \dfrac{6a^8b^3}{35} + \dfrac{9a^6b^4}{100}$;

k) $9x^4y^2 - 12x^2y^2 + 4y^2$; l) $4y^2 - 2x^3y^3 + 0.25x^6y^4$;

m) $4x^4y^4 - 0.8x^5y^3 + 0.04x^6y^2$; n) $2 + 2a\sqrt{2} + a^2$; o) $3a^2 + 2ab\sqrt{6} + 2b^2$;

p) $x^2 - 2x\sqrt{3} + 3$; q) $2x^2 - 2xy + \dfrac{y^2}{2}$; r) $\dfrac{4x^2}{3} + 4x + 3$; s) $\dfrac{9x^2}{2} + 6x + 2$;

t) $4y^2 - 4y\sqrt{3} + 3$.

III. 6. a) $x^2 - 4$; b) $4x^2 - 9$; c) $9x^2 - 1$; d) $9x^2 - 25$; e) $x^2 - 0.16$;

f) $-4x^2 + 9$; g) $0.04x^2 - 1$; h) $9x^2 - 16$; i) $\dfrac{x^2}{9} - \dfrac{1}{4}$; j) $\dfrac{a^2}{4} - \dfrac{4b^2}{9}$;

k) $9x^4 - \dfrac{1}{9}$; l) $\dfrac{y^2}{9} - \dfrac{9}{4}$; m) $\dfrac{a^2}{9} - 9a^4$; n) $\dfrac{25x^2}{9} - \dfrac{4y^2}{25}$; o) $\dfrac{16x^2}{9} - \dfrac{9y^2}{4}$;

p) $-x^2 + 3$; q) $5x^2 - 1$.

III. 7. a) $-3x^2 - 4x - 1$; b) $-2x^2 + 15x - 9$; c) $-6x^2 - 14x - 17$;

d) $-x^4 + 20x^2 - 4$; e) $8x^3$; f) $18x + 54$; g) $8x - 25$; h) $-7x + 57$;

i) $84x + 163$.

III.8. a) $x^2 + y^2 + z^2 + 2xy + 2yz + 2xz$ b) $x^2 + y^2 + 4 + 4x + 4y$;

c) $x^2 + y^2 + 0.01 + 0.2x + 0.2y$; d) $a^4 + 2a^3 + 3a^2 + 2a + 1$;

e) $x^4 + y^4 + z^4 + 2x^2y^2 + 2y^2z^2 + 2x^2z^2$; f) $6 + 2\sqrt{2} + 2\sqrt{3} + 2\sqrt{6}$;

g) $a^2 + b^2 + 2ab - 2b - 2a + 1$; h) $x^2 + 4y^2 + 4xy + 4y - 2x + 1$;

i) $x^2 + 4y^2 + 4z^2 - 4xy + 8zy - 4xz$; j) $4x^2 - 4x - 4x\sqrt{2} + 2\sqrt{2} + 3$.

III. 9. a) $(a + b)(x - y)$; b) $(x - y)(a - b)$; c) $(x + y)(m - n)$;

d) $(a - b)(x + 2)$; e) $(x + z)(a + y)$; f) $(a + d)(b - c)$; g) $(m + y)(x - n)$;

h) $(2 - 5a)(2x - 3)$; i) $(2a + b)(x - y)(x + y)$; j) $(y + \sqrt{3})(a - \sqrt{2})$;

k) $(x - y)(a^2 + b)$; l) $(x - 1)(x^2 + 1)$; m) $(x + 1)^2$; n) $(x + 1)(x^2 - 2)$;

o) $(4x - 3)(x - 3)$; p) $35x - 14$; q) $(x + 2\sqrt{3})(x + 1)$; r) $(x + \sqrt{3})(4 - 3x)$.

III. 10. a) $2m^2\left(m - 3m^3 + 3n + 1\right)$; b) $\left(a + 3x\right)\left(a^2 + x^2\right)$;

c) $n\left(m^2 + n^2\right)\left(m - 2n\right)$; d) $(a - 3b - 1)\left(2 - x^2\right)$; e) $\left(2m + 5q\right)\left(3a + b\right)$;

248

f) $(p+3m)(5a-7b)$; g) $(x^2+3z^2)(x+z)$; h) $(x-z)(x^2+2z^2)$;

i) $(m^2+4m-3)(m+3)(m+1)$; j) $(2a-5b)(2a-b)$;

k) $(5m+12p)(-11m+12p)$; l) $(m+6p)(19m-6p)$; m) $(5q-n)(3n-q)$;

n) $(2p^6-5z^4)^2$; o) $(5b^2-3a)^2$; p) $(3a^2b^4-1)^2$; q) $3x(1+3x^2y^3)^2$;

r) $(x-1)(x+1)(x-3)(x+3)$; s) $(a-b)(a+b)(3-a)(3+a)$;

t) $(4x+1)(3-y)(3+y)$; u) $-(a-2)^2(1-2c)(1+2c)$; v) $4a^{n-2}(3a^2+2b)^2$;

x) $(x+y+z)(x+y-z)$; y) $(3+y+5z)(3-y-5z)$;

z) $(3x-4)(x^2-8)(x^2+8)$.

III. 11. a) $\dfrac{x}{y}$; b) $\dfrac{15x^2}{y^2}$; c) $\dfrac{8}{y}$; d) $-\dfrac{5}{2x}$; e) $\dfrac{x}{y^2}$; f) $\dfrac{m}{x}$; g) 1; h) $\dfrac{b}{a}$; i) -11;

j) $\dfrac{3}{2}$. **III. 12.** 1) a) $\dfrac{5x}{3}$; b) $\dfrac{4y+x}{y}$; c) $\dfrac{24x+11}{12}$; d) $\dfrac{6x+5}{10}$; e) 1;

f) $\dfrac{6x+7}{15}$;

g) $\dfrac{12x-1}{6}$; h) $\dfrac{17x-3}{4}$; i) $\dfrac{17}{2x}$; j) $\dfrac{14x^2-4x-5}{14x^2}$; k) $\dfrac{1}{2(x+1)}$;

l) $-\dfrac{1}{15(x-2)}$. **III. 13.** a) $\dfrac{y}{x}$; b) $\dfrac{5}{x+4}$; c) $\dfrac{7}{9(x-3y)}$; d) $\dfrac{40x^2-40}{5x+5}$;

e) $\dfrac{1}{2x-3}$; f) $\dfrac{x^3-y^3}{x^6}$;

g) $\dfrac{x-1}{x+7}$; h) $\dfrac{x^2+2xy+y^2}{x+2y}$; i) $\dfrac{x-2y-2}{2y-x-1}$; j) $2x-2ax+2a^2$.

III. 14. a) $\dfrac{y}{4xz}$; b) $\dfrac{a}{y^2}$; c) $\dfrac{4}{3bx^2}$; d) $\dfrac{x-4}{x+3}$; e) $\dfrac{x}{2}$; f) $\dfrac{x-2}{x+2}$; g) $\dfrac{x}{2}$;

h) $3x-1$; i) $\dfrac{1}{5x-1}$; j) $\dfrac{x-3}{x+3}$; k) $a-5$; l) $\dfrac{6x-5}{6x+5}$; m) $\dfrac{2(x+4)}{x-4}$;

n) $\dfrac{2x+3}{4(x-1)}$; o) $\dfrac{3x+2}{2(2x+1)}$; p) $\dfrac{a^2+2}{a^2-2}$; q) $\dfrac{x+4}{x-4}$; r) $\dfrac{x+2}{x-2}$;

s) $\dfrac{x^2-5x+3}{x^2-5x-1}$; t) $\dfrac{x^2-4x-1}{x^2-4x-3}$; u) $\dfrac{2x^2-3x-2}{2x^2-3x-1}$.

III. 15. a) $\dfrac{2}{x}$; b) $\dfrac{3}{x-2}$; c) $\dfrac{x}{x+2}$; d) 1; e) $\dfrac{2}{(x^2-1)(x-1)}$;

f) $\dfrac{3a^2-x^2+2a}{6a^3x^2}$; g) $\dfrac{6}{x^4-1}$; h) $\dfrac{8}{x-1}$; i) $\dfrac{2(x+1)}{x(x-3)}$;

j) $\dfrac{2(x+9)}{(x-2)(x+5)}$; k) 2; l) $\dfrac{x-3}{x}$; m) $\dfrac{2(3x-4)}{(x-1)(x+1)}$; n) $\dfrac{x}{x^2-1}$;

III. 16. a) $\dfrac{1-3x}{(x-1)(x-2)}$; b) $\dfrac{-x}{x^3-1}$; c) $\dfrac{1}{x}$; d) -1; e) $\dfrac{x}{2}$; f) $4x$;

g) $\dfrac{3x^2+2}{2x}$; h) 1; i) $\dfrac{1}{x^2-3x}$; j) $\dfrac{1}{(x+1)(x-2)}$.

III. 17. a) $x \in \mathbf{R}\setminus\{-1,0,1\}$,

$$E = \left[\left(\frac{x}{x+1}-\frac{x^2}{(x+1)^2}\right) \div \left(\frac{x}{(x-1)(x+1)}-\frac{1}{x+1}\right)\right]\frac{x+1}{x} =$$

$$= \left[\left(\frac{x(x+1)-x^2}{(x+1)^2} \div \frac{x-x+1}{(x-1)(x+1)}\right)\right]\frac{x+1}{x} = \left(\frac{x^2+x-x^2}{x+1} \cdot \frac{x-1}{1}\right)\frac{x+1}{x} =$$

$$= \frac{x(x-1)}{x} = x-1;$$ b) $x \in \mathbf{R}\setminus\{-2,-1,0,1,2\}$, $E = 2x$;

c) $x \in \mathbf{R}\setminus\{-4,-1,1\}$; $E(x)=1-x$; d) $\dfrac{1}{x+2}$; e) -1;

f) $x-y+3$; g) $x+5$; h) $\dfrac{x-1}{x-3} = x-1, x=4$; i) $\dfrac{1}{y^2+y+1}$;

j) $x \notin \left\{-\dfrac{3}{2},-\dfrac{7}{5},-1,\ 1,\ 0\right\}$, 1; k) 1.

III. 18. a) $\left[\dfrac{x^2+2xy+y^2+2y^2-\left(x^2+xy+y^2\right)+x^2-y^2}{x^3-y^3}\right] \div \dfrac{x-y}{xy} =$

$$= \frac{x^2+2xy+y^2+2y^2-x^2-xy-y^2+x^2-y^2}{(x-y)(x^2+xy+y^2)} \cdot \frac{xy}{x-y} = \frac{x^2+xy+y^2}{x^2+xy+y^2} \cdot \frac{1}{xy} = \frac{1}{xy};$$

b) $\left\{\dfrac{x^2-y^2}{(x+y)^2}+\left(\dfrac{x^2-y^2}{(x+y)^2}\right)^3+2\left[\dfrac{x^2-y^2}{(x+y)^2}\right]^2\right\} \div \dfrac{x^2-y^2}{(x+y)^4} =$

$$= \frac{x^2 - y^2}{(x+y)^2} \left\{ 1 + \left[\frac{x^2 - y^2}{(x+y)^2} \right]^2 + 2\frac{x^2 - y^2}{(x+y)^2} \right\} \cdot \frac{(x+y)^4}{x^2 - y^2} =$$

$$= \left[1 + \frac{x^2 - y^2}{(x+y)^2} \right]^2 (x+y)^2 = \left[\frac{2x(x+y)}{(x+y)^2} \right]^2 (x+y)^2 = 4x^2; \quad \text{c)} \quad x-1;$$

d) $-\dfrac{m+n}{m^2 n^2}$; **e)** $\dfrac{x-y}{2xy}$; **f)** $-\dfrac{1}{4x}$; **g)** $\dfrac{1}{2a}$; **h)** $\dfrac{y}{x+y}$; **i)** $\dfrac{2}{x}$; **j)** $\dfrac{b-2}{2b}$.

III. 19. a) $E(x) = 1$; **b)** $x \in \{0, 1, 3, 4\}$;

III. 20. a) $x \in \{-2, -1, 0, 1\}$; **b)** $x = -2$, which is not a solution.

III. 21. a) $x \in \mathbf{R} \setminus \{-5, 5\}$; **b)** $\dfrac{9}{x+5}$; **c)** $x = 4$.

III. 22. a) $x \in \mathbf{R} \setminus \{-2, 0, 2\}$; **b)** $-\dfrac{x(x-2)}{x+2}$; **c)** $x \in \{0, 1\}$; **d)** $x = -6$.

III. 23. a) $x \in \mathbf{R} \setminus \{-1, 1\}$; **b)** $-\dfrac{x+1}{2(x-1)}$; **c)** $x = 2$.

III. 24. a) $x \in \mathbf{R} \setminus \left\{-2, -\dfrac{2}{3}, 2\right\}$; **c)** $x = 1$. **III. 25. a)** $11x - 48$; **b)** $4\dfrac{4}{11}$; **c)** 9.

III. 26. b) $x \in \{-11, -8, -7, -6, -4, -3, -2\}$.

III. 27. a) $x \in \mathbf{R} \setminus \{-1, 1\}$; **b)** $\dfrac{x^2 + 1}{x^2 - 1}$;

III. 28. $D = \mathbf{R} \setminus \{\pm 2\}$, $E(x) = \dfrac{3}{x-2}$, $x \in \{-1, 1, 3, 5\}$.

III. 29. a) $E_1(x)$: $x \in \mathbf{R} \setminus \{-1, 1\}$ and $E_2(x)$: $x \in \mathbf{R} \setminus \{-1, 0, 1\}$; **b)**

$E_1(x) = -2x$,

$E_2(x) = \dfrac{x-1}{x+1}$; **c)** -1;

III. 30. a) $x \in \mathbf{R} \setminus \left\{-\dfrac{1}{2}, -\dfrac{1}{3}, \dfrac{1}{3}, \dfrac{1}{2}\right\}$; **b)** $x \in \{0, 1\}$;

III. 31. a) $x \in \mathbf{R} \setminus \left\{-3, \dfrac{1}{2}, 3\right\}$; **c)** $x \in \{-5, -4, -2, -1\}$.

III. 32. a) $x = 3.5$; **b)** $x = \dfrac{1}{4}$; **c)** no solution; **d)** $x \in \mathbf{R} \setminus \{3\}$; **e)** no solution;

f) no solution; **g)** $x = 2$ (impossible); **h)** $x = \dfrac{15}{8}$; **i)** $x = 0$; **j)** $x = 3$;

k) $x = -5$; l) $x = \dfrac{a(a+b)}{b}$; m) $x = -\dfrac{2}{3}$; n) $x = -\dfrac{10}{21}$; o) $x = -3$; p) $x = \dfrac{5}{27}$;

q) $x = 24$; r) $x = \dfrac{1}{8}$; s) $x = 2\dfrac{a+b}{a-b}$; t) $x = 2$; u) $x = 2$.

III. 33. a) $x = 1$; b) $x = 1$; c) $x = 2$; d) $x = -1$; e) $x = 2$; f) $x = \dfrac{ab}{a+b}$;

g) $x = \dfrac{a}{a+1}$; h) $x = -a^2$; i) $x = \dfrac{2a^2}{3}$; j) $x = \dfrac{a+1}{4}$; k) $x = 1$; l) $x = 4$;

m) $x = \dfrac{16}{17}$. **III. 34.** $\dfrac{\overline{1x1} + \overline{2x2} + \overline{3x3} + ... + \overline{9x9}}{999} = x$ or

$\dfrac{101 + 202 + 303 + ... + 909}{999} = x$, which yields to $x = 5$.

III. 35. $\dfrac{1}{3x} = \dfrac{1}{6} - \dfrac{1}{5y} \Rightarrow \dfrac{1}{x} = \dfrac{1}{2} - \dfrac{3}{5y} \Rightarrow \dfrac{1}{x} = \dfrac{5y-6}{10y} \Rightarrow$

$x = \dfrac{10y}{5y-6} \Rightarrow x = 2 + \dfrac{12}{5y-6}$.

Therefore $5y - 6 = \{\pm 1; \ \pm 2; \ \pm 3; \ \pm 4; \ \pm 6; \ \pm 12\}$. y is an integer number,

then $y = 1$ and $x = -10$. **III. 36.** a) $x = 2$; b) $x = 6$; c) $x = \dfrac{3\sqrt{3}}{5}$; d) $x = \sqrt{2}$;

e) $x = \sqrt{3} - 1$; f) $x = -\dfrac{19}{36}$; g) $x = 14 - 2\sqrt{5}$; h) $x = 4\sqrt{2}$;

i) $x = -\dfrac{1}{20} - \sqrt{2}$; j) $x = 1$; k) $x = \dfrac{\sqrt{2}-1}{2}$; l) $x = \dfrac{10 + 3\sqrt{2}}{8}$; m) $x = 5$;

n) $x = \dfrac{4(1-\sqrt{3})}{9}$; o) $x = 3$; p) $z = -\dfrac{2(5+9\sqrt{2})}{21}$; q) $x = 3$; r) $x = 0$; s) $x = $

16; t) $x = 4$; u) $x = 0$; v) i) $a = -1$; ii) $a = 11$; iii) $a = -4$; iv) $a = 3$;
w) i) $m = 6$; ii) $m = 3$; iii) $m = 2$; iv) $m = -6$.

III. 37. a) $x_1 = -8$, $x_2 = 4$; b) $x_1 = -2$, $x_2 = 8$; c) $x_1 = -3$, $x_2 = \dfrac{5}{3}$;

d) $x_1 = -4$, $x_2 = 6$; e) $x = \pm 3$; f) Impossible; g) $x_1 = -\dfrac{11}{6}$, $x_2 = \dfrac{5}{6}$;

h) $x_1 = -\dfrac{2}{3}$, $x_2 = 0$; i) $x = 3$; j) $x = \pm m$; k) $x_1 = a-1$, $x_2 = -a-1$;

l) Impossible; m) $x = \pm 1$, $a \neq 0$; n) $x \in (-\infty, 0]$; o) $x = \pm 1$, $m \neq -\dfrac{1}{2}$;

q) $x_1 = m-1$, $x_2 = 1-m$, $m \neq 0$; r) $x = \pm 1$, $m \neq -2$; s) $x = \pm 1$, $m \neq 0$;

t) $x_1 = 1$, $m \neq -2$, $x_2 = \dfrac{m+2}{2-m}$, $m \neq 2$; u) If $a < 0$, $x_1 = \dfrac{2}{a}$, if $a > 0$, $x_2 = \dfrac{2}{3a}$,

and if $a = 0$, there is no solution; v) $x \in \left[-\dfrac{1}{2}, \dfrac{1}{2}\right]$; w) No solutions;

x) $x = -0.5$; y) $x_1 = -\dfrac{8}{3}$, $x_2 = -\dfrac{4}{5}$;

z) $|x| = \begin{cases} -x & \text{if } x < 0 \\ x & \text{if } x \geq 0 \end{cases}$, $|x+2| = \begin{cases} -x-2 & \text{if } x < -2 \\ x+2 & \text{if } x \geq -2 \end{cases}$; i) $x < -2$.

The equation becomes $-x - x - 2 = x + 5$. Thus $x = -\dfrac{7}{3}$; ii) $x \geq -2$. The

equation becomes $-x + x + 2 = x + 5$. Thus $x = -3$ (Impossible);

iii) $x \geq 0$. The equation becomes $x + x + 2 = x + 5$. Thus

$x = 3$. Therefore $x = -\dfrac{7}{3}$ and $x = 3$ are the solutions of the equation.

III. 38. a) $x = \dfrac{3}{2+m}$, $m \neq -2$; b) $x = 1$, $m \neq -1$; c) $x = \dfrac{4}{m-3}$, $m \neq 3$;

d) $x = \dfrac{m-1}{3}$; e) $x = \dfrac{3ab}{a^2+b^2}$, $a^2 + b^2 \neq 0$; f) $x = \dfrac{m-3}{m^2+m-2}$, $m \neq 1$,

$m \neq -2$; g) $x = \dfrac{1}{m+1}$, $m \neq -1$, $m \neq 3$; h) $x = \dfrac{1}{m^2-3m-4}$, $m \neq -1$, $m \neq 4$;

i) $x = \dfrac{4m+m^2}{m-4}$, $m \neq 4$; j) $x = \dfrac{2m}{m^2+2m}$, $m \neq 0$, $m \neq -2$;

k) $x = -1-m$, $m \neq 1$; l) $x = \dfrac{3a-7}{2a-1}$, $a \neq \dfrac{1}{2}$; m) $x = \dfrac{-1-2m}{m+2}$, $m \neq -2$;

n) $x = \dfrac{-a}{m}$, $m \neq 0$; o) $x = \dfrac{2am}{a+m}$, $m \neq -a$; p) $x = \dfrac{5}{m}$, $m \neq 0$;

q) $x = \dfrac{1}{a-m}$, $m \neq a$; r) $x = a+m$, $m \neq a$; s) $x = \dfrac{m\sqrt{2}}{4\sqrt{2}-4m}$, $m \neq \sqrt{2}$;

t) $x = \dfrac{2-3m}{m-1}$, $m \neq 1$; u) $x = \dfrac{4a^2-3m^2}{2a-m}$, $m \neq 2a$.

III. 39. a) $x_1 = -2$, $x_2 = 0$; b) $x = 3$; c) $x = \pm\sqrt{3}$; d) $x_1 = -\dfrac{5}{3}$, $x_2 = 0$;

e) $x = 2$, $x \neq 0$; f) $x_1 = \dfrac{5}{2}$, $x_2 = -\dfrac{5}{2}$; g) $x_1 = -17$, $x_2 = 15$; h) $x_1 = \dfrac{1}{5}$,

$x_2 = 5$; i) $x_1 = 1$, $x_2 = 2$, $x_3 = -3$; j) $x_1 = -2$, $x_2 = \dfrac{9}{4}$; k) $x = -4$; l)

$x_1 = 1$, $x_2 = -3$; m) $x_1 = 1$, $x_2 = -\dfrac{1}{8}$; n) $x = 1$; o) $x_1 = 3$, $x_2 = -\dfrac{3}{2}$;

p) $x = -3$; q) No solution; r) $x = 1$, $x \neq \pm 3$.

III. 40. a) $(\infty, 2]$; **b)** $\left(\dfrac{1}{3}, \infty\right)$; **c)** $\left(-\infty, -\dfrac{1}{7}\right]$; **d)** $\left(-\infty, -\dfrac{3}{2}\right]$; **e)** $[0, \infty)$;

f) $(-\infty, 3]$; **g)** $(-\infty, 1]$; **h)** $(-\infty, -2)$; **i)** $(-\infty, -2)$; **j)** $\left(\dfrac{2}{3}, \infty\right)$; **k)** $(-\infty, 1)$

; **l)** $\left[8\dfrac{1}{2}, \infty\right)$; **m)** $[0, \infty)$; **n)** $(-\infty, 5)$; **o)** $\left(-\infty, -3\dfrac{3}{7}\right]$; **p)** $\left(-\infty, -\dfrac{1}{6}\right]$;

r) $[4, \infty)$. **III. 41. a)** $(1, -1)$; **b)** $(2, -1)$; **c)** $(-1, -1)$; **d)** $(2, 1)$; **e)** $(3, 1)$;

f) $(1, 1)$; **g)** $(-1, -1)$; **h)** $(-5, -2)$; **i)** $(7, 6)$; **j)** $(6, 10)$; **k)** $(1, 1)$.

III. 42. a) $(1, -2)$; **b)** $(1, 0)$; **c)** $(2, 3)$; **d)** $(2, -1)$; **e)** $\left(\sqrt{2}, -\sqrt{3}\right)$;

f) $\left(\sqrt{5} - \sqrt{3}, \sqrt{5} + \sqrt{3}\right)$; **g)** $\left(\dfrac{1}{3}, -\dfrac{1}{3}\right)$; **h)** $\left(\dfrac{1}{2}, -\dfrac{1}{2}\right)$; **i)** $(28, 82)$; **j)** $(1, -3)$.

III. 43. a) $(6, -6)$; **b)** $(2, 1)$; **c)** $(4, -1)$; **d)** $(1, -1)$;

e) $\dfrac{x}{a} = 3 - \dfrac{y}{b}$, substitute in the second equation:

$$\dfrac{1}{3}\left(3 - \dfrac{y}{b}\right) + \dfrac{y}{4b} = \dfrac{5}{6}, \quad 1 - \dfrac{y}{3b} + \dfrac{y}{4b} = \dfrac{5}{6}, \quad \dfrac{-4y + 3y}{12b} = \dfrac{5}{6} - 1,$$

$\dfrac{-y}{12b} = \dfrac{-1}{6} \Rightarrow 12b = 6y \Rightarrow 2b = y$ and similarly $x = a$;

f) Multiply the second equation by $\dfrac{1}{a-b}$ and add the equations

$$x\left(\dfrac{1}{a+b} + \dfrac{1}{a-b}\right) = 2a + \dfrac{4ab}{a-b} \Rightarrow x\dfrac{2a}{(a+b)(a-b)} = \dfrac{2a^2 + 2ab}{a-b}, \text{ finally}$$

$x = (a+b)^2$. Replace x in the second equation, to find y: $(a+b)^2 - y = 4ab$,

and $y = (a-b)^2$;

g) $\dfrac{1}{x+y-1} = a$ and $\dfrac{1}{x-y+2} = b$ then the system becomes

$$\begin{cases} 2a + 3b = 5 \\ b - 2a = -1 \end{cases}, \quad x = \frac{1}{2}, \quad x = \frac{3}{2}; \quad \text{h) Denote} \quad \frac{1}{k} = \frac{3}{x+y} = \frac{4}{x+z} = \frac{5}{y+z},$$

$x + y = 3k, x + z = 4k, y + z = 5k$ and replacing in the third equation of the

system: $3k \cdot 4k \cdot 5k = 60 \Rightarrow k^3 = 1 \Rightarrow k = 1$.

Therefore $\begin{cases} x + y = 3 \\ x + z = 4 \\ y + z = 5 \end{cases}$, add these three equations:

$2(x + y + z) = 12 \Rightarrow x + y + z = 6,$

$x = 1, \quad y = 2, z = 3 \,; \text{i)} \quad y = 7, x = 5$.

III. 44. a, b, c be the sides of the triangle, then $a + b = 45, b + c = 52, a + c = 48$.

$a = \dfrac{41}{2}, \quad b = \dfrac{49}{2}, c = \dfrac{55}{2}$.

III. 45. b) $\dfrac{MN}{AC} = \dfrac{NP}{AB} = \dfrac{MP}{BC}$ and $AB + BC + AC = 70$. Applying the

proportion properties, we obtain $\dfrac{MN}{AC} = \dfrac{NP}{AB} = \dfrac{MP}{BC} = \dfrac{MN + NP + MP}{AC + AB + BC} = \dfrac{2}{5}$.

III. 46. $a + b + c = 222$, $b = 10c$ and $a = 10b = 100c$, $a = 200$, $c = 2$, $b = 20$.

III. 47. a) (9.5, 5.5); **b)** (6, 26); **c)** (7, 3); **d)** (2, 4); **e)** (6, 9); **f)** (1, 1); **g)**
(1, 2); **h)** (1, 1, 1); **i)** (1, 2, 3); **j)** (1, 5, 1);
k) (−1, −2, 5); **l)** (2, 2, 2); **m)** (−2, 2, 1); **n)** (−1, 8, 2); **o)** (1, 1, 1);

p) $\left(2^{-1}, 3^{-1}, 4^{-1}\right)$.

III. 48. $\dfrac{7}{2\left(2^3 - 1\right)} + \dfrac{13}{3\left(3^3 - 1\right)} + \dfrac{21}{4\left(4^3 - 1\right)} + ... + \dfrac{n^2 + n + 1}{n\left(n^3 - 1\right)} =$

$= \dfrac{1}{2(2 - 1)} + \dfrac{1}{3(3 - 1)} + \dfrac{1}{4(4 - 1)} + ... + \dfrac{1}{n(n - 1)} =$

$= \left(\dfrac{1}{1} - \dfrac{1}{2}\right) + \left(\dfrac{1}{2} - \dfrac{1}{3}\right) + ... + \left(\dfrac{1}{n-1} - \dfrac{1}{n}\right) = \dfrac{n-1}{n}$.

III. 49. a) The line AB is parallel to y-axis and AC is on x-axis;

b) $m_{AB} = 5$, $m_{BC} = -\dfrac{1}{5}$.

III. 50. a) $(2, 3)$; **b)** $(4, -1)$; **c)** $(1, 3)$;
d) $(2, 1.5)$; **e)** $(2, -3)$; **f)** $(1, 2)$.
III. 51. a) $(-6, 15)$; **b)** $(-8, -9)$; **c)** $(1, 14)$; **d)** $(3, -7)$.
III. 52. a) $(10, -9)$; **b)** ; **c)** $(13, -11)$; **d)** $(-7, -12)$.
III. 53. a) $B(2, 0)$ and $D(-4, 6)$; **b)** $B(-1, 1)$ and $D(-5, -5)$.
III. 54. $B(-2, 5)$ and $D(-4, 9)$. **III. 55.** $O(4.5, 4.5)$ and $C(7, 7)$.

III. 56. a) $m_{MN} = -3$, $m_{AC} = -3$, $M(3, 1)$, $N(4, -2)$; **b)** $AC = 2\sqrt{10}$,

$MN = \sqrt{10}$; **III. 57.** $M(4, 2.5)$ and $N(4.5, 4.75)$.

III. 58. a) $AB = 5$, $CD = 5$, $BC = 5$, $AD = 5$; **b)** $m_{AC} = -2$, $m_{BD} = \dfrac{1}{2}$.

III.59. a) $AB = 2\sqrt{5}$, $AC = 2\sqrt{5}$, $m_{AB} = \dfrac{1}{2}$, $m_{AC} = -2$; **b)** $M(4, 3)$;

c) $m_{AM} = -\dfrac{1}{3}$, $m_{BC} = 3$.

III. 60. a) $AB = 10$, $AM = \sqrt{k^2 + 81}$, $BM = \sqrt{k^2 + 1} \Rightarrow k = 3$;

b) $m_{AM} = \dfrac{1}{3}$, $m_{BM} = -3$. **III. 61. a)** $m_{AB} = 0$, $m_{CD} = 0$. $m_{AD} = \dfrac{3}{5}$,

$m_{BC} = \dfrac{3}{5}$; **b)** $AB = 9$, $DC = 9$, $BC = 3\sqrt{34}, AD = 3\sqrt{34}$;

c) $(5, 2.5)$. **III. 62. a)** $a = 2$; **b)** $a = 3$ or $a = -5$; **c)** $a = 1$ or $a = -5$.

III. 63. a) Denote $M(x_M, y_M)$ the middle point of $BC \Rightarrow x_M = \dfrac{x_B + x_C}{2}$ and

$y_M = \dfrac{y_B + y_C}{2} \Rightarrow x_M = 1$ and $y_M = 1$. The length of segment

$AM = \sqrt{(1-5)^2 + (1-4)^2} = 5$; **b)** $\left(-\dfrac{1}{2}, \dfrac{19}{4}\right)$, $\cdot \left(\dfrac{5}{2}, -\dfrac{11}{4}\right)$

III. 64. a) $m_{AB} = 5$, $m_{BC} = -\dfrac{1}{5}$; **b)** $AB = BC = \sqrt{26}$; **c)** right isosceles.
III. 65. $AB = \sqrt{x^2 + y^2}$.

III. 66. *Solution* **1.**
Let M be the intersection between the line AB and the perpendicular line on AB, passing through the point O.

$m_{AB} = \dfrac{y_B - y_A}{x_B - x_A} = \dfrac{1}{2}$, $y = mx + b$, $b = -\dfrac{5}{2}$. $AB : y = \dfrac{x}{2} + \dfrac{5}{2}$,

$$m_{OM} = \frac{y_B - y_A}{x_B - x_A} = -2,\ y = mx + b,\ b = 0.\ OM : y = -2x.$$

Solving the system $\begin{cases} y = \dfrac{x}{2} + \dfrac{5}{2} \\ y = -2x \end{cases}$, then $M(-1, 2)$.

$$OM = \sqrt{(x_M - x_O)^2 + (y_M - y_O)^2} = \sqrt{(-1-0)^2 + (2-0)^2} = \sqrt{5}.$$

Solution 2.

The equation of the line AB: $\dfrac{x - x_A}{x_B - x_A} = \dfrac{y - y_A}{y_B - y_A} \Leftrightarrow \dfrac{x+1}{3+1} = \dfrac{y-2}{4-2} \Leftrightarrow$

$x - 2y + 5 = 0$.

$$d(O, AB) = \frac{|ax_0 + by_0 + c|}{\sqrt{a^2 + b^2}} = \frac{|0 + 0 + 5|}{\sqrt{1+4}} = \sqrt{5}.$$

III. 67. a) $d = \dfrac{|4 \cdot 0 - 3 \cdot 3 - 1|}{\sqrt{4^2 + (-3)^2}} = 2$; **b)** $d = \sqrt{4.5}$; **c)** $d = 2\sqrt{5}$;

d) $d = 2\sqrt{5}$; **e)** $d = \dfrac{1}{\sqrt{29}}$.

III. 68. a) $d = 4\sqrt{2}$; **b)** $d = \dfrac{63}{\sqrt{61}}$.

III. 69. $d = \dfrac{|3 \cdot m - 4(m+1) - 1|}{\sqrt{3^2 + (-4)^2}} = 1$, then $|m + 5| = 5$.

Therefore $m = -10$ or $m = 0$.

III. 70. $m_{AC} = \dfrac{1}{\sqrt{3}}$, $m_{BC} = -\sqrt{3}$. **III. 71.** $AB = AC = BC = \sqrt{13}$.

III. 72. $P(1, 0)$.

III. 73. *Solution* 1: Let $P(a, 2)$ be the point on the line $y = 2$ such that

the angle APB be right. We have $m_{AP} = \dfrac{2-3}{a+3}$ and $m_{BP} = \dfrac{2-7}{a-3}$. Since

$m_{AP} = \dfrac{-1}{m_{BP}} \Rightarrow \dfrac{-1}{a+3} = \dfrac{a-3}{5}$. Hence $a = \pm 2$.

There are two points $(-2, 2)$ and $(2, 2)$.

Solution 2: Let $P(a, 2)$ be the point such that the triangle APB be right.

We have $AP = \sqrt{(a+3)^2 + (2-3)^2} = \sqrt{a^2 + 6a + 10}$,

$BP = \sqrt{(a-3)^2 + (2-7)^2} = \sqrt{a^2 - 6a + 34}$, and $AB = \sqrt{52}$.

Using the Pythagorean theorem, $AP^2 + PB^2 = AB^2$, we obtain

$a^2 + 6a + 10 + a^2 - 6a + 34 = 52$, with the solutions $a = \pm 2$.

III. 74. *Solution* 1: The point Q is the intersection between the lines MQ (the perpendicular bisector of the segment AB) and NQ (the perpendicular bisector of the segment BC).

We have $M\left(2, 3+\sqrt{3}\right)$ and $N\left(\dfrac{9}{2}, \dfrac{6-\sqrt{7}}{2}\right)$.

$m_{AB} = \dfrac{4-2\sqrt{3}}{2}$, then $m_{MQ} = \dfrac{-1}{2-\sqrt{3}}$. Therefore the equation of the line MQ

is (1) $y = -x\left(2+\sqrt{3}\right) + 7 + 3\sqrt{3}$.

$m_{BC} = \dfrac{4+\sqrt{7}}{-3}$, then $m_{NQ} = \dfrac{3}{4+\sqrt{7}}$. Therefore the equation of the line NQ is

(2) $y = \dfrac{\left(4-\sqrt{7}\right)x}{3} - 3 + \sqrt{7}$.

The intersection between the lines (1) and (2) is the point $Q(3,1)$.

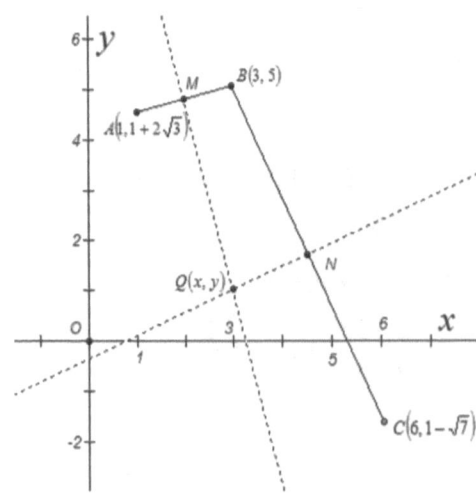

Solution 2: Let $Q(x, y)$ be the point such that $QA = QB = QC$. From $QA =$ QB we obtain $\sqrt{(x-1)^2 + (y-1-2\sqrt{3})^2} = \sqrt{(x-3)^2 + (y-5)^2}$ and from $QB = QC$ we obtain $\sqrt{(x-3)^2 + (y-5)^2} = \sqrt{(x-6)^2 + (y-1+\sqrt{7})^2}$.

Squaring both relations, we obtain a system of two linear equations with the solution $x = 3$ and $y = 1$.

III. 75. **a)** $f(x) = 2 - x$; **b)** $f(x) = 2x - 3$; **c)** $f(x) = 5 - 0.5x$;
d) $f(x) = -4x - 2$; **e)** $f(x) = x - 7$; **f)** $f(x) = 5x + 1$.

III. 76. **a)** $m = \dfrac{1}{3}$, $b = -\dfrac{2}{3}$; **b)** $m = -\dfrac{5}{3}$, $b = -35$;

c) $m = -\dfrac{14}{3}$, $b = -\dfrac{8}{3}$.

III. 77. **a)** $y = -\dfrac{1}{2}x + 4$; **b)** $y = 3x + 1$; **c)** $y = -\dfrac{3}{4}x + 3$;

d) $y = -\dfrac{5}{4}x - \dfrac{5}{2}$; **e)** $y = -2x - 2$; **f)** $y = \dfrac{1}{3}x - \dfrac{7}{3}$;

g) $y = \dfrac{5}{2}x - 5$; **h)** $y = -\dfrac{2}{3}x + \dfrac{7}{3}$.

III. 78. **a)** no; **b)** yes; **c)** no; **d)** yes. **III. 79.** **a)** $m = \dfrac{1}{4}$;

b) $m = -1$; **c)** $-\dfrac{3}{10}$; **d)** $-\dfrac{5}{4}$.

III. 80. **a)** $m_{AB} = \dfrac{a-4}{3}$; **b)** $a = 5$; **c)** $y = -3x + 15$.

III. 81. **a)** $b = 2$; **b)** $b = -\dfrac{11}{3}$.

III. 82. **a)** $a = -3$; **b)** $a = \dfrac{6}{5}$.

III. 83. $a = -3$. **III. 84.** $m = -4$.

III. 85.
The graph contains four points
lying on the line of equation

$y = 2x + 2$.

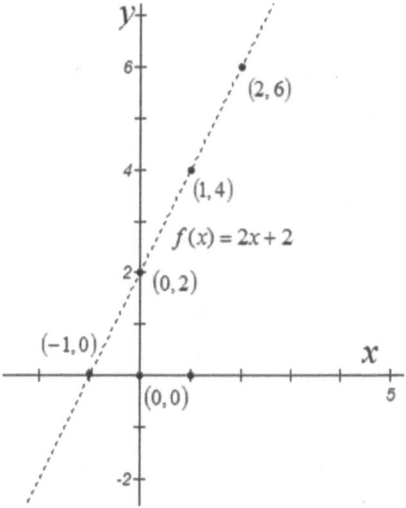

III. 86. a) $k = 8$, **b)** $k = 1$.

III. 87. $a = 1$. **III. 88. a)** $b = 5$; **b)** $b = -\dfrac{9}{4}$; **c)** $b = 1$; **d)** $b = \dfrac{5}{4}$; **e)** $b = 2$.

III. 89. $\left(\dfrac{1}{3}, \dfrac{5}{3}\right)$, yes.

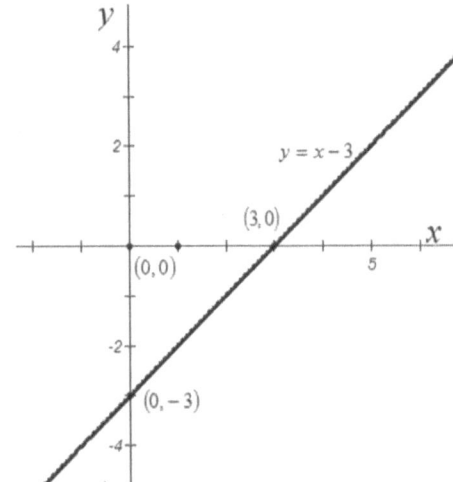

III. 90. $f(x) = -3x - 2\sqrt{6}$.

III. 91. a) 1;   **b)**

c) $M(3, 0)$ and $N(0, -3)$;

d) A; **e)** 4.5;

f) $\dfrac{3\sqrt{2}}{2}$;

g) $45°$;

h) $x = 1$;

i) $x \in (-\infty, 4]$.

III. 92. $f(x) = -2x + 4$.

III. 93. a) $(0, -8)$ and $(4, 0)$;

b)

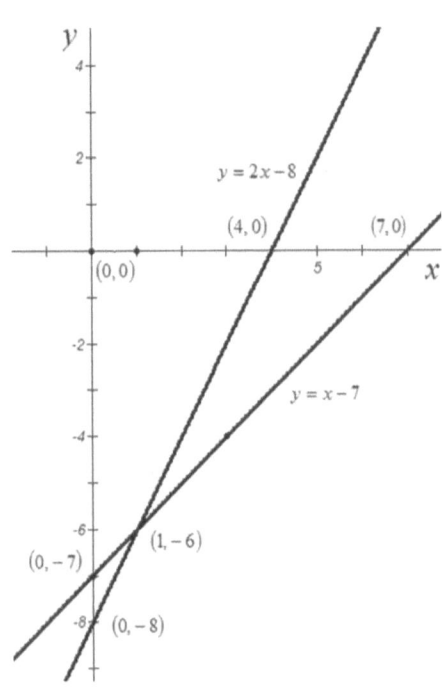

c) $(1, -6)$

d) $4\left(4 + \sqrt{5}\right)$ and 16;

e) $\dfrac{8\sqrt{5}}{5}$;

f) $2\left(3 + \sqrt{5}\right)$ and $2\sqrt{5}$;
g) $h(x) = x - 6$;

h) $x \in \left(-\infty, \ -\dfrac{7}{3}\right]$;

i) $(-1, 2)$; **k)** $C(11, 14)$.

260

III. 94. a) $M(-1, 1)$; **b)** $M(2, 0)$; **c)** $M(0, 1)$; **d)** $M(3, 2)$.

III. 95. $2m = \left(3\sqrt{6} - 5\sqrt{2}\right)$. **III. 96.** $m = 1, n = 2$.

III. 97. $m = 3, n = -1$. **III. 98.** $a = \dfrac{11}{18}$.

III. 99. $A_{\Delta ABC} = 3$

$P_{\Delta ABC} = 3 + 2\sqrt{2} + \sqrt{5}$.

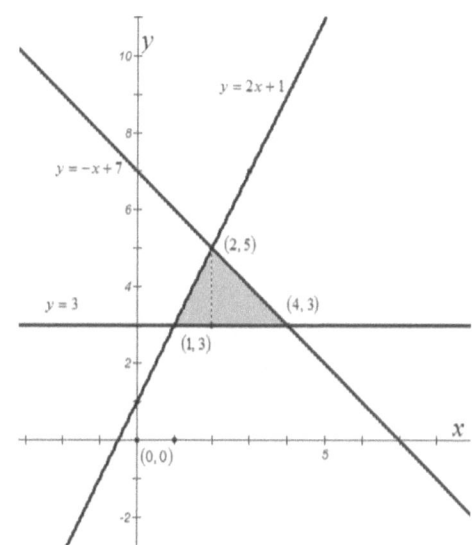

III. 100. a) $y = -\dfrac{3}{4}x + 5$;

b) $y = -\dfrac{3}{7}x - \dfrac{1}{7}$; **c)** $y = 3x - 6$;

d) $y = 3x - 15$; **e)** $y = 2x + 2$.

III. 101. b) Area = 6;

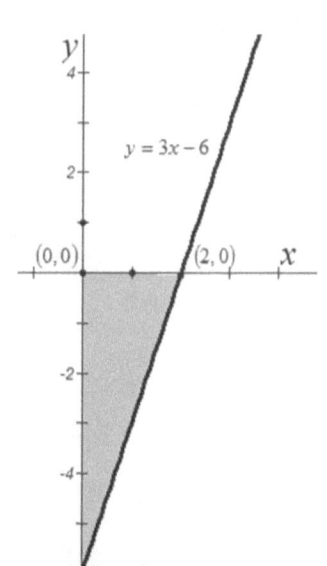

c) $a = 1$;

d) 105.

III. 102. a)
c) Area = 4.5;

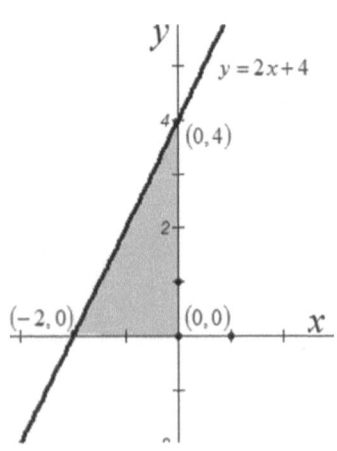

d) $\dfrac{3\sqrt{2}}{2}$;

e) $E = 6$;

III. 103. a) $f(x) = 2x + 3$; **b)** $f(x) = 2x + 7$.

III. 104. a) $a = 1$; **b)** $b = 3$; **c)** $a < 1$; **d)** $M(2, 2)$.

III. 105. $x_M = \dfrac{a + 3b}{10}$.

III. 106. a) $m = 1$; **b)**

c) Area = 4;

III. 107. b) (2, 0) and (0, – 4);

c) Area = 4;

d) Q is the

midpoint of the segment AB;

e) $\dfrac{4\sqrt{5}}{5}$.

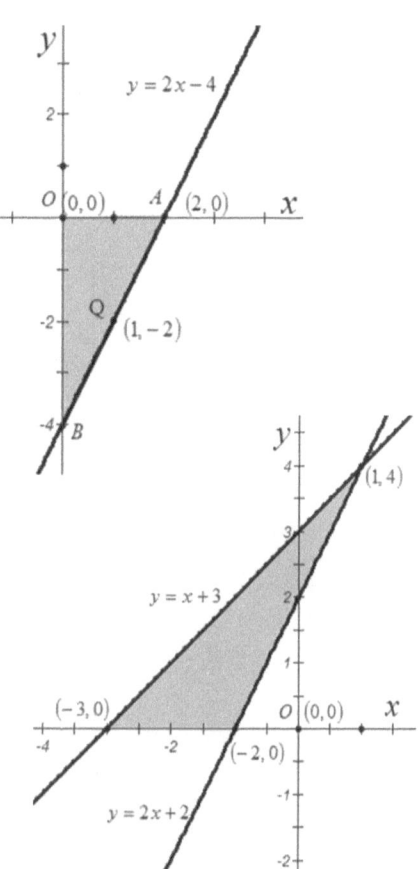

III. 108.

b) Area = 4;

III. 109. a) $a = 1$; **b)** Geometric mean = 1;

c) x intercept (2.5, 0) and y intercept (0, 5).

III. 110.

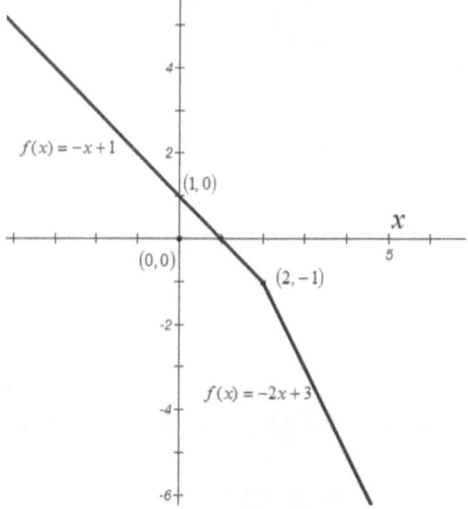

III. 111.

$x_1 = -3$, $x_2 = 3.5$.

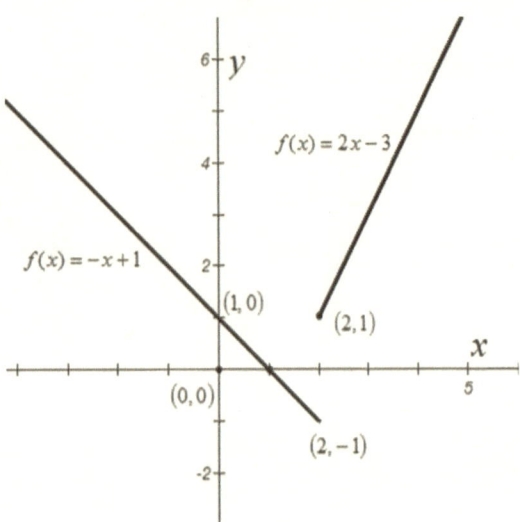

III. 112.

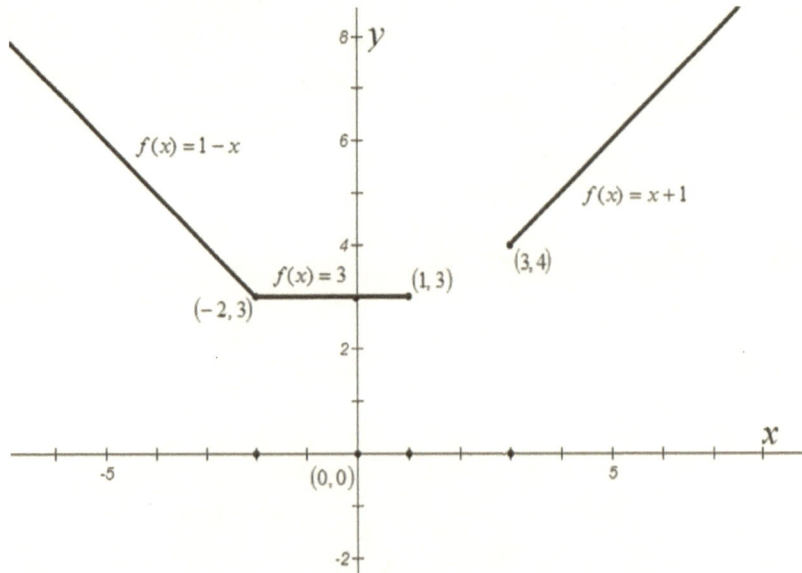

III. 113. $a = 1$.

III. 114. a) For $x = 1$ we have $f(0) + 3f(0) = -4$. Therefore $f(0) = -1$;

b) Denoting $t = x - 1$, we have $f(t) + 3f(0) = 2(t+1) - 6$, which yields $f(t) = 2t - 1$;

c) $\dfrac{1}{f(1)f(2)} + \dfrac{1}{f(2)f(3)} + \ldots + \dfrac{1}{f(2014)f(2015)} =$

$= \dfrac{1}{1 \cdot 3} + \dfrac{1}{3 \cdot 5} + \ldots + \dfrac{1}{4027 \cdot 4029} =$

$= \dfrac{1}{2}\left[\left(\dfrac{1}{1} - \dfrac{1}{3}\right) + \left(\dfrac{1}{3} - \dfrac{1}{6}\right) + \ldots + \left(\dfrac{1}{4027} - \dfrac{1}{4029}\right)\right] = \dfrac{1}{2}\left(\dfrac{1}{1} - \dfrac{1}{4029}\right) = \dfrac{2014}{4029}$.

III. 115. a) Let $M(\alpha, 0)$ be the mobile point on x-axis, where $\alpha \in \mathbf{R}$.

Line AB: $\quad y = \dfrac{1}{2}x + \dfrac{3}{2}$ (1).

Line LB: $\quad y = \dfrac{4}{5-\alpha}x - \dfrac{4\alpha}{5-\alpha}$ (2).

Line LA: $\quad y = \dfrac{2}{1-\alpha}x - \dfrac{2\alpha}{1-\alpha}$ (3)

The intersection between (2) and the y-axis is

$N\left(0, -\dfrac{4\alpha}{5-\alpha}\right)$.

The intersection between (2) and the $y = x$

is $Q\left(-\dfrac{4\alpha}{1-\alpha}, -\dfrac{4\alpha}{1-\alpha}\right)$.

The intersection between (3) and the y-axis

is

$M\left(0, -\dfrac{2\alpha}{1-\alpha}\right)$.

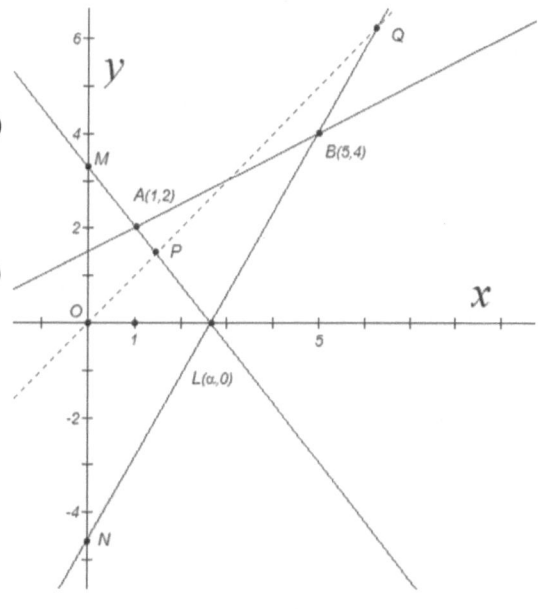

The intersection between (3) and the $y = x$ is $P\left(\dfrac{2\alpha}{1+\alpha}, \dfrac{2\alpha}{1+\alpha}\right)$.

The equation of the line NP is $y = \dfrac{\alpha+7}{5-\alpha}x - \dfrac{4\alpha}{5-\alpha}$, then the intersection

between NP and (1) is the point $\left(\dfrac{5}{3}, \dfrac{7}{3}\right)$.

b) $m_{MQ} = \dfrac{1}{2}$.

Chapter IV

POWERS AND RADICALS

IV. 1. a) x^4; b) y^4; c) $4a^3$; d) $-3b^3$; e) $-2x^2y$; f) $3a^2b^3$; g) 2^3a^3;

h) a^6b^6 ; i) a^{15}; j) $-3^3x^{12}y^9$; k) a^{21}; l) x^{11}; m) x^{n+1}; n) x^{2n}; o) a^{m+n+1} .

IV. 2. a) $6y^5$; b) $\left(3x^7y^5\right)$; c) $8y^3$; d) $16x^9$; e) $\dfrac{a}{b}$; f) $\dfrac{2}{b^4}$; g) $4r^7s$;

h) $\dfrac{8}{9}t^{10}$; i) $8y^3$; j) 12 ; k) $\dfrac{1}{x^3y^{12}}$; l) $\dfrac{d^7}{c^8}$; m) $\dfrac{y^2z}{x^2}$; n) x^3;

o) $x^3y^{12}z^{21}$; p) $\dfrac{s^4}{q^7r^6}$; q) $\dfrac{c^8}{3a^3}$; r) $\dfrac{1}{x^6y^4}$.

IV. 3. a) $6\sqrt{2}$; b) $2\sqrt{3}$; c) $-2\sqrt{5}$; d) $-\sqrt{7}$; e) $6\sqrt{3}-1$; f) $6-2\sqrt{5}$;

g) $7\sqrt{3}-7\sqrt{2}$; h) $10\sqrt{3}-8$.

IV. 4. a) 18; b) 12; c) 24; d) 20; e) 32; f) 28; g) 48; h) 27; i) -10 ; j) $10\sqrt{6}$;

k) -36; l) -24 .

IV. 5. a) $\sqrt{15}$; b) $\dfrac{1}{4}$; c) 3; d) $\dfrac{\sqrt{5}}{10}$; e) 6; f) $4\sqrt{3}$; g) 2; h) $\dfrac{\sqrt{14}}{2}$;

i) $\sqrt{2}$; j) $2\sqrt{2}$.

IV. 6. a) $5\sqrt{10}$; b) $8\sqrt{2}$; c) $2\sqrt{10}$; d) $3\sqrt{7}$; e) $6\sqrt{2}$; f) $5\sqrt{5}$;

g) $12\sqrt{2}$; h) $5\sqrt{7}$; i) $20\sqrt{2}$; j) $15\sqrt{2}$.

IV. 7. a) $\dfrac{3}{2}+\sqrt{2}$; b) $3-\sqrt{3}$; c) $1+\sqrt{2}$; d) $4-2\sqrt{2}$; e) $2+2\sqrt{2}$;

f) $\dfrac{3}{2}+\dfrac{\sqrt{3}}{2}$; g) $2-\dfrac{5\sqrt{2}}{2}$; h) $1-2\sqrt{2}$; i) $9-\sqrt{3}$.

IV. 8. a) $2\sqrt{3}+3$; b) $-4\sqrt{3}+3$; c) $4-2\sqrt{5}$; d) $-10+\sqrt{15}$;

e) $-2\sqrt{3}+12$; f) $-1-\sqrt{5}$; g) 5; h) $30+12\sqrt{3}$; i) $22-10\sqrt{5}$;

j) $7+7\sqrt{2}$. **IV. 9.** a) $50+7\sqrt{8}$; b) $9+6\sqrt{2}$; c) $17+12\sqrt{2}$; d) 8;

e) 3; f) 15; g) $4\sqrt{3}+2\sqrt{2}-4$.

IV. 10. a) $\sqrt{2}-1$; b) $6+4\sqrt{2}$; c) $\sqrt{5}-2$; d) $-\sqrt{2}-2$;

e) $-3-2\sqrt{2}$; f) $\dfrac{1}{2}+\dfrac{\sqrt{3}}{2}$; g) $\sqrt{3}+3$; h) $4+3\sqrt{3}$; i) $2+\sqrt{6}-\sqrt{2}$;

j) $3\left(\sqrt[3]{4}+\sqrt[3]{2}+1\right)$; k) $\sqrt[3]{2}-1$;

l) $\dfrac{1}{2\sqrt[3]{4}+3\sqrt[3]{2}+1}=\dfrac{1}{\left(\sqrt[3]{2}+1\right)\left(2\sqrt[3]{2}+1\right)}=$

$=\dfrac{\left(\sqrt[3]{2^2}-\sqrt[3]{2}+1\right)\left(\sqrt[3]{2^8}-\sqrt[3]{2^4}+1\right)}{51}=\dfrac{7\sqrt[3]{4}+5\sqrt[3]{2}-11}{51}$;

m) $\dfrac{-23}{\sqrt[3]{2}-\sqrt{3}}=\dfrac{-23\left(\sqrt[3]{4}+\sqrt{3}\sqrt[3]{2}+3\right)}{2-3\sqrt{3}}=\left(\sqrt[3]{4}+\sqrt{3}\sqrt[3]{2}+3\right)\left(2+3\sqrt{3}\right)$.

IV. 11. a) $6-4\sqrt{2}$; b) $5-3\sqrt{3}$; c) $-8\sqrt{5}$;

d) $-12\sqrt{2}$; e) $-4\sqrt{3}$; f) $\dfrac{20\sqrt{3}-9}{18}$; g) $-\dfrac{12+5\sqrt{2}}{12}$;

h) $\dfrac{-4-\sqrt{2}}{12}$; i) $\dfrac{-9+10\sqrt{3}}{18}$.

IV. 12. a) $\sqrt{14}$; b) 2; c) 5; d) $\sqrt{39}-\sqrt{26}$; e) $2\sqrt{3}$;

f) $2\sqrt{2}-2$; g) $\sqrt{2}$; h) $17\sqrt{5}$; i) 0; j) $7\sqrt{2}$; k) $-12\sqrt{2}-3\sqrt{5}$;

l) $12-18\sqrt{2}+12\sqrt{3}$; m) 10; n) $4\sqrt{5}$; o) $42\sqrt{10}-174$.

IV. 13. a) 4; b) 5; c) 5; d) 33; e) 1.

IV. 14. a) $-\sqrt{3}$; b) $2\sqrt{5}$; c) $3\sqrt{6}$; d) -72; e) $4\sqrt{3}+13\sqrt{2}$; f) 48;

g) $-3\sqrt{5}$; h) -1; i) 14; j) 172; k) -2.

IV. 15. a) $\dfrac{3\sqrt{2}+\sqrt{3}}{3}$; b) $\left(\sqrt{5}+\sqrt{2}\right)\cdot\left(\sqrt{2}+1\right)$; c) $\sqrt{2}+2$; d) 1; e) $10\dfrac{1}{5}$;

f) $3-\sqrt{3}$; g) $\sqrt{5}-\sqrt{2}$; h) $\sqrt{2}-1$; i) $\sqrt{2}-2\sqrt{3}$; j) 1; k) 9.

IV. 16. a) $\dfrac{2\sqrt{3}}{3}$; b) $\sqrt{3}$; c) $\dfrac{\sqrt{6}}{2}$; d) $-3+2\sqrt{2}$; e) $\sqrt{6}$;

f) $-3+2\sqrt{2}$; g) $7+2\sqrt{3}$; h) $\sqrt{7}-\sqrt{3}$; i) $2\sqrt{2}+\sqrt{10}-\sqrt{5}-1$; j) $\sqrt{3}-\sqrt{2}$.

IV. 17. a) $\sqrt[3]{9}-\sqrt[3]{21}+\sqrt[3]{49}$; b) $\sqrt{5}-1$; c) $\sqrt{6}-\sqrt{2}$; d) 4; e) $-5+3\sqrt{2}$.

IV. 18. a) $\dfrac{\left(\sqrt{7}-\sqrt{5}\right)^2}{\left(\sqrt{7}+\sqrt{5}\right)\left(\sqrt{7}-\sqrt{5}\right)}=\dfrac{12-2\sqrt{35}}{2}=6-\sqrt{35}$; b) $\dfrac{5+\sqrt{21}}{2}$;

c) $\sqrt{7}+\sqrt{5}$; d) $\sqrt{7}+\sqrt{6}$; e) $-\dfrac{1}{\sqrt{2}}$; f) $-3\sqrt{3}$; g) 4.

IV. 19. a) 4; b) 43^2, c) $\dfrac{-1-\sqrt{3}+\sqrt{5}+\sqrt{10}}{4}$; d) $-6-\sqrt{35}$.

IV. 20. a) $\sqrt{54}=3\sqrt{6}$; b) $\sqrt[3]{-108}=-\sqrt[3]{2^2\times3^3}=-3\sqrt[3]{4}$; c) $\sqrt[3]{4}$;

d) $\sqrt[4]{\dfrac{a^3}{c^3}}$; e) $26\cdot\sqrt[3]{2}$; f) $\dfrac{-23\cdot\sqrt[3]{3}}{6}$; g) 79; h) 1.

IV. 21. a) $x^{\frac{13}{15}}$; b) $-10a^{\frac{9}{4}}$; c) $2^4\cdot b^{\frac{9}{10}}$; d) $\dfrac{1}{2^2\cdot x^4}$; e) $\dfrac{1}{c^{\frac{2}{3}}d}$; f) $8x^9y^{12}$;

g) $y^{\frac{1}{2}}$; h) $\dfrac{1}{a^2}$; i) $\dfrac{2x^6}{y^2}$; j) $\dfrac{x^3}{y^9z^6}$; m) a^{31}.

IV. 22. a) $\dfrac{x^{15}}{y^5}$; b) $\dfrac{16x^3}{yz^2}$; c) $27a^{\frac{5}{4}}$; d) $12x^{\frac{5}{6}}$; e) $\dfrac{1}{3b}$; f) $x^{\frac{71}{120}}$;

g) $x^{\frac{2}{3}}\cdot y^{\frac{5}{2}}$; h) $x^{\frac{19}{2}}$; i) 64; j) $y^{\frac{5}{4}}$; k) $x^{\frac{3b}{2}}$.

IV. 23. a) $2-\sqrt{3}$; b) $-\sqrt{6}-\sqrt{2}+2$; c) $\dfrac{1}{3}$; d) $\dfrac{3}{2}$; e) $\sqrt{2}$; f) 2; g) a;

h) $\dfrac{x(a+b)}{ab}\Big|_{x=\sqrt{ab}}=\dfrac{a+b}{\sqrt{ab}}$.

IV. 24. a) $-\dfrac{b}{a}$; b) $\dfrac{a}{b}$; c) $\dfrac{x^2-\sqrt{x^4-a^4}}{a^2}$, $-\dfrac{m}{n}$; d) $\dfrac{a^n-b^n}{a^n+b^n}$.

IV. 25. a) $\left(\sqrt[3]{x}-1\right)\left(\sqrt[3]{x^2}+\sqrt[3]{x}+1\right)\left(\sqrt[3]{x}+1\right)\left(\sqrt[3]{x^2}-\sqrt[3]{x}+1\right)=(x-1)(x+1)$.

b) $\left(\sqrt{x^3}-2\cdot\sqrt[4]{y}\right)$; c) $2\cdot\sqrt[3]{xy}$; d) $3\sqrt{2}$.

IV. 28. a) $f^{-1}(x)=x^2-1$, $x\geq0$; b) $f^{-1}(x)=2-x^2$, $x\geq0$;

c) $f^{-1}(x)=\begin{cases}2x+6 & \text{if } x<3\\ x^2+3 & \text{if } x\geq3\end{cases}$; d) $f^{-1}(x)=\begin{cases}-x^2 & \text{if } x<0\\ x^2 & \text{if } x\geq0\end{cases}$;

e) $f(x)=2\sqrt{x-1}$, $x\geq2$, then $f^{-1}(x)=\dfrac{1}{4}\left(x^2+4\right)$, $x\geq2$;

f) $f(x)=\begin{cases}2\sqrt{1-x^2} & \text{if } 0\leq x\leq\dfrac{\sqrt{2}}{2}\\[2mm] 2x & \text{if } x>\dfrac{\sqrt{2}}{2}\end{cases}$, $x\geq0$ then

$$f^{-1}(x) = \begin{cases} \dfrac{1}{2}\sqrt{4-x^2} & \text{if } \sqrt{2} \le x \le 2 \\[2mm] \dfrac{1}{2}x & \text{if } x > 2 \end{cases}.$$

IV. 29. a) $y \in (-1,1)$; b) $f(x) = \begin{cases} 0 & \text{if } x < 1 \\ 2(x-1) & \text{if } x \ge 1 \end{cases}$, $y \ge 0$.

IV. 30. a) $2x - 1 + \sqrt{(x-4)^2} = 2x - 1 + |x-4| = \begin{cases} 3x - 5 & \text{if } x > 4 \\ x + 3 & \text{if } x \le 4 \end{cases}$;

b) $x - 3 + \sqrt{(x-3)^2} = x - 3 + |x-3| = \begin{cases} 2(x-3) & \text{if } x > 3 \\ 0 & \text{if } x \le 3 \end{cases}$;

c) $(x^2 - 4)\dfrac{|x-2| + 4}{|x-2|} = \begin{cases} x^2 - 4x - 12 & \text{if } x < 2 \\ (x+2)^2 & \text{if } x > 2 \end{cases}$;

d) $\dfrac{|x-1| \cdot |x|}{x^2 - |x| - x + 1} = \begin{cases} \dfrac{x^2 - x}{x^2 + 1} & \text{if } x < 0 \\[2mm] \dfrac{x}{1-x} & \text{if } 0 \le x < 1 \\[2mm] \dfrac{-x}{1-x} & \text{if } x > 1 \end{cases}$;

e) $\dfrac{|2x-3| + 6}{(2x-3)} \cdot \dfrac{|2x-3|}{|x|} = \begin{cases} 9x^{-1} - 2 & \text{if } x < 0 \\ 2 - 9x^{-1} & \text{if } 0 < x < 1.5 \\ 3x^{-1} + 2 & \text{if } x > 1.5 \end{cases}$;

f) $2x + |x| + |x+1| + 1 = \begin{cases} 0 & \text{if } x < -1 \\ 2x + 2 & \text{if } -1 \le x \le 0 \\ 4x + 2 & \text{if } x > 0 \end{cases}$;

g) $\dfrac{(x+1) \cdot |x-3|}{(x-1)^2 - 4|x-2|} = \begin{cases} \dfrac{3 + 2x - x^2}{x^2 + 2x - 7} & \text{if } x < 2 \quad x \ne -1 \pm 2\sqrt{2} \\[2mm] \dfrac{x+1}{3-x} & \text{if } 2 \le x < 3 \\[2mm] \dfrac{x+1}{x-3} & \text{if } x > 3 \end{cases}$;

h) $4 + \big||x-2| - 2\big| + |x-4| = \begin{cases} -2x + 8 & \text{if } x < 0 \\ 8 & \text{if } 0 \le x \le 2 \\ -2x + 12 & \text{if } 2 < x \le 4 \\ 2x - 4 & \text{if } x > 4 \end{cases}$;

i) $\dfrac{|x-3|+3(x-1)}{2|x-4|-|x-5|}=\begin{cases}\dfrac{2x}{3-x} & \text{if } x\in(-\infty,3)\\[2mm]\dfrac{4x-6}{3-x} & \text{if } x\in(3,4]\\[2mm]\dfrac{4x-6}{3x-13} & \text{if } x\in(4,5]\setminus\left\{\dfrac{13}{3}\right\}\\[2mm]\dfrac{4x-6}{x-3} & \text{if } x\in(5,\infty)\end{cases}.$

IV. 31. a) $x\in\left[\dfrac{1}{2},\infty\right)$; b) $x\in\mathbf{R}$; c) $x\in\left(-\infty,-\dfrac{1}{3}\right]$; d) $x\in\mathbf{R}$;

e) $x\in\left(-\infty,\dfrac{1}{3}\right]$; f) $x\in\mathbf{R}$; g) $x\in(-\infty,3]\cup[4,\infty)$; h) $x\in[-2,2]$;

i) $x\in[-4,-1]\cup[2,\infty)$; j) $x\in[1,3]$; k) $x\in(-\infty,-1]\cup[1,\infty)$;

l) $x\in[2,5]$; m) $x\in(-\infty,2]$; n) $x\in[0,4]$.

IV. 32. a) $\sqrt{32}>\sqrt{27}$; b) $\sqrt[4]{405}>\sqrt[4]{216}$; c) $\sqrt{2205}>\sqrt{2016}$;

d) $\sqrt{\dfrac{48}{14}}>\sqrt{\dfrac{47}{14}}$; e) $\sqrt{\dfrac{4}{75}}=-\sqrt{\dfrac{24}{450}}$; f) $-\sqrt[3]{378}<-\sqrt[3]{144}$;

g) $\sqrt[3]{\dfrac{2}{1}}>\sqrt[3]{\dfrac{7}{325}}$.

IV. 33. a) $\sqrt[12]{216},\ \sqrt[12]{625},\ \sqrt[12]{628}$;   b) $\sqrt{4},\ \sqrt[4]{15},\ \sqrt[3]{11}$;

c) $\sqrt[12]{777},\ \sqrt[3]{6},\ \sqrt{5}$; d) $\sqrt[12]{250},\ \sqrt[6]{16},\ 2\cdot\sqrt[3]{3}$;

e) $\sqrt[15]{448},\ \sqrt[10]{20},\ \sqrt[5]{50}$; f) $\sqrt[3]{7},\ \sqrt[12]{4000},\ \sqrt[4]{17}$; g) $\sqrt[5]{5},\ \sqrt{2},\ \sqrt[3]{3}$.

IV. 37. a) $\dfrac{7-4\sqrt{3}}{\sqrt[3]{26-15\sqrt{3}}}=\dfrac{\left(2-\sqrt{3}\right)^2}{\sqrt[3]{\left(2-\sqrt{3}\right)^3}}=\dfrac{\left(2-\sqrt{3}\right)^2}{2-\sqrt{3}}=2-\sqrt{3}.$

IV. 39. $S_1=-1+\sqrt{n}$; $S_2=1-\dfrac{1}{\sqrt{n+1}}$; $S_3=1-\dfrac{1}{\sqrt{n}}$;

$S_4=\dfrac{\sqrt{3}-\sqrt{1}}{\sqrt{2}}+\dfrac{\sqrt{5}-\sqrt{3}}{\sqrt{2}}+\ldots+\dfrac{\sqrt{2n+1}-\sqrt{2n-1}}{\sqrt{2}}=\dfrac{\sqrt{2n+1}-1}{\sqrt{2}}$;

$\dfrac{n}{\sqrt{n!}\cdot\left(n+1+\sqrt{n+1}\right)}=\dfrac{n\left(n+1-\sqrt{n+1}\right)}{\sqrt{n!}\cdot\left(n^2+n\right)}=$

$=\dfrac{n+1-\sqrt{n+1}}{(n+1)\cdot\sqrt{n!}}=\dfrac{1}{\sqrt{n!}}-\dfrac{\sqrt{n+1}}{(n+1)\cdot\sqrt{n!}}=\dfrac{1}{\sqrt{n!}}-\dfrac{1}{\sqrt{(n+1)!}}$,

Finally $S_5 = 1 - \dfrac{1}{\sqrt{(n+1)!}}$.

IV. 40. a) $\sqrt{x \cdot \sqrt[3]{\sqrt{x^3}}} = \sqrt{x \cdot \sqrt[6]{x^3}} = \sqrt{x \cdot \sqrt{x}} = \sqrt[4]{x^3}$, **b)** $\dfrac{1}{\sqrt[3]{a}}$;

c) $\sqrt[4]{\dfrac{x}{32}} \cdot \dfrac{\sqrt[4]{2}\left(\sqrt[4]{2^3 x}+2\right)}{\sqrt[4]{x}\left(\sqrt[4]{x}-\sqrt[4]{2}\right)} \times \dfrac{\sqrt[4]{2}\left(\sqrt[4]{2^3}-\sqrt[4]{x^3}\right)}{\sqrt{x}+\sqrt[4]{2x}+\sqrt{2}} =$

$= \sqrt[4]{\dfrac{x}{32}} \times \dfrac{\sqrt[4]{2^2}\left(\sqrt[4]{2^3 x}+2\right)\left(\sqrt[4]{2^3}-\sqrt[4]{x^3}\right)}{\sqrt[4]{x}\left(\sqrt[4]{x^3}-\sqrt[4]{2^3}\right)} = \sqrt[4]{\dfrac{x}{32}} \times \dfrac{-\sqrt[4]{2^2}\left(\sqrt[4]{2^3 x}+2\right)}{\sqrt[4]{x}} =$

$= \dfrac{\sqrt[4]{x}}{\sqrt[4]{2^5}} \times \dfrac{-\sqrt[4]{2^5}\left(\sqrt[4]{x}+\sqrt[4]{2}\right)}{\sqrt[4]{x}} = -\sqrt[4]{x}-\sqrt[4]{2}$; **d)** $a-b$; **e)** $x^2 \cdot \sqrt[4]{a}$.

IV. 41. a) $\sqrt{ab}$; **b)** $\dfrac{1}{2\sqrt{ab}}$; **c)** $\dfrac{\sqrt{a}}{\sqrt{2a+1}-\sqrt{a}}$;

d) $\left(\sqrt{x}-2\right)\left(\sqrt{x}-3\right)$; **e)** $2 \cdot \sqrt[4]{\dfrac{y}{x^2}}$. **IV. 42. a)** 1; **b)** $\dfrac{1}{\sqrt[8]{x-2}}$;

c) $\left(\sqrt[6]{1+\dfrac{1}{\sqrt{x}}} \cdot \sqrt[3]{\sqrt{x}+1} - \sqrt[3]{\sqrt{x}-1} \cdot \sqrt[6]{1-\dfrac{1}{\sqrt{x}}}\right)^{-2} \dfrac{\sqrt[12]{x}}{\sqrt{x}+\sqrt{x-1}} =$

$= \left(\dfrac{\sqrt[6]{\left(\sqrt{x}+1\right)^3}}{\sqrt[12]{x}} - \dfrac{\sqrt[6]{\left(\sqrt{x}-1\right)^3}}{\sqrt[12]{x}}\right)^{-2} \dfrac{\sqrt[12]{x}}{\sqrt{x}+\sqrt{x-1}} =$

$= \left(\dfrac{\sqrt[12]{x}}{\sqrt{\sqrt{x}+1}-\sqrt{\sqrt{x}-1}}\right)^2 \dfrac{\sqrt[12]{x}}{\sqrt{x}+\sqrt{x-1}} = \left(\dfrac{\sqrt[12]{x}\left(\sqrt{\sqrt{x}+1}+\sqrt{\sqrt{x}-1}\right)}{\left(\sqrt{\sqrt{x}+1}\right)^2-\left(\sqrt{\sqrt{x}-1}\right)^2}\right)^2 \dfrac{\sqrt[12]{x}}{\sqrt{x}+\sqrt{x-1}} =$

$= \dfrac{\sqrt[12]{x^2}\left(\sqrt{x}+1+2\sqrt{x-1}+\sqrt{x}-1\right)}{4} \dfrac{\sqrt[12]{x}}{\sqrt{x}+\sqrt{x-1}} = \dfrac{\sqrt[12]{x^3}\left(2\sqrt{x}+2\sqrt{x-1}\right)}{4\left(\sqrt{x}+\sqrt{x-1}\right)} = \dfrac{\sqrt[4]{x}}{2}$

d) $1-\sqrt{x^2-1}$; **e)** $\dfrac{\sqrt{a^2-4}}{\sqrt{a}}$.

IV. 43. a) $\dfrac{x}{2y-x}$; **b)** $\dfrac{-16x\sqrt{x}}{(1-x)\left(1-x^2\right)}$; **c)** 1; **d)** $x\sqrt{2}$;

e) $\dfrac{1}{\sqrt[4]{x^2-1}}$; **IV. 44. a)** $\dfrac{1}{x^2}$; **b)** 2; **c)** $y(x+y)$; **d)** 0; **e)** $\dfrac{\sqrt[24]{y}}{\sqrt{x}}$.

IV. 45. a) $\dfrac{2x^4}{x^8-16y^8}$, 3; b) ab, 1; c) $\dfrac{a-b}{2ab}$, $-\dfrac{7}{24}$;

d) $\sqrt{\left(a^3+b^3\right)^2}-a$, 0; e) $\dfrac{-\sqrt{ax}}{2x-a}$, 1; f) $-\dfrac{2\sqrt{x}}{3}$, -2;

g) $\dfrac{4x}{x-4}$, 20; h) $-x$, 1.

IV. 46. a) $-\sqrt{\dfrac{3x-2}{3x+2}}$; b) $x^3\cdot\sqrt[4]{a}$; c) $\sqrt[3]{\dfrac{2n}{1+n}}$; d) $\sqrt{\dfrac{a}{a+4b}}$;

IV. 47. a) 1; b) $\sqrt[4]{x}$; c) 1; d) $\pm\dfrac{1}{\sqrt{x}}$; e) $\dfrac{\sqrt{x}}{2}$.

IV. 48. a) $\sqrt{2}+\sqrt{3}$; b) $\sqrt{5}-1$; c) $\sqrt{7}-\sqrt{3}$;

d) $\sqrt{4\sqrt{2}+2\sqrt{6}}=\sqrt[4]{2}\sqrt{4+2\sqrt{3}}=\sqrt[4]{2}\left(\sqrt{3}+1\right)$; e) $3+2\sqrt{2}$;

f) $2-\sqrt{3}$; g) $5-\sqrt{3}$; h) $-2+\sqrt{5}$; i) $a-\sqrt{2-a^2}$.

IV. 49. a) $x\in[1,\infty)$, $x=9$; b) $x\in\left(-\infty,\dfrac{3}{4}\right]$, $x=-\dfrac{13}{4}$; c) $x=5$;

d) $x\in\left(-\infty,\dfrac{3}{4}\right]$; $x=-\dfrac{5}{8}$; e) $x\in\left[-3,-\sqrt{\dfrac{7}{2}}\right]\cup\left[\sqrt{\dfrac{7}{2}},\infty\right)$, $x_1=-2$, $x_2=8$;

f) $x\in\left[-\dfrac{1}{4},1\right]$, $x=\dfrac{4}{9}$; g) $x\in\mathbf{R}$, $x_1=-4$, $x_2=2$;

h) $x\in[2,\infty)$; $x_1=6$, $x_2=18$; i) $x\in[-1,\infty)$, $x=\dfrac{5}{4}$;

j) $x\in(-\infty,4]$, $x=1$; k) $x_1=-4$, $x_2=0$, $x_3=1$; l) $a=2$.

IV. 50. a) $x\in[2,\infty)$, $x-2+2\sqrt{x-2}\sqrt{2x-3}+2x-3=25$,

$2\sqrt{(x-2)(2x-3)}=30-3x$. $30-3x\geq0\Rightarrow x\in(-\infty,10]$ then the set of

permissible values is $x\in[2,\infty)\cap(-\infty,10]=[2,10]$. Squaring both sides of the last equation, we obtain

$4\left(2x^2-7x+6\right)=(30-3x)^2$, $x_1=6$ and $x_2=146$ (not solution).

b) $x\in\left[\dfrac{1}{2},2\right]$, $x_1=1$, $x_2=\dfrac{17}{9}$; c) $x\in\left[-\dfrac{1}{3},\dfrac{4}{3}\right]$; $x_1=0$, $x_2=1$;

d) $x\in[1,6]$; $x_1=2$, $x_2=5$; e) $x\in[0,\infty)$, $x=1$; f) $x\in\left[-\dfrac{1}{2},1\right]$, $x_1=0$,

$x_2 = \dfrac{8}{9}$; g) $x \in [-2, \infty)$, $x = 2$; h) $x \in [0, \infty)$, $x = \dfrac{1}{2}$; i) $x \in [3, \infty)$, $x = 3$.

IV. 51. a) $x \in \left[-\dfrac{1}{2}, 2\right]$, $x = \dfrac{3}{2}$; b) $x \in (-\infty, -4] \cup [4, \infty)$, $x_1 = 5$, $x_2 = -\dfrac{52}{5}$;

c) $x \in [0, \infty)$, $x = 0$; d) $x \in \left[\dfrac{6}{5}, \infty\right)$, $x = \dfrac{11}{5}$; e) $x \geq 4$, $x = 5$.

IV. 52. a) $x = 2$; b) $x = -1$; c) $x_1 = 7$, $x_2 = 8$; d) $x = 0$; e) $x = 4$.

IV. 53. a) $x = 3$; b) $x = 4$; c) $x = 14$.

IV. 54. a)) $x = -1$; b) $x_1 = 1$, $x_2 = 36$; c) $x_1 = 6$, $x_2 = -2$.

IV. 55. a) $x - 2 \geq 0 \Rightarrow x \in [2, \infty)$, $\sqrt{x-2} = t$, $t \geq 0$

$\sqrt{t^2 + 2t + 1} - \sqrt{t^2 - 2t + 1} = 1$, $|t + 1| - |t - 1| = 1$. I) $t \in (1, \infty)$ $t + 1 - t + 1 = 1$

impossible. II) $t \in [0,1)$ $t + 1 + t - 1 = 1 \Rightarrow t = \dfrac{1}{2}$, $x = 2.25$; b) $x + 3 \geq 0$,

$\sqrt{x+3} = t \geq 0$, $\sqrt{t^2 + 4t + 4} + \sqrt{t^2 - 4t + 4} = 6 \Leftrightarrow t + 2 + |t - 2| = 6$.

I) $t \in [2, \infty)$, $t + 2 + t - 2 = 6 \Leftrightarrow t = 3$, $\sqrt{x+3} = 3 \Rightarrow x = 6$.

II) $t \in [0, 2)$, $t + 2 - t + 2 = 6$ which is impossible; c) $x \in [-2, -1]$;

d) $x \in [0, 7]$; e) $x \in [5, 10]$; f) $x = 34$; g) $x \in [1, 2]$; h) $x \in [a, a+1]$.

IV. 56. a) $x \in [-1, \infty)$, $\sqrt[6]{x+1} = t$, $2\sqrt{5t + 4} = \sqrt{2t - 1} + \sqrt{20t + 5}$. Squaring

both sides we get $20t + 16 = 2t - 1 + 20t + 5 + 2\sqrt{(2t-1)(20t+5)}$ and collecting

terms, we obtain $\sqrt{(2t-1)(20t+5)} = 6 - t$. First side of this equation is

positive, this means $6 - t \geq 0 \Rightarrow t \in (-\infty, 6]$, then the set of permissible

values is $x \in [-1, \infty) \cap (-\infty, 6] = [-1, 6]$. Squaring both sides we obtain

$39t^2 + 2t - 41 = 0$ whose roots are $t_1 = -\dfrac{41}{39}$ (impossible) and $t_2 = 1$. From

(1) $\sqrt[6]{x+1} = 1 \Rightarrow x = 0$ is the unique solution of the initial equation;

b) $x = 1$; c) $x = 2$.

IV. 57. a) $x = \pm 4$; b) $x = a$; c) $x = \dfrac{a(a-1)}{a+1}$; d) $x_{1,2} = -a\left(1 \pm \sqrt{\dfrac{a}{a-1}}\right)$;

e) $\dfrac{1}{(\sqrt{x}-2)(\sqrt{x}-3)} + \dfrac{1}{(\sqrt{x}-1)(\sqrt{x}-2)} + \dfrac{1}{\sqrt{x}(\sqrt{x}-1)} = \dfrac{1}{\sqrt{x}(\sqrt{x}-3)}$.

$\sqrt{x}(\sqrt{x}-1) + \sqrt{x}(\sqrt{x}-3) + (\sqrt{x}-2)(\sqrt{x}-3) = (\sqrt{x}-1)(\sqrt{x}-2)$.

Finally $x = 1$ and $x = 4$. The equation doesn't have any solution (because the domain).

IV. 58. a) $\sqrt[3]{3x-1} = a$ and $\sqrt{3x+7} = b$, $a+b = 6$ and $a^3 - b^2 = -8$,

$(a-2)(a^2 + a + 14) = 0 \Rightarrow a = 2$, $x = 2$;

b) $x_1 = -2$, $x_2 = 7$, $x_3 = -1$; c) $x = 2$;

d) $\sqrt[3]{12 + x^2} = a$ and $\sqrt[3]{12 - x^2} = b$. Therefore $a + b = 2\sqrt[3]{3}$ and $a^3 + b^3 = 24$.

$x = \pm 2\sqrt{3}$; e) $x = 25$; f) $x = \square 1$.

IV. 59. a) $\sqrt[3]{x+1} + \sqrt[3]{x+2} = -\sqrt[3]{x+3}$. Cubing both sides of the last equation

$x + 1 + x + 2 + 3\sqrt[3]{x+1}\sqrt[3]{x+2}\left(\sqrt[3]{x+1} + \sqrt[3]{x+2}\right) = -x - 3$

$3 \cdot \sqrt[3]{x+1}\sqrt[3]{x+2}\left(-\sqrt[3]{x+3}\right) = -3x - 6 \Rightarrow \sqrt[3]{x+1}\sqrt[3]{x+2}\sqrt[3]{x+3} = x + 2$ Cubing

again both sides $(x+1)(x+2)(x+3) = (x+2)^3 \Rightarrow x = -2$;

b) $x_1 = -2$, $x_2 = -1.5$, $x_3 = -1$; c) $x_1 = 6, x_{2,3} = 6 \pm \dfrac{12}{7}\sqrt{21}$;

d) $x \in [0, \infty)$. Dividing both sides of the equation

by $\sqrt[3]{x-4} = \sqrt[3]{\left(\sqrt{x} - 2\right)\left(\sqrt{x} + 2\right)}$ we obtain

$2 \cdot \sqrt[3]{\dfrac{\left(\sqrt{x} - 2\right)^2}{\left(\sqrt{x} - 2\right)\left(\sqrt{x} + 2\right)}} - \sqrt[3]{\dfrac{\left(\sqrt{x} + 2\right)^2}{\left(\sqrt{x} - 2\right)\left(\sqrt{x} + 2\right)}} = 3 \Rightarrow$

$2 \cdot \sqrt[3]{\dfrac{\sqrt{x} - 2}{\sqrt{x} + 2}} - \sqrt[3]{\dfrac{\sqrt{x} + 2}{\sqrt{x} - 2}} = 3$, $\sqrt[3]{\dfrac{\sqrt{x} - 2}{\sqrt{x} + 2}} = t$, $2t - \dfrac{1}{t} = 3$, then $t_1 = 1$

and $t_2 = \dfrac{1}{2}$, $x = \left(\dfrac{18}{7}\right)^2$. IV. 60. a) $x = 3$; b) $x = 2$;

IV. 61. a) $x \geq b^2$ and $x \geq a^2$. We have $\dfrac{a^2 - b^2}{\sqrt{x - b^2} + \sqrt{x - a^2}} = a - b \Rightarrow$

$\sqrt{x - b^2} + \sqrt{x - a^2} = a + b$. Adding the last equation with the given equation,

we obtain $\sqrt{x - b^2} = a$. Therefore $x = a^2 + b^2$.

b) Multiply the given equation in both sides with the term

$\sqrt{4x^2 - 5x + 5} + \sqrt{4x^2 - 5x + 2}$. Therefore

$\left(4x^2 - 5x + 5\right) - \left(4x^2 - 5x + 2\right) = \sqrt{4x^2 - 5x + 5} + \sqrt{4x^2 - 5x + 2}$. Thus (1)

$\sqrt{4x^2 - 5x + 5} + \sqrt{4x^2 - 5x + 2} = 3$. Adding the given equation with (1) obtain

$\sqrt{4x^2 - 5x + 5} = 2$. The solutions are $x_1 = 1$, $x_2 = 0.25$;

c) Denote $\sqrt[3]{x^2 + 2} = m$, $\sqrt[3]{4x^2 + 3x - 2} = n$,

$\sqrt[3]{3x^2 + x + 5} = p$, $\sqrt[3]{2x^2 + 2x - 5} = q$, and we have $\begin{cases} m + n = p + q \\ m^3 + n^3 = p^3 + q^3 \end{cases} \Rightarrow$

$mn(m + n) = pq(p + q)$

$\begin{cases} m + n = p + q \\ (m + n)^3 - 3mn(m + n) = (p + q)^3 - 3pq(p + q) \end{cases} \Rightarrow$

I) If $m + n = p + q = 0 \Rightarrow p = -q \Rightarrow 5x^2 + 3x = 0 \Rightarrow \begin{cases} x_1 = 0 \\ x_2 = -\dfrac{3}{5} \end{cases}$

II) If $m + n = p + q \neq 0 \Rightarrow mn = pq \Rightarrow$

$\Rightarrow \dfrac{x^2 + 2}{2x^2 + 2x - 5} = \dfrac{3x^2 + x + 5}{4x^2 + 3x - 2} \Rightarrow (x^2 + 2x - 7) \cdot (2x^2 + x + 3) = 0$.

We have

1) $x^2 + 2x - 7 = 0 \Rightarrow x_{3,4} = -1 \pm 2\sqrt{2}$

2) $2x^2 - 3x + 4 = 0 \Rightarrow x_{5,6} \notin \mathbb{R}$.

Therefore $x \in \left\{ -1 - 2\sqrt{2}, -\dfrac{3}{5}, 0, -1 + 2\sqrt{2} \right\}$;   d) $x \in \left\{ -1, -\dfrac{3}{5}, -\dfrac{1}{3} \right\}$.

IV. 62. a) $\begin{cases} 28 + 5x \geq 0 \\ 54 - 5x \geq 0 \end{cases} \Leftrightarrow -\dfrac{28}{5} \leq x \leq \dfrac{54}{5}$.

Denote $\sqrt[4]{54 - 5x} = t \geq 0$ and $\sqrt[4]{28 + 5x} = z \geq 0$. We have

$\begin{cases} t + z = 4 \\ t^4 + z^4 = 82 \end{cases} \Leftrightarrow \begin{cases} t + z = 4 \\ \left[(t + z)^2 - 2tz \right]^2 - 2t^2 z^2 = 82 \end{cases} \Rightarrow$

$(16 - 2tz)^2 - 2t^2 z^2 = 82 \Rightarrow t^2 z^2 - 32tz + 87 = 0$. Therefore $tz = 3$ or

$tz = 29$. Now solve the systems $\begin{cases} t + z = 4 \\ tz = 3 \end{cases}$ and $\begin{cases} t + z = 4 \\ tz = 29 \end{cases}$. The second system

does not have real solutions; the first system is equivalent to the systems

276

$$\begin{cases} t = 1 \\ z = 3 \end{cases} \text{ and } \begin{cases} t = 3 \\ z = 1 \end{cases}. \text{ Since } x = \frac{t^4 - 28}{5}, \text{ the final solutions of the initial}$$

equation are

$x_1 = -\dfrac{27}{5}$ and $x_2 = \dfrac{53}{5}$; b) x = 23; c) x = 8; d) $x_1 = 1$, $x_2 = 2$;

e) $x_1 = 5$, $x_2 = 261$.

IV. 63. a) $x = 3^{10}$; **b)** x = 5; **c)** $x_1 = 32$, $x_2 = 243$; **d)** x = 8;

e) x = 8; **f)** x = 32; **g)** $x = \pm 1$; **h)** $x = \pm \dfrac{24}{25}$; **i)** x = 12;

j) $x > 0$, $\sqrt{x} = t > 0$,

$5t^3 + t - 42 = 0 \Rightarrow (t - 2)(5t^2 + 10t + 21) = 0$, $x = 4$;

k) $x^{\frac{2}{15}} = t$, $5t^3 + 3t^2 - 8 = 0 \Rightarrow (t - 1)(5t^2 + 8t + 8) = 0$, $x = \pm 1$.

IV. 64. a) $x \in (-1, 1)$. Dividing both members of the equation by $\sqrt[n]{1 - x^2}$,

$\sqrt[n]{\dfrac{1 + x}{1 - x}} + \sqrt[n]{\dfrac{1 - x}{1 + x}} = 4$, $\sqrt[n]{\dfrac{1 + x}{1 - x}} = t$, $t + \dfrac{1}{t} = 4$,

$t_{1,2} = 2 \pm \sqrt{3}$, $x = \dfrac{(2 \pm \sqrt{3})^n - 1}{(2 \pm \sqrt{3})^n + 1}$; **b)** $x = \dfrac{225}{289}$; **c)** $x = \dfrac{8}{27}$;

d) x = 3; **e)** x = 0; **f)** x = 1; **g)** $x_1 = \dfrac{314}{63}$, $x_2 = \dfrac{4369}{278}$.

IV. 65. a) Domain is $x \in (-1, \infty)$. The equation becomes

$\sqrt{x^2 + 6x} + \sqrt{x^2 + 6x + 5} = 4 + \sqrt{21}$. Denote $x^2 + 6x = t$, then

$\sqrt{t} + \sqrt{t + 5} = 4 + \sqrt{21}$. Squaring both sides of the equation and collecting the

terms, we get $\sqrt{t(t + 5)} - 4\sqrt{21} = 16 - t$. Squaring again and collecting the

terms, we have $8\sqrt{21\, t(t + 5)} = 80 + 37t$ and again squaring in both sides, it

yields $25t^2 - 800t + 6400 = 0$. The root is t = 16 and therefore only x = 2 is the

solution of the given equation; **b)** $x \in (0, 3]$, $x = 3\sqrt{\dfrac{2}{7}}$; **c)** $x_1 = 2$, $x_2 = 9$,

$x_{3,4} = \dfrac{11}{2} \pm \dfrac{7\sqrt{141}}{12}$; **d)** $x_1 = 5$, $x_2 = 24$; **e)** x = 1.

IV. 66. a) $x \geq 0$ and $2 - \sqrt{2 + \sqrt{2 + x}} \geq 0 \Rightarrow x \in [0, 2]$.

In this case denoting $x = 2\cos t$, $t \in \left[0, \dfrac{\pi}{2}\right]$, we obtain

$$\sqrt{2-\sqrt{2+\sqrt{2+x}}} = \sqrt{2-\sqrt{2+\sqrt{2+2\cos t}}} = 2\sin\frac{t}{8} \text{ and similarly}$$

$$\sqrt{6+3\sqrt{2+\sqrt{2+x}}} = 2\sqrt{3}\cos\frac{t}{8}. \text{ The initial equation becomes}$$

$$2\sqrt{3}\cos\frac{t}{8}+2\sin\frac{t}{8} = 4\cos t \Rightarrow \frac{\sqrt{3}}{2}\cos\frac{t}{8}+\frac{1}{2}\sin\frac{t}{8} = \cos t \Rightarrow \cos\left(\frac{\pi}{6}-\frac{t}{8}\right) = \cos t$$

with the solution $t = \dfrac{4\pi}{27}$ or $x = 2\cos\dfrac{4\pi}{27}$;

b) Use $\sqrt{1\pm\sqrt{x(2-x)}} = \sqrt{\dfrac{1}{2}(2-x)}\pm\sqrt{\dfrac{1}{2}x}$, $x \in [0,1]$.

IV. 67. a) We have $\left(\sqrt{5}-2\right)^n+\sqrt{m} = \sqrt{m+1}$, squaring both sides and

collecting the terms we obtain $2\left(\sqrt{5}-2\right)^n\sqrt{m} = 1-\left(\sqrt{5}-2\right)^{2n}$ or

$2\sqrt{m} = \left(\sqrt{5}+2\right)^n-\left(\sqrt{5}-2\right)^n$. Therefore $m = \dfrac{1}{4}\left[\left(\sqrt{5}+2\right)^n-\left(\sqrt{5}-2\right)^n\right]^2$;

b) $m = \dfrac{1}{4}\left[\left(\sqrt{17}+4\right)^n-\left(\sqrt{17}-4\right)^n\right]^2+1$; c) $m = \dfrac{1}{4}\left[\left(5-\sqrt{27}\right)^n-\left(5\sqrt{2}+7\right)^n\right]^2$;

d) $m = \dfrac{1}{4}\left[\left(\sqrt{2}+1\right)^n-\left(\sqrt{2}-1\right)^n\right]^2+1$; e) $m = 4n^2+4n$.

IV. 68. a) $x \in \mathbf{N}$, $x > 2$, we obtain the equation $\dfrac{3}{2x+2}+\dfrac{3}{(2x+2)x} = \dfrac{9}{2x^2}$,

$x = 3$; b) $x = 2$; c) $x = 3$; d) $x = 4$; e) $10x-4 \geq 0$, $11-x \in \mathbf{N}^*$, $2x-1 \in \mathbf{N}^*$,

$\dfrac{x+6}{4} \in \mathbf{N}^*$, therefore $x \in \{2,6\}$ and $x = 2$ verifies the equation;

f) $x = 2$; g) $x = 5$.

IV. 69. a) $x \in [1,2)$; b) $x \in (-\infty,-2]$; c) $x \in \left(\dfrac{3+2\sqrt{21}}{2},\infty\right)$; d) $x \in (2,\infty)$;

e) $x \in (-\infty,0]\cup(4.5,\infty)$; f) $x \in \left(\dfrac{-5-\sqrt{5}}{2},0\right]$; g) $x \in \left[0,\dfrac{3-\sqrt{13}}{2}\right)\cup(3,\infty)$;

h) $x \in \left[-2,\dfrac{7}{4}\right)$; i) $x \in [49,\infty)$; j) $x \in \left(-\infty,\dfrac{1}{3}\right)\cup(2,\infty)$; k) $x \in (11,\infty)$;

l) $x \in (2,8)$; m) $x \in (-\infty,0)\cup(1,2)$.

IV. 70. a) (4, 9), (9, 4); b) (1, 9), (9, 1); c) (1, 4), (4, 1); d) (3, 2);
e) (3, 1.5), (6, 3); f) (27, 8), (8, 27); g) (1, 16), (16, 1); h) (17/2, 5/3);

i) $(53, 28)$, $(317, -308)$.

IV. 71. a) $x_1 = 2$, $x_2 = 4$,...., $x_n = 2n$;

b) $(\sqrt{x_1 - 1} - 1)^2 + (\sqrt{x_2 - 2^2} - 2^2)^2 + ... + (\sqrt{x_n - n^2} - n^2)^2 = 0.$

IV. 72. $S = \dfrac{n \cdot (2n+1)}{4}$, $\left(\sqrt{x_1 - 1} - \dfrac{1}{2}\right)^2 + \left(\sqrt{x_2 - 2} - \dfrac{1}{2}\right)^2 + ... + \left(\sqrt{x_n - n} - \dfrac{1}{2}\right)^2 = 0,$

$x_1 = \dfrac{5}{4}, x_2 = \dfrac{9}{4}, ..., x_n = \dfrac{4n+1}{4}.$

IV. 73. For the given identity, adding the nominators to the denominators, we obtain

$$\frac{\sqrt{a}}{\sqrt{b} + \sqrt{c}} = \frac{\sqrt{b}}{\sqrt{c} + \sqrt{a}} = \frac{\sqrt{c}}{\sqrt{a} + \sqrt{b}} \quad \text{or}$$

$$\frac{\sqrt{a}}{\sqrt{b} + \sqrt{c}} = \frac{\sqrt{b}}{\sqrt{c} + \sqrt{a}} = \frac{\sqrt{c}}{\sqrt{a} + \sqrt{b}} = \frac{\sqrt{a} + \sqrt{b} + \sqrt{c}}{2(\sqrt{a} + \sqrt{b} + \sqrt{c})} = \frac{1}{2}.$$

From $\dfrac{\sqrt{a}}{\sqrt{b} + \sqrt{c}} = \dfrac{\sqrt{b}}{\sqrt{c} + \sqrt{a}}$ we get $(\sqrt{a} - \sqrt{c})(\sqrt{a} + \sqrt{b} - \sqrt{c}) = 0$. Since

$\dfrac{\sqrt{c}}{\sqrt{a} + \sqrt{b}} = \dfrac{1}{2}$, $\sqrt{a} + \sqrt{b} - \sqrt{c} = 0$ is impossible. Therefore $a = c$.

IV. 74. Squaring in both sides we obtain

$2(a + b) = a + c + a - c + 2\sqrt{(a+c)(a-c)}$ or $b = \sqrt{a^2 - c^2}$ and finally

$a^2 = b^2 + c^2$.

IV. 75. Denote $x = \sqrt{56 + \sqrt{56 + \sqrt{56 + ...}}}$,

then $x^2 = 56 + x$, with the solutions $x_1 = 8$ and $x_2 = -7$. Since x cannot be negative, $x = 8$ is the only solution. Similarly for

$y = \sqrt{12 + \sqrt{12 + \sqrt{12 + ...}}} = 3$ and $z = \sqrt{42 + \sqrt{42 + \sqrt{42 + ...}}} = 7.$

Then **A** $= \dfrac{8}{4} - 7 = -5$; **B** $= -5$.

Chapter V

QUADRATIC FUNCTIONS AND APPLICATIONS

V. 1. a) $(x-7)(x+7)$; **b)** $(5x-1)(5x+1)$; **c)** $(9x-2)(9x+2)$;

d) $5(2x+11)$; **e)** $(16x-35)(-2x+7)$; **f)** $(x+3)(x+4)$; **g)** $(x+3)(x-12)$;

h) $(x-2)(x+8)$; **i)** $(2x-1)(3x-4)$; **j)** $(7x-1)(x+2)$. **V. 2. a)** $(0,-6)$;

b) $\left(0,-5\sqrt{3}\right)$; **c)** $(0,4)$; **d)** $(0,0)$; **e)** $(0,-8)$; **f)** $(0,-3)$.

V. 3. a) $(-3,0)$, $(1,0)$, $x_V = -\dfrac{b}{2a} = \dfrac{x_1+x_2}{2} = -1$; **b)** $(-2,0)$, $(3,0)$, $x=-\dfrac{1}{2}$;

c) $(-7,0)$, $(3,0)$, $x=-2$; **d)** $(2,0)$, $(5,0)$, $x=\dfrac{7}{2}$; **e)** $(-5,0)$, $x=-5$;

f) $(x-2)^2 -11 = 0$, $x_{1,2} = 2\pm\sqrt{11}$, $x=2$; **g)** $x_1 = -\dfrac{b}{a}$, $x_2 = \dfrac{c}{a}$, $x = \dfrac{c-b}{2a}$.

V. 4. a) $f(x) = (x-1)^2 +7$; **b)** $f(x) = -(x-2.5)^2 +8.25$;

c) $f(x) = 0,\bar{6}(x-3)^2 -5$; **d)** $f(x) = 0.5(x-0.2)^2 -0.28$;

e) $f(x) = -3(x-0.1\bar{6})^2 +0.18\bar{3}$; **f)** $f(x) = \left(x-\dfrac{1}{15}\right)^2 +\dfrac{44}{45}$.

V. 5. a) $x=\dfrac{3}{4}$; **b)** $x=\dfrac{1}{6}$; **c)** $x=\dfrac{3}{8}$; **d)** $x=\dfrac{1}{5}$; **e)** $x=\dfrac{1}{8}$; **f)** $x=\dfrac{3}{4}$.

V. 6. a) $y = -(x+2)^2 +7$; **b)** $y = \dfrac{1}{2}(x-5)^2 +3$; **c)** $y = -(x+1)^2$.

V. 7. a) $f(x) = -\dfrac{1}{4}(x-4)^2 +3$; **b)** $f(x) = 5(-x+4)^2 -3$.

V. 8. a) $x_1 =1$, $x_2 =\dfrac{c}{a}$; **b)** $x_1 = -1$, $x_2 = -\dfrac{c}{a}$; **c)** $x_1 = 2$, $x_2 =\dfrac{c}{2a}$.

V. 9. a) $k=-4$; **b)** $l=-2$; **c)** $m=1$ or $m=-5$; **d)** $n=1$ or $n=-2.5$.

V. 10. a) $y_{max} = \dfrac{2}{3}$; **b)** $y_{min} = -\dfrac{25}{8}$; **c)** $y_{max} = 6$; **d)** $y_{max} = 5$;

e) $y_{max} = \dfrac{14}{5}$; **f)** $y_{max} = -4$.

V. 11. **a)**

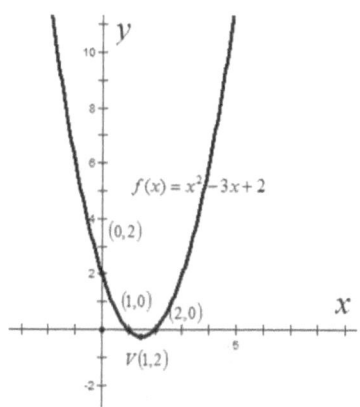

b)

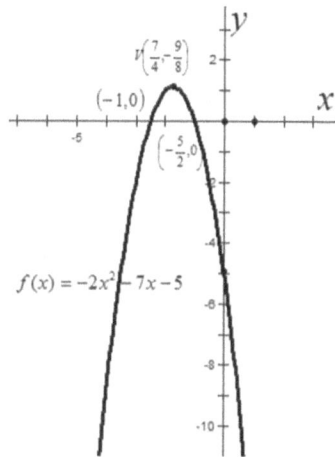

c)

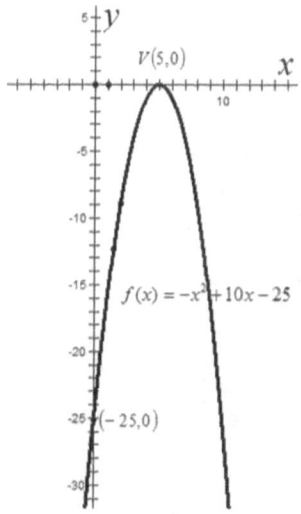

d)

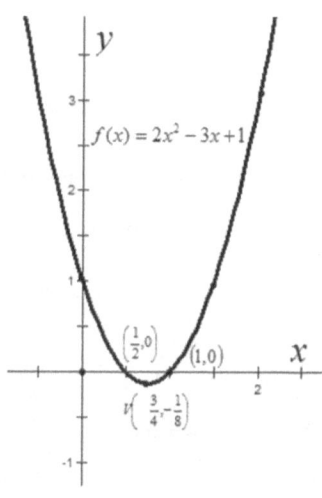

e)

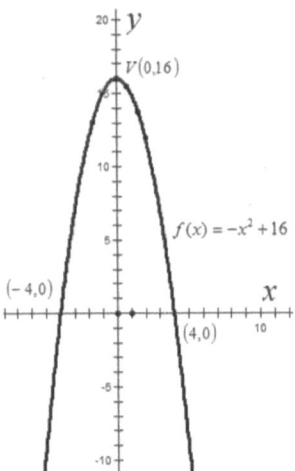

f)

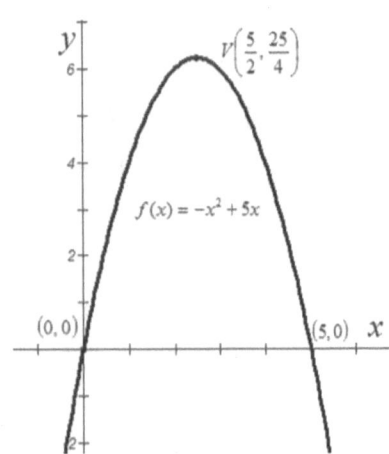

V. 12. $f(x) = -x^2 + 6x - 5$.

V. 13. $\Delta = 4ab\left(16\,ab(1-c)^2 - 1\right)$.

Using $0 < a < b < c$, then $\Delta < 4c^2\left(16c^2(1-c)^2 - 1\right)$.

Since $c(1-c) \le \dfrac{1}{4}$, we obtain $\Delta < 4c^2\left(16 \cdot \dfrac{1}{16} - 1\right) = 0$.

V. 14. $AB = \sqrt{(x_B - x_A)^2 + (y_B - y_A)^2} = \sqrt{(4 - a + 1)^2 + (4 + a - 2)^2} =$

$= \sqrt{(5-a)^2 + (2+a)^2} = 5$, $25 - 10a + a^2 + 4 + 4a + a^2 = 25$,

$2a^2 - 6a + 4 = 0 \Leftrightarrow a^2 - 3a + 2 = 0$, $a_1 = 1$ and $a_2 = 2$ are the solutions.

V. 15. a) $m_1 = -2$, $m_2 = 3$; b) $m_1 = -3$, $m_2 = 3$;

V. 16. a) $y = (x-2)(x-6)$; b) $y = 2(x-2)(x-6)$; c) $y = 3(x-2)(x-6)$.

V. 17. $p = -4$, $q = 1$, $V\left(-\dfrac{3}{2}, -\dfrac{25}{4}\right)$.

V. 18. a) $c = 1$, $-\dfrac{b}{2a} = 2$, and $-\dfrac{b^2 - 4ac}{4a} = -3$.

Thus, we obtain $a = 1$, $b = -4$, and $c = 1$. Therefore $f(x) = x^2 - 4x + 1$;

b) $\dfrac{1}{4}a + \dfrac{1}{2}b + c = 0$, $-\dfrac{b}{2a} = -\dfrac{5}{4}$, and $-\dfrac{b^2 - 4ac}{4a} = -\dfrac{49}{8}$. Thus,

282

we obtain $a = 2$, $b = 5$, and $c = -3$. Finally $f(x) = 2x^2 + 5x - 3$.

V. 19. $m = 4$. **V. 20. a)** $E(x) = \dfrac{2(x^2 - 4x + 8) + 1}{x^2 - 4x + 8} = 2 + \dfrac{1}{(x-2)^2 + 4}$.

The maximum is attained when $(x-2)^2 + 4$ is minimum, that is, for $x = 2$

and the maximum of the given expression is 2.25; **b)** $F_{max} = 3$.

V. 21. a) $(f \circ g)(x) = f(g(x)) = 18x^4$, $(g \circ f)(x) = g(f(x)) = 12x^4$;

b) $f(g(x)) = 9x^2 - 12x + 5$, $g(f(x)) = 3x^2 + 1$;

c) $f(g(x)) = (x^2 - 4x + 1)(x - 2)^2$, $g(f(x)) = (x^2 + 3x)(x^2 + 3x - 4) + 1$;

d) $f(g(x)) = \begin{cases} 12x - 20 & \text{if } x \le 1 \\ -8x + 12 & \text{if } x > 1 \end{cases}$, $g(f(x)) = \begin{cases} 12x - 14 & \text{if } x \le -2 \\ -8x - 6 & \text{if } x > -2 \end{cases}$;

e) $f(g(x)) = \begin{cases} (3x - 2)^2 + 1 & \text{if } 3x - 2 < 2,\ x \le -1 \\ (2x^2)^2 + 1 & \text{if } 2x^2 < 2,\ x > -1 \\ 2(2x^2) + 1 & \text{if } 2x^2 \ge 2,\ x > -1 \\ 2(3x - 2) + 1 & \text{if } x \ge 2,\ x \le -1 \end{cases}$,

$f(g(x)) = \begin{cases} 9x^2 - 12x + 5 & \text{if } x \le -1 \\ 4x^4 + 1 & \text{if } -1 < x < 1 \\ 4x^2 + 1 & \text{if } x \ge 1 \\ 2(3x - 2) + 1 & \text{if } x \in \varnothing \end{cases}$, $g(f(x)) = \begin{cases} 2x^4 + 4x^2 + 2 & \text{if } x < 2 \\ 8x^2 + 8x + 2 & \text{if } x \ge 2 \end{cases}$;

f) $f(g(x)) = \begin{cases} 9x^2 - 3x - 2 & \text{if } x < 0 \\ -6x & \text{if } 0 \le x \le 1 \\ (x^2 - 4x + 1)(x - 2)^2 & \text{if } 1 < x < 3 \\ -2x^2 + 8x - 6 & \text{if } x \ge 3 \end{cases}$

$g(f(x)) = \begin{cases} (x^2 + 3x)(x^2 + 3x - 4) + 1 & \text{if } x < \dfrac{-3 - \sqrt{13}}{2} \\ 3x^2 + 9x - 2 & \text{if } \dfrac{-3 - \sqrt{13}}{2} \le x < -2 \\ -6x - 14 & \text{if } x \ge -2 \end{cases}$;

g) $f(g(x)) = \begin{cases} 18x^4 & \text{if } 0 < x < \dfrac{1}{3} \\ 5 & \text{if } \dfrac{1}{3} \le x < 1 \end{cases}$, $g(f(x)) = \begin{cases} 12x^4 & \text{if } 0 < x < \dfrac{1}{3} \\ 5 & \text{if } \dfrac{1}{3} \le x < 1 \end{cases}$;

h) $f(x) = \begin{cases} 0 & \text{if } x \le 3 \\ 2(x-3) & \text{if } x > 3 \end{cases}$, $g(x) = \begin{cases} -2x+8 & \text{if } x < 0 \\ 8 & \text{if } 0 \le x \le 2 \\ -2x+12 & \text{if } 2 < x \le 4 \\ 2x-4 & \text{if } x > 4 \end{cases}$;

$f(g(x)) = \begin{cases} -4x+10 & \text{if } x < 0 \\ 10 & \text{if } 0 \le x \le 2 \\ -4x+18 & \text{if } 2 < x \le 4 \\ 4x-14 & \text{if } x > 4 \end{cases}$, $g(f(x)) = \begin{cases} 8 & \text{if } x \le 4 \\ 24-4x & \text{if } 4 < x \le 5 \\ 4x-16 & \text{if } x > 5 \end{cases}$.

V. 22. A function is invertible if it is a bijective function.

A function is bijective if it is one-to-one (injective) and onto (surjective).

A function $f : \mathbf{A} \to \mathbf{B}$, $f(x) = y$ is *one-to- one* (*injective*) if for any $x_1 \ne x_2$, $x_1, x_2 \in \mathbf{A}$ implies $f(x_1) \ne f(x_2)$.

A function $f : \mathbf{A} \to \mathbf{B}$, $f(x) = y$ is onto (surjective) if for any $y \in \mathbf{B}$ there is a $x \in \mathbf{A}$ such that $f(x) = y$.

a) *One-to-one* (*injective*): Let $x_1, x_2 \in (-\infty, 3]$, $x_1 < x_2 \le 3$.

We assume that (1) $f(x_1) = f(x_2)$ then we have $x_1^2 - 6x_1 + 8 = x_2^2 - 6x_2 + 8$ or $(x_1 - x_2)(x_1 + x_2 - 6) = 0$. Since $x_1 + x_2 < 6$ (according to the condition $x_1 < x_2 \le 3$) it implies $x_1 = x_2$, which is a contradiction.

Onto (*surjective*): Let any $y \in [-1, \infty)$, $y = x^2 - 6x + 8$, then

$x^2 - 6x + 8 - y = 0$ (2) with the solutions $x_{1,2} = 3 \pm \sqrt{y+1}$. We have to show that the equation (2) has only one solution in the interval $(-\infty, 3]$.

Consider $x_1 = 3 + \sqrt{y+1}$. It is obvious that $x_1 = 3 + \sqrt{y+1} \ge 3$, so x_1 cannot be a solution for (2). Finally $x_2 = 3 - \sqrt{y+1} \le 3$ is a solution for (2).

Indeed $f(x_2) = (3 - \sqrt{y+1})^2 - 6(3 - \sqrt{y+1}) + 8 =$

$= 9 - 6\sqrt{y+1} + y - 1 - 9 + 6\sqrt{y+1} + 8 = y$.

Therefore

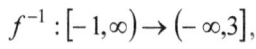

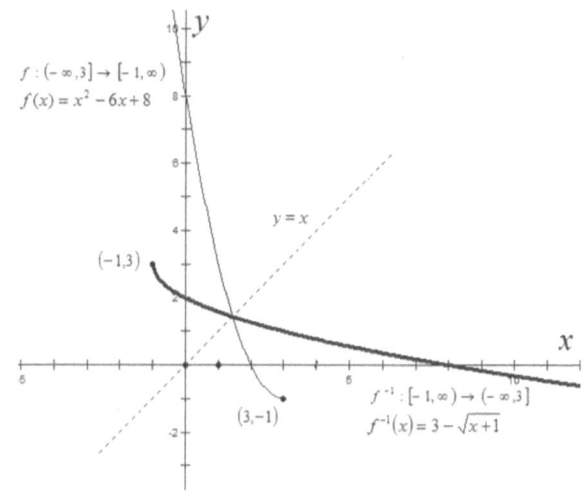

$f^{-1} : [-1,\infty) \to (-\infty,3]$,

$f^{-1}(x) = 3 - \sqrt{x+1}$

is the inverse of the
given function.

b) $f^{-1} : [-1,\infty) \to [3,\infty)$,
$f^{-1}(x) = 3 + \sqrt{x+1}$

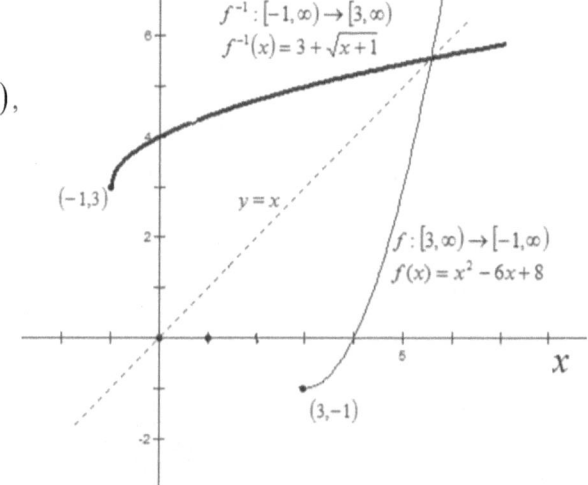

c) $f : \left(-\infty, \dfrac{9}{4}\right] \to \left(-\infty, \dfrac{5}{2}\right]$, $\quad f^{-1}(x) = \dfrac{5 - \sqrt{9-4x}}{2}$;

d) $f : \left(-\infty, \dfrac{9}{4}\right] \to \left[\dfrac{5}{2}, \infty\right)$, $\quad f^{-1}(x) = \dfrac{5 + \sqrt{9-4x}}{2}$.

V. 23. a)

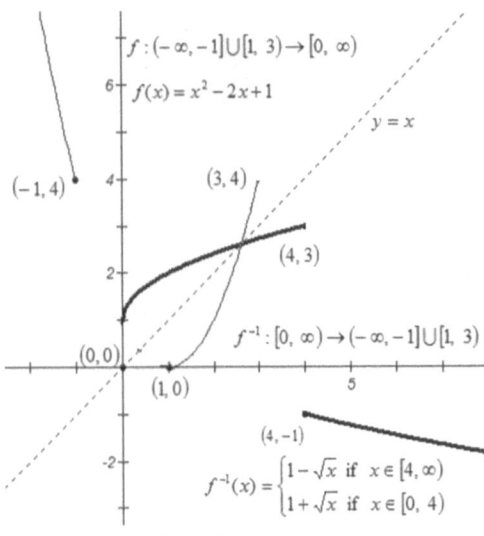

$f : (-\infty, -1] \cup [1, 3) \to [0, \infty)$

$f(x) = x^2 - 2x + 1$

$y = x$

$(-1, 4)$

$(3, 4)$

$(4, 3)$

$f^{-1} : [0, \infty) \to (-\infty, -1] \cup [1, 3)$

$(0, 0)$

$(1, 0)$

$(4, -1)$

$f^{-1}(x) = \begin{cases} 1 - \sqrt{x} & \text{if } x \in [4, \infty) \\ 1 + \sqrt{x} & \text{if } x \in [0, 4) \end{cases}$

b)

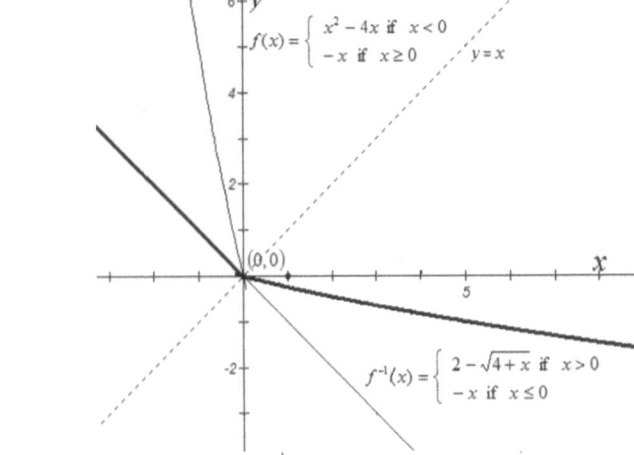

$f(x) = \begin{cases} x^2 - 4x & \text{if } x < 0 \\ -x & \text{if } x \geq 0 \end{cases}$

$y = x$

$(0, 0)$

$f^{-1}(x) = \begin{cases} 2 - \sqrt{4 + x} & \text{if } x > 0 \\ -x & \text{if } x \leq 0 \end{cases}$

c) $f^{-1}(x) = \begin{cases} 3 + \sqrt{1 + x} & \text{if } x \geq 0 \\ \dfrac{x + 8}{2} & \text{if } x < 0 \end{cases}$;

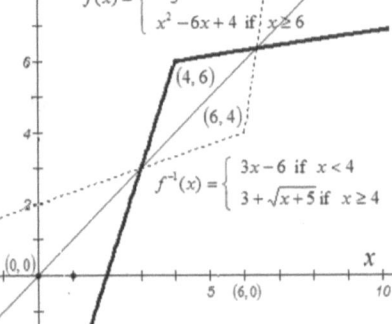

$f(x) = \begin{cases} \dfrac{x + 6}{3} & \text{if } x < 6 \\ x^2 - 6x + 4 & \text{if } x \geq 6 \end{cases}$

$y = x$

$(4, 6)$

$(6, 4)$

$f^{-1}(x) = \begin{cases} 3x - 6 & \text{if } x < 4 \\ 3 + \sqrt{x + 5} & \text{if } x \geq 4 \end{cases}$

$(0, 0)$

$5 \quad (6, 0)$

10

d) $f^{-1}(x) = \begin{cases} 3x - 6 & \text{if } x < 4 \\ 3 + \sqrt{x + 5} & \text{if } x \geq 4 \end{cases}$

286

e)

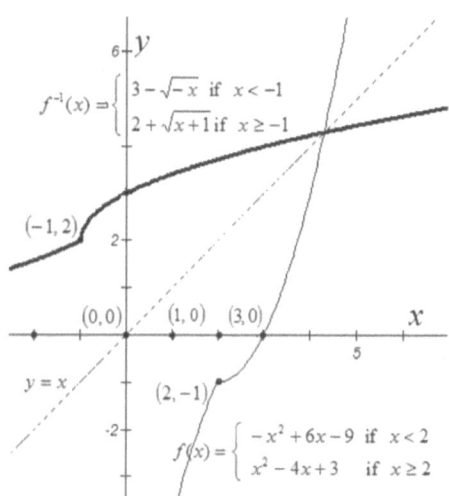

$$f^{-1}(x) = \begin{cases} 3 - \sqrt{-x} & \text{if } x < -1 \\ 2 + \sqrt{x+1} & \text{if } x \ge -1 \end{cases}$$

$(-1, 2)$

$(0, 0)$ $(1, 0)$ $(3, 0)$ x

$y = x$

$(2, -1)$

$$f(x) = \begin{cases} -x^2 + 6x - 9 & \text{if } x < 2 \\ x^2 - 4x + 3 & \text{if } x \ge 2 \end{cases}$$

V. 24. $AB = 8$.

V. 25. a) $f(x) = -(x+3)(x-1)$; **b)** $f(x) = \frac{3}{2}(x^2 + 2x - 7)$.

V. 26. a) $(6, 8)$, $(4, 0)$; **b)** $(5, 3)$; **c)** no point.

V. 27. a) $m > -2$; **b)** $m = -2$; **c)** $m < -2$. **28.** $a = 1, c = -2$.

V. 29. a) $f(x) = 2x^2 - 3x + 1$, **b)** $f(x) = x^2 - 3x + 2$; **c)** $f(x) = x^2 - 5x + 4$;

d) $f(x) = -\frac{2}{3}x^2 + \frac{4}{3}x + \frac{16}{3}$; **e)** $f(x) = -5x^2 + 10x - 3$;

f) $f(x) = \frac{1}{9}x^2 - \frac{10}{9}x + \frac{106}{9}$ or $f(x) = x^2 - 2x + 10$.

V. 30. $f(2) = 3$. **V. 31.** $a = -2$, $b = 0$.

V. 32. $f(x) = -x^2 + 6x - 4$. **V. 33.** $A_2(-3, 5)$.

V. 34. $P(1, 1)$, $Q(2, 4)$ the line passing through these points is

$y = 3x - 2$. **V. 35.** $b = -3$, $c = -14$.

V. 36. $a = 1$, $b = \frac{16}{3}$. **V. 37.** The equation $3x^2 + (a - 10)x + b + 11 = 0$

has the roots $x_1 = -1$ and $x_2 = 4$. $a = 1, b = -23$.

V. 38. $A(x_1, 0)$, $B(x_2, 0)$, $V\left(\frac{x_1 + x_2}{2}, f\left(\frac{x_1 + x_2}{2}\right)\right)$,

$A_{\triangle VAB} = \frac{1}{2}|x_1 - x_2| f\left(\frac{x_1 + x_2}{2}\right)$. **V. 39.** $\frac{a_1 + a_2 + ... + a_n}{n}$.

V. 40. $f(x) = -x^2 + 6x - 4$ or $f(x) = -9x^2 + 6x + 4$.

287

V. 41. a) $2m$; b) $m-1$; c) $4m^2 - 2m + 2$; d) $2m\left(4m^2 - 3m + 3\right)$;

e) $\pm 2\sqrt{m^2 - m + 1}$; f) $16m^4 - 16m^3 + 18m^2 - 4m + 2$; g) $\dfrac{2m}{m-1}$;

h) $\dfrac{4m^2 - 2m + 2}{m^2 - 2m + 1}$; i) $\dfrac{2m\left(4m^2 - 3m + 3\right)}{(m-1)^3}$; j) $\sqrt{2m + 2\sqrt{m-1}}$;

k) $\sqrt[4]{x_1 + x_2 + 6\sqrt{x_1 x_2} + 4 \cdot \sqrt[4]{x_1 x_2}\left(\sqrt{x_1} + \sqrt{x_2}\right)}$.

V. 42. a) $\dfrac{1}{7}$; b) $\dfrac{4}{3}$; c) -3, $\dfrac{1}{13}$; d) 14, -4; e) -1; f) $\dfrac{1}{3}$; g) -1;

h) -1, $\dfrac{5}{2}$; i) 1, $-\dfrac{111}{144}$; j) $m \in \left(-\infty, -\dfrac{13}{4}\right] \cup (-1, 3)$.

V. 43. a) $x_{1,2} = \dfrac{3 \pm \sqrt{13}}{2}$; b) $S_2 = 11$, $S_3 = 36$; c) $E = \dfrac{14}{3}$ and $F = -6$;

d) $m_1 = 1$, $m_2 = \dfrac{14}{9}$. **V. 44.** $m = \dfrac{2}{3}$.

V. 45. a) $x^2 - 9 = 0$; b) $x^2 + 2x = 0$; c) $x^2 + 2x - 8 = 0$;

d) $x^2 - x - 1 = 0$; e) $3x^2 - 2x + 1 = 0$.

V. 46. a) $y_1 + y_2 = \dfrac{x_1 + x_2}{x_1 x_2} = \dfrac{-m}{2m-1}$, $y_1 \cdot y_2 = \dfrac{1}{x_1 x_2} = \dfrac{1}{2m-1}$,

$(2m-1)y^2 + my + 1 = 0$. Another solution: use the substitution $x = \dfrac{1}{y}$ for the given equation at x.

b) $(2m-1)y^2 - (3m-2)y + m = 0$; c) $(2m-1)y^2 - m^2 y + m^2 = 0$;

d) $(2m-1)y^2 - \left(m^2 - 4m + 2\right)y + 2m - 1 = 0$;

e) $(2m-1)^2 y^2 - \left(m^2 - 4m + 2\right)y + 1 = 0$;

f) $(2m-1)^3 y^2 - \left(m^3 - 3m^2 + 3m\right)y + 1 = 0$.

V. 47. a) $6y^2 - 4y + 1 = 0$; b) $a = 3$; c) $a = \dfrac{1}{3}$.

V. 48. a) $x^2 - 6x + 1 = t$, then

$$\dfrac{t + (1-t)}{t} + \dfrac{(t+1) + (1-t)}{t+1} + \ldots + \dfrac{(t+2012) + (1-t)}{t+2012} = 2013 \Rightarrow$$

$$\left(1 + \dfrac{1-t}{t}\right) + \left(1 + \dfrac{1-t}{t+1}\right) + \left(1 + \dfrac{1-t}{t+2}\right) + \ldots + \left(1 + \dfrac{1-t}{t+2012}\right) = 2013 \Rightarrow$$

$$(1-t)\left(\dfrac{1}{t} + \dfrac{1}{t+1} + \dfrac{1}{t+2} + \ldots + \dfrac{1}{t+2012}\right) = 0 \Rightarrow t = 1 \text{ or }$$

$x^2 - 6x + 1 = 1 \Rightarrow x^2 - 6x = 0$. Therefore $x_1 = 0$ and $x_2 = 6$ are the roots of the initial equation.

b) $2x(2 + 4 + ... + 2012) - 2x(1 + 3 + ... + 2013) =$

$= (1^2 - 0^2) + (3^2 - 2^2) + ... + (2013^2 - 2012^2)$ then

$2x[(0 - 1) + (2 - 3) + ... + (2012 - 2013)] = (1 + 0) + (3 + 2) + ... + (2013 + 2012)$

$2x(-1006) = 1 + 2 + 3 + ... + 2013 \Rightarrow 2x(-1005) = \dfrac{(1 + 2013) \cdot 2013}{2}$.

Finally $x = -\dfrac{2014 \cdot 2013}{2012}$.

V. 49. a) 2; b) 2; c) $\dfrac{1}{9}$; d) $-\dfrac{7}{20}$; e) $\dfrac{9}{29}$; f) $\dfrac{16}{3}$; g) 1; h) $-\dfrac{1}{2}$.

V. 50. a) $a \neq 0$, $a = -1$; b) $a \neq 0$, $a = \dfrac{1}{2}$; c) $x \neq 0$, $x = -\dfrac{1}{5}$; d) $x \neq 0$,

$x = \pm 1$; e) $x \neq 0$ and $x \neq -5$, $x_1 = 1$, $x_2 = -\dfrac{15}{2}$; f) $p \neq 0$ and $p \neq -6$, $x_1 = 1$,

$x_2 = -\dfrac{21}{2}$; g) $v \neq 0$ and $v \neq 6$, $v = -\dfrac{15}{4}$; h) $x \neq -1$ and $x \neq 3$, $x_1 = 2$, $x_2 = \dfrac{6}{7}$;

i) $n \neq 0$, $n_1 = -1$, $n_2 = -\dfrac{37}{5}$; j) $x \neq 0$ and $x \neq 1$, $x = -5$;

k) $v \neq 1$ and $v \neq -2$, $v = 2$; l) $x \neq -5$, $x = 7$.

V. 51. a) $x_1 = 5$, $x_2 = \dfrac{8}{5}$; b) $x = \dfrac{1}{2}$; c) $x = \pm\sqrt{3}$;

d) Denoting $x^3 + 3 = t$, we obtain $\dfrac{1}{t} - \dfrac{1}{t+1} = \dfrac{1}{20}$.

The equation becomes $t^2 + t - 20 = 0$.

Then $t_1 = 4$, $t_2 = -5$. Therefore $x_1 = 1$, $x_2 = -2$;

e) The equation becomes

$\dfrac{(x - 2)(x + 1) + (x + 2)(x - 1)}{x^2 - 1} = -\dfrac{(x + 1)(x - 2) + (x + 2)(x - 1)}{x^2 - 4}$.

Expanding and collecting terms, we obtain $\dfrac{x^2 - 2}{x^2 - 1} = -\dfrac{x^2 - 2}{x^2 - 4}$,

then $x^2 = 2$ and $x^2 = \dfrac{5}{2}$. Therefore $x_{1,2} = \pm\sqrt{2}$ and $x_{3,4} = \pm\sqrt{\dfrac{5}{2}}$.

f) $x_1 = -1$, $x_2 = 3$, $x_3 = \dfrac{1}{3}$; **g)** $x_1 = -\dfrac{1}{2}$, $x_2 = 2$, $x_{3,4} = \dfrac{-7 \pm \sqrt{65}}{4}$;

h) The equation becomes $\dfrac{\left(x^2 - b^2\right) - \left(a^2 - b^2\right)}{(a+b)(x-b)} = \dfrac{\left(x^2 - c^2\right) - \left(a^2 - c^2\right)}{(a+c)(x-c)}$.

Then $x^2 - a^2 = 0$ and $(a+b)(x-b) = (a+c)(x-c)$. Therefore $x_{1,2} = \pm a$,

$x_3 = a + b + c$; **i)** $x_1 = a - 2b$, $x_2 = b - 2a$;

j) $\left(\dfrac{x+6}{x+3} - 2\right) + \left(\dfrac{x+10}{x+5} - 2\right) + \left(\dfrac{x+14}{x+7} - 2\right) = 0$, $x_1 = 0$, $x_{2,3} = \dfrac{-15 \pm 2\sqrt{3}}{3}$;

k) $x_1 = 2k$, $x_2 = -k$; **l)** $x_1 = a - b$, $x_2 = \dfrac{b^2 - a^2}{2b}$;

m) $x_1 = -a$, $x_2 = \dfrac{a^2 - 1}{a}$;

n) $x_1 = a - 1$, $x_2 = \dfrac{1-a}{a}$; **o)** Denote $x - 2a = t$, $6t^3 - 11at^2 - 3a^2t + 2a^3 = 0$

$(t - 2a)(3t - a)(2t + a) = 0$. Finally, $x_1 = 4a$, $x_2 = \dfrac{3a}{2}$, $x_3 = \dfrac{7a}{3}$.

V. 52. a) $4x^3 - 29x^2 + 33x + 54 = 0$, $x_1 = 3$, $x_{2,3} = \dfrac{17 \pm \sqrt{737}}{8}$;

b) $x^4 - 12x^3 + 34x^2 + 12x - 35 = 0$, $x_{1,2} = \pm 1$, $x_3 = 7$, $x_4 = 5$;

c) Denote $z^2 - z = k$. Then $z^2 - z = 0$, $z_1 = 0$, $z_2 = 1$;

d) $x^2 + 2x = 0$, $3x^2 + 6x - 34 = 0$, $x_1 = 0$, $x_2 = -2$, $x_{3,4} = \dfrac{-3 \pm \sqrt{111}}{3}$.

e) $x^4 + 6x^3 - 7x^2 - 50x + 12 = 0$; $x = 0$;

f) $4u^3 - 15u^2 - 15u + 44 = 0$, $u_1 = 4$, $u_2 = \dfrac{-1 \pm \sqrt{177}}{8}$;

g) $x^3 + 5x + 6 = 0$; $x = -1$; **h)** $x_1 = 1$, $x_2 = -1$, $x_3 = -a - 1$;

i) $x_1 = -1$, $x_2 = 1$, $x_3 = 2$, $x_4 = 4$;

j) $x^4 + 4x^3 + 3x^2 - 2x - 2 = 0$, $x_1 = -1$, $x_2 = -1$, $x_{3,4} = -1 \pm \sqrt{3}$;

k) Denote $\dfrac{x^2 + 1}{x} = m$, $x_1 = 2$, $x_2 = \dfrac{1}{2}$; **l)** $x_1 = 1$, $x_2 = -\dfrac{3}{5}$;

m) $x \neq \sqrt[3]{\dfrac{5}{2}}$, $x_1 = 2$, $x_2 = \sqrt[3]{\dfrac{3}{2}}$; **n)** Denote $x(x - 2) = t$.

V. 53. $2m + 6 < 0$ and $16 - m^2 > 0$. Therefore, $m \in (-4, -3)$.

290

V. 54. a) $x \in (-\infty, -6) \cup (-1, \infty)$;

b) $x \in [2,3]$; **c)** $x \in \left[-1, \dfrac{11}{3}\right]$; **d)** $x \in \left(-\infty, -\dfrac{2}{5}\right] \cup [1, \infty)$; **e)** $x \in \mathbf{R}$;

f)

x	$-\infty$		-1		1		2		9		∞
$-x+9$		$+$		$+$		$+$		$+$	$+$	0	$-$	$-$
$x+1$	$-$	$-$	0	$+$		$+$	$+$	$+$	$+$	$+$	$+$
$x-2$	$-$	$-$	$-$	$-$		$-$	$-$	0	$+$	$+$	$+$
$x-1$	$-$	$-$	$-$	$-$	0	$+$		$+$	$+$	$+$	$+$
$\dfrac{-x^2+8x+9}{x^2-3x+2}$	$-$	$-$	0	$+$	$+$	$\|$	$-$	$-$	$\|$	$+$	0	$-$	$-$

$x \in [-1, 1) \cup (2, 9]$;

g) $x \in (-\infty, -1) \cup \left(\dfrac{2}{3}, 1\right] \cup [3, \infty)$; **h)** $x \in (-3, -2) \cup \{0\}$; **i)** $x \in (5, 6.5)$;

j)

x	$-\infty$		$\left(-10-2\sqrt{22}\right)$		-2		$\left(-10+2\sqrt{22}\right)$		2	∞
$x^2+20x+12$		$+$	$+$	0	$-$	$-$	$-$	0	$+$	$+$
x^2-4		$+$	$+$	$+$	$+$	0	$-$	$-$	$-$	0	$+$
$\dfrac{x^2+20x+12}{2(x^2-4)}$		$+$	$+$	0	$-$	$-$	$-$	$\|$	$+$	0	$-$	$-$	$\|$	$+$

$x \in \left[-10-2\sqrt{2}, -2\right) \cup \left[-10+2\sqrt{2}, 2\right)$;

k) $x \in (-\infty, 0) \cup (1, \infty)$; **l)** $x \in [-5, -3) \cup [-1, 1)$;

m) $x \in \left(-\infty, \dfrac{11-\sqrt{105}}{2}\right) \cup \left(1, \dfrac{11+\sqrt{105}}{2}\right)$;

n) $x \in (-\infty, -3) \cup (-2, -1)$; **o)** $x \in [-3, -2) \cup (3, 3.5]$;

p) $\dfrac{3x(3x-2)}{(x-1)(x^2-4)} \le 0$, $x \in (-\infty, -2) \cup \left[0, \dfrac{2}{3}\right] \cup (1, 2)$;

q) $\dfrac{\sqrt{1+x^3}+x-2}{x-1} \ge x+1 \Leftrightarrow \dfrac{x^2-2x}{x-1} \le 0$, also $x+1 \ge 0$. Finally

$x \in [-1,0] \cup (1,2]$;   r) $x \in (-\infty,-2] \cup \{0\}$;   s) $x \in (3,4]$.

V. 55. a) $x \in (-\infty,-1) \cup (4,\infty)$; **b)** $x \in \left(3,\dfrac{7}{2}\right]$;   **c)** $(0,+\infty)$;   **d)** $x \in [0,1]$;

e) $x \in \left(-\infty,-2-\sqrt{10}\right] \cup \left(-2,-2+\sqrt{6}\right] \cup (2,\infty)$;

f) $x \in \left[\dfrac{3-\sqrt{11}}{2},\dfrac{17}{6}\right] \cup \left[\dfrac{3+\sqrt{11}}{2},\infty\right)$;   **g)** $x \in \left[0,\dfrac{8}{5}\right] \cup \left[\dfrac{5}{2},\infty\right)$.

V. 56. a) $x \in [-3,-1] \cup [1,3]$;   **b)** $x \in [-3,3]$;   **c)** $x \in [-1,1]$;

d) $x \in (-2,1] \cup (2,4]$;   **e)** $x \in (3,4]$;   **f)** $y \in [-1,0) \cup (0,1]$.

V. 57. a) $[-1,3]$;   **b)** $[-1,3)$;   **c)** $[-1,8]$;   **d)** $[3,\infty)$;   **e)** $[0,+\infty)$;

f) $[-1,\infty)$;   **g)** $[-1,+\infty)$.

V. 58.

$y \in [-4,\infty)$,

$f((\infty,-1) = (3,\infty)$,

$f([-1,1\ 2]) = [-1,3]$,

$f((3,7) = (-4,12)$.

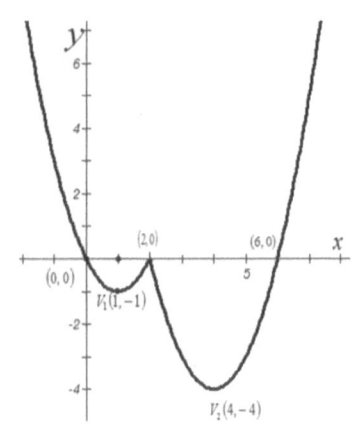

V. 59. a) $y \in \left[\dfrac{3}{4},+\infty\right)$;   **b)** $\left[-\dfrac{49}{4},+\infty\right)$;   **c)** $[-5,\infty)$;

d) Denote $m = \dfrac{x^2-x-3}{x^2+x+2}$, $(1-m)x^2 - (m+1)x - 2m - 3 = 0$, $\Delta \ge 0$,

$m \in \left[\dfrac{-1-\sqrt{92}}{7},\dfrac{-1+\sqrt{92}}{7}\right]$. Therefore, the range is $\left[\dfrac{-1-\sqrt{92}}{7},\dfrac{-1+\sqrt{92}}{7}\right]$.

e) $\left[\dfrac{1-\sqrt{3}}{2},\ \dfrac{1+\sqrt{3}}{2}\right];$

f) $y\in(-\infty,-1)$. **V. 60.** $a=\pm4\sqrt{3}$, $b=-1$.

V. 61. $m\in[2,\infty)\cup\{1\}$. **V. 62.** $m\in\left(3-2\sqrt{3},0\right)\cup\left(3+2\sqrt{3},\infty\right)$.

V. 63. $m\in[3,\infty)$. **V. 64.** $m\in\left(-1,\dfrac{3-\sqrt{17}}{2}\right]\cup\left[\dfrac{3+\sqrt{17}}{2},\ +\infty\right)$;

V. 65. a) $m_1=3$, $m_2=-\dfrac{15}{8}$; b) $x\in\left[3-\sqrt{14},0\right)\cup\left[3+\sqrt{14},\infty\right)$.

V. 66. $\Delta<0$, $m\in(1,2)$.

V. 67. a) $x_v=m+2$, $y_v=-m^2-3m-2$, $y=-x^2+x$;

b) $\Delta>0$, $m\in(-\infty,-2)\cup(-1,\infty)$. **V. 68.** $y=-x^2+x-1$.

V. 69. a) $x_v=\dfrac{-m-1}{m}$, $y_v=\dfrac{1}{m}$, $x_v=-1+y_v$; b) $AB=|x_1-x_2|$, $FV=\dfrac{1}{|m|}$;

c) $m\left(x^2+2x+1\right)+2x+2-y=0$, $(-1,0)$.

V. 70. a) $f_{-1}(x)=x^2-3$, $x_{\min}=0$, f_{-1} decreases on $(-\infty,\ 0]$ and

increases on $[0,\ \infty)$. b) $x_v=y_v$, $m=-6\pm\sqrt{13}$. c) $a>0$, $\Delta<0$, $m\in\varnothing$.

V. 71. $\Delta>0$, $12m+13>0$. **V. 72.** $\Delta=36>0$. **V. 73.** $m=-1$, $n=5$.

V. 74. $(-1,7)$, $(1,3)$. **V. 75.** $m=-\dfrac{1}{2}$, $n=-\dfrac{15}{2}$.

V. 76. a) 1) $m-1<0$ and $y_{\min}=\dfrac{4m+5}{1-m}<0$, $m\in\left(-\infty,\ -\dfrac{5}{4}\right)$;

2) $m=-\dfrac{5}{4}$; 3) $m\in(-\infty,\ -8)$; 4) $m=-2$.

b) $y=3x-1$; c) $m\left(x^2+2x+1\right)-x^2+4x+1-y=0$, $m\in(-1,-4)$.

V. 77. a) 1) $y_{\min}<0$, $y_{\min}=-(m+1)<0$, $m\in(-1,\infty)$; 2) $\Delta=0$, $m=-1$;

3) $m=-6$; 4) $m_1=2$, $m_2=3$; b) $x_{\min}=m-1$, $y_{\min}=-(m+1)<0$ and the
line is $y=-x-2$.

V. 78. $a>0$, $\Delta<0$, $m\in(2,\infty)$. **V. 79.** $\Delta<0$, $a<0$, $m\in\left(1,\dfrac{5}{3}\right)$.

V. 80. $a<0$, $\Delta<0$, $m\in(-\infty,\ -2)$.

V. 81. $(m-1)x^2+(m+2)x+m-1\geq0$, $a>0$, $\Delta\leq0$, $m\in(1,\ 4]$.

V. 82. $m\in[0,\infty)$, $x=3\pm\sqrt{3}$.

V. 83. a) $m\in\left(0,\ \dfrac{4}{3}\right)$; b) $m\in\left(0.25,\ 1+2\sqrt{2}\right]$; a) $m\in\left(-\dfrac{1}{5},\ 0\right)$.

V. 84. a) $(x_1 - 2)^2 + (x_2 - 2)^2 = (x_1 + x_2)^2 - 2x_1x_2 - 4(x_1 + x_2) + 8$, where

$x_1 + x_2 = m + 1$ and $2x_1x_2 = m^2 - 2m + 3$ (Vieta's formulae);

b) Denote $\dfrac{x_1}{x_2} = t$, then $t^2 - 4t + 1 = \dfrac{x_1^2 - 4x_1x_2 + x_2^2}{x_2^2} =$

$$= \frac{(x_1 + x_2)^2 - 6x_1x_2}{x_2^2} = \frac{m^2 + 2m + 1 - 3m^2 + 6m - 9}{x_2^2} = \frac{-2(m-2)^2}{x_2^2} \le 0$$

Therefore $t^2 - 4t + 1 \le 0$, which implies $t \in \left[2 - \sqrt{3}, 2 + \sqrt{3}\right]$;

c) From $s = x_1 + x_2 = m + 1$ and $p = x_1x_2 = \dfrac{m^2 - 2m + 3}{2}$, eliminate m and

get $2p = s^2 - 4s + 6$.

V. 85. a) $x_1 + x_2 = m + 3$, $2x_1x_2 = m^2 - 4m + 24$;

b) $(x_1 - 5)^2 = 5 - (x_2 - 5)^2 \le 5$, thus $|x_1 - 5| \le \sqrt{5}$, similarly for x_2;

c) $|x_1 - x_2| = \sqrt{(x_1 - x_2)^2} = \sqrt{(x_1 + x_2)^2 - 4x_1x_2} =$

$= \sqrt{10 - (m-7)^2} \le \sqrt{10}$; **d)** Since $x_1 + x_2 = m + 3$ and $\Delta = 4\left[10 - (m-7)^2\right]$,

from $10 - (m-7)^2 \ge 0$, we get $7 - \sqrt{10} \le m \le 7 + \sqrt{10}$, therefore

$-4 \le x_1 + x_2 \le 4$;

e) $x_1^2 + x_2^2 = (x_1 + x_2)^2 - 2x_1x_2 = 10m - 15$ and since $m \le 7 + \sqrt{10}$

then $x_1^2 + x_2^2 \le 55 + 10\sqrt{10}$; **f)** Denote $\dfrac{x_1}{x_2} = t$,

$$t^2 - \frac{5}{2}t + 1 = \frac{2x_1^2 - 5x_1x_2 + 2x_2^2}{2x_2^2} = \frac{-5(m-6)^2}{4x_2^2} \le 0.$$

Therefore $2t^2 - 5t + 2 \le 0$, which implies $t \in [0.5,\ 2]$;

g) $\Delta = 4\left[10 - (m-7)^2\right] \le 0$, $m \in \left(-\infty, 7 - \sqrt{10}\right] \cup \left[7 + \sqrt{10}, \infty\right)$;

h) Find the smallest possible value of $f_{\min} = \dfrac{1}{2}\left(m^2 - 14m + 39\right)$. The

minimum of $f_{\min}$ is holding for $m = 7$, therefore $f_{\min}(7) = -1.5$;

i) From $s = x_1 + x_2 = m + 3$ and $2p = 2x_1x_2 = m^2 - 4m + 24$, eliminate m and

obtain $2p = s^2 - 10s + 45$.

j) $f(5)f(6)<0$, $\left(m^2-14m+44\right)\left(m^2-16m+60\right)<0$.

V. 86. a) $(3, 1)$ and $(-3, -2)$; **b)** $(2, 1)$ and $\left(\dfrac{8}{5},\dfrac{1}{5}\right)$;

c) $(3, 2)$ and $\left(-\dfrac{5}{2},\dfrac{19}{4}\right)$; **d)** $(1, 3)$ and $\left(\dfrac{29}{2},-\dfrac{3}{2}\right)$;

e) $(-1, 2)$ and $\left(-\dfrac{13}{4},\dfrac{1}{2}\right)$; **f)** $(2, 1)$ and $\left(\dfrac{2}{3},\dfrac{11}{3}\right)$;

V. 87. a) $(7, 3)$ and $(-7, -3)$; **b)** Dividing the first equation by the second,

we get $\dfrac{x^2-y^2}{xy}=\dfrac{5}{2}$ or $2x^2-5xy+2y^2=0$. Dividing this by y^2, we have

$2\dfrac{x^2}{y^2}-5\dfrac{x}{y}+2=0$. Denote $\dfrac{x}{y}=t$. Therefore $2t^2-5t+2=0$. Then $t_1=2$,

$t_2=\dfrac{1}{2}$. Then we get the systems $\begin{cases} x=2y \\ x^4-y^4=15 \end{cases}$ and $\begin{cases} 2x=y \\ x^4-y^4=15 \end{cases}$. Finally,

the solutions are $(2, 1)$ and $(-2, -1)$. **c)** $(5,3)$. **d)** $(-4,-4),(-6,-2)$; **e)** $(2, 1)$;
f) $(2, 1), (-2, -1), (-2, 1), (2, -1)$;

g) Denote $a=\dfrac{1}{2x-y+1}$ and $b=\dfrac{1}{x-y-1}$, $(1, 2)$;

h) Denote $a=x^2+y^2-1$ and $b=\dfrac{x}{y}$, $(1, 1); (-1, -1); (1, -1); (-1, 1)$;

$\left(\dfrac{8\sqrt{85}}{425},\dfrac{3\sqrt{85}}{85}\right); \left(-\dfrac{8\sqrt{85}}{425},-\dfrac{3\sqrt{85}}{85}\right); \left(\dfrac{8\sqrt{85}}{425},-\dfrac{3\sqrt{85}}{85}\right); \left(-\dfrac{8\sqrt{85}}{425},\dfrac{3\sqrt{85}}{85}\right);$

i) $(2, -1)$; **j)** $(3, 1)$.

V. 88. a) Multiply the first equation by -6 and add it to the second equation.

We get $-5x^2+xy+18y^2=0$. Denote $\dfrac{x}{y}=t \Rightarrow 5t^2-t-18=0$. The initial

system is equivalent with two systems. I):$\begin{cases} x=2y \\ x^2+xy=6 \end{cases}$ with the solutions

$x_1=2$,

$y_1=1$ and $x_2=-2$, $y_2=-1$. II):$\begin{cases} x=-\dfrac{9}{5}y \\ x^2+xy=6 \end{cases}$

with the solutions $x_3=-\dfrac{3\sqrt{6}}{2}$,

$y_3 = \dfrac{5}{\sqrt{6}}$ and $x_4 = \dfrac{3\sqrt{6}}{2}$, $y_4 = -\dfrac{5}{\sqrt{6}}$; **b)** $(3, 2), (-3, -2)$; **c)** $(-1, 2), (1, -2)$,

$\left(\dfrac{5\sqrt{3}}{3}, \dfrac{4\sqrt{3}}{3}\right)$, $\left(-\dfrac{5\sqrt{3}}{3}, -\dfrac{4\sqrt{3}}{3}\right)$; **d)** $(1, 1), (-1, -1)$;

e) $(-1, 1), (1, -1)$; **f)** $(1, 1), (-1, -1), \left(-\dfrac{5}{\sqrt{37}}, \dfrac{4}{\sqrt{37}}\right), \left(\dfrac{5}{\sqrt{37}}, -\dfrac{4}{\sqrt{37}}\right)$.

V. 89. a) Denote $\begin{cases} x + y = s \\ xy = p \end{cases}$, the system becomes $\begin{cases} s^2 - 2p = 5 \\ p = 2 \end{cases}$, with the

solutions $\begin{cases} s = 3 \\ p = 2 \end{cases}$ and $\begin{cases} s = -3 \\ p = 2 \end{cases}$ or $\begin{cases} x + y = 3 \\ xy = 2 \end{cases}$ and $\begin{cases} x + y = -3 \\ xy = 2 \end{cases}$. Solving

the last two systems we get the solutions: $x_1 = 1$, $y_1 = 2$, $x_2 = 2$, $y_2 = 1$ and

$x_3 = -1$, $y_3 = -2$ and $x_4 = -2$, $y_4 = -1$; **b)** $(2, 3), (3, 2)$; **c)** $(0, 0), (2, 2)$;

d) $(1, 3), (3, 1), (1, -4), (-4, 1)$; **e)** $(1, 5), (5, 1), (3, 2), (2, 3)$;

f) *Solution* 1

Denote $x + y = s$ and $xy = p$ and the system becomes $\begin{cases} s^3 - 3ps = 26 \\ ps = -6 \end{cases}$.

Solution 2

Dividing the first equation by the second, we get $\dfrac{x^2 - xy + y^2}{xy} = -\dfrac{13}{3}$ or

$3x^2 + 10xy + 3y^2 = 0$. Dividing this by y^2 and denoting $\dfrac{x}{y} = t$, we get

$3t^2 + 10t + 3 = 0$ with the solutions $t_1 = -\dfrac{1}{3}$ and $t_2 = -3$. Then we get

the systems $\begin{cases} y = -3x \\ x^3 + y^3 = 26 \end{cases}$ and $\begin{cases} x = -3y \\ x^3 + y^3 = 26 \end{cases}$. Finally the solutions are

$(3, -1), (-1, 3)$; **g)** $(2, 3), (-2, -3), (3, 2), (-3, -2)$; **h)** $(-2, 3), (3, -2)$;

i) $(2, 4), (4, 2)$; **j)** $\left(\dfrac{1}{2}, \dfrac{1}{3}\right), \left(\dfrac{1}{3}, \dfrac{1}{2}\right)$; **k)** $(1, 4), (4, 1)$;

l) $(2, 5), (5, 2), (1, 8), (8, 1)$; **m)** From the second equation, we have $y^3 = \dfrac{8}{x^3}$.

Replacing this in the first equation, we get $x^3 + \dfrac{8}{x^3} = -9$ or $x^6 - 9x^3 + 8 = 0$

. Denoting $x^3 = t$ and applying quadratic formula in $t^2 - 9t + 8 = 0$ we get

$t_1 = 8$ and $t_2 = 1$. Therefore $x_1 = 2$, $x_2 = 1$ and $y_1 = 1$, $y_2 = 2$;

n) $(2, 2)$, $(-2, -2)$, $(-2, 2)$, $(2, -2)$; o) $(1, 2)$, $(2, 1)$, $(-1, -2)$, $(-2, -1)$;

p) $(1, 3)$, $(-1, -3)$, $(3, 1)$, $(-3, -1)$;

q) $(2, 3)$, $(2, -3)$, $(3, 2)$, $(3, -2)$, $(-2, 3)$, $(-2, -3)$, $(-3, 2)$, $(-3, -2)$.

V. 90. a) $(1.6, 1.3)$, $(1.4, 1.5)$; b) $(3, 1)$, $(-1, -3)$.

c) $\dfrac{x^3 + y^3}{3xy^2 - 2x^2 y} = \dfrac{9}{8}$ or $\dfrac{\left(\dfrac{x}{y}\right)^3 + 1}{3\dfrac{x}{y} - 2\left(\dfrac{x}{y}\right)^2} = \dfrac{9}{8}$.

Denoting $\dfrac{x}{y} = t$, we obtain the equation $(2t - 1)\left(4t^2 + 11t - 8\right) = 0$. One of the

solutions is $(1, 2)$; d) $(3, 5)$, and $(-3, -5)$;

e) $(4, 1)$, $(-4, -1)$; f) $(3, 2)$, $(-3, -2)$;

g) $(a, 2a)$, and $(2a, a)$; h) $y \neq 0$, Denote $\dfrac{x}{y} = t$, from the first

equation we get $t^3 + t^2 - 36 = 0 \Rightarrow (t - 3)\left(t^2 + 4t + 12\right) = 0$, then

$t = 3$. In the second equation denote $xy = z$, then it yields to

$z^2 + z = 12$ with the solutions $z_1 = 3$ and $z_2 = -4$. Therefore the initial system

is reduced to the systems $\begin{cases} \dfrac{x}{y} = 3 \\ xy = 3 \end{cases}$ and $\begin{cases} \dfrac{x}{y} = 3 \\ xy = -4 \end{cases}$. Only the first system has real

solutions, which are $(3, 1)$

and $(-3, -1)$; i) $(-1, 2)$ and $\left(\dfrac{17 \cdot \sqrt[3]{4}}{4}, \dfrac{19 \cdot \sqrt[3]{4}}{4}\right)$;

j) The system becomes $\begin{cases} \dfrac{x^3}{y} = 30 - xy \\ \dfrac{y^3}{x} = \dfrac{10}{3} - xy \end{cases}$. Multiplying both these equations we

get $x^2 y^2 = (xy - 30)\left(xy - \dfrac{10}{3}\right)$. The solutions are $(3, 1)$, $(-3, -1)$;

V. 91. a) $(2, -5, -4)$, $(-2, -1, 0)$; b), $(-2, -3, -4)$;

d) $(2, 3, 4)$, $(2, 6, 1)$, $(7, 3, -1)$, $(7, 6, -4)$; e) $(3, 3, 3)$;

f) The system becomes $\begin{cases} \dfrac{xy + yz + zx}{xyz} = 1 \\ \dfrac{x + y + z}{xyz} = \dfrac{1}{3} \\ xyz = 27 \end{cases}$ or $\begin{cases} xy + yz + zx = 27 \\ x + y + z = 9 \\ xyz = 27 \end{cases}$. From the

297

second equation $x + y = 9 - z$ (1) and from the third equation, we get $xy = \dfrac{27}{z}$

(2). Substituting (1) and (2) in the first equation, we obtain $\dfrac{27}{z} + z(9 - z) = 27$

or $z^3 - 9z^2 + 27z - 27 = (z - 3)^3 = 0$. Therefore

$z = 3$. Since the system is symmetric, the solution is (3, 3, 3);

g) Denote $a = \dfrac{1}{xy}$, $b = \dfrac{1}{yz}$, and $c = \dfrac{1}{xz}$, (1, 3, 4), ($-$ 1, $-$ 3, $-$ 4);

h) (a, a, a), $a \neq 0$; i) (1, 2, 3), (2, 1, 3);

j) $\begin{cases} (x + y - z)(x + y + z) = 9 \\ (y + z - x)(x + y + z) = 45 \\ (z + x - y)(x + y + z) = 27 \end{cases}$. Adding side by side, we obtain

$(x + y + z)(x + y + z) = 81 \Rightarrow (x + y + z) = \pm 9$.

I) $x + y + z = 9$, then $(x, y, z) = (2, 3, 4)$,

II) $x + y + z = -9$, then $(x, y, z) = (-2, -3, -4)$;

k) Subtracting the equations two-by-two, we obtain the system

$\begin{cases} (x - y)(1 - z) = 0 \\ (x - z)(1 - y) = 0 \\ (y - z)(1 - x) = 0 \end{cases}$. From this system we obtain two systems,

$\begin{cases} x = y \\ (x - z)(1 - y) = 0 \\ (y - z)(1 - x) = 0 \end{cases}$ and $\begin{cases} z = 1 \\ (x - z)(1 - y) = 0 \\ (y - z)(1 - x) = 0 \end{cases}$. The solution is (1, 1, 1);

l) (2, $-$3, 4), (2, $-$4, 3), (3, $-$2, 4), (3, $-$4, 2), (4, $-$2, 3), (4, $-$3, 2);

m) $\left(-\dfrac{7}{3}, -1, \dfrac{1}{3} \right)$, $\left(\dfrac{7}{3}, 1, -\dfrac{1}{3} \right)$.

V. 92. Taking $x = a^6$, $y = b^6$, $z = c^6$, the inequality becomes

$\left(a^2 + b^2 + c^2 \right)^3 > \left(a^3 + b^3 + c^3 \right)^2$. Expanding and collecting terms, we obtain

$3\left(b^2 + c^2 \right) a^4 - 2\left(b^3 + c^3 \right) a^3 + 3\left(b^2 + c^2 \right)^2 a^2 + b^2 c^2 \left[\left(2b^2 + 2c^2 \right) + (b - c)^2 \right] > 0$,

on the other hand $3\left(b^2 + c^2 \right) a^2 - 2\left(b^3 + c^3 \right) a + 3\left(b^2 + c^2 \right)^2 > 0$ is true because

$\Delta_a < 0$.

V. 93. 9 or -5. **V. 94.** 22, 24 or -22, -24. **V. 95.** 8 or -7.

V. 96. 3 or 6. **V. 97.** 7 and 8.

298

V. 98. $x^2 + (x+1)^2 - 421 = 0$, 14, 15 or -14, -15.

V. 99. 15 and 7. **V. 100.** 8. **V. 101.** $\dfrac{9}{10}$ or $\dfrac{10}{9}$. **V. 102.** 15 or -0.5.

V. 103. $(x-2)^2 + x^2 + (x+2)^2 = 116$.

The numbers are 4, 6, 8 or -4, -6, -8. **V. 104.** $BC = 10$ cm.

V. 105. $x + y = 21$ and $x^2 + y^2 = 225 \Rightarrow y^2 - 21y + 108 = 0$.

The numbers are 9 and 12.

V. 106. $\dfrac{x(2x+8)}{2} = 96 \Rightarrow x^2 + 4x - 96 = 0$. The base is 8cm and the height is

24cm. **V. 107.** 15 cm, 20 cm, and 25 cm.

V. 108. 12 cm. **V. 109.** 12 m and 15 m.

V. 110. 15. **V. 111.** $(x+30)(x+20) = 1200 \Rightarrow x^2 + 50x - 600 = 0$, $x = 10$.

V. 112. 10 cm and 14 cm.

V. 113. $(2x+60)(2x+40) = 4800 \Rightarrow x^2 + 50x - 600 = 0$, $x = 10$.

V. 114. a) Denote by x the price at which each CD will be sold. Number of
CD's sold $= 4,000 - 400$(number of \$1 increase) $= 4,000 - 400(x - 5)$.
Profit per CD is $x - 2$. Total profit $P(x) = $ (# CD's sold)(profit per CD)$= ($
$6,000 - 400x)(x - 2) = -400x^2 + 6,800x - 12,000$.
b) Profit is maximized for \$8.50 per CD and maximum monthly profit is
$P(8.5) = \$16,900$. **V. 115.** 325.

V. 116. The fare is reduced to $(8 - 0.25x)$ dollars per car, then the number of

cars is $(300 + 10x)$. The total daily revenue is $R(x) = (300 + 10x)(8 - 0.25x)$ or

$R(x) = -2.5x^2 + 5x + 2400$. The maximum for this function is 2402.50 and it is

attained for $x = 1$. The fare is \$7.75 per car.

V. 117. Let x be the number trees which maximize the yield per acre. The

number of bushes produced by each tree is $30 - 0.5(x - 40)$. The total yield

in bushels per acre is $f(x) = x\left(50 - \dfrac{x}{2}\right)$. The maximum of f is obtained for

$x = 50$.

V. 118. $h_{max} = 31.25$.

V. 119. a) 2 m; **b)** 41 m; **c)** 38 m; **d)** 3.125 s.

V. 120. Let x the width and y the length. We have $2x + y = 100$.

$A(x) = xy = x(100 - 2x)$. The function $A(x) = -x^2 + 100x$, $x \in [0, 50]$ has a maximum when $x = 25$ and $y = 50$.

V. 121. $AM = x$, the area of the triangles is

$$A(x) = \frac{x^2\sqrt{3}}{4} + \frac{(a-x)^2\sqrt{3}}{4},$$

$$A(x) = \frac{\sqrt{3}(2x^2 - 2ax + a^2)}{4}.$$

$$A_{min} = \frac{a^2\sqrt{3}}{8}.$$

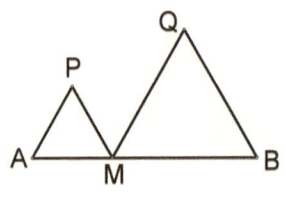

V.122. Let a, b, and c be the side lengths of $\triangle ABC$. Let $MD = x$. Applying Stewart's theorem in $\triangle MBC$, $MB^2 \cdot CD + MC^2 \cdot BD - MD^2 \cdot BC = BD \cdot DC \cdot BC$,

$$\frac{MB^2}{2} + \frac{MC^2}{2} - MD^2 = \frac{a^2}{4}.$$

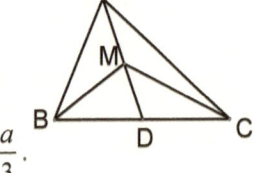

$$S = MA^2 + 2MD^2 + \frac{a^2}{2} = (m_a - x)^2 + 2x^2 + \frac{a^2}{2}$$

$$S = 3x^2 - 2m_a x + m_a^2 + \frac{a^2}{2}. \text{ Minimum of } S \text{ is for } x = \frac{a}{3}.$$

V.123. Let $AM = 2x$. $A(x) = r^2\pi - x^2\pi - (r-x)^2\pi$,

$$A(x) = -2x^2\pi + 2rx\pi. \quad A_{max} = \frac{r^2\pi}{2}.$$

V. 124. $AM = 2x$, the area of the circles is $A(x) = x^2\pi + (a-x)^2\pi$,

$$A_{min} = \frac{a^2\pi}{2} \text{ and } AM = a.$$

V. 125. Let y the width, x the length, and P the perimeter. We have $xy = 100$ and $P = 2x + 2y$ or $2x^2 - xP + 200 = 0$. This equation must have real roots, thus $\Delta = b^2 - 4ac \geq 0 \Rightarrow P \geq 40$,

$P = 40$ is minimum, then $x = 10$, and finally $y = 10$.

V. 126. $A_{\triangle ABC} = \dfrac{10h}{2}$. Let $AC = x$, $AD = y$, and $h = CD$, we have

$AC^2 - AD^2 = CB^2 - BD^2$ or $x^2 - y^2 = (10 - x)^2 - (8 - y)^2$,

$y = \dfrac{5x - 9}{4}$, $A_{\triangle ABC} = \dfrac{5}{2}\sqrt{-9x^2 + 90x - 81} = \dfrac{5}{2}\sqrt{144 - 9(x - 5)^2}$ thus

the maximum areas is 30 for $x = 5$, therefore $h = 3$cm.

V. 127. Let $AM = x$, then $DM = a - x$.

Since $MN \parallel CP$, then $\angle MND \equiv \angle PCN$.

Since $AB \parallel CD$, then $\angle PCN \equiv \angle CPB$, thus $\angle MND \equiv \angle CPB$, therefore the right triangles MND and CPB are similar. We have

$\dfrac{DN}{b} = \dfrac{a - x}{a} \Rightarrow DN = \dfrac{b}{a}(a - x)$.

The area of the quadrilateral $MNCP$ is

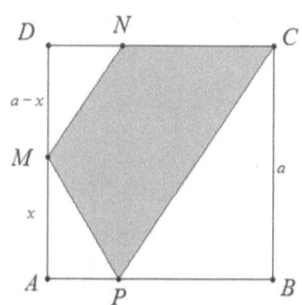

$f(x) = -\dfrac{b}{2a}x^2 + \dfrac{3b - a}{2}x + a^2 - ab$

and $x_0 = \dfrac{a}{2}\left(3 - \dfrac{a}{b}\right)$ is a maximum of f.

1) If $x_0 \le 0 \Leftrightarrow \dfrac{a}{b} \ge 3$, then the maximum

of f on the interval $[0, a]$ is $f(0) = a^2 - ab$.

2) If $0 < x_0 < a \Leftrightarrow 1 < \dfrac{a}{b} < 3$, then the maximum of f

on the interval $[0, a]$ is $f(x_0) = \dfrac{a}{8b}(a + b)^2$.

V. 128. Let x be the speed travelled at for the trip from Town A to Town B.

$$\frac{4000}{x} + \frac{4000}{x+300} = 13 \Rightarrow 13x^2 - 4100x - 1200000 = 0.$$ Therefore the speed

travelled at for the trip from Town A to Town B is 500km/h.

V. 129. Let x be the speed travelled at for the trip from B to C.

$$\frac{720}{x} + \frac{720}{x+10} = 17 \Rightarrow 17x^2 - 1270x - 7200 = 0.$$ Therefore the speed travelled

at for the trip from C to B is 90km/h.

V. 130. Let x be the speed travelled at for the trip from A to B.

$$\frac{4800}{x-200} - \frac{4800}{x} = 2 \Rightarrow x^2 - 200x - 480000 = 0.$$ Therefore the speed travelled

at for the trip from B to A is 600 km/h.

V. 131. a) $4 \le 2x - 1 < 5,\ \ x \in \left[\dfrac{5}{2}, 3\right)$; **b)** $x \in [17, 20)$;

c) $4 - x \le \dfrac{x-2}{4} < 5 - x$ and $x \in \mathbf{Z},\ \ x \in \left[\dfrac{16}{5}, \dfrac{22}{5}\right) \mathbf{IZ} = \{4\}$.

V. 132. a) Since $\left[\dfrac{2x+1}{3}\right] \le \dfrac{2x+1}{3} < \left[\dfrac{2x+1}{3}\right] + 1$, we have

$$\frac{x-4}{2} \le \frac{2x+1}{3} < \frac{x-4}{2} + 1 \Rightarrow -14 \le x < -8.$$

On the other hand $\dfrac{x-4}{2} = k \in \mathbf{Z}$, then $x = 2k + 4$. Therefore

$-14 \le 2k + 4 < -8$ which implies $k \in [-9, -6)$, since k is an integer number,

thus $k \in \{-9, -8, -7\}$.

The solutions of the equation are -16, -12, and -10;

b) -5 and -2; **c)** 7, 7.5, 8, 8.5, 9, 9.5; **d)** $5, \dfrac{19}{3}, \dfrac{23}{3}$; **e)** Denote $\left[3x - \dfrac{2}{5}\right] = t$,

$t \in \mathbf{Z}$. Then $t = 2x + 1$ or $t \le \dfrac{15t - 19}{10} < t + 1$ and the initial equation becomes

$\left[\dfrac{15t-19}{10}\right] = t$, therefore $t \le \dfrac{15t-19}{10} < t + 1$. Since $t \in \mathbf{Z}$, we obtain $t_1 = 4$

and $t_2 = 5$. The solutions of the equation are $x_1 = 1.5$ and $x_2 = 2$; f) $x_1 = \dfrac{3}{4}$,

$x_2 = 1$; g) $x_1 = -\dfrac{5}{2}$, $x_2 = -\dfrac{3}{2}$; h) $x \in \left\{ -\dfrac{1}{6}, 0, \dfrac{1}{6}, \dfrac{1}{3}, \dfrac{1}{2}, \dfrac{2}{3} \right\}$;

i) $1, 3, 5$.

V. 133. Denote $\dfrac{2x+5}{3} = t \in \mathbf{Z}$, then $x = \dfrac{3t-5}{2}$ and $\left[\dfrac{2t-3}{2t-5} \right] = t$. Thus

$0 \le \dfrac{2t-3}{2t-5} - t < 1$. Therefore $t \in \left(\dfrac{5-\sqrt{41}}{4}, \dfrac{1}{2} \right] \cup \left(\dfrac{5+\sqrt{41}}{4}, 3 \right]$. Since $t \in \mathbf{Z}$,

then $t = 0$ or $t = 3$. The solutions are $x = -\dfrac{5}{2}$ and $x = 2$.

V. 134. a) Denote $\left[x + \dfrac{2}{3} \right] = k$; $k \in \mathbf{Z}$, thus $\left[x + \dfrac{3}{5} \right] = 1 - k$. Then

$k - \dfrac{2}{3} \le x < k + \dfrac{2}{3}$ and respectively $-k + \dfrac{2}{5} \le x < -k + \dfrac{8}{5}$. The last

two inequalities become $-\dfrac{2}{3} < -x + k \le \dfrac{2}{3}$ and $\dfrac{2}{5} \le x + k < \dfrac{8}{5}$. Now

adding them side by side we get $-\dfrac{4}{30} < k < \dfrac{34}{30}$. Since k is an integer

number, then $k = 0$ or 1. Therefore $x \in \left[-\dfrac{2}{3}, \dfrac{2}{3} \right) \cap \left[\dfrac{2}{5}, \dfrac{8}{5} \right) = \left[\dfrac{2}{5}, \dfrac{2}{3} \right)$ or

$x \in \left[\dfrac{1}{3}, \dfrac{5}{3} \right) \cap \left[-\dfrac{3}{5}, \dfrac{3}{5} \right) = \left[\dfrac{1}{3}, \dfrac{3}{5} \right)$ and finally $x \in \left[\dfrac{1}{3}, \dfrac{3}{5} \right)$; **b)** $x \in \left[\dfrac{1}{3}, \dfrac{2}{3} \right)$.

V. 135. If $n \in \mathbf{Z}$, $n \le x < n + \dfrac{1}{2}$, then $[x] = \left[x + \dfrac{1}{2} \right] = n$ and $[2x] = 2n$.

If $n + \dfrac{1}{2} \le x < n + 1$, then $[x] = n$, $\left[x + \dfrac{1}{2} \right] = n + 1$ and $[2x] = 2n + 1$;

a) Denoting $\dfrac{2x+3}{3} = y$, then $x = \dfrac{3y-3}{2}$ and the equation becomes

$[y] + \left[y + \dfrac{1}{2} \right] = \dfrac{15y-3}{2}$ or $[2y] = \dfrac{15y-3}{2}$.

Taking $[2y] = \dfrac{15y-3}{2} = p \in \mathbf{Z}$, then $y = \dfrac{2p+3}{15}$, which yields to

$\left[\dfrac{4p+6}{15}\right] = p$. Therefore $p \le \dfrac{4p+6}{15} < p+1$, with the solution

$p \in \left(-\dfrac{5}{11}, \dfrac{10}{11}\right]$.

Since $p \in \mathbf{Z}$, then $p = 0$, $y = \dfrac{1}{5}$ and $x = -\dfrac{6}{5}$;

b) We have $\dfrac{3x-4}{12} + \dfrac{1}{2} = \dfrac{3x+2}{12} \Rightarrow$

$\left[\dfrac{3x+2}{12}\right] + \left[\dfrac{3x-4}{12}\right] = 2\left[\dfrac{3x-4}{12}\right] = \left[\dfrac{3x-4}{6}\right]$.

The equation becomes $\left[\dfrac{3x-4}{6}\right] + \left[\dfrac{3x-7}{6}\right] = \dfrac{3x+1}{2}$. Since

$\dfrac{3x-7}{6} + \dfrac{1}{2} = \dfrac{3x-4}{6}$,

the equation becomes $\left[\dfrac{3x-7}{3}\right] = \dfrac{3x+1}{2}$.

Taking $\dfrac{3x+1}{2} = k \in \mathbf{Z} \Rightarrow x = \dfrac{2k-1}{3}$ and substituting in the last equation, we

obtain $\left[\dfrac{2k-8}{3}\right] = k \Rightarrow k \le \dfrac{2k-8}{3} < k+1 \Rightarrow 3k \le 2k-8 < 3k+3 \Rightarrow$

$0 \le -k-8 < 3 \Rightarrow -3 < k+8 \le 0 \Rightarrow -11 < k \le -8$. Since k is an integer number,

then $k = -10$, $k = -9$, and $k = -8$.

Therefore $x \in \left\{-7, -\dfrac{19}{3}, -\dfrac{17}{3}\right\}$;

c) We successively have

$\left(\left[\dfrac{2x+3}{8}\right] + \left[\dfrac{2x+3}{8} + \dfrac{1}{2}\right]\right) + \left[\dfrac{2x+5}{4}\right] + \left[\dfrac{2x+4}{2}\right] = \dfrac{x+3}{2}$,

$\left[\dfrac{2x+3}{4}\right] + \left[\dfrac{2x+5}{4}\right] + \left[\dfrac{2x+4}{2}\right] = \dfrac{x+3}{2}$,

$\left(\left[\dfrac{2x+3}{4}\right] + \left[\dfrac{2x+3}{4} + \dfrac{1}{2}\right]\right) + \left[\dfrac{2x+4}{2}\right] = \dfrac{x+3}{2}$,

$\left[\dfrac{2x+3}{2}\right] + \left[\dfrac{2x+4}{2}\right] = \dfrac{x+3}{2}$,

$$\left[\frac{2x+3}{2}\right]+\left[\frac{2x+3}{2}+\frac{1}{2}\right]=\frac{x+3}{2},$$

$[2x+3]=\frac{x+3}{2}$. Next, we have $\frac{x+3}{2}\le 2x+3<\frac{x+3}{2}+1$ and $\frac{x+3}{2}\in \mathbf{Z}$.

Finally, $x=-1$.

V. 136. Since $\left[x+\frac{1}{2}\right]=[2x]-[x]$, we have

$$\left[\frac{n+1}{2}\right]=\left[\frac{n}{2}+\frac{1}{2}\right]=[n]-\left[\frac{n}{2}\right]$$

$$\left[\frac{n+2}{2^2}\right]=\left[\frac{n}{2^2}+\frac{1}{2}\right]=\left[\frac{n}{2}\right]-\left[\frac{n}{2^2}\right]$$

$$\left[\frac{n+2^k}{2^{k+1}}\right]=\left[\frac{n}{2^{k+1}}+\frac{1}{2}\right]=\left[\frac{n}{2^k}\right]-\left[\frac{n}{2^{k+1}}\right]$$

$$S=[n]-\left[\frac{n}{2^{k+1}}\right]=n$$

V. 137. a) $n<\sqrt{n^2+n+1}<n+1$, then $\left[\sqrt{n^2+n+1}\right]=n$;

b) $3n<\sqrt{9n^2+6n}<3n+1$, $\left[\sqrt{9n^2+6n}\right]=3n$.

V. 138. $\displaystyle\sum_{k=1}^{2013}\frac{1}{k^2}=1+\sum_{k=2}^{2013}\frac{1}{k^2}$, also $\dfrac{1}{k^2}<\dfrac{1}{k-1}-\dfrac{1}{k}$, for $k>2$. Therefore

$$\sum_{k=2}^{2013}\frac{1}{k^2}<\sum_{k=2}^{2013}\left(\frac{1}{k-1}-\frac{1}{k}\right)=1-\frac{1}{2013}<1.$$ Finally $\left[\displaystyle\sum_{k=2}^{2013}\frac{1}{k^2}\right]=0$ and

$$\left[\sum_{k=2}^{2013}\frac{1}{k^2}\right]=1.$$

V. 139. Adding (the equations of the system) side by side we get $x+3y=7.3$ (1)

On the other hand $3\{y\}=3.9-[x]\Rightarrow\{y\}=\dfrac{3.9-[x]}{3}\in[0,1)\Rightarrow 0.9\le[x]<3.9$.

Therefore $[x]=1$ or 2 or 3.

For $[x]=1$ $3\{y\}=3.9-[x] \Rightarrow \{y\}=\dfrac{2.9}{3}$ which is not possible (there are more decimals).

For $[x]=2$ $3\{y\}=3.9-[x] \Rightarrow \{y\}=\dfrac{1.9}{3}$ which is also not possible.

For $[x]=3$ $3\{y\}=3.9-[x] \Rightarrow \{y\}=0.3$ and $\{y\}=y-[y]=0.3$

Having in mind (1) and since $[x]=3 \Leftrightarrow 3 \leq x < 4$ we have

$1.1 < y \leq \dfrac{4.3}{3} \Rightarrow [y]=1$ and $\{x\}=0.4$. Then $\begin{cases} x=3.4 \\ y=1.3 \end{cases}$

V. 140. The equation becomes $5^{[x]}+5^{\{x\}}=6$. For $x<-1 \Rightarrow [x]<-1 \Rightarrow$ $5^{[x]}+5^{\{x\}}<5^{-1}+5<6$; For $x \geq 2 \Rightarrow [x] \geq 2 \Rightarrow 5^{[x]}+5^{\{x\}} \geq 5^2+1>6$; For $x \in [1;2) \Rightarrow [x]=1 \Rightarrow 5^{\{x\}}=1 \Rightarrow \{x\}=0 \Rightarrow x=1$;

For $x \in [0,1) \Rightarrow [x]=0 \Rightarrow 5^{\{x\}}=5 \Rightarrow \{x\}=1 \Rightarrow x \in \varnothing$;

For $x \in [-1,0) \Rightarrow [x]=-1 \Rightarrow 5^{\{x\}}=\dfrac{29}{5}>5 \Rightarrow \{x\}>1 \Rightarrow x \in \varnothing$;

Therefore the solution is $x=1$.

Chapter VI

COMPLEX NUMBERS

VI. 1. a) $z_1 = 5 + 2i$; **b)** $z_2 = 1 - 3i$; **c)** $z_3 = -5 + 14i$; **d)** $z_4 = \dfrac{7}{5} - \dfrac{1}{5}i$;

e) $z_5 = \dfrac{1-i}{1+2i} = \dfrac{(1-i)(1+2i)}{1^2 - (2i)^2} = \dfrac{3+i}{5}$; **f)** $z_6 = \dfrac{3}{5} - \dfrac{1}{5}i$.

VI. 2. a) $-32i$; **b)** $-1 - i$; **c)** 0 ; **d)** 0 ; **e)** 0 ;

f) $\left(i^4\right)^{502} \cdot i^2 + \left(i^4\right)^{502} \cdot i^3 + \left(i^4\right)^{502} \cdot i^4 + \left(i^4\right)^{502} \cdot i^5 = i^2 + i^3 + i^4 + i^5 = 0$;

g) $2^{2014}i$. **VI. 3.** 0. **VI. 4. a)** $9 + i$; **b)** $-1 - 5i$; **c)** $26 + 2i$; **d)** $\dfrac{2+i}{10}$;

e) $28 + 8i$. **VI. 5.** yes. **VI. 6. a)** $\operatorname{Re}(z_1) = -4$, $\operatorname{Im}(z_1) = 3$, $\left| z_1 \right| = 5$;

b) $\operatorname{Re}(z_2) = 3$, $\operatorname{Im}(z_2) = -2$, $\left| z_2 \right| = \sqrt{13}$;

c) $\operatorname{Re}(z_3) = -2$, $\operatorname{Im}(z_3) = 1$, $\left| z_3 \right| = \sqrt{5}$;

d) $\operatorname{Re}(z_4) = 1$, $\operatorname{Im}(z_4) = -2$, $\left| z_4 \right| = \sqrt{5}$.

VI. 8. a) $2i$; **b)** $\dfrac{1+3i}{6}$; **c)** $\dfrac{3+3i}{2}$. **VI. 9. a)** $z = -2 + 4i$; **b)** $|z| = 2\sqrt{5}$.

VI. 10. a) $\dfrac{7}{9} - \dfrac{3}{9}i$, **b)** $-\dfrac{3}{13} + \dfrac{11}{13}i$, **c)** $-\dfrac{4}{5} + \dfrac{7}{5}i$, **d)** $\dfrac{11i}{2}$,

e) $\dfrac{\left(1 + 2i + i^2\right)^2 + \left(1 - 2i + i^2\right)^2}{\left(1 + 4i + 4i^2\right)^2 + \left(1 - 4i + 4i^2\right)^2} = \dfrac{4i^2 + 4i^2}{(-3+4i)^2 + (-3-4i)^2} = \dfrac{4}{7}$;

f) $z = \dfrac{4}{3}(1+i)$; **g)** $z = -1 + i$; **h)** $z = -2i$; **i)** $\dfrac{3+7i}{10}$; **j)** $\dfrac{-15-8i}{17}$.

VI. 11. a) $\begin{cases} x + 2y = 3 \\ 3x - y = 2 \end{cases}$, $x = 1$, $y = 1$; **b)** $\begin{cases} x^2 + y = 2 \\ x + 3y^2 = 4 \end{cases}$, $x = 1$, $y = 1$;

c) $\begin{cases} 3\sqrt{x^2 - 2y} + x^2 = 2y \\ -x^2 = 4y - 12 \end{cases}$, $x = \pm 2$, $y = 2$; **d)** $x = -6$, $y = 7$; **e)** $x = 0$, $y = 3$;

f) $x - 3 + (y - 3)i = (x + 2)i - y - 4$, $\begin{cases} x - 3 = -y - 4 \\ y - 3 = x + 2 \end{cases} \Rightarrow y = 2$, $x = -3$.

g) $x = 1$, $y = 4$; **h)** $x = -1$, $y = -2$.

VI. 12. a) $a = 0$, $b = 1$; **b)** $a = 0$, $b = 1$; **c)** $a = \dfrac{1}{13}$, $b = -\dfrac{21}{13}$;

d) $a = \dfrac{6}{5}$, $b = 0$; **e)** $a = \dfrac{1}{2}$, $b = 0$.

VI. 13. a) $|z| = 1$; **b)** $|z| = \dfrac{\sqrt{2}}{100}$; **c)** $|z| = 3^{16}$; **d)** $|z| = 100$;

e) $|z| = 36$; **f)** $|z| = 16\sqrt{2}$. **VI. 14.** $z = -8$. **VI. 15.** $m = 4$.

VI. 16. $z_1\bar{z}_2 + z_2\bar{z}_1 = 4$. **VI. 17.** $z = 1 + i$. **VI. 18.** $z = -2 \pm i$.

VI. 19. $z_1 = 1 - 2i$, $z_2 = -1 + 2i$, $z_3 = -1 - 2i$, $z_4 = 1 + 2i$, $E_1 = E_2 = 0$.

VI. 20. a) $x_{1,2} = \dfrac{-1 \pm 3i}{3}$; **b)** $x_{1,2} = \dfrac{-5 \pm i\sqrt{3}}{2}$; **c)** $x_{1,2} = \dfrac{3 \pm i\sqrt{7}}{4}$;

d) $x_{1,2} = \dfrac{3 \pm i\sqrt{23}}{2}$. **VI. 21. a)** -6, 10; **b)** -8, 17; **c)** 0, 1; **d)** $-\dfrac{4}{25}$, 5.

VI. 22. The complex numbers $1 + 3i$ and $1 - 3i$ are the roots of the equation

$$x^2 - 2x + 10 = 0, \quad x^3 - 4x^2 + 14x - 20 = \left(x^2 - 2x + 10\right)\left(x - 2\right).$$

VI. 23. $\left(x + \dfrac{1}{2} + i\dfrac{\sqrt{3}}{2}\right)\left(x + \dfrac{1}{2} - i\dfrac{\sqrt{3}}{2}\right) = x^2 + x + 1$,

$$x^4 + x^2 + 1 = \left(x^2 + x + 1\right)\left(x^2 - x + 1\right).$$

VI. 24. We have $x_1^2 + x_1 + 1 = 0$ and $x_2^2 + x_2 + 1 = 0$; $x_1 + x_2 = -1$ and

$x_1 x_2 = 1$. On the other hand $x_1^3 - 1 = (x_1 - 1)\left(x_1^2 + x_1 + 1\right) = 0 \Rightarrow x_1^3 = 1$,

similarly $x_2^3 = 1$.

a) $\left(1 + x_1^2 - 1\right)^{2002} + \left(1 + x_2^2 - 1\right)^{2002} = \left(x_1^3\right)^{1334} x_1^2 + \left(x_2^3\right)^{1334} x_2^2 =$

$\qquad = x_1^2 + x_2^2 = (x_1 + x_2)^2 - 2x_1 x_2 = 1 - 2 = -1$; **b)** -1;

c) $\left[x_1^2\left(x_1^2 + x_1 + 1\right) + 1\right]^{2013} + \left[x_2^2\left(x_2^2 + x_2 + 1\right) + 1\right]^{2013} = 2.$

VI. 25. $\varepsilon^3 = 1$, $\varepsilon^3 - 1 = (\varepsilon - 1)\left(\varepsilon^2 + \varepsilon + 1\right) = 0 \Rightarrow \varepsilon^2 + \varepsilon + 1 = 0$

a) $z_1 = \varepsilon^4 + \varepsilon^2 + 1 = \varepsilon \cdot \varepsilon^3 + \varepsilon^2 + 1 = \varepsilon \cdot 1 + \varepsilon^2 + 1 = \varepsilon^2 + \varepsilon + 1 = 0$

b) $z_2 = \dfrac{1 + \varepsilon}{1 - 2\varepsilon + \varepsilon^2} + \dfrac{1 - \varepsilon}{1 + 2\varepsilon + \varepsilon^2} = \dfrac{1 + \varepsilon}{-3\varepsilon} + \dfrac{1 - \varepsilon}{\varepsilon} = \dfrac{-2 + 4\varepsilon}{-3\varepsilon}$.

If $\varepsilon_1 = \dfrac{-1 + i\sqrt{3}}{2} \Rightarrow z_2 = \dfrac{-2 + 4\varepsilon}{-3\varepsilon} = \dfrac{2}{3\varepsilon} - \dfrac{4}{3} = \dfrac{2}{3} \cdot \dfrac{2}{-1 + i\sqrt{3}} - \dfrac{4}{3} = \dfrac{-5 - i\sqrt{3}}{3}$,

If $\varepsilon_2 = \dfrac{-1-i\sqrt{3}}{2} \Rightarrow z_2 = \dfrac{-2+4\varepsilon}{-3\varepsilon} = \dfrac{2}{3\varepsilon} - \dfrac{4}{3} = \dfrac{2}{3} \cdot \dfrac{2}{-1-i\sqrt{3}} - \dfrac{4}{3} = \dfrac{-5+i\sqrt{3}}{3}$;

c) $z_3 = 0$; d) $z_4 = a^3 + b^3 + c^3 - 3abc$; e) $z_5 = 4$; f) $z_6 = 9$.

VI. 27. $z^2 - z + 1 = 0$ then $z^3 + 1 = (z+1)(z^2 - z + 1) = 0 \Rightarrow z^3 = -1$,

$$z^{13} + \dfrac{1}{z^{13}} = \left(z^3\right)^4 z + \dfrac{1}{\left(z^3\right)^4 z} = z + \dfrac{1}{z} = 1.$$

VI. 28. $i^4 = 1$, $\dfrac{1+i}{1-i} = \dfrac{(1+i)(1+i)}{1^2 - i^2} = i$,

$z = \left(i^4\right)^{502} \cdot i^2 + \left(i^4\right)^{502} \cdot i^3 + \left(i^4\right)^{503} + \left(i^4\right)^{503} \cdot i = -1 - i + 1 + i = 0$.

VI. 29. $x_1{}^2 + x_1 + 1 = 0$ and $x_1{}^3 - 1 = (x_1 - 1)(x_1{}^2 + x_1 + 1) = 0 \Rightarrow x_1{}^3 = 1$,

similarly $x_2{}^3 = 1$, $E = \dfrac{\left(-x_1{}^2\right)^{2014}}{-x_1} + \dfrac{\left(-x_2{}^2\right)^{2014}}{-x_2} =$

$= -\left(x_1{}^3\right)^{1342} x_1 - \left(x_2{}^3\right)^{1342} x_2 = -x_1 - x_2 = -1$.

VI. 30. *Two solutions:* **i)** A complex number is a real number, then $z = \bar{z}$.
From $|z_1| = |z_2| = 1$ this means $z_1 \bar{z}_1 = z_2 \bar{z}_2 = 1$.

$$\overline{\left(\dfrac{z_1 + z_2}{1 + z_1 z_2}\right)} = \dfrac{\bar{z}_1 + \bar{z}_2}{1 + \bar{z}_1 \bar{z}_2} = \dfrac{\bar{z}_1 + \bar{z}_2}{1 + \bar{z}_1 \bar{z}_2} = \dfrac{\dfrac{1}{z_1} + \dfrac{1}{z_2}}{1 + \dfrac{1}{z_1} \dfrac{1}{z_2}} = \dfrac{z_1 + z_2}{1 + z_1 z_2}.$$

ii) Consider $z_1 = \cos t_1 + i\sin t_1$, $z_2 = \cos t_2 + i\sin t_2$, $|z_1| = |z_2| = 1$.

We have $\dfrac{z_1 + z_2}{1 + z_1 z_2} = \dfrac{\cos t_1 + \cos t_2 + i(\sin t_1 + \sin t_2)}{1 + \cos(t_1 + t_2) + i\sin(t_1 + t_2)} =$

$$= \dfrac{2\cos\dfrac{t_1 + t_2}{2}\cos\dfrac{t_1 - t_2}{2} + 2i\sin\dfrac{t_1 + t_2}{2}\cos\dfrac{t_1 - t_2}{2}}{2\cos^2\dfrac{t_1 + t_2}{2} + 2i\sin\dfrac{t_1 + t_2}{2}\cos\dfrac{t_1 + t_2}{2}} = \dfrac{\cos\dfrac{t_1 - t_2}{2}}{\cos\dfrac{t_1 + t_2}{2}} \in \mathbf{R}.$$

VI. 31. $|z| = 1 \Rightarrow \bar{z} = \dfrac{1}{z}$. Then $\overline{\left(\dfrac{1 + z^{2n}}{z^n}\right)} = \dfrac{1 + \bar{z}^{2n}}{\bar{z}^n} = \dfrac{1 + \left(\dfrac{1}{z}\right)^{2n}}{\left(\dfrac{1}{z}\right)^n} = \dfrac{1 + z^{2n}}{z^n}$.

VI. 32. $\overline{\dfrac{1 - z + z^2}{1 + z + z^2}} = \dfrac{1 - z + z^2}{1 + z + z^2} \Rightarrow \dfrac{1 - \bar{z} + \bar{z}^2}{1 + \bar{z} + \bar{z}^2} = \dfrac{1 - z + z^2}{1 + z + z^2} \Rightarrow$

$(z - \bar{z})(1 - z \cdot \bar{z}) = 0$, $z \neq \bar{z}$, then $\bar{z} \cdot z = 1$, which means $|z| = 1$.

VI. 33. $z_2 = \dfrac{1 - \bar{z}_1}{1 + \bar{z}_1}$, $x_2 + iy_2 = \dfrac{1 - x_1^2 - x_2^2}{(1 + x_1)^2 + x_2^2} + \dfrac{2y_1 i}{(1 + x_1)^2 + x_2^2}$,

$y_2 = \dfrac{2y_1}{(1 + x_1)^2 + x_2^2}$. Therefore $y_1 y_2 > 0$. **b)** $z_1 + z_2 \in \mathbf{R} \Rightarrow y_1 + y_2 = 0$,

from $y_1 y_2 > 0 \Rightarrow y_1 = 0$ and $y_2 = 0$.

VI. 34. $\bar{z} = \overline{(3 + 2i)^{4n} + (2 + 3i)^{4n}} = \overline{(3 + 2i)^{4n}} + \overline{(2 + 3i)^{4n}} =$

$= \overline{(3 + 2i)}^{4n} + \overline{(2 + 3i)}^{4n} = (3 - 2i)^{4n} + (2 - 3i)^{4n} =$

$= i^{4n}(2 + 3i)^{4n} + i^{4n}(3 + 2i)^{4n} = z$

VI. 35. Use $z \in \mathbf{R} \Leftrightarrow \bar{z} = z$.

$$(1 + i)^m (1 - i)^n = (1 - i)^m (1 + i)^n = \left(\dfrac{1 + i}{1 - i}\right)^{m - n} = i^{m - n}.$$

VI. 36. a) $\bar{z}_n = z_n$;

b) $2z_{n+1} - 2z_n = 2(1 + i)^{n+1} + 2(1 - i)^{n+1} - 2(1 + i)^n - 2(1 - i)^n =$

$= 2i(1 + i)^n - 2i(1 - i)^n = (1 + i)^2 (1 + i)^n + (1 - i)^2 (1 - i)^n = z_{n+2}$.

VI. 37. a) point $(-2, 3)$; **b)** point $(3, -2)$; **c)** point $(-2, 0)$;
d) point $(0, 2)$; **e)** point $(4, -8)$; ; **f)** point $(0.5, 2.5)$.

VI. 38. $OA = |z_1| = \sqrt{10}$, $OB = |z_2| = \sqrt{10}$, $AB = |z_1 - z_2| = 2\sqrt{5}$,

VI. 39. $|z_1 - z_2| = 3\sqrt{10}, |z_1 - z_3| = 2\sqrt{10}, |z_3 - z_2| = \sqrt{10}$.

VI. 40. $|z_1 - z_2| = 3\sqrt{2}$, $|z_1 - z_3| = 3\sqrt{2}$.

VI. 41. $x^2 + y^2 = 5$. **VI. 42. a)** $z = -\dfrac{7}{3} - \dfrac{4}{3}i$; **b)** $z = -\dfrac{7}{6} + \dfrac{5}{6}i$;

c) $z_1 = 2 - 3i$, $z_2 = -4 - 3i$; **d)** $z = a + bi$, $|z| = \sqrt{a^2 + b^2}$,

$$\left|\dfrac{z + 1}{z}\right| = \dfrac{|z + 1|}{|z|} = \dfrac{\sqrt{(a + 1)^2 + b^2}}{\sqrt{a^2 + b^2}},$$

$$\left|\dfrac{z - i}{1 - i}\right| = \dfrac{|z - i|}{|1 - i|} = \dfrac{|z - i|}{\sqrt{2}} = \dfrac{\sqrt{a^2 + (b - 1)^2}}{\sqrt{2}},$$

$$\begin{cases} \dfrac{\sqrt{(a + 1)^2 + b^2}}{\sqrt{a^2 + b^2}} = 2 \\ \dfrac{\sqrt{a^2 + (b - 1)^2}}{\sqrt{2}} = 1 \end{cases} \Rightarrow \begin{cases} 3(a^2 + b^2) = 2a + 1 \\ a^2 + b^2 = 2b + 1 \end{cases}, \quad z_1 = 1,$$

$z_2 = \dfrac{-1-2i}{5}$; e) $z = 2 - i$.

VI. 43. a) $x^2 + (y-1)^2 = 4$; $C(0,\ 1)$; $r = 2$;

b) $\dfrac{x^2}{4} + \dfrac{y^2}{3} = 1$ (ellipse); c) Perpendicular bisector of the segment AO, where

$A(-2, 0)$ and $O(0, 0)$; d) $x^2 + (y+1)^2 < 4$;

e) $\left(\dfrac{x}{\sqrt{2}}\right)^2 - \left(\dfrac{y}{\sqrt{2}}\right)^2 - 1 < 0$; f) $\dfrac{1}{4} < \left(x + \dfrac{3}{2}\right)^2 + (y-1)^2 \le 1$;

g) $x^2 + (y-1)^2 > 16$; h) $z = x + iy$,

$z + i - 2 = (x-2) + i(y+1) \Rightarrow |z + i - 2| = \sqrt{(x-2)^2 + (y+1)^2}$

$\sqrt{(x-2)^2 + (y+1)^2} \le 2 \Rightarrow (x-2)^2 + (y+1)^2 \le 4$.

VI. 44. a) $x_1 = 1$, $x_{2,3} = \dfrac{-1 \pm i\sqrt{3}}{2}$; b) $x_1 = 2$, $x_{2,3} = -1 \pm i\sqrt{3}$;

c) $x_1 = -\dfrac{1}{3}$, $x_{2,3} = \dfrac{1 \pm i\sqrt{3}}{6}$; d) $x_1 = \dfrac{3}{4}$, $x_{2,3} = \dfrac{-3 \pm 3i\sqrt{3}}{8}$; e) ± 2 and $\pm 2i$;

f) ± 3, $\pm 3i$; g) $x_1 = 1$, $x_{2,3} = \dfrac{-1 \pm i\sqrt{3}}{2}$; $x_4 = \sqrt[3]{2}$, $x_{5,6} = \dfrac{-\sqrt[3]{2} \pm i \cdot \sqrt[6]{108}}{2}$;

h) $x_1 = -1$, $x_{2,3} = \dfrac{1 \pm i\sqrt{3}}{2}$; $x_4 = -\sqrt[3]{2}$, $x_{5,6} = \dfrac{\sqrt[3]{2} \pm i \cdot \sqrt[6]{108}}{2}$.

VI. 45. c) $|a + b| \le |a| + |b|$, $|(1 + 2i)z + 2i|z|| \le |1 + 2i||z| + |2i||z| < \sqrt{5} \cdot \dfrac{1}{2} + 2 \cdot \dfrac{1}{2} < 3$;

f) Let $z + \dfrac{1}{z} = u \Rightarrow z^3 + \dfrac{1}{z^3} = u^3 - 3u$. Since $\left|z^3 + \dfrac{1}{z^3}\right| \le 2$, then

$|u^3 - 3u| \le 2 \Rightarrow |u|\left(|u|^2 - 3\right) \le |u^3 - 3u| \le 2$.

The inequality $|u|\left(|u|^2 - 3\right) \le 2$ is equivalent to $\left(|u| - 1\right)^2 \left(|u| - 2\right) \le 0$. Therefore

$|u| \le 2$.

VI. 46. c) $\varepsilon^2 + \varepsilon + 1 = 0$, $\varepsilon^3 = 1$, $\bar\varepsilon = \varepsilon^2$, $\left(\bar\varepsilon\right)^2 = \varepsilon$,

(1) $|z - 1|^2 = (z - 1)(\bar z - 1) = |z|^2 - z - \bar z + 1$,

(2) $|z - \varepsilon|^2 = |z|^2 - z\varepsilon^2 - \bar z\varepsilon + 1$,

(3) $\left|z - \varepsilon^2\right|^2 = |z|^2 - z\varepsilon - \bar z\varepsilon^2 + 1$. Add (1), (2) and (3).

VI. 47. $z = \left(\sqrt[4]{a} + i\sqrt[4]{b}\right)\left(\sqrt[4]{a} + i^2\sqrt[4]{b}\right)\ldots\left(\sqrt[4]{a} + i^{4k}\sqrt[4]{b}\right)$.

VI. 48. a) $a = \pm 2$; b) $a = \pm\sqrt{6}$.

VI. 49. a) $\dfrac{z-2}{z-6} = \dfrac{x-2+iy}{x-6-iy} = \dfrac{(x-2+iy)(x-6+iy)}{(x-6)^2 - (iy)^2} =$

$= \dfrac{(x-2)(x-6)+y^2}{(x-6)^2 + y^2} + i\dfrac{y(x-2)-y(x-6)}{(x-6)^2 + y^2}$,

$\dfrac{(x-2)(x-6)+y^2}{(x-6)^2 + y^2} = \dfrac{y(x-6)-y(x-2)}{(x-6)^2 - (iy)^2} = 0 \Rightarrow (x-4)^2 + (y+2)^2 = 8.$

A circle, of center $C(4,-2)$, $r = 2\sqrt{2}$; b) $\left(x - \dfrac{3}{2}\right)^2 + y^2 = \dfrac{1}{2}$, $C\left(\dfrac{3}{2}, 0\right)$;

$r = \dfrac{1}{2}$;

c) The points from the line $x = 0$;

d) From $\left|z^2 - i\right| = \left|z^2 - 2i\right| \Rightarrow \left|x^2 - y^2 + i(2xy-1)\right| = \left|x^2 - y^2 + i(2xy-2)\right| \Rightarrow$

$(2xy-1)^2 = (2xy-2)^2$. Finally $y = \dfrac{3}{4x}$, which is a hyperbola;

e) $\dfrac{z+1+i}{2-iz} = \dfrac{x+1+i(y+1)}{2+y-ix} = \dfrac{[x+1+i(y+1)](2+y+ix)}{(2+y)^2 + x^2} =$

$= \dfrac{(x+1)(y+2)-x(y+1)+i(x^2+x+y^2+3y+2)}{(2+y)^2 + x^2}$.

Since $\operatorname{Im}\left(\dfrac{z+1+i}{2-iz}\right) = 0$, then

$x^2 + x + y^2 + 3y + 2 = 0 \Rightarrow \left(x + \dfrac{1}{2}\right)^2 + \left(y + \dfrac{3}{2}\right)^2 = \dfrac{5}{2}$, a circle with the center

$C\left(-\dfrac{1}{2}, -\dfrac{3}{2}\right)$, $r = \sqrt{\dfrac{5}{2}}$. **VI. 50.** $\dfrac{4}{9} < (x-3)^2 + y^2 < 100$.

VI. 51. $x \in \left(-\infty, \dfrac{-5-\sqrt{28}}{3}\right) \cup \left(\dfrac{-5+\sqrt{28}}{3}, \infty\right)$.

VI. 52. a) $x_1^2 + x_2^2 = m^2 - m$, $x_1^3 + x_2^3 = m^2(m-1.5)$, $\dfrac{1}{x_1} + \dfrac{1}{x_2} = 2$;

b) $m \in (0, 2)$; c) 4 and $1 - \sqrt{17}$.

VI. 53. $a = \dfrac{8-3i}{2}$. **VI. 54.** $a = -3$, $b = 6$.

$\left(x + \dfrac{1}{2} + i\dfrac{\sqrt{3}}{2}\right)\left(x + \dfrac{1}{2} - i\dfrac{\sqrt{3}}{2}\right) = x^2 + x + 1$,

$$x^4 + x^2 + 1 = \left(x^2 + x + 1\right)\left(x^2 - x + 1\right).$$

VI. 55. **a)** $x(x-1)(x+1)\left(x + \dfrac{1}{2} + i\dfrac{\sqrt{3}}{2}\right)\left(x + \dfrac{1}{2} - i\dfrac{\sqrt{3}}{2}\right)$;

b) $(x+2)\left(x - 2 - i\sqrt{3}\right)\left(x - 2 + i\sqrt{3}\right)$; **c)** $x^2(x-i)(x+i)$;

d) $x^2(x+1)\left(x + \dfrac{3}{2} + i\dfrac{3\sqrt{3}}{2}\right)\left(x + \dfrac{3}{2} - i\dfrac{3\sqrt{3}}{2}\right)$.

VI. 56. **a)** $4 - 3i$; **b)** $1 - 2i$; **c)** $-3 - 4i$; **d)** $z = x + yi$, $x, y \in \mathbf{R}$ the equation

becomes $x^2 - y^2 + 2ixy + 4x + 4 = 0 \Rightarrow x^2 - y^2 + 4x + 4 = 0$ and $xy = 0$.

Finally $z_1 = -2i$; $z_2 = 2i$, $z_3 = -2$; **e)** $z = -0.5i$; **f)** $z = 2 + 1.5i$;

g) $z = -2 + 0.5i\sqrt{2}$; **h)** $z_1 = 3 - 4i$, $z_2 = -3 + 4i$, $z_3 = 4 - 3i$, $z_4 = -4 + 3i$;

i) $z = \dfrac{-1 + i\sqrt{3}}{2}$. **VI. 57.** **a)** $0 + 0i$, $1 + 0i$, $-\dfrac{1}{2} \pm \dfrac{i\sqrt{3}}{2}$;

b) $z = 0$ is a solution. For $z \neq 0$, $r^2 = r \Rightarrow r = 1$. Then $z^3 = \bar{z}z \Rightarrow z^3 = i$

or $z^3 = -i^3$ or $(z+i)\left(z^2 - iz - 1\right) = 0$. The solutions are $z = 0$, $z = -i$,

$z = \dfrac{\pm\sqrt{3} + i}{2}$; **c)** $z = 0$, $z = \pm i$; **d)** $z_1 = 0$, $z_2 = -3i$, $z_3 = 3i$;

e) If $z = a + bi$, then the solutions are $z = a\left(1 \pm i\sqrt{3}\right)$, where $a \leq 0$.

VI. 58. **a)** $z = \dfrac{\left(1 - i\sqrt{3}\right)\left[a - (a+1)i\right]}{a^2 - \left[(a+1)i\right]^2} = \dfrac{a - (a+1)\sqrt{3} - \left(a + 1 + a\sqrt{3}\right)i}{a^2 + (a+1)^2}$,

$a + 1 + a\sqrt{3} = 0 \Rightarrow a = \dfrac{-1}{1 + \sqrt{3}} = \dfrac{1 - \sqrt{3}}{2}$;

b) $b = x + yi$, $\left(x - \dfrac{1}{2}\right)^2 + \left(y - \dfrac{1}{2}\right)^2 = \dfrac{9}{2}$.

VI. 59. **a)** $z_1 = 1 - 2i$, $z_2 = -1 + 2i$; **b)** $z_1 = 2 - \sqrt{3}i$, $z_2 = -2 + \sqrt{3}i$;

c) $z_1 = 2 + i$, $z_2 = -2 - i$; **d)** $z_1 = 4 + 3i$, $z_2 = -4 - 3i$.

VI. 60. **a)** $z_1 = i$, $z_2 = 3i$;

b) $z_{1,2} = \dfrac{1 - i \pm \sqrt{-8 - 6i}}{2} = \dfrac{1 - i \pm \sqrt{(1 - 3i)^2}}{2} = \dfrac{1 - i \pm (1 - 3i)}{2}$, $z_1 = i$,

$z_2 = 1 - 2i$;

c) $z_{1,2} = \dfrac{2 - 6i \pm \sqrt{-48 + 14i}}{2 - i} = \dfrac{2 - 6i \pm (1 + 7i)}{2 - i}$, $z_1 = 1 + i$, $z_2 = \dfrac{-11 + 27i}{5}$;

d) $z_1 = -1+3i$, $z_2 = 1+2i$; e) $z_1 = -1-i$, $z_2 = i$;

f) $z_1 = 2$, $z_2 = 3-2i$; g) $z_1 = 1-i$, $z_2 = 1+2i$; h) $z_1 = 2-i$, $z_2 = 1+2i$;

i) $z_1 = 3-i$, $z_2 = -1+2i$; j) $z_1 = 2+3i$, $z_2 = 3-i$; k) $z_1 = 5+i$, $z_2 = 3+2i$.

VI. 61. a) $(1+i)(x+iy) + (5-3i)(x-iy) = 20 - 4i \Rightarrow 6x - 4iy - 2ix - 4y = 20 - 4i$.

Then $\begin{cases} 6x - 4y = 20 \\ -4y - 2x = -4 \end{cases}$ whose solution is $\begin{cases} x = 3 \\ y = -\dfrac{1}{2} \end{cases}$. Therefore $z = 3 - \dfrac{1}{2}i$;

b) $z = 2+8i$; c) $z = 1+i$; d) $z = -1-i$;

e) $(1+3i)z - (3-2i)z = 11i + 13 \Rightarrow z(-2+5i) = 11i + 13 \Rightarrow z = 1 - 3i$.

f) $2(a+bi) + i(a-bi) = 9 + 3i,\ \begin{cases} 2a+b = 9 \\ 2b + a = 3 \end{cases},\ z = 5 - i$.

VI. 62. a) $z_1 = -\sqrt{2} - i\sqrt{2}$, $z_2 = \sqrt{2} + i\sqrt{2}$, $z_3 = -i\sqrt{2}$, $z_4 = i\sqrt{2}$;

b) $z_1 = 1-i$, $z_2 = -1+i$, $z_3 = 2-i$, $z_4 = -2+i$;

c) $(z^2 + 1)(z^2 - 2z + 3) = 0$, $z_1 = -i$, $z_2 = i$, $z_3 = 1 - i\sqrt{2}$, $z_4 = 1 + i\sqrt{2}$;

d) $z_1 = -1-i$, $z_2 = 1+i$, $z_3 = -1-2i$, $z_4 = 1+2i$;

e) Denote $\dfrac{5z - 2i}{2z - 5i} = a$, $a^3 + a^2 + a + 1 = 0$. Thus $a_1 = -1$,

$a_2 = -i$, and $a_3 = i$.

Finally $z_1 = i$, $z_2 = \dfrac{-29 + 20i}{29}$, and $a_3 = \dfrac{29 + 20i}{29}$.

VI. 63. a) From $(a+bi)^3 = -11 + 2i$ we obtain

$(a^3 - 3ab^2) + (3a^2b - b^3) = -11 + 2i$ or $\begin{cases} a^3 - 3ab^2 = -11 \\ 3a^2b - b^3 = 2 \end{cases}$. In the

equation $2(a^3 - 3ab^2) = -11(3a^2b - b^3)$ setting $b = ta$, yields

$2(1 - 3t^2) = -11(3t - t^3)$ or $(t+2)(11t^2 - 16t - 1) = 0$.

The only convenient solution of this equation is $t_1 = -2$; hence

$z = 1 - 2i$; b) $z = 3 - 2i$; c) $z_1 = 2i$, $z_2 = \sqrt{3} - i$, $z_3 = -\sqrt{3} - i$;

d) $z_1 = \sqrt{3} + 3i$, $z_2 = -\sqrt{3} - 3i$; e) $z_1 = -\sqrt{3} + i$, $z_2 = \sqrt{3} - i$, $z_3 = 1 + i\sqrt{3}$,

$z_4 = -1 - i\sqrt{3}$; f) $z_1 = -\sqrt{3} + i$, $z_2 = \sqrt{3} - i$, $z_3 = \sqrt{3} + i$, $z_4 = \sqrt{3} - i$,

$z_5 = 2i$, $z_6 = -2i$.

VI. 64. a) $\dfrac{z-5+12i}{z-3+4i}$; **b)** $\dfrac{z+1}{z-1+i}$; **c)** $\dfrac{z-i}{z+i}$.

VI. 65. a) $x = 1 - i$, $y = i$; **b)** $x = i$, $y = 2 + i$; **c)** $x = 1 + i$, $y = 1 - i$;

d) $x = 1 + i$, $y = 2 + i$.

VI. 66. a) $z_1 = 1 - i$, $z_2 = i$; **b)** $z_1 = 1 + i$, $z_2 = -2i$;

c) $z_1 = 3 - i$, $z_2 = -1 + 2i$; **d)** $z_1 = 2 + 3i$, $z_2 = 3 - i$;

e) $z_1 = i$, $z_2 = 1 + 3i$.

VI. 67. a) $x_1 = \dfrac{3 + i\sqrt{7}}{2}$, $y_1 = \dfrac{3 - i\sqrt{7}}{2}$, $x_2 = \dfrac{3 - i\sqrt{7}}{2}$, $y_2 = \dfrac{3 + i\sqrt{7}}{2}$;

b) $x_1 = \dfrac{\sqrt{2}}{2} + i$, $y_1 = -\dfrac{\sqrt{2}}{2} + i$; $x_2 = -\dfrac{\sqrt{2}}{2} + i$, $y_2 = \dfrac{\sqrt{2}}{2} + i$; **c)** $x_1 = i\sqrt{2}$,

$y_1 = 1; x_2 = i\sqrt{\dfrac{41}{15}}, y_2 = \dfrac{4}{15}$;

d) $x_1 = \dfrac{1}{2} + i\dfrac{\sqrt{39}}{6}$, $y_1 = 1 - i\dfrac{\sqrt{39}}{3}$; $x_2 = \dfrac{1}{2} - i\dfrac{\sqrt{39}}{6}$, $y_2 = 1 + i\dfrac{\sqrt{39}}{3}$.

VI. 68. a) $f(-1) = -3$, $f(i) = -1 + 4i$, $f(2 - i) = 7 - 6i$,

$f(\bar{z}) = \bar{z}^2 + 3\bar{z} - z$, $f(|z|) = |z|^2 + 2|z|$.

VI. 69. $A = \{(x, 0) \mid x \in \mathbb{R} \setminus \{1\}\}$.

VI. 70. Switching x and $\varepsilon^2 x$ we obtain $f(x) = \varepsilon^2 x$.

VI. 71. $f(x)f(ix) = x^2$ also $f(ix)f(-x) = -x^2$, hence $f(-x) = -f(x)$.

We have $g(x) + g(\varepsilon x) = x$ (1), $g(\varepsilon x) + g(\varepsilon^2 x) = \varepsilon x$ (2), $g(\varepsilon^2 x) + g(x) = \varepsilon^2 x$

(3). Adding (1), (2), and (3) we obtain $g(x) = -\varepsilon x$.

VI. 72. Clearly $b = \dfrac{b}{1 + \varepsilon} + \varepsilon \cdot \dfrac{b}{1 + \varepsilon}$, therefore

the relation (1) becomes (2) $f(x + a) - \dfrac{b}{1 + \varepsilon} = -\varepsilon \left(f(x - a) - \dfrac{b}{1 + \varepsilon} \right)$.

Denoting $g(x) = f(x - a) - \dfrac{b}{1 + \varepsilon}$, (2) becomes $g(x + 2a) = -\varepsilon \cdot g(x)$ and

in general (3) $g(x + 4an) = (-\varepsilon)^{2n} \cdot g(x)$. Since $\varepsilon^n = 1$, (3) becomes (4)

$g(x + 4an) = g(x)$. Using (2) and (4)

$f(x + 4an - a) - \dfrac{b}{1 + \varepsilon} = f(x - a) - \dfrac{b}{1 + \varepsilon} \Leftrightarrow f(x + a(4n - 1)) = f(x - a)$. Taking

$x - a = t$, we obtain $f(t + 4an) = f(t)$, $\forall t \in \mathbb{R}$.

Therefore the period is $T = 4an$.

VI. 73. For $\alpha = 1$, $E = 1 + 2 + 3 + ... + n = \dfrac{n(n+1)}{2}$. For $\alpha \neq 1$,

$$E = \left(\alpha + \alpha^2 + \alpha^3 + ... + \alpha^n\right) + \left(\alpha^2 + \alpha^3 + ... + \alpha^n\right) + \left(\alpha^3 + \alpha^4 + ... + \alpha^n\right) +$$

$$... + \alpha^n = \alpha\frac{1-\alpha^n}{1-\alpha} + \alpha^2\frac{1-\alpha^{n-1}}{1-\alpha} + \alpha^3\frac{1-\alpha^{n-2}}{1-\alpha} + ... + \alpha^n\frac{1-\alpha}{1-\alpha} =$$

$$= \frac{\left(\alpha - \alpha^{n+1}\right) + \left(\alpha^2 - \alpha^{n+1}\right) + \left(\alpha^3 - \alpha^{n+1}\right) + ... + \left(\alpha^n - \alpha^{n+1}\right)}{1-\alpha} =$$

$$= \frac{\left(\alpha + \alpha^2 + \alpha^3 + ... + \alpha^n\right) - \left(n\alpha^{n+1}\right)}{1-\alpha} = \frac{\left(\alpha + \alpha^2 + \alpha^3 + ... + \alpha^n\right) - \left(n\alpha^{n+1}\right)}{1-\alpha} =$$

$$= \frac{\alpha\dfrac{1-\alpha^n}{1-\alpha} - n\alpha^{n+1}}{1-\alpha} = \frac{-n\alpha}{1-\alpha}.$$

VI. 74. If $z = \cos t + i\sin t$, $z^n = \cos nt + i\sin nt$, $n \in N$

(De Moivre's theorem) also $z^{-n} = \cos nt - i\sin nt$. So

$$\left(\cos nt + i\sin nt\right) + \left(\cos nt - i\sin nt\right) = 2\cos nt = z^n + z^{-n} = \frac{z^{2n}+1}{z^n} \text{ and}$$

$$\left(\cos nt + i\sin nt\right) - \left(\cos nt - i\sin nt\right) = 2i\sin nt = z^n - z^{-n} = \frac{z^{2n}-1}{z^n}.$$

b) Let $z = \cos 20^0 + i\sin 20^0$, we get $z^9 = \cos 180^0 + i\sin 180^0 = -1$.

Using a) we have $\sin 70^0 = \cos 20^0 = \dfrac{z^2+1}{2z}$,

$\sin 50^0 = \cos 40^0 = \dfrac{z^4+1}{2z^2}$ and $\sin 10^0 = \cos 80^0 = \dfrac{z^8+1}{2z^4}$. So

$$A = \frac{z^2+1}{2z}\frac{z^4+1}{2z^2}\frac{z^8+1}{2z^4} = \frac{1+z^2+z^4+z^6+z^8+z^{10}+z^{12}+z^{14}}{8z^7} =$$

$$= \frac{\dfrac{1-z^{16}}{1-z^2}}{8z^7} = \frac{1-z^7z^9}{8\left(z^7-z^9\right)} = \frac{1+z^7}{8\left(z^7+1\right)} = \frac{1}{8}.$$

VI. 76. $z = \cos t + i\sin t$ then

$$Z_k = \sqrt[n]{z} = \sqrt[n]{\cos t + i\sin t} = \cos\frac{t+2k\pi}{n} + i\sin\frac{t+2k\pi}{n}, k = 0,1,...,n-1 . \text{ We have}$$

$$z^3 = \frac{3+i}{2-i} = 1+i = \frac{\sqrt{2}}{2}\left(\cos\frac{\pi}{4} + i\sin\frac{\pi}{4}\right), \text{ then } Z_k = \sqrt[3]{\frac{\sqrt{2}}{2}}\left(\cos\frac{\pi}{4} + iso\frac{\pi}{4}\right) \Rightarrow$$

$$Z_k = \frac{1}{\sqrt[6]{2}}\left[\cos\frac{1}{3}\left(\frac{\pi}{4}+2k\pi\right)+i\sin\frac{1}{3}\left(\frac{\pi}{4}+2k\pi\right)\right], \quad k=1,2,3.$$

VI. 77. Let be $\dfrac{1+ia}{1-ia} = r(\cos t + i\sin t)$.

But $\left|\dfrac{1+ia}{1-ia}\right| = \dfrac{|1+ia|}{|1-ia|} = \dfrac{\sqrt{1+a^2}}{\sqrt{1+a^2}} = 1 \Rightarrow r = 1$.

So $\dfrac{1+ia}{1-ia} = \cos t + i\sin t \Rightarrow a = \tan\dfrac{t}{2}$. From $\left(\dfrac{1+iz}{1-iz}\right)^n = \cos t + i\sin t$ we get

$$\frac{1+iz}{1-iz} = \sqrt[n]{\cos t + i\sin t} = \cos\frac{t+2k\pi}{n} + i\sin\frac{t+2k\pi}{n}, k = 0,1,\ldots,n-1 \quad \text{and}$$

$z = \tan\dfrac{t+2k\pi}{2n}, k = 0,1,\ldots,n-1$ Since $t = 2\arctan a$, therefore $z \in \mathbf{R}$.

VI. 78. Let be (1) $\dfrac{x-i}{x+i} = z$, then the initial equation becomes

$$z^{n-1} + z^{n-2} + \ldots + z + 1 = 0.$$

$z^n - 1 = (z-1)(z^{n-1} + z^{n-2} + \ldots + z + 1) = 0 \Rightarrow z^n = 1$, but $z \neq 1$. If $z^n = 1$ we

obtain, $z^n = \cos 0 + i\sin 0 \Rightarrow z = \sqrt[n]{\cos 0 + i\sin 0} = \cos\dfrac{0+2k\pi}{n} + i\sin\dfrac{0+2k\pi}{n}$,

From (1) we get $\dfrac{x-i}{x+i} = \cos\dfrac{2k\pi}{n} + i\sin\dfrac{2k\pi}{n} \Rightarrow x = \dfrac{i(1+z)}{1-z}$. So

$$x_k = i\left(\frac{1+\cos\dfrac{2k\pi}{n}+i\sin\dfrac{2k\pi}{n}}{1-\cos\dfrac{2k\pi}{n}-i\sin\dfrac{2k\pi}{n}}\right) = \frac{2i\cos^2\dfrac{2k\pi}{2n}-2i\sin\dfrac{2k\pi}{2n}\cos\dfrac{2k\pi}{2n}}{2\sin^2\dfrac{2k\pi}{2n}-2i\sin\dfrac{2k\pi}{2n}\cos\dfrac{2k\pi}{2n}} = -\cot\frac{k\pi}{n}$$

$k = 1,2,\ldots,n-1$.

VI. 79. We have $\left(\dfrac{z+i\sqrt{1-z^2}}{z-i\sqrt{1-z^2}}\right)^n = -1 = \cos\pi + i\sin\pi \Rightarrow$

$$\frac{z+i\sqrt{1-z^2}}{z-i\sqrt{1-z^2}} = \cos\frac{\pi+2k\pi}{n} + i\sin\frac{\pi+2k\pi}{n} \Rightarrow$$

$$\frac{z}{\sqrt{1-z^2}} = \frac{i\left[\cos\dfrac{(2k+1)\pi}{2n}+i\sin\dfrac{(2k+1)\pi}{2n}+1\right]}{\cos\dfrac{(2k+1)\pi}{2n}+i\sin\dfrac{(2k+1)\pi}{2n}-1} =$$

$$= \frac{2i\cos^2\frac{(2k+1)\pi}{2n} - 2\sin\frac{(2k+1)\pi}{2n}\cos\frac{(2k+1)\pi}{2n}}{-2\sin^2\frac{(2k+1)\pi}{2n} + 2i\sin\frac{(2k+1)\pi}{2n}\cos\frac{(2k+1)\pi}{2n}} = \cot\frac{(2k+1)\pi}{2n}.$$

Finally $z = \cos\frac{(2k+1)\pi}{2n}$ $k = 0,1,...,n-1$.

VI. 81. Let $M(f(z_1)), N(f(z_2))$, and $P(f(z_3))$ be the geometric images of the numbers $f(z_1)$ $f(z_2)$ and $f(z_3)$, then we have

$$\frac{f(z_3) - f(z_1)}{f(z_2) - f(z_1)} = \frac{az_3 + b - az_1 - b}{az_2 + b - az_1 - b} = \frac{a(z_3 - z_1)}{a(z_2 - z_1)} = \frac{z_3 - z_1}{z_2 - z_1} \in \mathbf{R}^*,$$

therefore M, N and P are collinear.

VI. 82. a) It is known that a function is a *bijection*, if it is a injection (one-to one) and a *surjection* (onto).

A function $f : A \rightarrow B$ is said to be *one-to-one* (*injective*), if and only if $f(z_1) = f(z_2)$ implies that $z_1 = z_2$ for all z_1 and z_2 in the domain of f.

A function $f : A \rightarrow B$ is called *onto* (*surjective*), if and only if for every element y in B there is an element x in A with $f(x) = y$ ($\forall y \in B \Rightarrow \exists x \in A$ such as $f(x)=y$).

For injectivity, let $z_1, z_2 \in C$, $z_1 = x_1 + iy_1$ and $z_2 = x_2 + iy_2$

if $f(z_1) = f(z_2) \Rightarrow 3z_1 + |z_1| = 3z_2 + |z_2| \Rightarrow$

$$3(x_1 + iy_1) + {}_-\sqrt{x_2^2 + y_2^2} = 3(x_2 + iy_2) + \sqrt{x_2^2 + y_2^2} \Rightarrow$$

$$\begin{cases} 3x_1 + \sqrt{x_2^2 + y_2^2} = 3x_2 + \sqrt{x_2^2 + y_2^2} \\ y_1 = y_2 \end{cases} \Rightarrow \begin{cases} x_1 = x_2 \\ y_1 = y_2 \end{cases} \Rightarrow z_1 = z_2.$$

b) For surjectivity $\forall u \in C$, $u = a + ib \Rightarrow \exists z \in C$ such as $f(z)=u$, $z=x+iy$. Since

$f(z) = 3x + 3iy + \sqrt{x^2 + y^2}$. Therefore the function f is a surjection (onto) if

the system (1) $\begin{cases} 3x + \sqrt{x^2 + y^2} = a \\ 3y = b \end{cases}$ has a unique solution. Substituting $y = \frac{b}{3}$

in the first equation we have

$\sqrt{9x^2 + b^2} = 3a - 9x$. This is an irrational equation, $3a - 9x \geq 0$ and

we get $x \leq \frac{a}{3}$. Squaring the both sides of the irrational equation we get

$9x^2 + b^2 = (3a - 9x)^2 \Rightarrow 72x^2 - 54ax + 9a^2 - b^2 = 0$ whose roots are

$$x_1 = \frac{9a + \sqrt{9a^2 + 8b^2}}{24} \quad \text{and}$$

$$x_2 = \frac{9a - \sqrt{9a^2 + 8b^2}}{24}. \text{ Notice that } x_1 = \frac{9a + \sqrt{9a^2 + 8b^2}}{24} > \frac{9a + 3|a|}{24} > \frac{a}{3}$$

and $x_2 = \dfrac{9a - \sqrt{9a^2 + 8b^2}}{24} < x_1 = \dfrac{9a - 3|a|}{24} < \dfrac{a}{3}$. Finally

$$x_2 = \frac{9a - \sqrt{9a^2 + 8b^2}}{24}.$$

So the solution of the system (1) is $x = \dfrac{9a - \sqrt{9a^2 + 8b^2}}{24}$ and $y = \dfrac{b}{3}$.

Since the function is injective and surjective , it is a bijection. Therefore the function f is invertible and its inverse is $f^{-1}(u) = z$. Finally

$$f^{-1}(u) = \frac{9\,\mathrm{Re}\,u - \sqrt{9(\mathrm{Re}\,u)^2 + 8(\mathrm{Im}\,u)^2}}{24} + i\frac{\mathrm{Im}\,u}{3}.$$

Chapter VII

EXPONENTIAL AND LOGARITHMIC FUNCTIONS

VII. 1. a) $x \in (-\infty, -4]$; **b)** $x \in (-\infty, -2]$; **c)** $x \in [-2, +\infty)$; **d)** $x \in \mathbf{R}$;
e) If $a > 1$, $x \in [0, +\infty)$, if $0 < a < 1$, $x \in (-\infty, 0]$, if $a = 1$, $x \in \mathbf{R}$; **f)** $x \in \mathbf{R}$.

VII. 2. a) $f : [-1, -1] \to [1, 2]$; **b)** $f_1 : (-\infty, -1] \cup [1, \infty) \to (0, 1]$; ;

c) $f_2 : [3, +\infty) \ [0, +\infty)$; **d)** $f_3 : \mathbf{R} \ (0, 1]$;

e) $f_4(x) = (2^x - 4)^2 + 1$, $f_4 : \mathbf{R} \ [1, +\infty)$.

VII. 3. d) Vertical stretch by factor 2; Horizontal translation of 3 units to the right; Vertical translation by 1.5 units down.

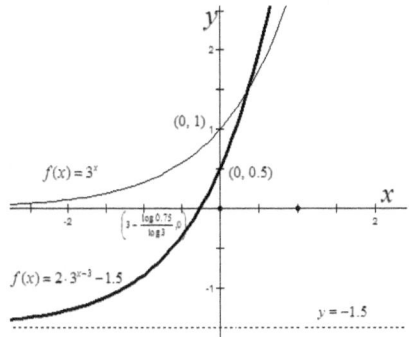

e) Reflection in y-axis; Vertical stretch by factor 3; Vertical translation by 2 units up; $(x, y) \to (x, -3y + 2)$.

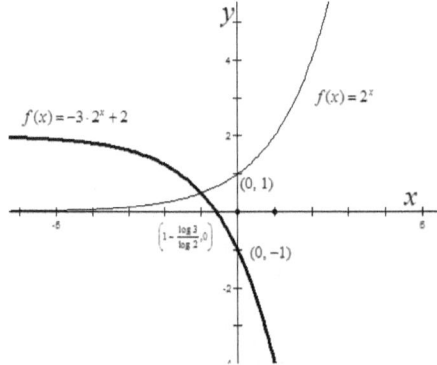

f) Reflection in x-axis and y-axis; Vertical stretch by factor of 2; Horizontal translation by a factor of 3 to the right; Vertical translation by 4 units down; $(x, y) \to (-x - 3, -2y - 4)$.

VII. 4. a) $x \in \left(-\dfrac{2}{3}, \infty\right)$; **b)** $x \in \left(-\infty, \dfrac{1}{3}\right)$; **c)** $x \in (-\infty, -2) \cup (2, \infty)$;

d) $x \in (0, 1)$; **e)** $x \in (0, 3)$; **f)** $x \in [4, 10)$. **VII. 5. a)** $x \neq 0$; **b)** $x \in (0, \infty)$.

VII. 6. a) $x \neq -1$; **b)** $x \in (-1, 3) \backslash \{0\}$; **c)** $x \in (-\infty, -1) \cup (6, \infty)$;

d) $10 - x^2 > 0$, $\log_3\left(10 - x^2\right) > 0$, $x + 1 > 0$ and $x + 1 \neq 0$, $x \in (-1, 3) \backslash \{0\}$; **e)**

$x \in (1, 16)$; **f)** $x \in (1, \infty)$; **g)** $x \neq 0$ **h)** $x \in (-\infty, -3) \cup (1, \infty)$.

VII. 7.

	$y = 2^x$	$y = -2^{x-3} + 4$
Shape	increasing	decreasing
Shape factor	1	-1
point	$(0, 1)$	$(3, 3)$
domain	$x \in \mathbf{R}$	$x \in \mathbf{R}$
range	$y \in (0, \infty)$	$y \in (-\infty, 4)$
asymptote	$y = 0$	$y = 4$
mapping transformations	(x, y)	$(x + 3, -y + 4)$

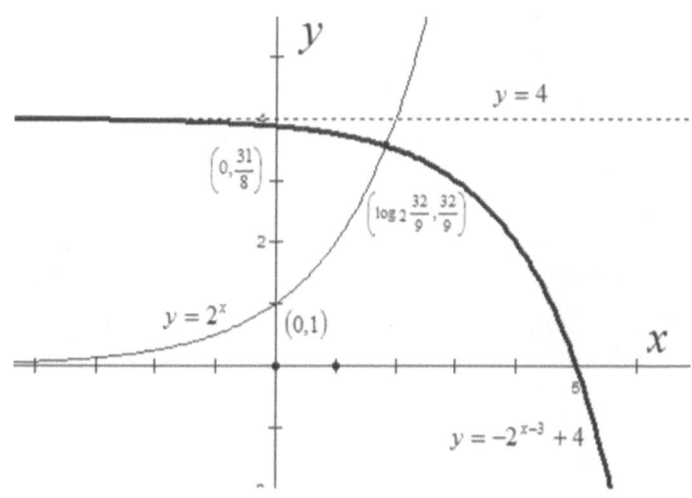

VII. 8. The basic function is $y = \left(\dfrac{1}{2}\right)^x$. It is translated 3 units to the right, a vertically stretched by a factor 3, a reflection in the x-axis, and a translation up 2 units.

Function	$y = \left(\dfrac{1}{2}\right)^x$	$y = -3 \cdot \left(\dfrac{1}{2}\right)^{x-3} + 2$
shape	decreasing	increasing
domain	$x \in \mathbf{R}$	$x \in \mathbf{R}$
range	$y \in (0, \ \infty)$	$y \in (\infty, \ 2)$
asymptote	$y = 0$	$y = 2$
y-intercept	$(0, 1)$	$(0, -22)$
mapping transformations	(x,y)	$(x + 3, -3y + 2)$

VII. 9.

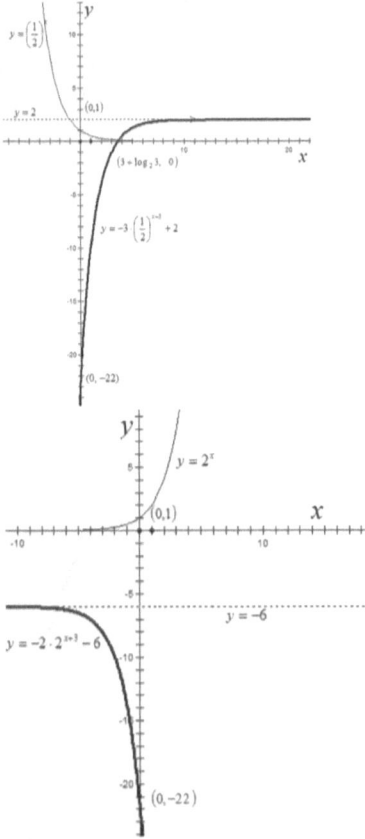

VII. 10.

Function	$y = \log_{\frac{1}{3}} x$	$y = 2\log_{\frac{1}{3}} (x-1) - 3$
Shape	decreasing	decreasing
Shape factor	1	2
point	$(1, 0)$	$(2, -3)$
domain	$x \in (0, \infty)$	$y \in (1, \infty)$
range	$y \in \mathbf{R}$	$y \in \mathbf{R}$
asymptote	$x = 0$	$x = 1$
mapping transformations	(x, y)	$(x+1, 2y-3)$

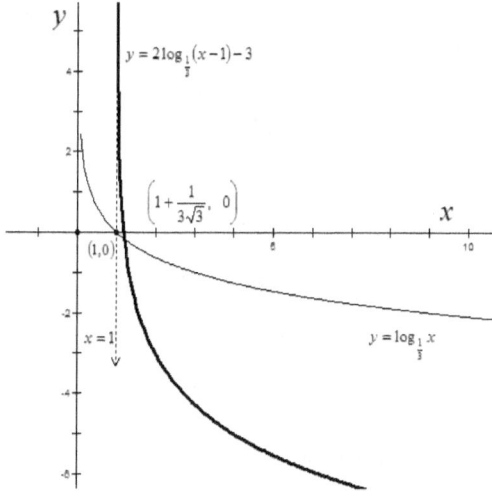

VII. 11.
The basic function is $y = \log_3 x$. It is translated 2 units to the left, vertically stretched by a factor 2, reflected in the x-axis, and translated down 1 unit.

Function	$y = \log_2 x$	$y = -2\log_2 (x+2) - 1$
shape	increasing	decreasing
domain	$x \in (0, \infty)$	$x \in (-2, \infty)$
range	$y \in \mathbf{R}$	$y \in \mathbf{R}$
asymptote	$x = 0$	$x = -2$
x-intercept	$(1, 0)$	$(-1, -3)$
mapping transformations	(x, y)	$(x-2, -2y-1)$

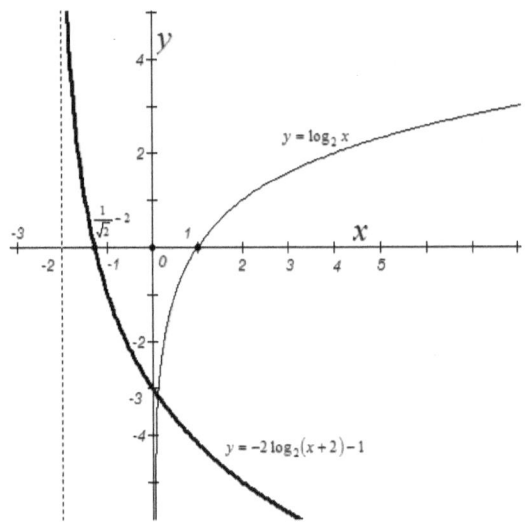

$y = \log_2 x$

$y = -2\log_2(x+2) - 1$

VII. 12. $y = -2\log_2(-x+3) + 5$.

VII. 13. f) A reflection in the x-axis, a vertical stretch by a factor of 5, translation by 2 units to the right and 4 units down, a dilatation (stretch) by a factor 3.

VII. 14. a) 2.0922; **b)** 2.4812; **c)** 0.9244; **d)** 1.0300.

VII. 15. a) 3; **b)** 4; **c)** 2; **d)** 32; **e)** 0.5; **f)** 0; **g)** $-\dfrac{3}{4}$; **h)** $\dfrac{4}{5}$; **i)** $\dfrac{5}{4}$.

VII. 16. a) $\dfrac{1}{4}$; **b)** 10; **c)** $\dfrac{4}{3}$; **d)** -2; **e)** $-3 + \log_2 7$; **f)** 2.

VII. 17. a) $\ln\dfrac{\sqrt[3]{x(x+1)}}{\sqrt{x-1}}$; **b)** $\ln\dfrac{x^2(x^2+1)^3}{\sqrt{x^2-1}}$.

VII. 18. a) 4; **b)** -7; **c)** -1; **d)** 1; **e)** -3; **f)** $-\dfrac{5}{2}$; **g)** $\log_2 5$; **h)** $-0.5\log_2 3$;

i) $\dfrac{1}{2}$; **j)** 1.

VII. 19. a) $\log_6 5$; **b)** $\log_a b = x$, $\left(x + \dfrac{1}{x} + 2\right)\cdot\left(x - \dfrac{x}{x+1}\right)\dfrac{1}{x} - 1 = x$, $\log_a b$;

c) $\dfrac{9}{8}$; **d)** $\log_a 2$; **e)** $\dfrac{\dfrac{1}{3}\log_a(a^2-1)\cdot\dfrac{1}{-4}\log_a(a^2-1)}{\dfrac{1}{4}\log_a(a^2-1)\cdot\dfrac{6}{4}\log_a(a^2-1)} = -\dfrac{2}{9}$;

f) $\begin{cases} 1 & if \ \ 1 < a \le b \\ \log_a b & if \ \ 1 < b < a \end{cases}$

VII. 20. a) $\left(b^{\frac{1}{\frac{1}{2}\log_{10} a}{\log_{10} a}} \cdot a^{\frac{\frac{1}{2}\log_{10} b}{\log_{10} b}} \right)^{2\log_{ab}(a+b)} = \left(\sqrt{b} \cdot \sqrt{a} \right)^{2\log_{ab}(a+b)} = (ab)^{\log_{ab}(a+b)} = a+b \,;$

b) $\sqrt{x^{1+\frac{1}{\log_2 x}} + 8^{\frac{1}{3}\log_x 2}} + 1 = \sqrt{x^{1+\log_x 2} + \left(2^{\log_2 x}\right)^2} + 1 = \sqrt{2x + x^2} + 1 = x+1\,;$

c) $\left[\left(9^{\log_9 5}\right)^2 + \left(3^{\log_3 6}\right)^{\frac{3}{2}} \right]\left[25 - \left(5^{\log_5 6}\right)^{\frac{3}{2}} \right] = \left(25 + 6\sqrt{6}\right)\left(25 - 6\sqrt{6}\right) = 409 \quad;$

d) $\dfrac{\left[\left(3^{\log_3 2}\right)^3 + 5^{\log_5 7}\right]\cdot\left[2\cdot\left(3^{\log_3 2}\right)^4 - \left(2^{\log_2 3}\right)^3\right]}{\left(3 + 5^{\log_5 4}\right)\cdot 5^{\log_5 3}} = \dfrac{\left(2^3 + 7\right)\left(2\cdot 2^4 - 3^3\right)}{(3+4)\cdot 3} = \dfrac{25}{7}\,;$

e) $\dfrac{\log_a b}{\log_a b - 1}$; **f)** $\left(\log_2 x + 1\right)^3$.

VII. 23. a) $\dfrac{4(3-a)}{3+a}$; **b)** $\dfrac{a+3}{2}$; **c)** $4(1-a-b)$; **d)** $\log_6 18 = \dfrac{a+2b}{a+b}$;

e) $abc + 1$.

VII. 24. a) $\log_{30} 8 = \dfrac{\log_{10} 8}{\log_{10} 30} = \dfrac{3\log_{10} 2}{\log_{10} 2 + \log_{10} 3 + \log_{10} 5}$,

$\log_{10} 2 = \log_{10}\dfrac{10}{5} = \log_{10} 10 - \log_{10} 5 = 1 - \log_{10} 5 = 1 - a,$

$\log_{30} 8 = \dfrac{3(1-a)}{b+1}$; **b)** $\dfrac{(1-a-b)(2+a-2b)}{2a(1-b)}$; **c)** $\dfrac{6-6a+3b}{4a-2b}$;

d) $\dfrac{1-2\log_7 c}{1-\log_7 c}$. **VII. 26. a)** Denoting $\log a = x$, $\log b = y$, $\log c = z$, we

obtain $\log_b ab = \dfrac{x+y}{y}$, $\log_c bc = \dfrac{y+z}{z}$,

$\log_a ac = \dfrac{x+z}{x}$. Applying log in (1), we obtain

$\dfrac{x(x+y)}{y} = \dfrac{y(y+z)}{z} = \dfrac{z(x+z)}{x} \Leftrightarrow xyz(x+y) = xyz(z+y) = xyz(x+z).$

Finally $a = b = c$;
c) We have

$(ab)^x = A^2 \Rightarrow x\log_A(ab) = 2 \Rightarrow \dfrac{1}{x} = \dfrac{1}{2}\log_A(ab),$

325

$$(bc)^y = A^2 \Rightarrow y\log_A(bc) = 2 \Rightarrow \frac{1}{y} = \frac{1}{2}\log_A(bc),$$

$$(ac)^z = A^2 \Rightarrow z\log_A(ac) = 2 \Rightarrow \frac{1}{z} = \frac{1}{2}\log_A(ac).\ \text{Adding these equalities, we}$$

obtain

$$\frac{1}{x} + \frac{1}{y} + \frac{1}{z} = \frac{1}{2}\log_A(abc)^2 = \frac{1}{\log_a A} + \frac{1}{\log_b A} + \frac{1}{\log_c A}.$$

VII. 27. a) We have $2 = 3^{\frac{1}{a}}$ and $2^{b+1} = 10$. Therefore

$$\left(3^{\frac{1}{a}}\right)^{b+1} = 10 \Rightarrow 3^{\frac{b+1}{a}} = 10, \text{which yields } \log_3 10 = \frac{b+1}{a}.$$

VII. 28. a) $E = (2 + 4 + ... + 512)^{\frac{1}{9}} = 2$;

b) $F = 3^{\log_3 2} + 2\log_2 \dfrac{4}{\sqrt{3}+\sqrt{2}} - \log_2 \dfrac{1}{\left(\sqrt{3}+\sqrt{2}\right)^2} =$

$$= 2 + \log_2 16 - \log_2\left(\sqrt{3}+\sqrt{2}\right)^2 + \log_2\left(\sqrt{3}+\sqrt{2}\right)^2 = 6.$$

VII. 29. $\dfrac{1}{\log_{\sqrt{n}} 3} = \dfrac{1}{\frac{1}{2}\log_n 3} = 2\log_3 n$ and $\left(\sqrt{3}\right)^{\frac{1}{\log_{\sqrt{n}} 3}} = 3^{\log_3 n} = n$.

Therefore $E = 1 + 2 + 3 + ... + n = \dfrac{n(n+1)}{2} - 1$.

VII. 30. $\dfrac{3}{2}$.

VII. 33. b) $\log_n 3 + \log_n 5 + ... + \log_n(2n-1) < n \Leftrightarrow 1\cdot 3\cdot 5\cdot ... \cdot(2n-1) < n^n$.
Using the inequality between arithmetic mean and geometric mean, we get

$$\sqrt[n]{1\cdot 3\cdot 5\cdot ... \cdot(2n-1)} < \frac{1+3+5+...+(2n-1)}{n} = \frac{n^2}{n} = n.$$

VII. 34. a) $x = 2 + \log_6 3$; **b)** $x = 6.321928$; **c)** $x = 1 + 4\log_{15} 3$; **d)** $x \in \mathbf{R}$,

$x = \ln 3$; **e)** $x \in \mathbf{R}$, $x = \dfrac{4 + \ln 10}{3}$; **f)** $x = 0.173286$; **g)** $x = 0.339482$;

h) $x \in \mathbf{R}$, $x = \ln(\ln 7)$; **i)** $x = 5 + \log_7^2 5$.

VII. 35. a) $x = 1$; **b)** $x = 1$; **c)** $x = -3$; **d)** $x = 1$; **e)** $x_1 = 3$, $x_2 = -\dfrac{7}{2}$.

VII. 36. a) 1; **b)** $x_1 = 0$, $x_2 = 25$; **c)** $x_1 = -2$, $x_2 = 1$; **d)** $x \in (0, \infty) \setminus \{1\}$,

$$\frac{1}{\sqrt{x}-1} - \frac{1}{\sqrt{x}+1} = \frac{2}{\sqrt{x}+1} \Leftrightarrow 4 = 2\sqrt{x} \Rightarrow x = 4 ;$$

e) $x \in [5, \infty)$, $3^{\sqrt{x-5}} = a$, $a^2 - 6a - 27 = 0$, $a_1 = -3$ (impossible),

$a_2 = 9$, $x = 9$; **f)** $x < 9$, $\left(\dfrac{5}{2}\right)^{\frac{4+\sqrt{9-x}}{\sqrt{9-x}}} \cdot \left(\dfrac{5}{2}\right)^{-1+\sqrt{9-x}} = \left(\dfrac{5}{2}\right)^5$,

$$\frac{4+\sqrt{9-x}}{\sqrt{9-x}} - 1 + \sqrt{9-x} = 5, \quad 13 - x = 5\sqrt{9-x}, \quad x_1 = -7, \quad x_2 = 8 ;$$

g) $5^{\frac{1}{3}-\frac{1}{2x}} = 5^{2x-2}$, $x \neq 0$ $\quad x_{1,2} = \dfrac{7 \pm \sqrt{13}}{12}$.

VII. 37. a) 2; **b)** 1.5; **c)** $\dfrac{1}{3}$; **d)** 0.

VII. 38. a) 2; **b)** 4; **c)** $x = 3$; **d)** 4; **e)** 3; **f)** 1; **g)** $6^{2x} \cdot 6 - 6^{2x} \div 6 = 210$ or

$6^{2x}\left(6 - \dfrac{1}{6}\right) = 210$, $x = 1$; **h)** $x = 7$. **VII. 39. a)** 3; **b)** 4; **c)** 4; **d)** 1; **e)** 3.

VII. 40. a) $x = 2$; **b)** $x = 1$ (no solution); **c)** $x = \log_5 2$; **d)** $x = 3$ (no

solution); **e)** $x \in \mathbf{R} \setminus \{2\}$.

VII. 41. a) $x_1 = 2$, $x_2 = -2^{-1}$; **b)** $x = 1$;

c) $5^{x^2} \cdot \dfrac{1}{5} \cdot 4^{x^2} = 4 \cdot 3^{x^2} \cdot \dfrac{1}{3}$, $\dfrac{5^{x^2} 4^{x^2}}{3^{x^2}} = \dfrac{5 \cdot 4}{3}$, $\left(\dfrac{20}{3}\right)^{x^2} = \dfrac{20}{3} \Rightarrow x^2 = 1$, $x = \pm 1$;

d) $x_1 = 1$, $x_2 = \dfrac{1}{4}$; **e)** $x_1 = 3$, $x_2 = -5^{-1}$; **f)** $x_1 = 3$, $x_2 = -2.5$.

VII. 42. a) $x_1 = -1$, $x_2 = 0$; **b)** $2^x\left(2^x + 10\right) = 144$, $2^x = a > 0$, $x = 3$;

c) $x = 1$; **d)** $x = 1$; **e)** $x = 0$; **f)** $5^{2x} \cdot 5^{-1} + 2 \cdot 5^x \cdot 5^{-1} = 5^x + 2$,

$5^x = t > 0$, $t^2 - 3t - 10 = 0 \Rightarrow t_1 = 5$ $t_2 = -2$, $x = 1$.

VII. 43. a) $9 \cdot 9^{x^2} - \dfrac{9}{9^{x^2}} = 80$, $9^{x^2} = t > 0$, $9t^2 - 80t - 9 = 0$,

$t_1 = 9$, $t_2 = -\dfrac{1}{9}$, $x = \pm 1$; **b)** $x_1 = 1$, $x_2 = 2$; **c)** $x = \log_6 8$;

d) $9^{x^2} - 12 \cdot 3^{x^2} + 27 = 0$, $3^{x^2} = a$; $a^2 - 12a + 27 = 0$, $a_1 = 3 \Leftrightarrow x_{1,2} = \pm 1$

and $a_2 = 9 \Leftrightarrow x_{3,4} = \pm \sqrt{2}$; **e)** $x_1 = 0$, $x_2 = 8$.

VII. 44. a) $x = 3$; **b)** $x_1 = 1$, $x_2 = -\log_3 2$; **c)** $x_1 = 1$, $x_2 = 2$;

d) $x_1 = -0.5$, $x_2 = 0$; **e)** $x_1 = -1$, $x_2 = 8$.

VII. 45. a) $x = 1$; **b)** $x = \dfrac{5 + \sqrt{5}}{2}$; **c)** $x = 1$; **d)** $x = 1$.

VII. 46. a) $x = 0$; **b)** $x = -1$; **c)** $x = 2$; **d)** $x = 0$.

VII. 47. a) $x = -\dfrac{1}{2}$; **b)** $3^{3x} - 13 \cdot 3^{2x} + 39 \cdot 3^x - 27 = 0$, $3^x = a$,

$(a - 3)(a^2 - 10a + 9) = 0$ $a_1 = 3 \Rightarrow x = 1$, $a_2 = 9 \Rightarrow x = 2$ and

$a_3 = 1 \Rightarrow x = 0$;

c) $3^{3x} + \dfrac{1}{3^{3x}} + 3 \cdot 3^x + \dfrac{3}{3^x} = 8 \Rightarrow \left(3^x + \dfrac{1}{3^x}\right)^3 = 2^3 \Rightarrow$

$3^{2x} - 2 \cdot 3^x + 1 = 0$, $\left(3^x - 1\right)^2 = 0$, $3^x = 1$, $x = 0$;

d) $3^x - \dfrac{2}{3^x} = t$, $x_1 = -\dfrac{1}{2}$, $x_2 = \dfrac{1}{2}$;

e) $\dfrac{5}{3^x - \dfrac{3}{3^x} + 3} + \dfrac{3}{3^x - \dfrac{3}{3^x} - 1} = 4$, $3^x - \dfrac{3}{3^x} = t$, $\dfrac{5}{t + 3} + \dfrac{3}{t - 1} = 4$.

Finally, $x_1 = 0$, $x_2 = 1$.

VII. 48. a) $x = 20$; **b)** $x = 1$; **c)** $x \in [-5, \infty)$, $3^{\sqrt{x+5}} = t$, $t^2 - 6t - 27 = 0$,

$t_1 = -3$ (impossible) and $t_2 = 9$, $\sqrt{x + 5} = 2 \Rightarrow x + 5 = 4 \Rightarrow x = -1$;

d) $x \in (-\infty, -2] \cup [2, \infty)$; $2^{x + \sqrt{x^2 - 4}} = a$; $2a^2 - 5a - 12 = 0$; $x = 2.5$;

e) $\left(1 + \sqrt{3}\right)^x = t > 0$, $x = \log_{1 + \sqrt{3}} 2$.

VII. 49. a) $\left(2 + \sqrt{3}\right)^x = t$, $\left(2 - \sqrt{3}\right)^x = \dfrac{1}{t}$, $t^2 - 14t + 1 = 0$,

$t_{1,2} = 7 \pm 4\sqrt{3} = \left(2 \pm \sqrt{3}\right)^2$, $x = \pm 2$; **b)** ± 1; **c)** ± 2; **d)** ± 2.

VII. 50. a) $x = 1$; **b)** $x_{1,2} = 1 \pm \sqrt{2}$, $x_3 = 1$; **c)** ± 1; **d)** 1; **i)** ± 1.

VII. 51. a) $0 < x^2 - x - 1 \neq 1$ and $x^2 - 1 = 0$. $x_1 = -1$, $x_2 = 1$, $x_3 = 2$;

b) Since the LHS is a positive number, it follows that $x \in [2, \infty)$. From

328

$x^2 + 3x - 4 = 0$, we obtain the solutions $x_1 = -4$ (impossible), $x_2 = 1$ (impossible), $x = 3$ is a solution; c) $x \in \{-4, 1, 2, 4\}$;

d) If $x^2 - 5x + 6 = 1$ the solutions are $x_{1,2} = \dfrac{5 \pm \sqrt{5}}{2}$ and if

$0 < x^2 - 5x + 6 \neq 1$ the solutions are $x_3 = -3$, $x_4 = 0$; e) $x_1 = 0$, $x_2 = 3$, $x_3 = 6$.

VII. 52. a) $x \in [1, \infty) \cup \{-1\}$; b) $x \in [1, 2]$; c) $x \in [1, 2]$.

VII. 53. a) ± 1; b) -2; c) -1, 1, and 0;

d) $\dfrac{5^{2x}}{5} + 2^{2x} - 5^{2x} + 2^{2x} \times 2^2 = 0 \Leftrightarrow 5^{2x}\left(\dfrac{1}{5} - 1\right) + 2^{2x} \times 5 = 0$,

$\left(\dfrac{5}{2}\right)^{2x} = \left(\dfrac{5}{2}\right)^2 \Rightarrow x = 1$; e) Dividing by $6^{\frac{1}{x+1}}$ we get

$6 \cdot \left(\dfrac{3}{2}\right)^{\frac{1}{x+1}} + 6 \cdot \left(\dfrac{2}{3}\right)^{\frac{1}{x+1}} - 13 = 0$,

$\left(\dfrac{3}{2}\right)^{\frac{1}{x+1}} = t$, $6t^2 - 13t + 6 = 0$, $x_1 = 0$ and $x_2 = -2$; f) 0; g) -1; h) 0.5; i) 0 and $\log_{1.5} 2$.

VII. 54. a) $5^{x-2012}\left(3 \cdot 5^4 + 5^3 + 2\right) = 2002$, since 2002 does not have 5 as a

factor, it follows that $3 \cdot 5^4 + 5^3 + 2 = 2002$, and $5^{x-2012} = 1$, $x = 2012$;

b) $6^{x-2003}\left(9 \cdot 6^3 + 6^2 + 3 \cdot 6 + 4\right) = 2002$. Through a similar reasoning as in a), it follows that $x = 2003$.

VII. 55. Suppose that $\log_5 10$ is a rational number. $\log_5 10 = \dfrac{a}{b}$, where a, b are

integer positive numbers and relatively prime. The equality $5^a = 10^b$ is false,

left side odd number, right side even number, contradiction.

VII. 56. a) $x \in (0, \infty)$, $x = e^6$; b) $x \in (1.5, \infty)$, $x = 0.5(e + 3)$;

c) $x = 15.085536$; d) $x \in \mathbb{R}$, $x = 2 - \ln 8$; e) $x \in (0, \infty)$, $x = 20$;

f) $x \in (1, \infty)$, $x = e^{e^2}$.

VII. 57. a) $x \in (0, \infty) \setminus \{1\}$, $x = 9$; b) $x \in (-0.5, \infty)$, $x = 3$; c) $x \in (0, \infty)$, $x = 1$;

d) $x \in (-\infty, 10)$, $x = 1$; e) $x \in (2.5, \infty)$, $x = 3$; f) $x \in (2.\overline{3}, \infty)$, $x = 34$;

g) $x \in (-1.5, \infty)$, $x_1 = -1$, $x_2 = 2$; h) $x = 0$; i) $x = 3$.

VII. 58. a) $x = 15$; **b)** $x = 27$; **c)** $x = 27$; **d)** $x \in (-0.5, \infty) \setminus \{0\}$, $x = 1$; **e)**
$x \in (-0.6, \infty) \setminus \{0.5\}$, $x = 3$; **f)** $x = 3$.

VII. 59. a) $x \in (3, \infty)$, $x^2 - 3x = x$, $x_2 = 4$; **b)** $x = 1$; **c)** $x \in (2, \infty)$; $x = 5$;

d) $\begin{cases} x^3 + 1 > 0 \\ x^2 + 2x + 1 > 0 \end{cases}$, $\begin{cases} x \in (-1, \infty) \\ x \neq -1 \end{cases}$, $x \in (-1, \infty)$, $\log_3 (x^2 + x + 1) = \log_3 7$,

$x_1 = 2$ and $x_2 = -3$; **e)** $x = \sqrt[10]{10}$; **f)** $x \in (0, \infty)$;

$\log_4 x + \dfrac{1}{2} \log_4 x = -\log_4 \sqrt{3} \Rightarrow x = \dfrac{1}{\sqrt[3]{3}}$;

g) $x \in (-1, 0)$; $x = -\dfrac{7}{8}$; **h)** $\log_2 \sqrt{\log_2 x} + \log_2 \dfrac{\log_2 x}{2} = 2 \Leftrightarrow$

$\log_2 x \sqrt{\log_2 x} = 8 \Leftrightarrow \log_2 x = 4 \Rightarrow x = 16$.

VII. 60. a) $x_1 = 2$; $x_2 = 2^{-2}$; **b)** $x_1 = 2\sqrt{2}$; $x_2 = -2\sqrt{2}$; **c)** $\pm \sqrt{7}$;

d) $x_1 = 25$; $x_2 = 5^{-1}$; **e)** $x_1 = 5$; $x_2 = 10^{-2}$; **f)** $x = 0.99$; **g)** $x \in (0, \infty)$,

$\log_a x = t$, $3t - 2 = t^2 (2t - 3)$, $(t + 1)(2t^2 - 5t + 2) = 0$, $x_1 = a^{-1}$, $x_2 = a^2$,

$x_3 = \sqrt{a}$; **h)** $x_1 = 7, x_2 = \dfrac{11}{7}$.

VII. 61. a) $x \in (-2, \infty)$,

$\log_6 (2 + x) + \log_6 (3 + x) = 1 \Rightarrow \log_6 (2 + x)(3 + x) = 1$, $x_1 = 0$ and $x_2 = -5$
(impossible); **b)** $x \in (4, \infty)$, $\log_2 (2 + x)(x - 4) = 0$, $x_1 = 1 + \sqrt{10}$,

$x_2 = 1 - \sqrt{10}$ (impossible); **c)** $x \in (0, \infty)$, $\log_2 x + 2 \log_2 x = 6$ $\log_2 x = 2$,
$x = 4$;

d) $x \in (2, \infty)$, $\log_2 (x - 2) = \log_2 \sqrt{2x - 1} \Rightarrow x - 2 = \sqrt{2x - 1}$, $x = 5$;

e) $x = 2.5$; **f)** $x = 3.5$.

VII. 62. a) $x \in (0, \infty) \setminus \{1\}$, $\dfrac{1}{2} \log_x 5 + \log_x 5 + 1 = \dfrac{1}{4} (\log_x 5)^2 + \dfrac{9}{4}$,

$\log_x 5 = t$, $t^2 - 6t + 5 = 0$, $x = \sqrt[5]{5}$, $x = 5$; **b)** $x \in (0, \infty)$,

$(-\log_3 9 - \log_3 x)^2 + 2 \log_3 x - \log_3 27 = 8$, $\log_3 x = t$, $t^2 + 6t - 7 = 0$,

$x_1 = 3$, $x_2 = 3^{-7}$; c) $\sqrt{\log_a x} = y$, $y + \dfrac{1}{y} = \dfrac{10}{3}$, $y_1 = 3 \Rightarrow x_1 = a^9$ and

$y_2 = \dfrac{1}{3} \Rightarrow x_2 = a^{\frac{1}{9}}$; d) $x = 6$; e) $x_1 = 3$ (impossible), $x_2 = 7$; f) $x = 4$;

g) $x = 3$.

VII. 63. a) $x = 2$; b) $x = \pm\sqrt{\log_2 3}$; c) $x_1 = 3$, $x_2 = 5$;

d) $\log_8\left(9^{x+1} - 1\right) - \log_8\left(3^{x+1} + 1\right) = \dfrac{1}{3}$, $\log_8 \dfrac{\left(3^{x+1} + 1\right)\left(3^{x+1} - 1\right)}{3^{x+1} + 1} = \dfrac{1}{3}$, $3^{x+1} - 1 = 8^{\frac{1}{3}}$,

$x = 0$; e) $x = 5$; f) $x_1 = 2$, $x_2 = 4$; g) $x = 0$.

VII. 64. a) $\begin{cases} x \in (0,\infty) \setminus \{1\} \\ \log_x \sqrt{4x} \geq 0 \end{cases}$, $x \in \left(0, \dfrac{1}{4}\right] \cup (1,\infty)$. From $\sqrt{\log_x \sqrt{4x}} = -\log_x 4$,

$-\log_x 4 \geq 0$, $x \in \left(0, \dfrac{1}{4}\right]$, $\sqrt{\dfrac{1}{2}\left(\log_x 4 + \log_x x\right)} = -\log_x 4$, $\log_x 4 = t < 0$,

$\sqrt{\dfrac{1}{2}(t+1)} = -t$ $2t^2 - t - 1 = 0$, $x = \dfrac{1}{16}$; b) $x \in \left(0, \dfrac{1}{3}\right]$,

$\sqrt{\dfrac{1}{2}\log_x 3 + \dfrac{1}{2} \cdot \dfrac{1}{\log_x 3}} = -1$,

$\log_x 3 = -\dfrac{1}{2} \Rightarrow x = \dfrac{1}{9}$ and $\log_x 3 = 1$ (impossible); c) $x = 16$; d) $x_1 = 2^{-8}$,

$x_2 = 2^{27}$; e) 50, $\dfrac{1}{200}$; f) $\log_5 x = a$, $x = 5$.

VII. 65. a) $x_1 = 2^{-\frac{4}{27}}$, $x_2 = 2^{\frac{1}{3}}$; b) $\log_6 x = t$, $y^2 + \dfrac{1}{y^2} + 2\cdot\left(y + \dfrac{1}{y}\right) + \dfrac{3}{4} = 0$,

$y + \dfrac{1}{y} = z$, $4z^2 + 8z - 5 = 0$, $x_1 = \dfrac{1}{36}$, $x_2 = \dfrac{1}{\sqrt{6}}$; c) $x_1 = \sqrt[5]{5}$; $x_2 = 5$;

d) $x = \dfrac{a^4 + 1}{1 - a^4}$; e) $x_1 = \dfrac{1}{3}$, $x_2 = 9$; f) $x = a^6$;

g) $x \in (0,\infty) \setminus \{1\}$, $x_1 = \sqrt{2}$, $x_2 = \sqrt[4]{2}$.

VII. 66. a) $x \in (0,\infty) \setminus \left\{\dfrac{1}{3}\right\}$, $\dfrac{1}{\log_3 3 + \log_3 x} - \dfrac{1}{\log_x 3 + \log_x x} + \log_3^2 x = 1$,

$\log_3 x = t$, $t^3 + t^2 - 2t = 0$, $x_1 = 1$, $x_2 = 3^{-2}$, $x_3 = 3$;

b) $x \in (0, \infty) \setminus \left\{ \dfrac{1}{16}, \dfrac{1}{4}, 1 \right\}$. $x_2 = \dfrac{1}{2}$, $x_3 = \dfrac{1}{8}$; **c)** $x_1 = 3$, $x_2 = 3^5$;

d) $x \in (0, \infty) \setminus \left\{ \dfrac{1}{81}, \dfrac{1}{9}, 3 \right\}$. We observe that $x = 1$ verifies the initial equation.

We have $\dfrac{10}{\log_x 9x} + \dfrac{21}{\log_x 81x} - \dfrac{6}{\log_x \dfrac{x}{3}} = 0$,

$\dfrac{10}{2\log_x 3 + \log_x x} + \dfrac{21}{4\log_x 3 + \log_x x} - \dfrac{6}{\log_x x - \log_x 3} = 0$,

$\log_x 3 = t$, $26t^2 - 3t - 5 = 0$, $x_1 = 3^3$, $x_2 = \dfrac{1}{\sqrt[5]{3^{13}}}$; **e)** $x_1 = a$, $x_2 = a^{-\frac{16}{9}}$;

f) $x \in (0, \infty) \setminus \left\{ \dfrac{1}{2}, \sqrt{2}, 2 \right\}$, $\dfrac{1}{\log_x \dfrac{2}{x^2}} + \dfrac{2}{\log_x \dfrac{2}{x}} - \dfrac{8}{\log_x 2x} = 0$,

$\dfrac{1}{t-2} + \dfrac{2}{t-1} - \dfrac{8}{t+1} = 0$, $5t^2 - 22t + 21 = 0$, $\log_x 2 = t$, $x_1 = \sqrt[3]{2}, x_2 = \sqrt[7]{32}$.

Notice: The initial equation has $x_3 = 1$ as a solution, too;

g) $x \in (0, \infty) \setminus \left\{ \dfrac{1}{27}, \dfrac{1}{3} \right\}$, $x_1 = 3$, $x_2 = 3^{-4}$;

h) $x \in (0, \infty) \setminus \left\{ \dfrac{1}{a^3}, \dfrac{1}{a} \right\}$, $x_1 = a$, $x_2 = a^{-4}$.

VII. 67. a) $\dfrac{1}{81}$, 9; **b)** $x^{\frac{\log_9 16}{\log_3 x}} = x^{\frac{\log_3 4}{\log_3 x}} = x^{\log_x 4} = 4$, $\log_2 3 + \log_2 x = 4$, $x = \dfrac{16}{3}$;

c) $x \in (0, \infty) \setminus \{1\}$, $x^{\log_{10} \sqrt{x}} = 100 \Leftrightarrow \log_{10}\left(x^{\log_{10} \sqrt{x}} \right) = \log_{10} 10^2 \Leftrightarrow$

$\log_0 \sqrt{x} \log_{10} x = 2 \Leftrightarrow \dfrac{1}{2}(\log_{10} x)\log_{10} x = 2$, $x = 10^{\pm 2}$;

d) $x_1 = 10^{-4}$, $x_2 = 10$; **e)** $\left(2\log^3 x - 1.5\log x \right)\log x = 0.5$, $\log^2 x = t \geq 0$,

$4t^2 - 3t - 1 = 0$; $x = 10^{\pm 1}$; **f)** $x_1 = 7, x_2 = 14$; **g)** $x = \sqrt{3}$.

VII. 68. a) The equation becomes $(7^x)^3 + \left(8^x\right)^3 + \left(9^x\right)^3 = 3 \cdot (7 \cdot 8 \cdot 9)^x$.

$\dfrac{a+b+c}{3} \geq \sqrt[3]{abc}$ is true for any $a, b, c \geq 0$, the equality holds for $a = b = c$.

Therefore $343^x = 512^x = 729^x$, thus $x = 0$;

b) The arithmetic-geometric mean inequality gives

$$9^x + 36^x + 32^x + 32^x \geq 4\sqrt[4]{9^x \cdot 36^x \cdot 32^x \cdot 32^x} = 4\sqrt[4]{3^{2x} \cdot 3^{2x} \cdot 2^{2x} \cdot 2^{5x} \cdot 2^{5x}} =$$

$$= 4\sqrt[4]{3^{4x} \cdot 2^{12x}} = 4\sqrt[4]{3^{4x} \cdot 8^{4x}} = 4 \cdot 24^x.$$ Equality holds when the numbers are

equal, then $9^x = 36^x = 32^x \Rightarrow x = 0.$

VII. 69. a) $(1, 1)$, $\left(\dfrac{1}{\sqrt[3]{3}}, \sqrt[3]{9}\right)$; **b)** $(2, 4)$, $\left(-2, \dfrac{1}{4}\right)$; **c)** Multiplying both

equations, $6^{x+y} = 6^3 \Rightarrow x + y = 3$. Dividing both equations side by side,

$\left(\dfrac{2}{3}\right)^{x-y} = \dfrac{3}{2} \Rightarrow x - y = -1$, $(1, 2)$; **d)** $(3, 1)$, $(-1, -3)$; **e)** $(-17, 2)$;

f) $\begin{cases} x^y \cdot x = 27 \\ \dfrac{(x^y)^2}{x^5} = \dfrac{1}{3} \end{cases} \Rightarrow \begin{cases} x^y = \dfrac{27}{x} \\ (x^y)^2 = \dfrac{x^5}{3} \end{cases}$, $(3, 2)$; **g)** $(1, 2)$, $(2, 1)$;

h) $2^{x+y} = t$, $3^{x-y} = z$, $\begin{cases} t^2 - z^2 = -65 \\ 5t + 6z = 18 \end{cases}$, $t_1 = 4$, $t_2 = \dfrac{664}{77}$ and $z_1 = 9$, $z_2 < 0$.

Therefore $\begin{cases} x + y = 2 \\ x - y = 2 \end{cases}$, $(2, 0)$.

VII. 70. a) $x, y \in (0, \infty)$, $\begin{cases} \left(\sqrt[4]{x} + \sqrt{y}\right)\log x = \dfrac{8}{3}\log y \\ \left(\sqrt[4]{x} + \sqrt{y}\right)\log y = \dfrac{2}{3}\log x \end{cases}$,

$\left(\sqrt[4]{x} + \sqrt{y}\right)^2 = \dfrac{16}{9} \Rightarrow \sqrt[4]{x} + \sqrt{y} = \dfrac{4}{3}$, substituting the last relation in the first

equation $x^{\frac{4}{3}} = y^{\frac{8}{3}}$, $x = y^2$, $\begin{cases} x = y^2 \\ \sqrt[4]{x} + \sqrt{y} = \dfrac{4}{3} \end{cases}$, $x = \dfrac{16}{81}$, $y = \dfrac{4}{9}$, $x = y = 1$ is a

solution for the system, too.

b) $(-2, \infty)$ $x, y \in (0, \infty)$, $a = 4\sqrt[3]{x}$, $b = 3\sqrt{y}$, $\begin{cases} ab = 144 \\ a^2 + b^2 = 337 \end{cases} \Leftrightarrow \begin{cases} ab = 144 \\ a + b = 25 \end{cases}$,

$(8, 4)$, $\left((\log_4 9)^3, (\log_3 16)^2\right)$;

c) $(8, 3)$, $\left((\log_3 5)^3, 2 + (\log_5 9)^4\right)$;

d) $(1, 1, 1)$, $(4, 2, \sqrt{2})$; **e)** $(2, 3, 4)$; **f)** $(0, 0, 0)$.

VII. 71. a) $x \in (0, \infty) \setminus \{1\}$, $\begin{cases} x + y = x^2 \\ 4x - 2y = x^2 \end{cases}$, $(2, 2)$; **b)** $x, y \in (0, \infty) \setminus \{1\}$, $(4, 4)$;

c) $(3, 9)$; **d)** $(2, 4)$; **e)** $(3, 2)$; **f)** $(3, 9)$, $(\frac{1}{9}, \frac{1}{3})$; **g)** $(3, 27)$, $(27, 3)$;

h) $\begin{cases} \sqrt{y}\left(2\sqrt[4]{x}-1\right)=3 \\ \sqrt[4]{x}\left(\sqrt{y}+1\right)=4 \end{cases}$, $(1, 9)$, $(6, 1)$.

VII. 72. a) $(3, 4)$, $(4, 3)$; **b)** $(3, 9)$; **c)** $(3, 5)$, $(5, 3)$;

d) $\begin{cases} \log_3 x + \log_3 y = 3 \\ \log_3 y \log_3 x = 2 \end{cases}$, $(3, 9)$ or $(9, 3)$; **e)** $x, y > 0$, $\begin{cases} x^2 + y^2 = 10^2 \\ x \cdot 2^2 = y \cdot 3 \end{cases}$, $(6,8)$;

f) $(2, 6)$, $(6, 2)$; **g)** $\log_a\left(x^{\log_a x} y^{\log_a y}\right) = 2m$, $x^{\log_a x} y^{\log_a y} = a^{2m}$, $u = x^{\log_a x}$,

$v = y^{\log_a y}$, $uv = a^{2m}$, $u + v = 2\dfrac{p^2 + q^2}{p^2 - q^2} a^m$; **h)** $\left(10^{-1}, 1, 10\right)$.

VII. 73. For $x > 1$, let $f(x) = 5^{\log_{10}\left(x^3 - x^2\right)}$. The three equations are $f(x) = y^2$,

$f(y) = z^2$ and $f(z) = x^2$. Since $x^3 - x^2 = x^2(x - 1)$ is increasing, f is an
increasing function. If, say, $x < y$, then $y < z$ and $z < x$, yielding a contradiction.
Thus, we can only have that $x = y = z$ and so $\log_D\left(x^3 - x^2\right) = \log_5 x^2$. Let

$2t = \log_5 x^2$ so that $t > 0$, $x^2 = 5^{2t}$ and so

$x = 5t$. Therefore $5^{3t} - 5^{2t} = 10^{2t} \Rightarrow 5^t - 1 = 4^t \Rightarrow 5^t - 4^t = 1$.

Since $5^t - 4^t = 4^t\left[\left(\dfrac{5}{4}\right)^t - 1\right]$ is an increasing function of t, we see

that the equation for t has a unique solution, namely $t = 1$. Therefore $x = 5$.
VII. 74. a) $\left(2^x - 1\right)\left(3^x - 1\right) \geq 0$, $x \in \mathbf{R}$; **b)** $x \in [1, 2]$; **c)** $x \in (0, \infty)$;

d) $x \in [0, 4]$; **e)** $x \in \left(-0.\overline{3}, \infty\right)$; **f)** $3^{\sqrt[4]{x} - \sqrt{x}} = t$, $9t^2 + 8t - 1 \geq 0$, $x \in [0, 16]$;
g) $x \in (0, 1) \cup (1, 2)$; **h)** $x \in [-1, 3)$; **i)** $x \in (-\infty, 0) \cup (1, \infty)$;

j) $x \in (-\infty, 1]$; **k)** $x \in \left(0, \log_{\frac{2}{3}} \dfrac{1}{2}\right]$; **l)** $x^{3x} < x^{2x^2 - 2x + 1}$ or $x^{(2x-1)(x-2)} > 1$;

I) $x \in (0, 1)$, $(2x - 1)(x - 2) < 0$, then $x \in (0.5, 1)$;

II) $x \in (1, \infty)$, $(2x - 1)(x - 2) > 0$, then $x \in (2, \infty)$. Finally, $x \in (0.5, 1) \cup (2, \infty)$.

VII. 75. a) $x \in \left(-\dfrac{5 + \sqrt{13}}{2}, -4\right) \cup \left(-1, \dfrac{-5 + \sqrt{13}}{2}\right)$;

b) $x \in \left(2^{-28}, 1\right)$; **c)** $x \in (-1 + \log_2 3, 2)$; **d)** $x \in (-1, 0) \cup [1, \infty)$;

e) $x \in (-\infty, 0) \cup \left[0, \dfrac{1}{2}\right) \cup \left(1, \dfrac{3}{2}\right] \cup (3, \infty)$.

VII. 76. a) $\dfrac{3}{\log_2 x - 1} - \dfrac{3^2}{\log_2 x - 2} < \dfrac{2\log_2 x}{\log_2 x - 2}$; $x \in (0, 2) \cup (4, \infty)$;

b) $2\log_3 \log_4 \dfrac{2x-1}{x-1} < 0$, $x \in (-\infty, 0) \cup \left(\dfrac{3}{2}, \infty\right)$;

c) $x \in (3, 4] \cup [6, \infty)$; i) $x = -1$; d) $x \in \left(\dfrac{1}{8}, \dfrac{1}{4}\right) \cup (4, 8)$.

VII. 77. For $x, y > 1$, then $\log_x y > 0$. $\left(\log_c a + \log_c b\right) \geq 2\sqrt{\log_c a \cdot \log_c b} \Leftrightarrow$

$\sqrt{\log_a c \cdot \log_b c}\left(\log_a c + \log_b c\right) \geq 2 \Leftrightarrow (ab)^{\sqrt{\log_a c \cdot \log_b c}} \geq c^2$.

VII. 78. a) Notice that $x = 2$ is a solution of the equation.

$\left(\dfrac{3}{5}\right)^x + \left(\dfrac{4}{5}\right)^x = 1$. Assume that there is another solution $x_1 < 2$, therefore

$1 = \left(\dfrac{3}{5}\right)^{x_1} + \left(\dfrac{4}{5}\right)^{x_1} > \left(\dfrac{3}{5}\right)^2 + \left(\dfrac{4}{5}\right)^2 = 1$, contradiction. Similarly for the interval

$(2, \infty)$; b) $x = 4$;

c) $7 \cdot \left(\dfrac{4}{5}\right)^{2x-1} + 38 \cdot \left(\dfrac{1}{5}\right)^{2x-1} = 6$, $x = \dfrac{3}{2}$; d) $f(x) = \left(\dfrac{\sqrt{3}}{2}\right)^x + \left(\dfrac{1}{2}\right)^x$ is a

decreasing function. The unique solution is $x = 2$;

e) $f(x) = 2^x - \left(\dfrac{2}{3}\right)^x$ is a decreasing function. The unique solution is $x = 2$;

f) $1 = 3 \cdot \sqrt[3]{\left(\dfrac{1}{3}\right)^x} + \sqrt[3]{\left(\dfrac{5}{8}\right)^x}$, $x = 3$.

VII. 79. a) Consider the function $f : \mathbf{R} \to (0, \infty)$

$f(x) = \log_3 \left(2^x + 1\right)$. The function is invertible (bijection), $f^{-1} : (0, \infty) \to \mathbf{R}$

$f^{-1}(x) = \log_2 \left(3^x - 1\right)$. The initial equation is equivalent to the equation (1)

$f(x) = f^{-1}(x)$. But the two graphs, being symmetric to the first bisector

$y = x$, the equation (1) is equivalent to one of the equations $f(x) = x$ or

$f^{-1}(x) = x$, $\log_3 \left(2^x + 1\right) = x \Rightarrow 2^x + 1 = 3^x \Rightarrow$

(2) $\left(\dfrac{2}{3}\right)^x + \left(\dfrac{1}{3}\right)^x = 1$. The functions $\left(\dfrac{2}{3}\right)^x$ and $\left(\dfrac{1}{3}\right)^x$ are monotonically

decreasing also the sum $\left(\dfrac{2}{3}\right)^x +\left(\dfrac{1}{3}\right)^x$ is monotonically decreasing. So the

equation $\left(\dfrac{2}{3}\right)^x +\left(\dfrac{1}{3}\right)^x =1$ has maximum one solution, $x = 1$;

b) $x = 1$; c) $f : \mathbf{R}\ \mathbf{R}$, $f(x)=\dfrac{e^x - e^{-x}}{2}$ is an odd function and

$f^{-1}(x)= \ln\left(x + \sqrt{x^2 +1}\right)$; $x = 0$.

VII. 80. a) Denote $\sqrt{x} +\dfrac{1}{\sqrt{x}} = t \ge 2 \Rightarrow x+\dfrac{1}{x}=t^2 -2$. The equation becomes

$2^{t^2 -2} +2^t =8$. The function $f(t)= 2^{t^2 -2} +2^t$ is an increasing function on the

domain $t \ge 2$. Since $f(2)=8$, the only solution is $t = 2$. Therefore $x = 1$.

b) We have $2^{\log_5\left(2^{\log_5 x}+3\right)} +3 = x$. Consider the function $f :(0,\infty)\to \mathbf{R}$

$f(x)= 2^{\log_5 x} +3$ which, obviously is a strictly increasing function. Then

the given equation becomes $f(f(x))= x$. Denoting $t = \log_5 x$ we get

$2^t +3 = 5^t \Rightarrow \left(\dfrac{2}{5}\right)^t +3 \cdot \left(\dfrac{1}{5}\right)^t =1$ with the solution $t = 1$, because the left side

function is decreasing, thus injective. Finally, $x = 5$ is the unique solution;

c) Denoting $t = \cos 2x$, we obtain $2^t =1+ \log_2 (1+t)$.

For $t \in (0,1) \Rightarrow 2^t <t+1<1+ \log_2 (1+t)$.

For $t \notin (0,1) \Rightarrow 2^t >t+1>1+ \log_2 (1+t)$. Therefore $t = 0$ or $t = 1$ and finally ,

$x = \dfrac{\pi}{2}$ is the solution of the given equation.

VII. 81. Domain $\Rightarrow x \in (0,\infty)\setminus \{1\}$. Denoting $\sqrt[4]{x} = t$ then $x = t^4$ and

$\log_x\left(\sqrt{x} + \sqrt[4]{x}\right)= \dfrac{1+ \log_t (t +1)}{4}$, $\log_3 12 =1+ \log_3 4$. Inequation becomes:

$\log_t (t +1) \le \log_3 4 \Rightarrow \dfrac{\log_3 (t +1)}{\log_3 t} \le \log_3 4$. With the substitution t

$= 3^y \Rightarrow \dfrac{\log_3 (3^y +1)}{y} - \log_3 4 \le 0 \Rightarrow \dfrac{\log_3 (3^y +1)- \log_3 4^y}{y} \le 0$ or (1)

$$\frac{\log_3\left[\left(\dfrac{3}{4}\right)^y+\left(\dfrac{1}{4}\right)^y\right]}{y}\le 0,$$ the function $f(y)=\left(\dfrac{3}{4}\right)^y+\left(\dfrac{1}{4}\right)^y$ is decreasing and

$f(y)=1$ has a unique solution $y=1$. Therefore $y\in(-\infty,0)$ $[1,\infty)$, $y=\log_3 t\Rightarrow$ $t\in(0,1)$ $[3,\infty)$ and $x=t^4\Rightarrow x\in(0,1)$ $[81,\infty)$.

VII. 82. Since $f(x)=\left|\sqrt{x-1}-2\right|+\left|\sqrt{x-1}-3\right|=\begin{cases}5-2\sqrt{x-1} & if\ x\in[1,5)\\ 1 & if\ x\in[5,10]\\ 2\sqrt{x-1}-5 & if\ x\in(10,\infty)\end{cases}$

from $f(\log x)=1$ we obtain $x\in\left[10^5,10^{10}\right]$.

VII. 83. (f - invertible) $\Leftrightarrow$ (the ecuation $f(x)=m, m\in\mathbf{R}$ has a unique solution in $D(f)=\mathbf{R}$).

$$f(x)=m\ \Leftrightarrow\ 2^x-2^{-x}+1=m\ \Leftrightarrow\ \begin{cases}2^x=t,\ t>0,\\ t-\dfrac{1}{t}+1=m\end{cases}\Leftrightarrow$$

$\begin{cases}2^x=t,\ t>0,\\ t^2+(1-m)t-1=0\end{cases}$. $t_1=\dfrac{1}{2}\left(m-1-\sqrt{m^2-2m+5}\right)<0$ is not

convenient, $t_2=\dfrac{1}{2}\left(m-1+\sqrt{m^2-2m+5}\right)>0$.

$$2^x=\frac{m-1+\sqrt{m^2-2m+5}}{2}\ \Leftrightarrow\ x=\log_2\frac{m-1+\sqrt{m^2-2m+5}}{2}$$

The inverse is $f^{-1}:\mathbf{R}\to\mathbf{R}$, $f^{-1}(x)=\log_2\dfrac{x-1+\sqrt{x^2-2x+5}}{2}$.

VII. 84. Denote $\alpha=\lg a, \beta=\lg b, \gamma=\lg c$ and we have

$\alpha\le\beta\le\gamma<0$, $A=\dfrac{\alpha}{\beta}+\dfrac{\beta}{\gamma}+\dfrac{\gamma}{\alpha}$, and $B=\dfrac{\beta}{\alpha}+\dfrac{\gamma}{\beta}+\dfrac{\alpha}{\gamma}$ Therefore

$A-B=\dfrac{(\alpha-\beta)(\beta-\gamma)(\gamma-\alpha)}{\alpha\beta\gamma}\le 0\Rightarrow A\le B$.

VII. 85. a) $f\left(a,a^3\right)=\log_a a^3+\log_{a^3}a=3+\dfrac{1}{3}=\dfrac{10}{3}$.

b) $f\left(x,\dfrac{1}{y}\right)=\log_x\dfrac{1}{y}+\log_{\frac{1}{y}}x=-\log_x y-\log_y x=f(x,y)$ and

$$f\left(\frac{1}{x}, y\right) = \log_{\frac{1}{x}} y + \log_y \frac{1}{x} = -\log_x y - \log_y x = f(x, y)$$

c) $\log_{a^x} b + \log_b a^x - \log_a b^x - \log_{b^x} a = \dfrac{15}{4}(\log_b a - \log_a b) \Rightarrow$

$\left(x - \dfrac{1}{x} - \dfrac{15}{4}\right)(\log_b a - \log_a b) = 0$, $x_1 = 4$, $x_2 = -\dfrac{1}{4}$.

Chapter VIII

ARITHMETIC AND GEOMETRIC SEQUENCES

VIII. 1. a) 2, 4, 6, 8, 10; **b)** 9, 10, 11, 12, 13; **c)** 3, 6, 9, 12, 15;

d) 2, 4, 8, 16, 32; **e)** 3, 5, 7, 9, 5; **f)** 11, 22, 33, 44, 55; **g)** 4, 7, 10, 13, 16;

h) $-1, 2, 7, 14, 23$; **i)** $-1, \dfrac{1}{2}, \dfrac{-1}{6}, \dfrac{1}{24}, \dfrac{-1}{120}$; **j)** $0, \dfrac{3}{2}, \dfrac{-8}{3}, \dfrac{15}{4}, \dfrac{-24}{5}$;

k) $\dfrac{1}{2}, \dfrac{2}{5}, \dfrac{3}{10}, \dfrac{4}{17}, \dfrac{5}{26}$; **l)** $\dfrac{3}{2}, \dfrac{9}{4}, \dfrac{16}{9}, \dfrac{25}{16}, \dfrac{36}{25}$; **m)** $1, \dfrac{1}{2}, \dfrac{1}{3}, \dfrac{1}{4}, \dfrac{1}{5}$.

VIII. 2. a) $(-1)^{n+1}\dfrac{1}{n}$; **b)** $a_n = \dfrac{1}{n^2+1}$; **c)** $a_n = \dfrac{2n \cdot 2(n+1)}{(2n-1)(2n+1)}$;

d) $a_n = \dfrac{2 \cdot 3^{n-1}}{5^n}$; **e)** $b_1 = 1,\ b_2 = \dfrac{1}{2}, b_3 = \dfrac{1}{3}, b_4 = \dfrac{1}{4}, \ldots, b_n = \dfrac{1}{n}$; **f)** $x_n = n!$;

g) $x_n = \dfrac{n+1}{n+2}$; **h)** $x_n = an$.

VIII. 3. a) $t_n = 2n-1$; **b)** $t_n = 6n$; **c)** $t_n = 5n-1$; **d)** $t_n = 10^n$; **e)** $t_n = \dfrac{1}{2} \cdot 10^n$;

f) $t_n = 7 + 6(n-1)$.

VIII. 4. $a_1 = 349, a_n = a_{n-1} - 28$. **VIII. 5.** $a_1 = -\dfrac{1}{2},\ a_n = a_{n-1} + \dfrac{1}{2}$.

VIII. 6. The first arithmetic sequence has the general term

$x_n = 6n$, $n \geq 1$, the second $y_k = 5k-1$, $k \geq 1$. The common terms

$x_n = y_k \Leftrightarrow 6n = 5k-1$, where n, k are natural numbers.

From $k = \dfrac{6}{5}(n+1) - 1$ we obtain $n+1 = 5, 10, 15, 20, 25, \ldots$, therefore the new

sequence is 24, 54, 84, 114, 134,... The general term is $a_n = 24 + 30(n-1)$,
which is also an arithmetic sequence.

VIII. 7. 13. **VIII. 8.** $x_{13} = 130$.

VIII. 9. $a = 2, b = -4, c = -1$.

VIII. 10. a) 32, 36; **b)** 19, 37; **c)** 12, 77; **d)** 4, -32; **e)** $\dfrac{4}{7}, \dfrac{7}{13}$; **f)** $\dfrac{28}{27}, \dfrac{82}{81}$

VIII. 11. $\dfrac{\sqrt{5n-3}}{n^2-2}$, $n \geq 2$. **VIII. 12. a)** $a_n = n(2n-1)$;

b) $a_{n+1} = S_{n+1} - S_n = 6n - 3$, $a_{n+1} - a_n = 6$. **VIII. 13.** $S_n = 2^{n+1} - n^2 - 2$.

VIII. 14. a) 3, 3, 3; **b)** 2, 8, 14; **c)** –1, 3, 7; **d)** 2, 5, 5; **e)** 3, 5, 7.

VIII. 15. a) –4; **b)** 16; **c)** 32; **d)** 40.

VIII. 16. a) i) $n^2 - 3n + 2$; **ii)** $2n - 2$; **b) i)** $3n^2 - 11n + 8$; **ii)** $6n - 8$;

c) i) $2^{n-1} - 1$; **ii)** 2^{n-1}; **d) i)** $2n^2 - 7n + 5$; **ii)** $4n - 5$;

e) i) $2(3^{n-1} - 1)$; **ii)** $4(3^{n-1})$; **f) i)** $n^2 + 1$; **ii)** $2n - 1$.

VIII. 17 a) $a_n = 2n + 2$, yes, it is $a_{1005} = 2012$.

b) $(a_n)_{n\geq 1}$ is an arithmetic sequence, $a_1 = 4$ and the ratio is 2.

VIII. 18. a) i) 79, $4n - 1$; **ii)** 820, $2n^2 + n$; **b)** 125. **VIII. 19. a)** 72, 152; **b)** t_{51}.

VIII. 20. a) 20, 14, – 4; **b)** 2, 23, 30; **c)** 17, 27, 32; **d)** 10, 17, 31; **e)** 1, – 9, – 14;

f) 6, –3, – 12.

VIII. 21. –10 ,–6, –2. **VIII. 22. a)** no; **b)** no; **c)** a_{12}; **d)** a_{37}.

VIII. 23. $a_n = 2^n + 1$, for any $n > 1$. **VIII. 24.** $a_n = n^2$.

VIII. 25. $a_2 = 3$, $a_3 = 6$, $a_4 = 10$. Setting $b_n = \sqrt{8a_n + 1}$ we have

$$a_n = \frac{b_n^2 - 1}{8} \quad \text{or} \quad (1) \; a_{n+1} = \frac{b_{n+1}^2 - 1}{8}, \text{ on the other hand,}$$

(2) $a_{n+1} = \dfrac{b_n^2 - 1}{8} + \dfrac{1}{2}b_n + \dfrac{1}{2}$. Equating (1) and (2), we obtain $b_{n+1}^2 = (b_n + 2)^2$.

Since $b_n > 0$, then $b_{n+1} = b_n + 2$. Therefore $b_n = 2n + 1$ and $a_n = \dfrac{n(n+1)}{2}$.

VIII. 26. $a_n = n^2 - n$. **VIII. 27. a)** ii; **b)** iv; **c)** v; **d)** vi.

VIII. 28.

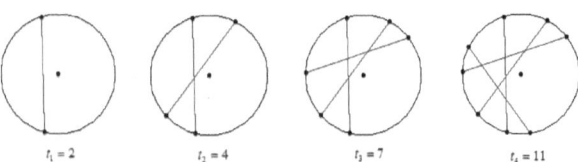

$t_1 = 2$ $t_2 = 4$ $t_3 = 7$ $t_4 = 11$

The second difference is 1, therefore $t_n = an^2 + bn + c$. Solving the system

$$\begin{cases} a + b + c = 2 \\ 4a + 2b + c = 4 \\ 9a + 3b + c = 7 \end{cases}, \text{ we obtain } t_n = \frac{1}{2}\left(n^2 + n + 2\right)$$

VIII. 29. $a_n = 7n - 10$. **VIII. 30.** $a_{20} = -\dfrac{57}{2}$.

VIII. 31.

$n = 1:$ $a_2 - a_1 = 1^2 - 1 + 1$

$n = 2:$ $a_3 - a_2 = 2^2 - 2 + 1$

....

$n = 99:$ $a_{100} - a_{99} = 99^2 - 99 + 1$. Adding all these 99 equalities

$a_{100} - a_1 = 1^2 + 2^2 + ... + 99^2 - (1 + 2 + ... + 99) + 99 = 323499;$

$1 + 2 + ... + n = \dfrac{n(n+1)}{2}$; $1^2 + 2^2 + ... + n^2 = \dfrac{n(n+1)(2n+1)}{2}$. $a_{100} = 323{,}500$.

VIII. 32. a) $2, -1, -4, -7, -10,$ $a_n = -3n + 5$;

b) $a_1 = 2$, $a_2 = 4$, $a_3 = 6$, $a_4 = 8$, $a_5 = 10$, $a_n = 2n$;

c) $a_1 = 21$, $a_2 = 22\dfrac{1}{2}$, $a_3 = 24$, $a_4 = 25{,}5$, $a_5 = 27$, $a_n = \dfrac{3}{2}(n + 13)$;

d) $a_1 = 4$, $a_2 = 9$, $a_3 = 14$, $a_4 = 19$, $a_5 = 24$, $a_n = 5n - 1$.

VIII. 33. a) 31; b) 37; c) 41; d) 21; e) 32; f) 100. **VIII. 34.** 4210

VIII. 35. 1380. **VIII. 36.** a) 179; b) 2760.

VIII. 37. $a_{10} = 30$, S12=570. **VIII. 38.** $a_1 = 1$, $d = 4$, $n = 26$, $S_{26} = 1326$.

VIII. 39. 735. **VIII. 40.** a) 2; b) $2x^2 = (x - 2) + (x + 6)$, $x_1 = -1$, $x_2 = 2$.

VIII. 41. a) $a = \pm 1$, b) $a = 10$. **VIII. 42.** $x - 5y + 2 = 0$.

VIII. 43. $m = 15$, $n = 63$.

VIII. 44. $a_{50} = 2a_{49} - a_{48} = 2\sqrt{27 - 10\sqrt{2}} - \sqrt{18 - 8\sqrt{2}} =$

$= 2(5 - \sqrt{2}) - (4 - \sqrt{2}) = 6 - \sqrt{2}$.

VIII. 45. a) 345; b) 105; c) $- 2670$.

VIII. 46. $\begin{cases} a_1 + a_5 = 8 \\ a_3 \cdot a_5 = 80 \end{cases} \Rightarrow \begin{cases} 2a_1 + 4d = 8 \\ (a_1 + 2d)(a_1 + 4d) = 80 \end{cases}$, $d = 8$,

$a_1 = -12$, $a_{20} = 140$, $S_{20} = 1280$. **VIII. 47.** $3 + 10 + 17 + ...$

VIII. 48. $a_1 = 0$, $d = 14/3$, and S10= 210.

VIII. 49. $a_1 = 4$, $d = 3$. **VIII. 50.** a) $a_1 = 3$, $d = 5$;

b) $a_1 = 3$, $d = 5$ or $a_1 = 13$, $d = -5$; c) $a_1 = 1$, $d = 1$;

d) $a_1 = \pm\sqrt{2}$, $d = 0$ or $a_1 = 1$, $d = 1.5$ or $a_1 = \dfrac{2}{7}$, $d = \dfrac{3}{7}$.

VIII. 51. 5, 7, 9, ..., 23 and reverse.

VIII. 52. a) $a_1 = 2$, $d = 2$ or $a_1 = 6$, $d = -2$ or $a_1 = 2 - \sqrt{2}$, $d = \sqrt{22}$ or

$a_1 = 2 + \sqrt{22}$, $d = -\sqrt{22}$;

b) $a_1 = a - 3x$, $a_2 = a - x$, $a_3 = a + x$, $a_4 = a + 3x$, $a = 9$, $x = \pm 3$.

VIII. 53. 5. **VIII. 54.** 1.5, 405. **VIII. 55.** 11, 15, 19.

VIII. 56. a) $a_1 = 11$, $d = 10$; **b)** $a_1 = 7$, $d = 6$.

VIII. 57. Let $a - d$, a, and $a + d$ be the numbers, where d is the

ratio of the arithmetic sequence. Then we have $\begin{cases} (a-d) + a + (a+d) = 6 \\ (a-d)^2 + a^2 + (a+d)^2 = 62 \end{cases}$.

From the first equation we obtain $a = 2$ and the second equation becomes

$d^2 = 25$. Therefore the numbers are $-3, 2, 7$ or $7, 2, -3$.

VIII. 58. $x_1 = x_2 = 1$. **VIII. 60.** $n = 10$. **VIII. 61.** $r = 2\alpha$, $a_n = \alpha(2n - 1)$.

VIII. 62. From $S_m = S_n \Rightarrow \dfrac{(a_1 + a_m)m}{2} = \dfrac{(a_1 + a_n)n}{2} \Rightarrow \dfrac{a_1 + a_m}{a_1 + a_n} = \dfrac{n}{m} \Rightarrow$

$\dfrac{a_m - a_n}{a_1 + a_n} = \dfrac{n - m}{m} \Rightarrow \dfrac{(m - n)d}{a_1 + a_n} = \dfrac{n - m}{m} \Rightarrow a_1 + a_n = -dm$.

Then $a_1 + a_n + dm = 0 \Rightarrow a_1 + a_{n+m} = 0$. Therefore $S_{n+m} = 0$.

VIII. 63. $2(a^2 + b^2) = (a + b)^2 + (a - b)^2$.

VIII. 65. $m_a^2 = \dfrac{2(b^2 + c^2) - a^2}{4}$. **VIII. 67.** Factoring the given equality we

obtain $(a + b - 2c)(c + b - 2a)(a + c - 2b) = 0$.

VIII. 68. $m = 3$. **VIII. 70.** $\dfrac{1 + 2 + \dots + (x - 1)}{x} = 10$. $1 + 2 + \dots + (x - 1)$ is an

arithmetic sequence with $x - 1$ terms, where $a_1 = 1$, $d = 1$ and $a_n = x - 1$ then

$S_n = \dfrac{(a_1 + a_n)n}{2} = \dfrac{(1 + x - 1)(x - 1)}{2}$; $\dfrac{x(x - 1)}{2} = 10x$, $x = 21$.

VIII. 71. Let $a_1 = 3$, $d = 4$; $x = a_n$, we have

$a_n = a_1 + (n - 1)d \Rightarrow x = 3 + (n - 1)4$, $S_n = \dfrac{(a_1 + a_n)n}{2}$,

then $210 = \dfrac{(3 + x)n}{2}$; $x = 39$; **b)** $x = 0$; **c)** $x = \pm 2$.

VIII. 72. $a_{2k} = \dfrac{a_{2k-1} + a_{2k+1}}{2} \geq \sqrt{a_{2k-1}a_{2k+1}} \Rightarrow \dfrac{a_{2k-1}}{a_{2k}} \leq \dfrac{\sqrt{a_{2k-1}}}{\sqrt{a_{2k+1}}}$, $k \geq 1$ and

multiplying these inequalities side by side we obtain the result.

VIII. 73. a) Use the identity $\dfrac{1}{\sqrt{a_k}+\sqrt{a_{k+1}}}=\dfrac{\sqrt{a_{k+1}}-\sqrt{a_k}}{d}$, or by induction.

VIII. 74. $S_1=\dfrac{1}{d}\left(\dfrac{1}{a_1}-\dfrac{1}{a_n}\right)$, $S_2=\dfrac{1}{d}\left(\dfrac{1}{a_1^2}-\dfrac{1}{a_n^2}\right)$, $S_3=\dfrac{a_1a_2}{a_{n-1}a_n}\cdot\dfrac{a_{n-1}a_n+d^2}{a_1a_2+d^2}$;

$S_4=\dfrac{n-1}{\sqrt{a_1}+\sqrt{a_n}}$.

VIII. 75. a) Setting $x_n=a_n a_{n+1}...a_{n+k}\Rightarrow x_{n+1}-x_n=(k+1)da_{n+1}a_{n+2}...a_{n+k}$, d as the ratio.

$A_n=a_1a_2...a_k+\left[(x_2-x_1)+(x_3-x_2)+...+(x_{n+1}-x_n)\right]\dfrac{1}{(k+1)r}$ therefore

$A_n=a_1a_2...a_k+\dfrac{a_{n+1}a_{n+2}...a_{n+k-2}-a_1a_2...a_{k+1}}{(k+1)r}$;

b) $y_n=\dfrac{1}{a_n a_{n+1}...a_{n+k-2}}$ using y_n-y_{n+1} we obtain

$B_n=1\dfrac{1}{(k-1)d}\left(\dfrac{1}{a_1a_2...a_{k-1}}-\dfrac{1}{a_{n+1}a_{n+2}...a_{n+k-1}}\right)$, d is the ratio of the sequence.

VIII. 78. $q_n=b_{n+1}-b_n=b_1r^n-b_1r^{n-1}=b_1r^{n-1}(r-1)=kr^{n-1}$.

VIII. 79. a) $b_1=2$, $b_2=6$, $b_3=18$, $b_4=54$, $b_5=162$;

b) $b_1=3$, $b_2=-6$, $b_3=12$, $b_4=-24$, $b_5=48$; **c)** $b_1=\sqrt{2}$, $b_2=\sqrt{6}$,

$b_3=3\sqrt{2}$, $b_4=3\sqrt{6}$, $b_5=9\sqrt{2}$; **d)** $b_1=5$, $b_2=10$,

$b_3=20$, $b_4=40$, $b_5=80$.

VIII. 80. $b_1=7$. **VIII. 81.** 2. **VIII. 82.** $b_n=6\cdot7^{n-1}$.

VIII. 83. $b_1=144$, $b_n=\dfrac{1}{2}b_{n-1}$. **VIII. 84. a)** $r_{a_n}=\dfrac{1}{2}$ and $r_{b_n}=3$;

b) $y_n=2^{n+1}$. **VIII. 85.** $y_n=\dfrac{10}{3}2^{n-1}$ and $x_n=\dfrac{5\cdot2^n+3}{5\cdot2^n-3}$.

VIII. 86. $A_n(-n,\ n)$, A_1A_2: $y=-x$.

VIII. 87. $b_4=b_1\cdot r^3\Rightarrow r=\sqrt{3}$.

Therefore $b_2=\sqrt{6},b_3=3\sqrt{2},b_5=9\sqrt{2},b_6=9\sqrt{6}$.

VIII. 88. $\begin{cases}x^2=y\sqrt{3}\\ y^2=x\cdot18\sqrt{2}\end{cases}$, $x=3\sqrt{2}$ and $y=6\sqrt{3}$. **VIII. 90.** 11.

VIII. 91. $b_5 = -81$, $S_{10} = \dfrac{1-3^{10}}{2}$. **VIII. 92.** $\dfrac{b_1 r + b_1 r^2}{b_1 r^2 + b_1 r^3} = \dfrac{14}{28}$,

$r = 2$, $b_1 = \dfrac{7}{3}$, $S_{10} = \dfrac{7}{3}\left(2^{10} - 1\right)$. **VIII. 93.** 15, 45. **VIII. 94.** 3.

VIII. 95. a) $b_1 = 1$, $r = 3$; $b_1 = -81$, $r = \dfrac{1}{3}$; **b)** $b_1 = 2$, $r = 3$.

VIII. 96. $\dfrac{1+\sqrt{5}}{2}$. **VIII. 97.** $\dfrac{-1+\sqrt{5}}{2}$. **VIII. 100.** $m = 18$, $n = 288$.

VIII. 101. $\begin{cases} b_1 + b_2 + b_3 = 21 \\ b_1 b_2 b_3 = 64 \end{cases}$; $\begin{cases} b_1\left(1 + r + r^2\right) = 21 \\ b_1^3 r^3 = 64 \end{cases}$; $1, \dfrac{1}{4}, \dfrac{1}{16}$ or $16, 4, 1$.

VIII. 102. $\dfrac{1}{2}$. **VIII. 103.** $b_1, b_2, b_3, \ldots, b_n$

$\begin{cases} b_1 b_2 b_3 = 3\sqrt{3} \\ b_1^3 + b_2^3 + b_3^3 = 28 + 3\sqrt{3} \end{cases} \Rightarrow \begin{cases} b_1^3 r^3 = 3\sqrt{3} \\ b_1^3\left(1 + r^3 + r^6\right) = 28 + 3\sqrt{3} \end{cases}$;

$3\sqrt{3} r^6 - 28 r^3 + 3\sqrt{3} = 0$,

I) $r = \sqrt{3}$ and $b_1 = 1$, $b_2 = \sqrt{3}$, $b_3 = 3$;

II) $r = \dfrac{1}{\sqrt{3}}$ and $b_1 = 3$, $b_2 = \sqrt{3}$, $b_3 = 1$.

VIII. 104. a) $(2, 10, 18)$, $(20, 10, 0)$; **b)** $(0, -2, -4)$, $(2, 4, 6)$,

$\left(\dfrac{1}{2}, -\dfrac{7}{2}, -\dfrac{15}{2}\right)$.

VIII. 105. a) $a = 5$, $b = 10$, $c = 20$, $d = 40$;

b) $a = 1$, $b = 5$, $c = 9$, $d = 13$. **VIII. 106.** $a = 3$, $b = 6$, $c = 12$.

VIII. 107. $b^2 = ac$, $a + b + c = 13$, $(b+1)^2 = (a+1)(c-1)$; $1, 3, 9$;

$\dfrac{-55}{7}, \dfrac{121}{7}, \dfrac{25}{7}$. **VIII. 108.** 4. **VIII. 109.** $\dfrac{2}{3}$.

VIII. 110. a) 4; **b)** -3; **c)** 5; **d)** $-\dfrac{8}{3}$; **e)** 119; **f)** $\dfrac{500}{67}$.

VIII. 111. $a_1 = 4$, $d = 4$. **VIII. 112.** $b^2 = ac$, $b + 8 = \dfrac{a+c}{2}$,

$(b+8)^2 = a(c+64)$; $(4, 12, 64)$ or $\dfrac{4}{9}, -\dfrac{20}{9}, \dfrac{100}{9}$.

VIII. 113. $\begin{cases} b_1 - b_4 = b_2 \\ b_3 - 6 = b_4 \end{cases} \Rightarrow \begin{cases} b_1(1-r) = 54 \\ b_1 q^2(1-r) = 6 \end{cases} \Rightarrow \dfrac{b_1(1-r)}{b_1 q^2(1-r)} = \dfrac{54}{6} \Rightarrow r = \pm\dfrac{1}{3}.$

Therefore $q = \dfrac{1}{3}$, $b_1 = 81$, or $q = -\dfrac{1}{3}$, $b_1 = \dfrac{81}{2}$

VIII. 114. Let $a_1, a_2, a_3, ..., a_n$; d and $b_1, b_2, b_3, ..., b_p$; r the arithmetic and geometric sequences, respectively. We have $a_1 = b_1 = 1$, $S_9 = 369$, $a_9 = b_9$.
Finally $r^8 = 81$, therefore $r = \pm\sqrt{3}$. **VIII. 115.** 32.

VIII. 116. $a_1, a_2, a_3, ..., a_n$; d, $b_1, b_2, b_3, ..., b_n$; r.

$a_1 + a_2 + a_3 + a_4 + a_5 = 62$ and $a_5 = b_1$, $a_8 = b_2$, and $a_{11} = b_{10}$.

$a_1 + (a_1 + d) + (a_1 + 2d) + (a_1 + 3d) + (a_1 + 4d) = 62$ and $a_1 + 4d = b_1$,

$a_1 + 7d = b_1 r$, $a_1 + 10d = b_1 r^9$. Therefore $b_1 = 2$ or $b_1 = \dfrac{62}{5}$.

VIII. 117. $b_1 + b_2 + b_3 = 65$ and $\log_{15} b_1 + \log_{15} b_2 + \log_{15} b_3 = 3$, (5, 3) and

(45, 1/3).

VIII. 118. $b_3 = 2$, $b_5 = 6$, $S_9 = \dfrac{2\left(1 - 81\sqrt{3}\right)}{3\left(1 - \sqrt{3}\right)}$.

VIII. 119. $2a_1 + 9b_1 = 31$, $b_1(1 + a_1) = 9$, $a_1 = \dfrac{25}{2}$, $b_1 = \dfrac{2}{3}$ or $a_1 = 2$, $b_1 = 3$.

VIII. 120. $\begin{cases} b_1 + b_5 = 164 \\ b_3 \cdot b_5 = 2916 \end{cases}$; $r = 3$, $b_1 = 2$, $b_{10} = 39366$.

VIII. 125. a) $a_{n+1} = \dfrac{3}{2} a_n$; **b)** $n = 5$. **VIII. 126. a)** $a_{n+1} - a_n = 2$, $\dfrac{b_{n+1}}{b_n} = 4$;

b) $a_1 + a_2 + ... + a_n = \dfrac{(a_1 + a_n)n}{2} = n^2$, $n = 2014$.

VIII. 127. a) We have: $S = 1 + 3 + 3^2 + 3^3 + ... + 3^{98} + 3^{99} +$

$+ 3 + 3^2 + 3^3 + ... + 3^{98} + 3^{99} +$

$+ 3^2 + 3^3 + ... + 3^{98} + 3^{99} +$

$..$

$+ 3^{98} + 3^{99} +$

$+ 3^{99}.$

Therefore

$$S = \frac{1\left(3^{100}-1\right)}{3-1} + \frac{3\left(3^{99}-1\right)}{3-1} + \frac{3^2\left(3^{98}-1\right)}{3-1} + ... + \frac{3^{98}\left(3^2-1\right)}{3-1} + \frac{3^{99}\left(3^1-1\right)}{3-1} =$$

$$= \frac{1}{2}\left[\left(3^{100}-1\right)+\left(3^{100}-3\right)+\left(3^{100}-3^2\right)+...+\left(3^{100}-3^{98}\right)+\left(3^{100}-3^{99}\right)\right] =$$

$$= \frac{1}{2}\left[100 \cdot 3^{100} - \left(1+3+3^2+...+3^9\right)\right] = \frac{1}{4}\left(199 \cdot 3^{100}+1\right).$$

b) $3 = \dfrac{10-1}{3}$, $\quad 3 = \dfrac{10^2-1}{3},..., \quad \underbrace{33...3}_{n \text{ times}} = \dfrac{10^n-1}{3}$; $\quad S_n = \dfrac{10\left(10^n-1\right)}{27} - \dfrac{n}{3}$.

VIII. 128. $S_1 = \dfrac{nq^{n+1}}{q-1} - \dfrac{q^{n+1}q}{(q-1)^2}$; $\quad S_2 = 2\dfrac{q^{n+1}-q}{(q-1)^3} - \dfrac{(2n-1)q^{n+1}+q}{(q-1)^2} - \dfrac{n^2q^{n+1}}{q-1}$;

$S_3 = \dfrac{2}{3}\dfrac{2^n-1}{2^n}$. **VIII. 129.** $S_1 = \dfrac{\left(a^{2n}-1\right)\left(a^{2n+2}-1\right)}{a^{2n}\left(a^2-1\right)} + 2n$;

$S_2 = \dfrac{\left(a^{2n}-1\right)\left(a^{2n+2}-1\right)}{a^{2n}\left(a^2-1\right)} - 2n$. **VIII. 130. a)** $A_n = \dfrac{n-1}{\sqrt{r}-1}$, **b)** $B_n = b_1\dfrac{r^{n+1}-r}{r^2-1}$.

VIII. 131. $A = b_1\dfrac{r^n-1}{r-1}$, $\quad B = \dfrac{1}{b_1}\dfrac{r^n-1}{r^{n-1}(r-1)}$, $\quad \dfrac{A}{B} = b_1^2 r^{n-1}$,

$C = \left(b_1 r^{\frac{n-1}{2}}\right)^n = \left(b_1^2 r^{n-1}\right)^{\frac{n}{2}} = \left(\dfrac{A}{B}\right)^{\frac{n}{2}}$.

VIII. 132. $b_1 = x^4+1$, $r = \sqrt{x^4+1}$, $S = \dfrac{n^2+3n}{4}$.

VIII. 133. $b_1^2 \cdot (1-r^2+r^4-r^6+...+r^{4n-4}-r^{4n-2})(1-r^{4n}) =$

$b_1^2 \cdot \dfrac{r^{4n}-1}{-r^2-1} \cdot (1-r^{4n}) = b_1^2 \cdot \dfrac{(1-r^{4n})^2}{1+r^2} \geq 0$.

VIII. 135. $S_1 = \dfrac{1}{b_1^p\left(r^p+1\right)}\left(1+\dfrac{1}{r^p}+...+\dfrac{1}{r^{(n-1)p}}\right)$,

$S_2 = \dfrac{1}{b_1^p\left(r^p-1\right)}\left(1+\dfrac{1}{r^p}+...+\dfrac{1}{r^{(n-1)p}}\right)$, therefore $\dfrac{S_1}{S_2} = \dfrac{r^p-1}{r^p+1}$, where r is the

ratio of the geometric sequence. **VIII. 136.** Use $S_n = b_1\dfrac{r^n-1}{r-1}$.

VIII. 138. For $n = 2$ we have $2\left(\sqrt{x_1} + \sqrt{x_2}\right) = 3\sqrt{x_2} \Rightarrow x_2 = 4a$, for $n = 3$

we have $2\left(\sqrt{x_1} + \sqrt{x_2} + \sqrt{x_3}\right) = 4\sqrt{x_3} \Rightarrow x_3 = 9a$. Next, using mathematical

induction $x_n = n^2 a$. Therefore $\displaystyle\sum_{k=1}^{n} x_k = a\left(1^2 + 2^2 + \ldots + n^2\right) = \frac{n(n+1)(2n+1)a}{2}$.

VIII. 139. $|x_0| \leq 2p \Rightarrow \left|\dfrac{x_0}{2p}\right| \leq 1$. If we denote $y_0 = \dfrac{x_0}{2p}$ in this case there exists

a real number $\alpha \in \left[-\dfrac{\pi}{2}, \dfrac{\pi}{2}\right]$ such that $y_0 = \sin\alpha$. Denoting $y_1 = \dfrac{x_1}{2p}$ we get

$y_1 = \sin 3\alpha$ and using mathematical induction, we get $y_n = \sin 3^n \alpha$. Finally

$x_n = 2p\sin 3^n \alpha$ where $\alpha = \arcsin\dfrac{x_0}{2p}$.

VIII. 140. The equation $r^2 - \dfrac{5}{7}r - \dfrac{2}{7} = 0$ has the roots $r_1 = 1$ and $r_2 = -\dfrac{2}{7}$.

Therefore $x_n = A \cdot 1^n + B \cdot \left(-\dfrac{2}{7}\right)$, $\forall n \geq 2$.

For $n = 0$ and 1 $\quad \begin{cases} A - \dfrac{2}{7}B = b \\ A + B = a \end{cases}$. Thus $A = \dfrac{2a + 7b}{9}$ and

$B = \dfrac{7(a-b)}{9}$. Consequently $x_n = \dfrac{2a + 7b}{9} + \dfrac{7(a-b)}{9} \cdot \left(-\dfrac{2}{7}\right)^n$.

Chapter IX

POLYNOMIALS AND ALGEBRAIC EQUATIONS

IX. 1.

	x^4	x^3	x^2	x^1	x^0
-1	1	1	2	0	1
	1	0	2	-2	3

$P(x) = x^4 + x^3 + 2x^2 + 1 = (x^3 + 2x - 2)(x + 1) + 3$.

IX. 2. a) $q(x) = x^2 + 5x + 3$, $r = 0$;

b) $q(x) = 2x^3 + 9x^2 + 22x + 63$, $r = 190$;

c) $q(x) = 3x^3 + 9x^2 + x + 4$, $r = 1$;

d) $q(x) = x^3 - 2x^2 + 7x - 13$, $r = 16$;

e) $q(x) = 2x^2 + 2x - 4.5$, $r = 12.5$;

f) $r(x) = x^4 - 3x^3 + 6x^2 - 9x + 21$, $r = -44$;

g) $r(x) = x^4 - x^3 - x^2 + 2x + 2$, $r = -3$;

h) $r(x) = x^7 - x^6 - 3x^5 + 3x^4 + 3x^3 - 3x^2 + x - 1$, $r = 3$.

IX. 3. a) $q(x) = 2x + 4$, $r = -6x - 7$; **b)** $q(x) = x^3 - x^2 + 6x - 11$, $r = 34$;

c) $q(x) = x + 6$, $r(x) = 9x^2 + 5x - 5$; **d)** $q(x) = x^2 + 2x + 8$, $r = 0$;

e) $q(x) = x^3 - x - 4$, $r = 8$; **f)** $q(x) = x^3 + 3$, $r(x) = 3x - 4$;

g) $q(x) = x^5 + x^3 - x^2 + 9x - 5$, $q(x) = 33x^2 - 29x + 6$.

IX. 4. $f(1) = 0$. **IX. 5.** 13. **IX. 6. a)** $P(1) = 0$; **b)** The quotient is $x^2 - x - 6$ and the remainder is zero;

c) $x_1 = -2$, $x_2 = 1$, $x_3 = 3$.

IX. 7. $2x - 1$. **IX. 8.** $a(a + 1)(a + 2) = -1320$, $-12, -11, -10$. **IX. 9.** 3, 2, 7.

IX. 10. $V(r) = \frac{4}{3}\pi(r^3 + 12r^2 + 48r + 64) = \frac{4}{3}\pi(r + 4)^3$. Therefore the radius is $(r + 4)$.

IX. 11. a) $f(1) = -1, f(-1) = -1$; **b)** The quotient is $x^2 - 5x + 5$ and the remainder is -1; **c)** $x_1 = -1$, $x_2 = 1$, $x_{3,4} = \frac{5 \pm \sqrt{5}}{2}$.

IX. 12. $P(x) = ax + b$, $P(-1) = -a + b = 2$, $P(1) = a + b = -3$, $P(x) = -\dfrac{5}{2}x - \dfrac{1}{2}$.

IX. 13. a) $(x^2 - x - 2)(x - 5)$. **IX. 14. a)** -1; **b)** 1; **c)** -60.

IX. 15. $f(2) = 0$, $m = 5$. **IX. 16.** $m = -1$. **IX. 17.** $m = 1$.

IX. 18. $a_1 = 1$, $a_2 = -\dfrac{1}{3}$.

IX. 19. $P(1) = 1 - m + m^2 + 4 - 2 = 5 \Rightarrow m^2 - m - 2 = 0$. Therefore $m = -1$ or 2.

IX. 20. $a = -5$, $m = 2$, $n = 5$. **IX. 21.** $f(m) = 0$ or $(m^2 - m + 6)(m - 1)$, $m = 1$.

IX. 22. $a = 1.5$, $b = 5.5$. **IX. 23. a)** $f(x) = 2x^2 + x - 1$; **b)** $f(x) = 3x^2 - 9x + 6$.

IX. 24. $a = -1$, $b = 20$, $c = -12$.

IX. 25. a) $f(-1) = 0$, $f(1) = 4$, $a = 1$ and $b = 2$; **b)** $x_1 = -1$, $x_2 = \dfrac{1 + i\sqrt{7}}{2}$,

$x_3 = \dfrac{1 - i\sqrt{7}}{2}$; **c)** $x \in (-\infty, -1)$.

IX. 26. $k = 7$. **IX. 27.** We have $\begin{cases} P(-2) = 4(-2)^3 + a(-2)^2 - 2b + 11 = -7 \\ P(1) = 4(1)^3 + a(1)^2 + b + 11 = 14 \end{cases}$.

$\begin{cases} 4a - 2b = 14 \\ a + b = -1 \end{cases}$. From the second equation $a = -1 - b$. Substituting this in the

first equation, we obtain $4(-1 - b) - 2b = 14 \Rightarrow b = -3$ and $a = 2$.

IX. 28. $a = -\dfrac{7}{3}$, $b = \dfrac{1}{3}$.

IX. 29 $\begin{cases} f(1) = 1 + 6 + k - 4 = 3 + k \\ f(-2) = -8 + 24 - 2k - 4 = 12 - 2k \end{cases}$. They have the same remainder,

so they are equal to each other. $3 + k = 12 - 2k \Rightarrow k = 3$. **IX. 30.** $m = -6$,

$n = -5$. **IX. 31.** $m = -21$ and $n = -3$. **IX. 32.** 6.

IX. 33. $P(x) = -\dfrac{1}{3}x^3 + \dfrac{4}{3}x$. **IX. 34.** The sum of the coefficients is $f(1)$,

$f(1) = 1 + (1 + a) + (1 + a)^2 + ... + (1 + a)^n = \dfrac{(1 + a)^{n+1} - 1}{a}$.

IX. 35. For $m \neq -1$ and $m \neq 2$ the degree of f is four; for $m = -1$ the degree of f is zero; for $m = -2$ the degree of f is two.

IX. 36. $P(x) = (x - a)(x - b)q(x) + mx + n$.

IX. 37. $P(x) = (x-1)(x+1)(x-2)Q(x) + ax^2 + bx + c$, $P(1) = 0$, $P(-1) = 1$,

$P(2) = -3$, $r(x) = -\dfrac{5}{6}x^2 - \dfrac{1}{2}x + \dfrac{4}{3}$.

IX. 38. $m = -5$. **IX. 39.** $a = -3$.

IX. 40. $a = -40$, $b = 288$, $c = -32$.

IX. 41. $P(-1) = 2$, $P(2) = -1$, $P(x) = (x^2 - x - 2)q(x) + ax + b$, $-a + b = 2$ and $2a + b = -1$. The remainder is $-x + 1$.

IX. 42.

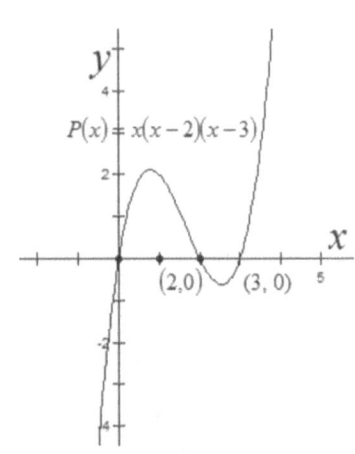

$x_1 = 0$, $x_2 = 2$, $x_3 = 3$.

IX. 43. $f(x) = 3(x^2 - 1)(x + 2)$. **IX. 44.** $f(x) = 2(x^2 - 1)(x + 3)(x - 2)$.

IX. 45. $f(x) = 2(x - 2)^2(x + 2)(x - 4)$. **IX. 46.** $P(x) = -2(4x - 1)(x - 2)(x + 1)$.

IX. 47. a)

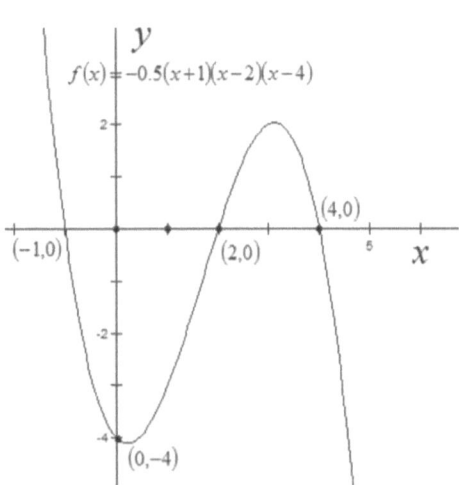

350

b)

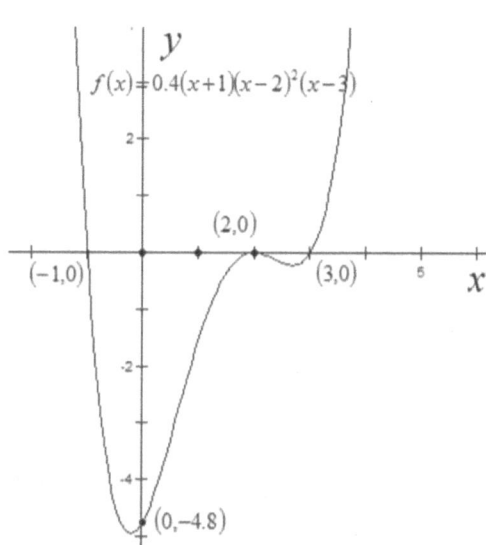

c)

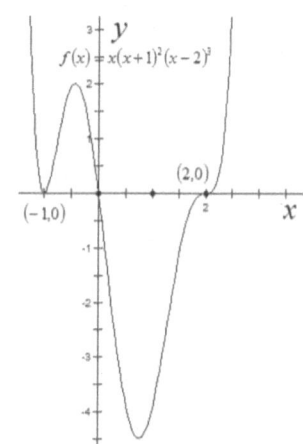

IX. 48.

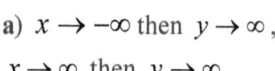

a) $x \to -\infty$ then $y \to \infty$,
$x \to \infty$ then $y \to \infty$

b) $x \in \left(-\infty, 1\right] \cup \left[4, \infty\right)$;

c) $x \in \left(1, 4\right)$.

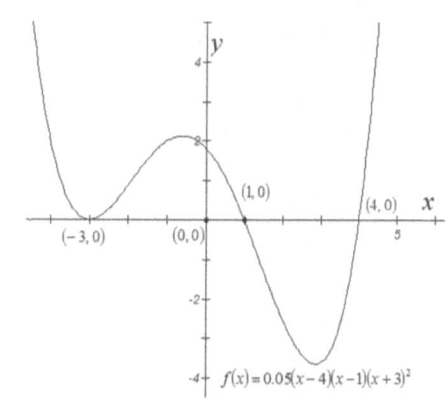

IX. 49.

$$f(x) = \frac{7}{15}(x-1)(x-2)(x+2).$$

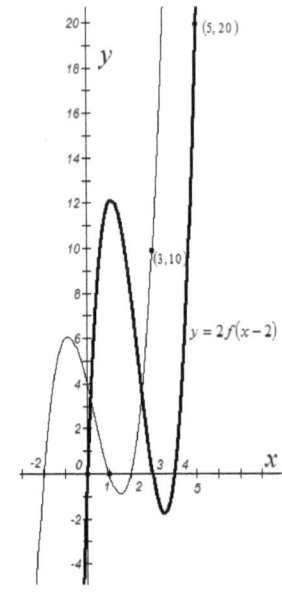

IX. 50.
a) Vertical stretch by a factor of 2;
horizontal translation
2 units to the right.

b) Reflection in the y-axis;
Vertical translation 3 units up.

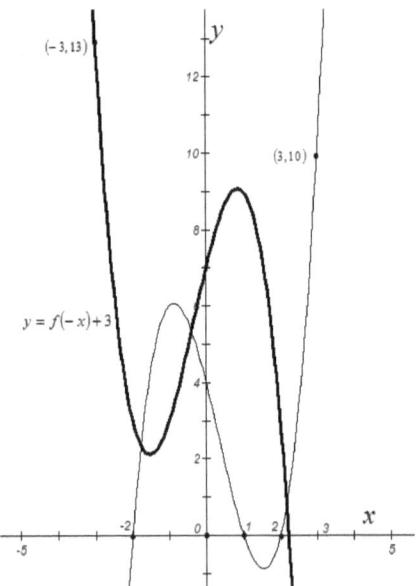

c) Reflection in the x-axis;
Vertical stretch by a factor of 2;
Horizontal stretch
by a factor of 3.

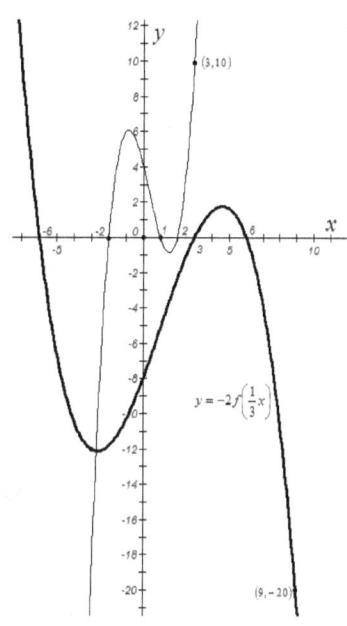

c) Vertical compression

by a factor of $\dfrac{1}{3}$;

Horizontal stretch
by a factor of 2;
Vertical translation 1 unit down.

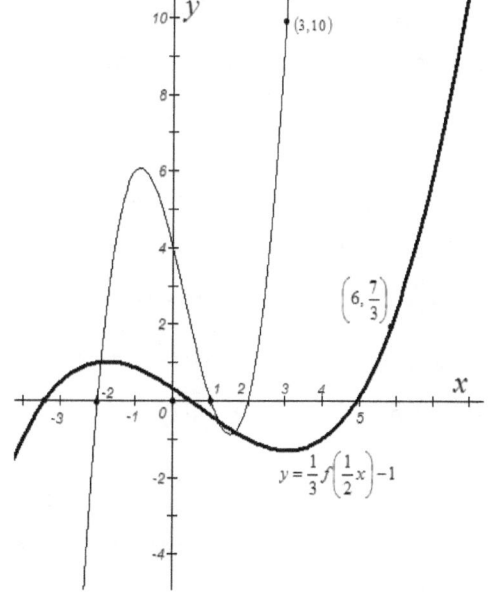

IX. 51. $f(x) = 2(x-4)(x-1)(x+3)$.

IX. 52. a) $f(x) = ax(x-4)(x-2)(x+3)$; **b)** $f(x) = -2x(x-4)(x-2)(x+3)$.

IX. 53 a) $f(x) = x(x+2)(x-1)(x+1)$;

b)

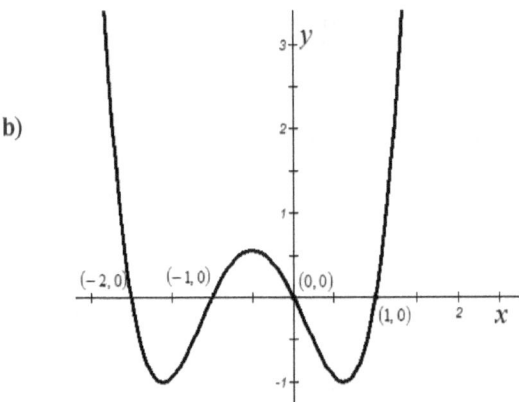

c) $x \in (-\infty, -2] \cup [-1, 0] \cup [1, \infty)$.

IX. 54. a) $f(2) = 0$ and $f(4) = 160$, $a = 2$ and $b = 48$;

b) $f(x) = 2(x-3)(x-2)(x+1)(x+4)$; c) $f(0) = 48$.

IX. 55. *Solution* 1. We have $f(-3) = 0$, $f(-1) = 0$, and $f(2) = 0$, which is

equivalent to the system $\begin{cases} -27a + 9b - 3c - 12 = 0 \\ -a + b - c - 12 = 0 \\ 8a + 4b + 2c - 12 = 0 \end{cases}$, with the solution $a = 2$,

$b = 4$, $c = -10$.

Solution 2. $f(x) = k(x+3)(x+1)(x-2)$, also $f(0) = -12$, therefore

$f(x) = 2(x+3)(x+1)(x-2) = 2x^3 + 4x^2 - 10x - 12$.

IX. 56. a) $f(1) = 0$ and $f'(1) = 0$. *Another solution* using synthetic division

(Horner's Algorithm)

	x^{n+1}	x^{n-1}	x^{n-1}	.	.	.	x^2	x^1	x^0
1	1	0	0				0	$-n-1$	n
1	1	1	1				1	$-n$	0
	1	2	3				n	0	

In this case the quotient is $q(x) = x^{n-1} + 2x^{n-2} + 3x^{n-3} + \ldots + (n-1)x + n$.

354

IX. 57. Let ε be a root of the equation $g(x) = 0 \Rightarrow$

$\varepsilon^2 + \varepsilon + 1 = 0$, $\varepsilon^3 - 1 = (\varepsilon - 1)(\varepsilon^2 + \varepsilon + 1) = 0 \Rightarrow \varepsilon^3 = 1$ and

$\varepsilon \neq 1$. $f(\varepsilon) = 0$, $f(\varepsilon) = \varepsilon(\varepsilon^2 + 2\varepsilon + 1)^n (\varepsilon + 1) + (m-1)\varepsilon^n =$

$= (\varepsilon^2 + \varepsilon)\varepsilon^n + (m-1)\varepsilon^n = (m-2)\varepsilon^n = 0 \Rightarrow m = 2$.

IX. 59. a) The roots of the equation $3x^2 + 3x + 1 = 0$ are x_1 and x_2, therefore

$3x_1^2 + 3x_1 + 1 = 0$, and $3x_2^2 + 3x_2 + 1 = 0$.

$f(x_1) = (3x_1 + 1)^{6n+1} + 3x_1^2 = \left[9x_1(3x_1^2 + 3x_1 + 1) + 1\right]^{2n}(3x_1 + 1) + 3x_1^2 =$

$= 3x_1 + 1 + 3x_1^2 = 0$, thus f is divisible by $x - x_1$. Similarly f is divisible by

$x - x_2$. Therefore f is divisible by $(x - x_1)(x - x_2) = 3x^2 + 3x + 1$.

IX. 60. $f(x) = ax^5 + bx^4 + cx^3 = (x+1)^3(ax^2 + px + q) + 2$,

$f(x) = -12x^5 - 30x^4 - 20x^3$.

IX. 61. $f(x) = (x-1)^3\left(-\dfrac{5}{16}x^2 - \dfrac{13}{16}x - \dfrac{3}{8}\right) - 1$.

IX. 62. $b + d - c = a \Rightarrow -\dfrac{b}{a} + \dfrac{c}{a} - \dfrac{d}{a} = -1 \Rightarrow S_1 + S_2 + S_3 + 1 = 0 \Rightarrow$

$x_1 + x_2 + x_3 + x_1x_2 + x_2x_3 + x_3x_1 + x_1x_2x_3 + 1 = 0 \Rightarrow$

$x_1(1 + x_2 + x_3 + x_2x_3) + (1 + x_2 + x_3 + x_2x_3) = 0 \Rightarrow$

$(1 + x_2 + x_3 + x_2x_3)(1 + x_1) = 0 \Rightarrow (1 + x_1)(1 + x_2)(1 + x_3) = 0$. Therefore

$x_1 = x_2 = x_3 = -1$.

IX. 63. $x_1 + x_2 = 0 \Leftrightarrow (x_1 + x_2)(x_2 + x_3)(x_3 + x_1) = 0 \Leftrightarrow$

$(x_1 + x_2 + x_3)(x_1x_2 + x_2x_3 + x_3x_1) - x_1x_2x_3 = 0 \Leftrightarrow$

$S_1S_2 = S_3 \Leftrightarrow -\dfrac{b}{a} \cdot \dfrac{c}{a} = -\dfrac{d}{a} \Leftrightarrow bc = ad$.

IX. 64. a) $x_1 + x_2 + x_3 = \dfrac{1}{2}$, $x_1 + x_2 = 1 \Rightarrow x_3 = -\dfrac{1}{2}$; $m = -3$;

b) $m = 11$ $x_1 = 2$, $x_2 = -5$, $x_3 = 1$, $x_4 = -1$;

c) $m = 0$, $x_1 = 1 + 2i$, $x_2 = 1 - 2i$, $x_3 = 1 + \sqrt{2}$, $x_4 = 1 - \sqrt{2}$;

d) $x_1 = -4$, $x_2 = -2$, $x_3 = 1$; **e)** $x_1 = 3$, $x_2 = 1$, $x_3 = 2$, $x_4 = 4$;

f) $m = 5$, $x_1 = 2$, $x_2 = 1$, $x_3 = 5$; **g)** $m = 7$, $x_1 = 1$, $x_2 = 4$, $x_3 = 2$, $x_4 = 3$;

h) $m = -8$, $x_1 = 1 + \sqrt{3}$, $x_2 = 1 - \sqrt{3}$, $x_3 = 1 + 2\sqrt{2}$, $x_4 = 1 - 2\sqrt{2}$;

i) $x_1 = 7$, $x_2 = 4$, $x_3 = -2$; j) $x_1 = -2$, $x_2 = 1$, $x_3 = \sqrt{3}$;

k) $m = 4$; l) $m = -5$; m) $x_1 = -\dfrac{2}{3}$, $x_{2,3} = -1 \pm i\sqrt{2}$.

IX. 65. $4a^3 - 27b^2$. IX. 66. $2 - \sqrt{2}$. IX. 67. a) $S - S' = -5$;

b) $E = \dfrac{(x_1 x_2 + x_3 x_1 + x_2 x_3)^2 - 2(x_1 x_2 x_3)(x_1 + x_2 + x_3)}{(x_1 + x_2 + x_3)^2} = \dfrac{39}{25}$,

we have $x_1^3 + 3x_1^2 + 3x_1 + 5 = 0$, $x_1^4 + 3x_1^3 + 3x_1^2 + 5x_1 = 0$ and

$x_1^5 + 3x_1^4 + 3x_1^3 + 5x_1^2 = 0$, similarly for x_2 and x_3. Thus we obtain

$$\sum_{k=1}^{3} x_k = -3, \quad \sum_{k=1}^{3} x_k^2 = 3, \quad \sum_{k=1}^{3} x_k^3 = -15, \quad \sum_{k=1}^{3} x_k^4 = 51.$$

Therefore $\displaystyle\sum_{k=1}^{3} x_k^5 + 3\sum_{k=1}^{3} x_k^4 + 3\sum_{k=1}^{3} x_k^3 + 5\sum_{k=1}^{3} x_k^2 = 0$,

$$F = x_1^5 + x_2^5 + x_3^5 = \sum_{k=1}^{3} x_k^5 = -3\sum_{k=1}^{3} x_k^4 - 3\sum_{k=1}^{3} x_k^3 - 5\sum_{k=1}^{3} x_k^2 = -123,$$

$G = -24$. IX. 68. $m = \dfrac{1}{5}$, $E = -\dfrac{7}{2}$.

IX. 69. a) $S(m) = m^3 + 6m - 6$; b) $m \in (-\infty, 1]$; c) $m = \dfrac{4}{3}$. IX. 70. $m = 1$.

IX. 72. a) $f(-1) \cdot f(1) = -(m+1)^2 < 0$; b) $m = -1$, $x_1 = x_2 = -1$, $x_3 = 1$;

c) $(x_1 x_2 + x_3 x_1 + x_2 x_3)^2 - 2x_1 x_2 x_3 (x_1 + x_2 + x_3) = 4x_1 x_2 x_3$, $m = \pm\sqrt{2}$.

IX. 73. $0 < \displaystyle\sum_{i=1}^{4} x_i^2 = \left(\sum_{i=1}^{4} x_i\right)^2 - 2\sum_{1 \le i \le j \le 4} x_i x_j =$

$= [-2(a+1)]^2 - 2(2a^2 + 4a + 3) = -2 < 0$.

IX. 74. $x_6 = -3$, $b = 720$. IX. 75. $a = -3$, $b = -6$. IX. 76. $a = 3.5$, $b = 39$.

IX. 77. a) $m = 5$, $x_1 = 1$, $x_2 = 4$, $x_3 = 7$;

b) $x_1 = -5$, $x_2 = -1$, $x_3 = 3$, $x_4 = 7$, $m = 6$;

c) $x_1 = -1$, $x_2 = 2$, $x_3 = 5$, $m = 2$;

d) $x_1 = -5$, $x_2 = -2$, $x_3 = 1$, $x_4 = 4$, $m = 3$.

IX. 78. $x_2^2 = x_1 x_3$ and Vieta's Formulas $x_1 + x_2 + x_3 = -a$,

$x_1 x_2 + x_2 x_3 + x_3 x_1 = b$, $x_1 x_2 x_3 = -c$.

Solving this system we obtain $b^3 - a^3 c = 0$.

IX. 79. a) Setting $x_1 = a$, $x_2 = qa$, $x_3 = q^2 a$, and $x_4 = q^3 a$.

Using Vieta's formulae
$$\begin{cases} x_1 + x_2 + x_3 + x_4 = 15 \\ x_1 x_2 + x_1 x_3 + x_1 x_4 + x_2 x_3 + x_2 x_4 + x_3 x_4 = 70 \\ x_1 x_2\, x_3 + x_1 x_2 x_4 + x_1 x_3 x_4 + x_2\, x_3 x_4 = 120 \\ x_1 x_2\, x_3 x_4 = m \end{cases}$$

we obtain $q = 2$, $x_1 = 1$, $x_2 = 2$, $x_3 = 4$, $x_4 = 8$, $m = 64$;

b) x_1, x_2, x_3, and x_4 are in geometric sequence $\Rightarrow x_2^2 = x_1 x_3$ and $x_3^2 = x_2 x_4$

$\Rightarrow x_2 x_3 = x_1 x_4$. Denote $x_2 + x_3 = s_1$, $x_2 x_3 = p_1$ and $x_1 + x_4 = s_2$, $x_1 x_4 = p_2$.

Then $s_1 + s_2 = \dfrac{15}{2}$, $s_1 s_2 + p_1 + p_2 = \dfrac{35}{2}$, $s_1 p_2 + s_2 p_1 = 15$, $p_1 p_2 = \dfrac{m}{2}$, and

$p_1 = p_2$. Therefore $m = 8$ and the roots are $\left(\dfrac{1}{2}, 1, 2, 4\right)$ or $\left(4, 2, 1, \dfrac{1}{2}\right)$;

c) $m = 2$, $x_1 = 1$, $x_2 = 2$, $x_3 = 4$;

d) $m = 2$, $x_1 = 1$, $x_2 = -2$, $x_3 = 4$, $x_4 = -8$.
IX. 80. $a = 4$, $b = -6$, $c = 4$, $d = -1$.

IX. 81. a) $\alpha = \sqrt[3]{7 + \sqrt{41}} + \sqrt[3]{7 - \sqrt{41}}$,

$\alpha^3 = \left(\sqrt[3]{7 + \sqrt{41}}\right)^3 + \left(\sqrt[3]{7 - \sqrt{41}}\right)^3 + 3\left(\sqrt[3]{7 + \sqrt{41}} \cdot \sqrt[3]{7 - \sqrt{41}}\right)\alpha = 14 + 6\alpha$,

$P(\alpha) = \alpha^3 - 6\alpha + 5 = 14 + 5 = 19$; b) 13; c) 17.

IX. 82. $\alpha^3 = \left(\sqrt[3]{\sqrt{5} + 2}\right)^3 + \left(\sqrt[3]{\sqrt{5} - 2}\right)^3 + 3\left(\sqrt[3]{\sqrt{5} + 2} \cdot \sqrt[3]{\sqrt{5} - 2}\right)\alpha = 2\sqrt{5} + 3\alpha$

IX. 83. $(x + 1)^2 = \left(\sqrt{2} + \sqrt{3}\right)^2$, then $\left(x^2 + 2x - 4\right)^2 = 24$ or

$x^4 + 4x^3 - 4x^2 - 16x - 8 = 0$. **IX. 84.** Denote $a = \sqrt[5]{27}$ and $b = \sqrt[5]{9}$.

$a^3 - b^3 = (a - b)^3 - 3ab(a - b) = x^3 - 9x$,

$(a - b)^5 = a^5 - b^5 - 5ab(a^3 - b^3) + 10a^2 b^2 (a - b)$ or

$x^5 = 18 - 15\left(x^3 - 9x\right) + 90x$, thus $x^5 + 15x^3 - 225x - 18 = 0$.

IX. 85. a) Let α be the real root, then $\alpha^3 - (4+i)\alpha^2 + (7+3i)\alpha - 2i - 6 = 0$

or $\alpha^3 - 4\alpha^2 + 7\alpha - 6 - i\left(\alpha^2 - 3\alpha + 2\right) = 0$. Therefore $\alpha = 2$ and

$z_2 = 1 - i$, $z_3 = 1 + 2i$.

b) Let $x_1 = \alpha$ be a real root, then $\alpha^3 - 4\alpha^2 + (6-i)\alpha - 3 + i = 0$ or

$\alpha^3 - 4\alpha^2 + 6\alpha - 3 + i(-\alpha + 1) = 0 \Rightarrow -\alpha + 1 = 0$ and $\alpha^3 - 4\alpha^2 + 6\alpha - 3 = 0 \Rightarrow$

$\alpha = 1$. Therefore $f(x) = (x-1)\left(x^2 - 3x + 3 - i\right)$, $x_2 = 2 + i$ and $x_3 = 1 - i$.

IX. 86. (*Three solutions*) **i)** $x_2 = \dfrac{1 + i\sqrt{3}}{2}$ is also a root, therefore there is p

$\in \mathbf{R}$ such that $f(x) = \left(x^2 - x + 1\right)\left(x + p\right) = x^3 + (p-1)x^2 + (1-p)x + p$, then
equating the coefficients $p - 1 = a$, $1 - p = 4$ and $p = b$.
Finally, $p = -3$, $a = -4$ and $b = -3$.

ii) Using long division

$x^3 + ax^2 + 4x + b = \left(x^2 - x + 1\right)\left(x + a + 1\right) + x(a + 4) + a + b + 1$. Since the
remainder is zero, $a = -4$ and $b = -3$.

iii) x_1 is a root, then $x_1^2 - x_1 + 1 = 0 \Rightarrow x_1^3 + 1 = \left(x_1 + 1\right)\left(x_1^2 - x_1 + 1\right) = 0 \Rightarrow$

$x_1^3 = -1$. Therefore $f(x_1) = x_1^3 + ax_1^2 + 4x_1 + b = -1 + a(x_1 - 1) + 4x_1 + b = 0 \Rightarrow$

$-1 + a \cdot \left(\dfrac{1 - i\sqrt{3}}{2} - 1\right) + 4 \cdot \dfrac{1 - i\sqrt{3}}{2} + b = 0$ or

$b + 1 - \dfrac{a}{2} + i \cdot \left(\dfrac{-\sqrt{3}}{2} a - 2\sqrt{3}\right) = 0 \Rightarrow a = -4$ and $b = -3$.

IX. 87. $a = -6$ and $b = -20$.

IX. 88. $x_2 = 1 + i$ is also a root, therefore $f(x) = \left(x^2 - 2x + 2\right)\left(x^2 + ax + b\right)$.
Equating the coefficients $a = -4$, $b = 3$, $m = -6$, $n = 6$, $x_3 = 1$, $x_4 = 3$.

IX. 89. $-1 - i\sqrt{3}$, $-1 - i$, $f(x) = x^4 + 4x^3 + 10x^2 + 12x + 8$.

IX. 90. (*Four solutions*) **i)** Solving the system $x_1 = x_2$, $x_1 + x_2 + x_3 = -\dfrac{m}{2}$,

$x_1 x_2 + x_1 x_3 + x_2 x_3 = 2$, and $x_1 x_2 x_3 = -2$, we get $m = -7$ and $x_1 = x_2 = 2$.

ii) Obviously $x = 0$ is not a solution of $f(x) = 0 \Rightarrow 2x + m + \dfrac{4}{x} + \dfrac{4}{x^2} = g(x)$ has the same roots as f. Therefore g has a double root and $g'(x)=0$, $g(x)=0$.

$$g'(x)=2-\frac{4}{x^2}-\frac{8}{x^3}, \ g'(x)=0 \Rightarrow x^3-2x-4=0 \Rightarrow x=2.$$

From $g(2)=0 \Rightarrow m=-7$.

iii) Let $\alpha \in \mathbf{R}$ be a double root. Using synthetic division (Horner's algorithm) two times

	x^3	x^2	x^1	x^0
α	2	m	4	4
α	2	$2\alpha+m$	$2\alpha^2+m\alpha+4$	$2\alpha^3+m\alpha^2+4\alpha+4$
	2	$4\alpha+m$	$6\alpha^2+2m\alpha+4$	

we get the system $2\alpha^3+m\alpha^2+4\alpha+4=0$, $6\alpha^2+2m\alpha+4=0$.

Therefore $\alpha=2$ and $m=-7$. **iv)** Let $\alpha \in \mathbf{R}$ be a double root.

$f(x)=(x-\alpha)^2(2x+p)=2x^3+(p-4\alpha)x^2+(2\alpha^2-2p\alpha)x+\alpha^2p$. Identifying the coefficients we obtain the system $p-4\alpha=m$, $2\alpha^2-2p\alpha=4$ and

$\alpha^2p=4$. Finally $\alpha=2$ and $m=-7$.

IX. 91. $x_1=x_2=x_3=\dfrac{1}{3}$, $a=-\dfrac{2}{3}$, $b=\dfrac{8}{27}$, $c=-\dfrac{1}{27}$.

IX. 92. $a=-2$, $b=1$, $c=-2$. The roots are $x_2=\dfrac{1-i\sqrt{3}}{2}$, $x_3=2$, $x_4=-1$.

IX. 93. $x_2=2+i$, $x_3=1$, $x_4=0$, $a=9$, $b=-5$, $c=0$.

IX. 94. a) $x_2=2+\sqrt{3}$, $x_3=-1$; **b)** $x_2=1+2\sqrt{2}$, $x_3=1$;

c) $x_2=1+\sqrt{2}$, $x_3=i$, $x_4=-i$; **d)** $x_2=3-i$, $x_3=1$;

e) $x_2=2-3i$, $x_3=\sqrt{2}$, $x_4=-\sqrt{2}$; **f)** $x_2=2-i$, $x_3=1$, $x_4=2$,

$m=-23$, $n=10$; **g)** $x_2=-1-i$, $x_3=1$, $x_4=-1$, $m=1$, $n=-2$

h) $x_2=-\sqrt{3}+\sqrt{5}$, $x_3=\sqrt{3}-\sqrt{5}$, $x_4=\sqrt{3}+\sqrt{5}$, $x_5=-1$;

i) $x_2=-\sqrt{3}+\sqrt{7}$, $x_3=\sqrt{3}-\sqrt{7}$, $x_4=\sqrt{3}+\sqrt{7}$, $x_5=-\sqrt{3}$, $x_6=\sqrt{3}$.

IX. 95. a) $x_1=-1, x_2=\dfrac{-1+i\sqrt{15}}{4}$, $x_3=\dfrac{-1-i\sqrt{15}}{4}$;

b) $x_1=-1, x_2=-2, x_3=-\dfrac{1}{2}$; **c)** $x_1=\dfrac{1+i\sqrt{3}}{2}$, $x_2=\dfrac{1-i\sqrt{3}}{2}$,

$x_3 = \dfrac{-3+i\sqrt{7}}{4}, x_4 = \dfrac{-3-i\sqrt{7}}{4}$; d) $x_1 = 2, x_2 = \dfrac{1}{2}, x_3 = i, x_4 = -i$;

e) $x_1 = -1, x_2 = i, x_3 = -i, x_4 = \dfrac{-1+i\sqrt{3}}{2}$,

$x_5 = \dfrac{-1-i\sqrt{3}}{2}$; f) $x_1 = -1, x_2 = -1, x_3 = -1, x_4 = \dfrac{1+i\sqrt{3}}{2}, x_5 = \dfrac{1-i\sqrt{3}}{2}$.

IX. 96. $P(x) = (x-1)^2 q(x) + ax + b$. Then $P(1) = a + b$ and $P'(1) = a$.

Therefore $r(x) = (n+3)x - n - 1$.

IX. 97. $f(x) = g(x)x^2(x-1) + r(x)$ where $r(x) \le 2$. Let $r(x) = ax^2 + bx + c$

be the remainder with a, b, $c \in \mathbf{R}$ and $a \ne 0$. For $x = 0 \Longrightarrow c = 1$; for $x = 1$
$\Longrightarrow a + b + c = n + 1$. Differentiating in both members of the equation

of the division theorem, we obtain $nx^{n-1} + (n-1)x^{n-1} + ... + 2x + 1 =$

$= g'(x)x^2(x-1) + g(x)(3x^2 - 2x) + 2ax + b$. For $x = 0 \Longrightarrow b = 1$.

Finally $r(x) = (n-1)x^2 + x + 1$.

IX. 98. a) $f(x) = \left(x^2 + x + 1\right)\left[(m-1)x^2 - 2(m+1)x + m - 4\right]$.

b) $(m-1)x^2 - 2(m+1)x + m - 4 = 0$ has a double root

$\Delta = 4(m+1)^2 - 4(m-1)(m-4) = 0$, thus $m = \dfrac{3}{7}$.

c) Using Vieta's formulae, we obtain $\dfrac{m+6}{m-4} \ge \dfrac{5}{4}$, $4 < m \le 44$.

d) $x_0 = -1 + i\tan\beta$ cannot be a root for $x^2 + x + 1$.

$\overline{x_0} = -1 - i\tan\beta$ is also a root for $(m-1)x^2 - 2(m+1)x + m - 4 = 0$.

Therefore $-2 = \dfrac{2(m+1)}{m-1}$ and $1 + \tan^2\beta = \dfrac{m-4}{m-1}$, $m \ne 1$.

Finally $m = 0$ and $\beta = \pm\dfrac{\pi}{3} + 2k\pi$, $k \in \mathbf{Z}$ or $\beta = \pm\dfrac{2\pi}{3} + 2k'\pi$, $k' \in \mathbf{Z}$. The

roots are $x_0 = -1 - i\sqrt{3}$ and $x_0' = -1 + i\sqrt{3}$.

IX. 99. $f(x) = x^2 + 5x + 5$, $f(x^2 - 1) = x^4 + 3x^2 + 1$.

IX. 100. a) If a is a root of the equation $x^3 - x - 5 = 0$, then $a^3 = a + 5$

(1). On the other hand $y = a \cdot a^3 - 5a + 3 = a(a+5) - 5a + 3 = a^2 + 3$ (2). From

(2) $a^2 = y - 3$ substituting in (1) $a(y-3) = a + 5$ we obtain

$a = \dfrac{5}{y-4}$ (3). Now, substituting (3) in (2), we obtain the equation

$y^3 - 11y^2 + 40y - 23 = 0$;

b) $4y^3 - 40y^2 + 244y - 1220 = 0$; c) $y^3 - 20y^2 + 57y - 559 = 0$;

d) $y^3 - 13y^2 + 48y - 3161 = 0$; e) $y^3 + y^2 - 205y + 35478 = 0$.

IX. 101. a) If x_1 is a root of the equation $x^3 - 2x - 1 = 0$, then $x_1^3 = 2x_1 + 1$, similarly $x_2^3 = 2x_2 + 1$ and $x_3^3 = 2x_3 + 1$.

Also $\sum x_1 = 0$, $\sum x_1 x_2 = -2$.

$P(x_1) = x_1^5 - x_1^4 - x_1 - 4 = x_1^2(x_1^3) - x_1(x_1^3) - x_1 - 4 =$
$= x_1^2(2x_1 + 1) - x_1(2x_1 + 1) - x_1 - 4 = -x_1^2 + 2x_1 - 2$

$P(x_1) + P(x_2) + P(x_3) = \sum P(x_1) = \sum(-x_1^2 + 2x_1 - 2) = -\sum x_1^2 + 2\sum x_1 - \sum 2 =$
$= -\left(\sum x_1\right)^2 + 2\left(\sum x_1 x_2\right) + 2\sum x_1 - \sum 2 = 0 - 4 + 0 - 6 = -10$;

b) If $\sum P(x_1) = -3$.

IX. 102. Let $f(x) = x^{n-1} + x^{n-2} + ... + 1 = (x - x_1)(x - x_2)...(x - x_{n-1})$.

a) $f(1) = n$; b) $\displaystyle\sum_{i=1}^{n} \frac{1}{1-x_i} = \frac{f'(1)}{f(1)}$; c) $f(-1) = \dfrac{1-(-1)^n}{2}$.

IX. 103. $\displaystyle\sum_{k=1}^{n} \frac{x_k^2+1}{x_k-1} = \sum_{k=1}^{n} \frac{(x_k-1)(x_k+1)+2}{x_k-1} = \sum_{k=1}^{n}(x_k+1) - 2\sum_{k=1}^{n}\frac{1}{1-x_k}$

Since $x_1 + x_2 + ... + x_n = 0$ and $\displaystyle\sum_{k=1}^{n}\frac{1}{1-x_k} = \frac{P'(1)}{P(1)} = \frac{n-4}{2}$. Therefore

$\displaystyle\sum_{k=1}^{n}\frac{x_k^2+1}{x_k-1} = 4$.

IX. 104. a) From the given equation, we have $P_n(i) - P_n(i-1) = (i-1)^n$, $(\forall) i = 1, 2...k$; b). From $P_n(1) - P_n(0) = 0 \Rightarrow P_n(1) = 0$.

IX. 105. Let α be a real root of the given equation. We have

$\alpha^{2n+1} - \alpha^{2n} + \alpha^{2n-1} + 2n\alpha^n - n^2 = 0 \Longrightarrow \alpha^{2n-1}(\alpha^2 + 1) = (\alpha^n - n)^2$ or

$\alpha^{2n-1} = \dfrac{(\alpha^2 - n)^2}{\alpha^2 + 1} \geq 0$. Then $\alpha \geq 0$. Since 0 is not a solution for the given equation, it follows that $\alpha > 0$.

IX. 106. We assume that there is $\lambda \in \mathbf{Z}$ such that $f(\lambda) = 0$. We have the cases $\lambda = 2k$ and $\lambda = 2k + 1$ and we obtain $f(2k)$ and $f(2k + 1)$ odd numbers. Therefore f does not have any integer root.

IX. 107. The degrees of the polynomial on the left side is n^2, but the degree of the polynomial on the right side is $2n$. Then $n^2 = 2n \Rightarrow n = 2$. Finally

$$f(x) = a\left(x^2 - x\right) \text{ where } a \in \mathbf{R}^*.$$

IX. 108. Let $x_1, x_2, \ldots, x_n$ be the roots of the polynomial P. Therefore

$$x_1 x_2 \ldots x_n = 1 \text{ and } |x_1| \cdot |x_2| \cdot \ldots \cdot |x_n| = 1 \Rightarrow |x_1| = |x_2| = \ldots = |x_n| = 1, \text{ since}$$

$$P(x) = (x - x_1)(x - x_2) \ldots (x - x_n) \Rightarrow P(-1) = (-1)^n (1 + x_1)(1 + x_2) \ldots (1 + x_n) \text{ as}$$

$$|x_i| = 1, \ (\forall) i = 1, 2 \ldots k \Rightarrow x_i \cdot \overline{x_i} = 1 \Rightarrow \overline{x_i} = \frac{1}{x_i}. \text{ Then}$$

$$\overline{P(-1)} = (-1)^n \left(1 + \overline{x_1}\right)\left(1 + \overline{x_2}\right) \ldots \left(1 + \overline{x_n}\right) = (-1)^n \frac{(1 + x_1)(1 + x_2) \ldots (1 + x_n)}{x_1 \cdot x_2 \cdot \ldots \cdot x_n} = P(-1)$$

therefore $P(-1)$ is real.   **IX. 109.** It is obvious that $\varepsilon^n = 1$.

Therefore $\displaystyle\sum_{k=1}^{n} f\left(\frac{1}{\varepsilon^k}\right) = \sum_{k=1}^{n} \left[\sum_{j=1}^{n} a_j \left(\frac{1}{\varepsilon^k}\right)^j\right] = \sum_{k=1}^{n} a_j \sum_{j=1}^{n} \frac{1}{\varepsilon^{kj}} =$

$$= na_0 + a_1 \frac{1 - \left(\dfrac{1}{\varepsilon}\right)^n}{\varepsilon \ \ 1 - \dfrac{1}{\varepsilon}} + a_2 \frac{1 - \left(\dfrac{1}{\varepsilon}\right)^{2n}}{\varepsilon^2 \ \ 1 - \dfrac{1}{\varepsilon^2}} + \ldots + na_n = n\left(a_0 + a_n\right).$$

Chapter X

PERMUTATIONS AND COMBINATIONS.
NEWTON'S BINOMIAL THEOREM

X. 1. $5! = 120$. **X. 2.** $6! = 720$. **X. 3.** a) $6! - 5! = 600$; b) $5! = 120$; c) $5! - 4! = 96$;

d) $4! = 24$; e) $600 - 24 = 576$. **X. 4.** $\dfrac{7!}{3! \cdot 2! \cdot 1! \cdot 1!}$, $\dfrac{14!}{(3!)^2 (2!)^2}$, $\dfrac{11!}{2! (4!)^2}$.

X. 5. $\dfrac{15!}{6! \cdot 5! \cdot 4!} = 630,630$.

X. 6. $P(14,3) = 2184$. **X. 7.** $P(7,5) = 2520$, $7! = 5040$. **X. 8.** Five-digit

numbers are $P(5,5) - P(4,4) = 96$, four-digit numbers are $P(5,4) - P(4,3) = 96$,

three-digit numbers are $P(5,3) - P(4,2) = 48$, two-digit numbers are

$P(5,2) - P(4,1) = 16$, and one-digit numbers are five. In total 261 numbers.

X. 9. $\binom{7}{4} + \binom{7}{5} + \binom{7}{6} + \binom{7}{7} = 35 + 21 + 7 + 1 = 64$. **X. 10.** $\binom{20}{4} = 4845$.

X. 11. a) $\binom{4}{1} \cdot \binom{8}{5}$; b) $\binom{12}{6} - \binom{8}{6}$; c) $\binom{11}{5}$. **X. 12.** $\binom{4}{2} \cdot \binom{10}{5} = 1512$.

X. 13. a) $\binom{25}{5} = 53,130$, b) $\binom{24}{4} = 10,626$. **X. 14.** $\binom{6}{2} = 15$. **X. 15.** $\binom{5}{2} = 10$.

X. 16. $\binom{9}{5} = 126$; $\binom{10}{5} = 252$.

X. 17. $\binom{6}{3} \cdot \binom{9}{4} + \binom{6}{4} \cdot \binom{9}{3} + \binom{6}{5} \cdot \binom{9}{2} + \binom{6}{6} \cdot \binom{9}{1} = 4.005$. **X. 18.** $\binom{7}{4} \cdot P(9,4)$.

X. 19. a) $\binom{15}{4} \cdot \binom{10}{6}$; b) $785,565$. **X. 20.** See the previous problem.

X. 21. $\binom{9}{7} = 36$. **X. 22.** a) $\dfrac{(x+9)!}{(x+4)! \cdot 5!} = 5 \dfrac{(x+7)!}{(x+4)!} \Rightarrow \dfrac{(x+9)(x+8)}{5!} = 5$

$\Rightarrow x^2 - 17x - 528 = 0$, $x = 33$; b) $x = 7$; c) $x = 10$; d) $x = 5$;

e) $x = 19$; f) $x = 5$; g) $x = 2$; h) $x = 10$; i) $x = 4$; j) $x = 3$;

k) $x = 2$; l) $n = 10$; m) $n = 12$. **X. 24.** a) 15; b) 84; c) $2\dfrac{6}{7}$;

d) $8\dfrac{13}{14}$; e) $4k^2$; f) $\dfrac{(n-1)!}{(k-1)!}$; g) $\dfrac{k+1}{n+1}$; h) $\dfrac{2(n-k)}{n-1}$; i) $\dfrac{k}{(n+1)(n+2)}$.

X. 25. Use $\binom{n}{k} = \binom{n}{n-k}$, $\left[\binom{2n+1}{1} \cdot \binom{2n+1}{2} \cdot \ldots \cdot \binom{2n+1}{n}\right]^2$.

X. 26. a) $n \geq 6 \, (n \in \mathbf{N})$; b) $n \in [0,5]$, $n \in \mathbf{N}$; c) $n \in [4,10]$, $n \in \mathbf{N}$;

d) $1 \leq n \leq 9$ $(n \in \mathbb{N})$; e) $4 \leq n \leq 13$ $(n \in \mathbb{N})$; f) $9 \leq n \leq 15$ $(n \in \mathbb{N})$;

g) $5 \leq n \leq 20$ $(n \in \mathbb{N})$; h) $0 \leq n \leq 5$ $(n \in \mathbb{N})$; i) $1 \leq n \leq 4$ $(n \in \mathbb{N})$.

X. 27. a) $x = 7$, $y = 6$; b) $x = 3$, $y = 8$; c) $x = 5$, $y = 5$; d) $n = 10$, $k = 4$;

e) $x = 3$, $y = 5$ or $x = 5$, $y = 3$; f) $x = 6$, $y = 3$; g) $x = 5k - 1$, $y = 3k$.

X. 28. a) Use $\binom{2n}{n} = 2\binom{2n-1}{n}$.

X. 29. a) $S_1 = \sum_{k=1}^{n} \left[(k+1)! - k! \right] = (2! - 1!) + (3! - 2!) + \ldots + (n+1)! - n! = (n+1)! - 1$;

b) $S_2 = \sum_{k=2}^{n} \left(\frac{1}{k!} - \frac{1}{(k+1)!} \right) = 1 - \frac{1}{(k+1)!}$; c) $S_3 = \frac{1}{2} - \frac{1}{n!}$;

d) $S_4 = \sum_{k=2}^{n} \frac{k(k+1)}{k!} - \sum_{k=2}^{n} \frac{k+2}{k!} = \sum_{k=2}^{n} \frac{k+1}{(k-1)!} - \sum_{k=2}^{n} \frac{k+2}{k!} = 3 - \frac{n+2}{n!}$;

e) $S_5 = \sum_{k=1}^{n} \frac{1}{k(k+1)} = 1 - \frac{1}{n+1}$;

f) $S_6 = \sum_{k=2}^{n} \left(\frac{1}{k! k} - \frac{1}{(k+1)!(k+1)} \right) = \frac{1}{4} - \frac{1}{(n+1)!(n+1)}$;

g) $S_7 = \sum_{k=1}^{n} \frac{k+1}{(k-1)! + k! + (k+1)!} = \sum_{k=1}^{n} \frac{k+1}{(k-1)! [1 + k + k(k+1)]} =$

$= \sum_{k=1}^{n} \frac{k+1}{(k-1)!(k+1)^2} = \sum_{k=1}^{n} \frac{k}{(k+1)!} = \sum_{k=1}^{n} \left(\frac{1}{k!} - \frac{1}{(k+1)!} \right) = \frac{n}{n+1}$.

X. 30. Use the formula $\binom{n}{k} = \binom{n-1}{k} + \binom{n-1}{k-1}$.

X. 31. a) $\dfrac{S_1}{3!} = \dfrac{1 \cdot 2 \cdot 3}{3!} + \dfrac{2 \cdot 3 \cdot 4}{3!} + \ldots + \dfrac{n \cdot (n+1) \cdot (n+2)}{3!} =$

$= \binom{3}{3} + \binom{4}{3} + \ldots + \binom{n+2}{3} = \binom{n+3}{4}$. $S_1 = \dfrac{n \cdot (n+1) \cdot (n+2) \cdot (n+3)}{4}$;

b) Similarly $\dfrac{S_2}{k!} = \binom{k}{k} + \binom{k+1}{k} + \ldots + \binom{n+k-1}{k} = \binom{n+k}{k+1}$. Therefore

$S_2 = \dfrac{(n+k) \cdot (n+k-1) \ldots (n+1) n}{k+1}$.

X. 32. a) $\dfrac{n(n+1)(n-1)}{6}$;

b) $\sum_{k=2}^{n} \dfrac{n^2 (n-1)^2}{4} = \dfrac{1}{4} \sum_{k=2}^{n} k^4 - \dfrac{1}{2} \sum_{k=2}^{n} k^3 + \dfrac{1}{4} \sum_{k=2}^{n} k^2 = \dfrac{n \cdot (n^2 - 1)(3n^2 - 2)}{60}$.

We used the equalities $\displaystyle\sum_{k=1}^{n}k=\frac{n\cdot(n+1)}{2}$; $\displaystyle\sum_{k=1}^{n}k^2=\frac{n\cdot(n+1)(2n+1)}{6}$;

$$\sum_{k=1}^{n}k^3=\frac{n^2\cdot(n+1)^2}{4}; \quad \sum_{k=1}^{n}k^4=\frac{n\cdot(n+1)(2n+1)(3n^2+3n-1)}{30}.$$

X. 33. $a=2$, $b=2$, $c=4$. **X. 34.** Use $\binom{n-1}{k}=\frac{n-k}{k}\binom{k-1}{n-1}\binom{n-1}{k-1}$

and $\binom{n}{k}=\frac{n}{k}\binom{n-1}{k-1}$. **X. 36.** $\dfrac{\binom{2n}{n}}{n+1}=\dfrac{(2n)!}{n!\cdot(n+1)!}=\dfrac{(2n)!}{n!\cdot(n-1)!}\cdot\dfrac{1}{n(n+1)}=$

$=\dfrac{(2n)!}{n!\cdot(n-1)!}\left(\dfrac{1}{n}-\dfrac{1}{n+1}\right)=\binom{2n}{n}-\binom{2n}{n-1}$. **X. 37.** Use the identity $\dfrac{k\binom{n}{k}}{\binom{n}{k-1}}=n-k$.

X. 38. $\binom{n+1}{n-1}=\dfrac{n(n+1)}{2}=1+2+3+\ldots+n$. Therefore

$$\frac{\binom{n+1}{n-1}!}{\prod\limits_{k=1}^{n}k!}=\frac{(1+2+3+\ldots+n)!}{1!\cdot2!\cdot3!\cdot\ldots\cdot n!}=\binom{1+2+3+\ldots+n}{n}\cdot\binom{1+2+3+\ldots+(n-1)}{n-1}\cdot\binom{1+2+3+\ldots+(n-2)}{n-2}\cdot\ldots\cdot\binom{1+2}{2}.$$

X. 39. a) $\dfrac{1}{\binom{2n}{k}}=\dfrac{2n+1}{2(n+1)}\left(\dfrac{1}{\binom{2n+1}{k}}+\dfrac{1}{\binom{2n+1}{k+1}}\right)$; $\displaystyle\sum_{k=0}^{2n}(-1)^k\cdot k!(2n-k)!=\dfrac{(2n+1)!}{n+1}$.

$$\sum_{k=0}^{2n}(-1)^k\cdot k!(2n-k)!=(2n)!\cdot\sum_{k=0}^{2n}(-1)^k\cdot\frac{k!(2n-k)!}{(2n)!}=$$

$$=(2n)!\cdot\sum_{k=0}^{2n}\frac{(-1)^k}{\binom{2n}{k}}=(2n)!\cdot\sum_{k=0}^{2n}(-1)^k\cdot\frac{2n+1}{2(n+1)}\left(\frac{1}{\binom{2n+1}{k}}+\frac{1}{\binom{2n+1}{k+1}}\right)=$$

$$=\frac{(2n+1)!}{2(n+1)}\cdot\sum_{k=0}^{2n}(-1)^k\cdot\left(\frac{1}{\binom{2n+1}{k}}+\frac{1}{\binom{2n+1}{k+1}}\right)= \quad\text{(we used a))}$$

$$=\frac{(2n+1)!}{2(n+1)}\cdot\left\{\left[\frac{1}{\binom{2n+1}{0}}+\frac{1}{\binom{2n+1}{1}}\right]-\left[\frac{1}{\binom{2n+1}{1}}+\frac{1}{\binom{2n+1}{2}}\right]+\ldots+\left[\frac{1}{\binom{2n+1}{2n}}+\frac{1}{\binom{2n+1}{2n+1}}\right]\right\}=$$

$$=\frac{(2n+1)!}{n+1}.$$

X. 40. a) $\binom{4}{0}+\binom{4}{1}\cdot(2x)+\binom{4}{2}\cdot(2x)^2+\binom{4}{3}\cdot(2x)^3+\binom{4}{4}\cdot(2x)^4$;

b) $2\binom{6}{0}x^6+2\binom{6}{2}\cdot x^4\left(x^2-1\right)+2\binom{6}{4}\cdot x^2\left(x^2-1\right)^2+2\binom{6}{6}\cdot\left(x^2-1\right)^3$.

X. 41. $t_5 = \binom{9}{4}\left(\sqrt[3]{x}\right)^4 \cdot \left(\dfrac{1}{\sqrt[3]{x}}\right)^5 = \dfrac{126}{\sqrt[3]{x}}$.

X. 42. $\dfrac{x^5}{32} + \dfrac{5x^4 y}{8} + 5x^3 y^2 + 20x^2 y^3 + 40xy^4 + 32y^5$

X. 43. a) $256 - 3072x + 16128x^2$; **b)** 24.

X. 44. $t_{k+1} = \binom{2010}{k} x^k \left(\dfrac{2}{\sqrt[5]{x}}\right)^{2010-k} = \binom{2010}{k} \cdot 2^{2010-k} \cdot x^{k+\frac{k-2010}{5}}$,

$k + \dfrac{k-2010}{5} = 0 \Rightarrow k = 335$. Therefore $t_{336} = \binom{2010}{335} \cdot 2^{1675}$.

X. 45. a) 198; **b)** $1\dfrac{2}{7}$. **X. 46.** 152. **X. 47.** $\dfrac{1}{x}\left[(1+x)^{21} - 1\right]$, 210 each.

X. 48. $x = \dfrac{1}{3}$, $n = 9$, $\dfrac{28}{9}$. **X. 49.** 2300. **X. 50.** $n = 10$, $a = 2$, $b = 960$.

X. 51. $-\dfrac{448}{3\sqrt{3}} x^3$. **X. 52.** 11. **X. 53.** $n = 15$. **X. 54.** $n = 20$.

X. 55. $\dfrac{1}{2}$. **X. 56.** $a = \pm 1$. **X. 57.** $n = 3$. **X. 58.** $t_5 = 792$.

X. 59. $t_{n+1} = (-1)^n \dfrac{(3n)!}{n! \cdot (2n)!}$. **X. 60.** $t_6 = 210x^2 y^3$. **X. 61.** $t_3 = 153x^{\frac{13}{2}}$.

X. 62. $2^{n-1} = 2048$, $t_7 = -264x^3 y^7$. **X. 64. a)** $n = 6$; **b)** $n = 10$.

X. 65. $t_{k+1} = \binom{21}{k} \cdot \left(\sqrt[3]{\dfrac{x}{\sqrt{y}}}\right)^k \left(\sqrt{\dfrac{y}{\sqrt[3]{x}}}\right)^{21-k} = \binom{21}{k} \cdot x^{\frac{k}{3} - \frac{21-k}{6}} \cdot y^{\frac{21-k}{2} - \frac{k}{6}}$. Since x and y

have equal powers, we obtain $k = 12$, therefore $t_{13} = \binom{21}{12} \sqrt{x^5 y^5}$.

X. 66. $x_1 = \dfrac{1}{100\sqrt{10}}$, $x_2 = 10$. **X. 67.** $m = 12$.

X. 68. $\binom{n}{0} + \binom{n}{1} + \binom{n}{2} = 22$, $n = 6$.

$t_3 + t_5 = \binom{6}{2}\left(\sqrt{2^x}\right)^4 \left(\sqrt{2^{1-x}}\right)^2 + \binom{6}{4}\left(\sqrt{2^x}\right)^2 \left(\sqrt{2^{1-x}}\right)^4$,

$15 \cdot 2^{x+1} + 15 \cdot 2^{2-x} = 135$, $x_1 = -1$, $x_2 = 2$.

X. 69. $t_3 = \binom{9}{2}\left(x^{\log\sqrt{x}}\right)^2 \cdot \left(\dfrac{1}{\sqrt[7]{x^2}}\right)^7 = 36x^{\log x - 2}$. Therefore

$36x^{\log x-2} = 36000 \Rightarrow x^{\log x-2} = 1000 \Rightarrow (\log x - 2)\log x = 3$, $x_1 = 10^{-1}$ and

$x_2 = 10^3$. **X. 70.** $n = 6$, $x_1 = -2$, $x_2 = 1$.

X. 71. $x = 6$. **X. 72.** $x \in (0, \infty) \setminus \{10^{-1}\}$, $x_1 = 10$, $x_2 = 10^{-4}$.

X. 73. $n = 8$, t_0, t_4, t_8.

X. 74. $10 - 3^x > 0$, $\binom{n}{1} + \binom{n}{3} = 2\binom{n}{2} \Rightarrow n = 7$.

From $t_6 = 21 \Rightarrow x_1 = 0$ and $x_2 = 2$. **X. 75.** $n = 8$; t_6. **X. 76.**

$\binom{n}{1} + \binom{n}{3} + \binom{n}{5} + \ldots = 2^{n-1}$, $n = 9$, $t_7 = \binom{9}{3}x^{-3}$.

X. 77. a) $t_{k+1} = \binom{5}{k}\left(\sqrt[3]{3}\right)^{5-k} \cdot \left(\sqrt{2}\right)^k = \binom{5}{k} \cdot 3^{\frac{5-k}{3}} \cdot 2^{\frac{k}{2}}$ where $\dfrac{5-k}{3} \in Z$, $\dfrac{k}{2} \in Z$ and

$0 \le k \le 5$ we get $k = 2$ and $t_3 = \binom{2}{5} \cdot 3 \cdot 2 = 60$;

b) t_6 and t_{21}; **c)** t_{14k+1}, $0 \le k \le 14$, $k \in N$; **d)** $t_{15} = -\binom{24}{10} \cdot 36$.

X.78. a) 161,700; **b)** t_{91}; **c)** $(2.5)^{100}$; **d)** t_{50}; **e)** 7.

X. 79. a) 1; **b)** 2^{22}.

X. 80. a) $t_{k-1} < t_k > t_{k+1}$, $t_{50} = \binom{100}{50} \cdot \left(\dfrac{1}{2}\right)^{100}$, **b)** t_{10}; **c)** $314{,}925 \cdot 10^5$.

X. 81. $20a^2 - 5a$. **X.82. a)** $(x-1)^9 = 1$, $x = 2$; **b)** $x = 4$;

c) $\sin x = -\dfrac{1}{2}$, thus $x_1 = -\dfrac{\pi}{6} + 2k\pi$ $k \in Z$, $x_2 = \dfrac{\pi}{6} + (2k'+1)\pi$ $k' \in Z$.

X. 83. a) 1; **b)** $\left(\cosh^2 x - \sinh^2 x\right)^n = 1$.

X. 85. a) $z = 1 + i$, $z^n = 2^{\frac{n}{2}} \cdot \left(\cos\dfrac{n\pi}{4} + i\sin\dfrac{n\pi}{4}\right)$,

$(1+i)^n = 1 - \binom{n}{2} + \binom{n}{4} - \binom{n}{6} + \ldots + i\left(\binom{n}{1} - \binom{n}{3} + \binom{n}{5} - \binom{n}{7} + \ldots\right)$;

d) $z = 1 + i\sqrt{3}$, $z^n = 2^n \cdot \left(\cos\dfrac{n\pi}{3} + i\sin\dfrac{n\pi}{3}\right)$.

X. 86. a) $\varepsilon = \dfrac{-1 + i\sqrt{3}}{2}$ is the root of the equation $\varepsilon^2 + \varepsilon + 1 = 0$, also $\varepsilon^3 = 1$.

$\varepsilon = \cos\dfrac{2\pi}{3} + i\sin\dfrac{2\pi}{3}$, $1 + \varepsilon = \cos\dfrac{\pi}{3} + i\sin\dfrac{\pi}{3}$,

$1 + \varepsilon^2 = \cos\dfrac{\pi}{3} - i\sin\dfrac{\pi}{3}$. Finally, add the equalities

$$2^n = \binom{n}{0} + \binom{n}{1} + \binom{n}{2} + \binom{n}{3} + \binom{n}{4} + \dots ,$$

$$(1+\varepsilon)^n = \binom{n}{0} + \binom{n}{1}\varepsilon + \binom{n}{2}\varepsilon^2 + \binom{n}{3}\varepsilon + \binom{n}{4}\varepsilon + \dots ,$$

$$\left(1+\varepsilon^2\right)^n = \binom{n}{0} + \binom{n}{1}\varepsilon^2 + \binom{n}{2}\varepsilon + \binom{n}{3}\varepsilon + \binom{n}{4}\varepsilon^2 + \dots .$$

X. 87. $z^n = \left(1 + i\dfrac{1}{\sqrt{3}}\right)^n = \left(\dfrac{2}{\sqrt{3}}\right)^n \left(\cos\dfrac{n\pi}{6} + i\sin\dfrac{n\pi}{6}\right).$

X. 88. $z^{6n} = \left(1 - i\sqrt{3}\right)^{6n} = \left[2\left(\cos\dfrac{5\pi}{3} + i\sin\dfrac{5\pi}{3}\right)\right]^{6n} = 2^{6n}.$

X. 89. Use $z = (1+i)^n$. **X. 90.** **a)** $S_n = 2^n$; **b)** Denote

(1) $S_n = \binom{n}{1} + 2\binom{n}{2} + 3\binom{n}{3} + \dots + n\binom{n}{n}$. From $\binom{n}{k} = \binom{n}{n-k}$, we get

(2) $S_n = n\binom{n}{0} + (n-1)\binom{n}{1} + (n-2)\binom{n}{2} + \dots + \binom{n}{n-1}$. Adding (1) and (2) we obtain

$2S_n = n\left[\binom{n}{0} + \binom{n}{1} + \binom{n}{2} + \dots + \binom{n}{n}\right]$. Finally $S_n = 2^{n-1}n$.

Another solution: Using $\binom{n}{k} = \dfrac{n}{k}\binom{n-1}{k-1}$,

$$S_n = n\binom{n-1}{0} + n\binom{n-1}{1} + n\binom{n-1}{2} + \dots + n\binom{n-1}{n-1} = 2^{n-1}n ;$$

c) $S_n = 2^{n-1}(n+2)$; **d)** $S_n = (n-2)2^{n-1} + 1$; **e)** $S_n = (n+1)2^n$;

f) $S_n = 2^{n-1}(2k+n)$; **g)** $S_n = 0$; **h)** $S_n = 2^{n-1}n - 2^n + 1$;

i) Use $\binom{n}{k} = \binom{n-1}{k-1} + \binom{n-1}{k}$, 0; **j)** Use $\binom{n}{k} = \dfrac{k+1}{n+1}\binom{n+1}{k+1}$, $S_n = \dfrac{2^{n+1}-1}{n+1}$;

k) Use $\binom{n-1}{k} = \binom{n}{k} - \binom{n-1}{k-1}$,

$$S_n = \sum_{k=0}^{n} \frac{\binom{n}{k}}{k+2} = \sum_{k=0}^{n-1}\frac{1}{k+2}\binom{n+1}{k+1} - \sum_{k=0}^{n-1}\frac{1}{k+2}\binom{n}{k+1} + \frac{1}{n+1} =$$

$$= \sum_{k=0}^{n-1}\frac{1}{k+2}\frac{k+2}{n+2}\binom{n+2}{k+2} - \left[\sum_{k=0}^{n-1}\frac{1}{k+2}\frac{k+2}{n+1}\binom{n+1}{k+2}\right] + \frac{1}{n+1} =$$

$$= \frac{1}{n+2}\sum_{k=0}^{n-1}\binom{n+2}{k+2} - \left[\frac{1}{n+1}\sum_{k=0}^{n-1}\binom{n+1}{k+2}\right] + \frac{1}{n+1} = \frac{2^{n+1}n+1}{(n+1)(n+2)} ;$$

l) Use $\binom{n}{k} = \dfrac{(k+2)(k+1)}{(n+2)(n+1)}\binom{n+2}{k+2}$, $S_n = \dfrac{2^{n+2}-n-3}{n+2}$; **m)** $S_n = \dfrac{1}{n+1}$.

X. 91. a) The coefficient of x^n in the identity

$$x^n(1-x)^n + x^{n-1}(1-x)^n + \ldots + x^{n-m}(1-x)^n = x^{n-m}(1-x)^{n-1} - x^{n+1}(1-x)^{n-1}$$ is

$(-1)^m C_{n-1}^m$ when $m \le n-1$ and it is zero if $m = n$;

b) Similarly, the coefficient of x^k from the expression

$$(1+x)^n + (1+x)^{n+1} + \ldots + (1+x)^{n+m} = \frac{1}{x}\left[(1+x)^{n+m+1} - (1+x)^n\right]$$ is

$\binom{n+m+1}{k+1} - \binom{n}{k+1}$ when $k \le n-1$ and it is $\binom{n+m+1}{n+1}$ when $k = n$.

X. 92. a) The coefficient of x^n in the identity $(1+x)^n(1+x)^n = (1+x)^{2n}$ is

$\binom{2n}{n}$; **b)** The coefficient of x^n in the identity $(1+x)^n(1-x)^n = (1-x^2)^n$ is zero

if $n = 2k+1$ and $(-1)^k\binom{2k}{k}$ if $n = 2k$; **c)** Use $\binom{n}{k} = \binom{n-1}{k} + \binom{n-1}{k-1}$.
The sum is 1 for $n = 3k$, 0 for $n = 3k+1$, and -1 for $n = 3k-1$, $k \in \mathbb{N}$; **d)**

2^{2n}; **e)** $\dfrac{(2n-1)!}{[(n-1)!]^2}$.

X. 93. Consider the equality $(1+x)^p(x+1)^{n-p} = (1+x)^n$.

$$\sum_{k=0}^{p}\binom{p}{k}P(m,k)P(n-p,m-k) = \binom{n}{m} \text{ or } m!\sum_{k=0}^{p}\binom{p}{k}P(m,k)P(n-p,m-k) = P(n,m).$$

X. 94. The number $\binom{3n}{n}$ is the coefficient of x^n from the expansion $(1+x)^{3n}$.

In the equality $(1+x)^{3n} = (1+x)^n(1+x)^{2n}$ we have $(1+x)^{2n} = \sum_{k=0}^{2n}\binom{2n}{k}x^k$ and

$(1+x)^n = \sum_{l=0}^{n}\binom{n}{l}x^l$. Therefore $(1+x)^{3n} = \sum_{k=0}^{2n}\sum_{l=0}^{n}\binom{2n}{k}\binom{n}{l}x^{k+l}$ and the coefficient

of x^n in $(1+x)^{3n}$ is $\sum_{k+l=n}\binom{2n}{k}\binom{n}{l} = \sum_{k=0}^{n}\binom{2n}{k}\binom{n}{k}$, thus $\binom{3n}{n} = \sum_{k=0}^{n}\binom{2n}{k}\binom{n}{k}$.

X. 95. $\left(1+\sqrt{2}\right)^{2n} = \sum_{k=0}^{n}2^k\binom{2n}{2k} + \sqrt{2}\cdot\sum_{k=0}^{n}2^{k+1}\binom{2n}{2k+1}$ (1),

$\left(1-\sqrt{2}\right)^{2n} = \sum_{k=0}^{n}2^k\binom{2n}{2k} - \sqrt{2}\cdot\sum_{k=0}^{n}2^{k+1}\binom{2n}{2k+1}$ (2).

Multiplying (1) and (2) member by member, we get the initial identity.

X. 96. Use $\dfrac{k\binom{n}{k}}{\binom{n}{k-1}} = n-k$. **X. 97.** Use $[(n-1)+1]^n$.

X. 98. $S_1 + iS_2 = 1 + \binom{n}{1}(\cos x + i\sin x)^1 + \binom{n}{2}(\cos x + i\sin x)^2 + ... +$

$+ \binom{n}{n}(\cos x + i\sin x)^n = (1 + \cos x + i\sin x)^n = \left(2\cos^2\dfrac{x}{2} + 2i\sin\dfrac{x}{2}\cos\dfrac{x}{2}\right)^n =$

$= 2^n \cos^n\dfrac{x}{2}\left(\cos\dfrac{nx}{2} + i\sin\dfrac{nx}{2}\right).$

Bibliography

*Canadian Mathematics Competition, Problems Problems Problems, Waterloo, On, Canada, Volume 1, 2, 3, 4, 5.

*James Stewart, Lothar Redlin, Saleem Watson, *Precalculus Mathematics for calculus 4th*, Brooks/Cole.

*Wesner/Nustad, *Elementary Algebra with applications*, Iowa1983.

*Elliot Mendelson, *Theory and problems of beginning calculus,* 2nd edition, McGra-Hill.

*Titu Andreescu, Dorin Andrica, *Complex Numbers from A to...Z*, Birkhäuser, 2006.

*A. Tsypkin, A. Pinsky, *Methods of solving problems in high school math*, "Mir" Publishers Moscow, 1983.

*Călugăriţa Gh., Mangu V., *Probleme de matematică pentru treapta I şi a II-a de liceu*, Ed. Albatros, Bucureşti, 1977.

*Comissaire H., Anzemberger E., *Exercises d'Algebre et de Trigonometrie*, Paris, 1923.

* Iaglom I. M., Iaglom A. M., *Challenging Mathematical Problems with elementary Solutions*, Dover Publications, 1964.

*Ioachimescu A. G., *Culegere de probleme de algebră*, ediţia a V-a, Ed. Didactică şi Pedagogică, Bucureşti, 1968.

*Ikramov H. D., *Zadacnik po lineinoi algebre*, Moskva, 1975.

*Kuterov A., Rubanov A., *Zadacnik po algebre i elementarnîm funcţiiam*, Moskva, 1974.

*M. I Skanavi, Sbornik Zadach po matematike, Moskva, 1996.

*Vîşenski V. A., Kartaşov N. V., Mihailovski B. I., Iardenko M. I., *Sbornik zadac kievskih matematiceskih olimpiad*, Kiev, 1984.

*Nesterenko I. V., Olenik S. N., Potapov M. K., *Zadaci vstupitelnîh eczamenov po matematike*, Moskva, 1986.

*Dorofeev G. V., Potapov M. K., Rozov N. H., *Posobie po matematike dlia postupaiuşcih v vuzî*, Moskva, 1976.

*Sklearski D. O., Cenţov N. N., Iaglom I. M., *Izbrannîe zadacii teoremî elementarnoi matematiki (Arifmetica i algebra)*, Moskva, 1965.

*Stamate I., Stoian I., *Culegere de exerciţii şi probleme de algebră*, Ed. Didactică şi Pedagogică, Bucureşti, 1979.

*C. Nastasescu, M. Brandiburu, C. Nita, D. Joita, *Exercitii si probleme de algebra*, Editura Didactica si pedagogica, Bucuresti.

Matematică - Algebră (high school textbooks, grades IX - X, 1980 edition), Ed. Didactică şi Pedagogică, Bucureşti, 1980.

*Collection of "Gazeta Matematică - seria B", Bucharest.

www.ingramcontent.com/pod-product-compliance
Lightning Source LLC
Chambersburg PA
CBHW020724180526
45163CB00001B/97